Ecological Studies, Vol. 148

Analysis and Synthesis

Ecological Studies

Volumes published since 1995 are listed at the end of this book.

Springer

Berlin
Heidelberg
New York
Barcelona
Hong Kong
London
Milan
Paris
Singapore
Tokyo

F.V. Wulff L.A. Rahm P. Larsson (Eds.)

A Systems Analysis of the Baltic Sea

With 151 Figures, 6 in Color, and 60 Tables

Springer

Prof. Dr. Fredrik V. Wulff
University of Stockholm
Department of Systems Ecology
10691 Stockholm
Sweden

Prof. Dr. Lars A. Rahm
Linköping University
Department of Water and Environmental Studies
58183 Linköping
Sweden

Prof. Dr. Per Larsson
Lund University
Ecotoxicology
Department of Ecology
Ecology Building
22362 Lund
Sweden

ISSN 0070-8356
ISBN 3-540-67769-0 Springer-Verlag Berlin Heidelberg New York

Library of Congress Cataloging-in-Publication Data.

A systems analysis of the Baltic Sea / Fredrik V. Wulff, Lars A. Rahm, Per Larsson (eds.).
p. cm. – (Ecological studies ; v. 148)
Includes bibliographical references (p.).
ISBN 3540677690 (alk. paper)
1. Marine ecology–Baltic Sea. 22. Pollution–Environmental aspects–Baltic Sea. I. Wulff, Fredrik. II. Rahm, Lars A., 1948- III. Larsson, Per. IV. Series

QH92.7 .S97222001
577.7'334–dc21 00-045707

Springer-Verlag Berlin Heidelberg New York
a company of BertelsmannSpringer Science+Business Media GmbH

http://www.springer.de

Printed in Germany

Cover design: *design & production* GmbH, Heidelberg
Typesetting: Bader • Damm • Kröner, Heidelberg

SPIN 10680195 31/3130 YK – 5 4 3 2 1 0 – Printed on acid free paper

Preface

During recent decades, it has become apparent that human influences on aquatic ecosystems are no longer restricted to freshwater and marine coastal regions, but are affecting entire seas like the Baltic and Mediterranean. Widespread anoxia in deep water and sediments, increased sedimentation and massive plankton blooms are some effects caused by large-scale processes.

Eutrophication, caused by increased inputs of nutrient or organic matter to the sea, is the ultimate process behind many of these changes. Other effects, like disturbed reproduction in marine mammals and predatory birds, are caused by toxic substances. An important group of these compounds are the persistent organic pollutants (POP), like PCBs, DDTs and HCHs. The effects of nutrients and POP have been documented on a long time scale in the Baltic Sea, not only locally, but also in the large, open sea areas.

Nutrients and POP have at least one factor in common. In order to take measures, it is vital to identify the sources and transportation pathways. Budget calculations and identification of the important physical and biogeochemical pathways that control the observed distributions of nutrients and some selected toxic compounds have also been the prime objectives of the studies presented in this book. Many of the articles presented here stem from studies funded by the Swedish Environmental Protection Agency in a project called "Large- scale Environmental Effects and Ecological Processes in the Baltic Sea". Additional funding agencies in Sweden and from the European Union, which are acknowledged in the separate chapters of the book, have also contributed.

We have deliberately not described in detail the effects of eutrophication and persistent toxic pollutants on the biota in the Baltic. They are well known, at least qualitatively. Less is known quantitatively about the underlying causes behind the changes that have occurred. This book is highly interdisciplinary and uses a systems approach for analysing and describing this ecosystem. Budgets and dynamic simulation models in the book are supplemented by extensive reviews, where our understanding is validated against the extensive field studies available for this, maybe, most intensively studied marine area in the world.

We hope that this book will serve as a valuable reference as well as provide an understanding of the large-scale dynamics of the Baltic Sea. Hopefully, our scientific approach will be valuable to researchers challenged with understanding the causes behind changes that have occurred in other marine ecosystems.

Fredrik Wulff, Stockholm
Lars Rahm, Linköping,
and *Per Larsson*, Lund

December 2000

Contents

Contributors

Agrell, Cecilia

Ecotoxicology, Department of Ecology, Ecology Building,
223 62 Lund, Sweden

Alexandersson, Hans

Swedish Meteorological and Hydrological Institute,
601 76 Norrköping, Sweden

Axelman, Johan

Institute of Applied Environmental Research,
University of Stockholm, 106 91 Stockholm, Sweden

Azam, Facouk

Marine Biology Research Division, Scripps Institution
of Oceanography, University of California San Diego, La Jolla,
California 92093-0202, USA

Bandh, Cecilia

Institute of Applied Environmental Research,
University of Stockholm, 106 91 Stockholm, Sweden

Bergström, Sten

Swedish Meteorological and Hydrological Institute,
610 76 Norrköping, Sweden

Blomqvist, Sven

Department of Systems Ecology, University of Stockholm,
106 91 Stockholm Sweden

Bremle, Gudrun

School of Engineering, Jönköping University, P.O. Box 1026,
551 11 Jönköping, Sweden

Broman, Dag

Institute of Applied Environmental Research (ITM),
Laboratory for Aquatic Ecotoxicology, University of Stockholm, 106 91
Stockholm, Sweden

Carlsson, Bengt

Swedish Meteorological and Hydrological Institute,
601 76 Norrköping, Sweden

Carman, Rolf

Department of Geology and Geochemistry,
University of Stockholm, 106 91 Stockholm, Sweden

Cederwall, Hans

Department of System Ecology, University of Stockholm
106 91 Stockholm, Sweden

Danielsson, Åsa

Department of Water and Environmental Studies,
Linköping University, 581 83 Linköping, Sweden

Granat, Lennart

Department of Geology and Geochemistry,
University of Stockholm, 106 91 Stockholm, Sweden

Grimvall, Anders

Department of Statistics, Linköping University,
58 183 Linköping, Sweden

Hagström, Åke

Department of Marine Sciences, Kalmar University,
391 82 Kalmar, Sweden

Hallin, Ann-Katrin

Department of Environmental Assessment, Swedish University of Agricultural Sciences, Box 7062, 750 07 Uppsala, Sweden

Heiskanen, Anna-Stiina

Einvironment Institute, TP 290, Joint Research Centre, European Comission, 21020 Ispra (VA), Italy

Ishaq, Rasha

Institute of Applied Environmental Research, University of Stockholm, 106 91 Stockholm, Sweden

Jansson, Mats

Department of Physical Geography, University of Umea, 901 87 Umea, Sweden

Johansson, Niclas

Ecotoxicology, Department of Ecology, Ecology Building, 223 62 Lund, Sweden

Josefsson, Weine

Swedish Meteorological and Hydrological Institute, 601 76 Norrköping, Sweden

Karlsson, Karl-Göran

Swedish Meteorological and Hydrological Institute, 601 76 Norrköping, Sweden

Klavins, Maris

Department of Environmental Science, University of Latvia, Raina Bullevard 19, 1586 Riga, Latvia

Kuparinen, Joma

Finnish Institute of Marine Research, P.O. Box 33, 009 31 Helsinki, Finland

Larsson, Per

Ecotoxicology, Department of Ecology, Ecology Building,
223 62 Lund, Sweden

Okla, Lennart

Ecotoxicology, Department of Ecology, Ecology Building,
223 62 Lund, Sweden

Näf, Carina

J&W Energy and Environment, Stjärnv. 2, 181 83 Lidingö, Sweden

Pettersen, Harald

Amersham Pharmacia Biotech AB, 751 84 Uppsala, Sweden

Rahm, Lars

Department of Water and Environmental Studies,
Linköping University, 581 83 Linköping, Sweden

Roots, Ott

Ministry of the Environment, Environment Information Centre
EE0006, Mustamäe Tee Str. 33, 106 16 Tallinn, Estonia

Sandberg, Johannes

Department of Systems Ecology, University of Stockholm,
106 91 Stockholm, Sweden

Savchuk, Oleg

Department of Systems Ecology, Stockholm University,
106 91 Stockholm, Sweden

Stålnacke, Per

Jordforsk – Centre for Soil and Environmental Research,
Frederik A. Dahls vei 20, 1432 Aas, Norway

Stigebrandt, Anders

Department of Oceanography, Earth Sciences Center,
Göteborg University, 413 81, Sweden

Wania, Frank

Department of Chemistry and Division of Physical Sciences, University of Toronto at Scarborough, 1265 Military Trail, Scarborough, Ontario, Canada M1C 1A4

Westring, Gustaf

Marsep International, P.O. Box 130 Kokstad, 5863 Bergen, Norway

Wulff, Fredrik

Department of Systems Ecology, University of Stockholm, 106 91 Stockholm, Sweden

Zebühr, Yngve
Institute of Applied Environmental Research, University of Stockholm, 106 91 Stockholm, Sweden

Zelechowska, Alexandra
Institute of Meteorology and Water Management, Department of Water Pollution Control, Jaskowa Dolina 29, 80 252 Gdansk, Poland

Zweifel, Ulla Li
Department of Marine Science, Kalmar University, 391 82 Kalmar, Sweden

1 Introduction

F. WULFF, L. RAHM, and P. LARSSON

1.1 Background

During recent decades, large-scale effects of pollution on the entire Baltic Sea have become apparent. The prevalence of anoxic deep basins, decreases in the large populations of grey seals and increased frequency of toxic phytoplankton blooms (Horstmann 1975; Kahru et al. 1994) are examples of phenomena governed by large-scale processes. Marine resources, like fish for consumption, are endangered due to overall high levels of toxic substances and overfishing. Measures must be implemented to reduce the loads of pollutant to such levels that the environmental quality of the Baltic is restored. At the same time, the costs of different measures should be weighted against the results achieved (see HELCOM 1991; Wulff and Niemi 1992; Gren et al. 1997). Consequently, demand for strong scientific argument increases when international measures need to be motivated (see e.g. Wikner et al. 1996).

So far, most scientific studies have concentrated on local recipients in the coastal zone, hence our understanding of the factors controlling large-scale effects is limited. It is now obvious that these effects arise as a result of a complex interplay between pollutant loads from many sources. They originate from all the countries in the Baltic drainage basin and, in the case of atmospheric deposition, from sources even further away. A complex interplay between physical, biogeochemical and ecological processes governs concentration levels and effects. Large-scale interdisciplinary efforts on an international level are needed to understand and remedy these effects.

Assessments (e.g. HELCOM 1996) show that concentrations of both eutrophication and toxic substances are still high in the Baltic Sea although for several substances the increasing trends have ceased during the last decades. Although there are clear relationships between increasing anthropogenic inputs and increasing concentrations and ecological effects on a long-term (50-year) time scale, these relationships are less obvious and

Ecological Studies, Vol. 148
F. Wulff et al. (eds.)
A Systems Analysis of the Baltic Sea

even absent on a decade scale or less. In fact, the concentration of nutrients in the sediments of the Baltic proper did not start to increase until the middle of the nineteenth century (e.g. Carman et al. 2000) parallel with increases in population, agriculture and industrialization. There is now a growing understanding that variations in concentration and effects of various pollutants are caused by a complex interplay between inputs, climate forcing and processes in the sea. This book is an attempt to synthesize our current understanding of the processes governing large-scale effects and ecological processes, critical for the distribution and effects of pollutants in the Baltic Sea. It is also an attempt to summarize some of the results gained in a Baltic Sea research project 'Large-scale environmental effects and ecological processes in the Baltic Sea' which was initiated by the Swedish Environmental Protection Agency (SEPA) with connections with several Baltic states. Some of the work presented in this book has subsequently been funded within the Baltic Sea System Study (BASYS), an international EU/MAST (Marine Science and Technology) program.

Models synthesize and evaluate our understanding of external forcing and internal biogeochemical and physical processes controlling distributions, concentrations and effects of eutrophication and toxic substances. Models are also tools to evaluate various management strategies. Therefore we have organized the chapters in this book by starting with descriptions of physical transports and external forcing by climate and by loads of nutrients and some selected toxic substances. We then describe various critical processes governing plankton growth, sedimentation and transformation in sediments. The next chapters describe quantities and distributions of nutrients, organic matter and toxic substances. The book concludes with chapters in which both budgets and dynamic models of nutrient and toxic substances are developed, based on the information presented in previous chapters.

1.2 Physical Transports and External Forcing

In Chapter 2, Stigebrandt describes the physical transport systems of the Baltic, and the various current and mixing processes. The Baltic has strong salinity gradients both vertically and horizontally. This chapter provides an extensive overview of the physical oceanography of the Baltic Sea and the processes governing the distribution of sea salt and thus many other dissolved substances in this large estuary. In addition, the overall distributions of temperature, currents, oxygen and optical properties are described and related to forcing functions.

With a drainage basin area that is four times larger than the surface of the sea, the Baltic is highly affected by inputs from the surrounding land. The freshwater outflow through the Danish straits is large, comparable to the Mississippi River. Variations in weather affect to a high degree the supply of nutrients, seasonally, interannually and on a climate scale. In Chapter 3, Bergström et al. describe the processes governing freshwater inputs and spatial and temporal characteristics of climate variations. They illustrate the large regional variations in climate and also in hydrology. The latter depends on climate as well as on variations in topography and soils in the different subregions.

In Chapter 4, Grimvall and Stålnacke have assembled an extensive database of nutrient concentrations in rivers from all Baltic countries and have estimated riverine inputs for the period 1970–1993. They show that the large variations in freshwater runoff to the Baltic, reported by Bergström et al. (Chap. 3), are also reflected in large variations in nutrient inputs. If the variations in freshwater flows were accounted for, there were no clear trends in riverine nutrient inputs during the studied period. The major increases in nutrient loads occurred earlier, probably in the 1950–1960s, when drastic changes in land use and increases in fertilizer applications occurred. Despite significant changes in land use, point source emissions and atmospheric emissions in certain parts of the Baltic drainage basin, the total annual riverine load of both nitrogen and phosphorus to the sea has been fairly constant between 1970 and 1993.

Runoff from agricultural land has been estimated to account for about half the total N load and a fifth the P load to the Baltic. After independence in 1990, fertilizer use in Estonia, Latvia and Lithuania dropped drastically to 1950 levels and the yield of most agricultural products decreased by almost 50 %. In spite of these drastic changes, there have been almost no changes in nutrient runoff to the Baltic, at least until 1998. Löfgren et al. (1999) attribute this to large nutrient pools still remaining in the previously heavily fertilized soils that will continue to leak nutrients for many decades, before substantial decreases in runoff will be seen.

The clear gradient in nutrient inputs, with an increase from north to south, is explained by the increasing population densities and the area of agricultural land, as shown by Sweitzer et al. (1996). Poland contributes most to the nutrient load but is inhabited by only approximately half of the 85 million people that live in the area of the Baltic drainage basin. The per capita load is not dramatically different between, for example, Sweden and Poland, in spite of much more advanced sewage treatment in the former country. Today, the major source of nutrients to the Baltic is, as mentioned above, leakage from agricultural land; Stålnacke (1996) estimates this to be roughly 50 % of the total load. Presently, the former East-block countries are using less fertilizer per unit of land. They have soil properties and more wetlands that favor nutrient retention to a higher degree than elsewhere in the drainage basin, including the more 'developed' Baltic EU countries.

Although the riverine input is the major source of nutrients, the atmospheric deposition of nitrogen directly over the sea could represent about a third of the total input. Consequently, better estimates are urgently needed for nitrogen (see e.g. Rosenberg et al. 1990). Only a minor fraction of the phosphorus load comes from atmospheric deposition. In Chapter 5, Granat presents a thorough examination of existing data and interpolated wet and dry depositions of nitrate and ammonium with a high resolution over the Baltic. The nitrogen concentrations are averaged over 1986–1990 for the entire Baltic Sea. He reports large regional and seasonal variations and even doubled nitrogen deposition in the last decades, compared to data from before 1965. Recent calculations of nitrogen deposition to the Baltic with the EMEP model for the period 1986–1996 show a downward trend in deposition after 1989. The calculated deposition for the last 3 years is thus about 20 % lower than during 1986–1990 (Barret et al. 1998).

There has been a decrease in the levels of some toxic persistent halogenated pollutants in the Baltic biota [Bignert et al. 1998; for example herring (*Clupea harengus*), cod (*Gadus morhua*) and guillemot (*Uria aalgae*)] during the last decades (Fig. 1.1), although the levels are still high. Health authorities are still cautioning extensive consumption of Baltic fish. There has been a recovery in top predator populations, e.g. the breeding success of white-tailed eagles is almost back to the level of the 1950s before the extensive contamination by persistent pollutants of the Baltic started (Fig. 1.1). Some of this population recovery in these birds of prey is also attributed to winter feeding on food free of persistent pollutants. The rapid recovery of the seal populations, which previously showed a high degree of reproduction disturbance due to PCBs, has now urged fishery organizations to legalize hunting again.

Although trends in persistent pollutants in some Baltic biota are well documented (Bignert et al. 1998), little has been known about the origin of these pollutants and their flows between different components of the ecosystem. Inputs of chlorinated hydrocarbons, specifically atmospheric and river inputs of PCBs, DDTs and HCHs to the Baltic, are described by Agrell et al. in Chapter 6. An extensive sampling program was set up, covering the entire Baltic region, and with participation of scientists from many countries around the Baltic. The same techniques were used at all sites to obtain comparable data. Agrell et al. (Chap. 6) found that the rivers and the atmosphere contribute almost equally to the PCB load, while atmospheric deposition of HCHs and DDTs was about five to seven times more important. Large-scale, even global, redistribution of pesticides via

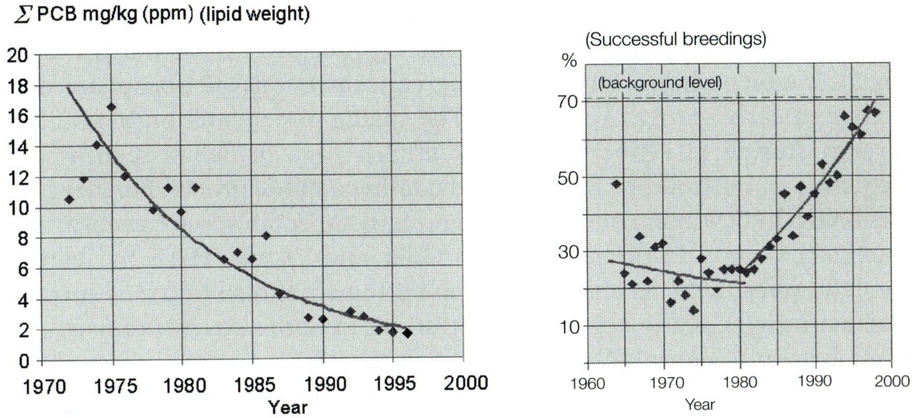

Fig. 1.1. ΣPCB in herring muscle (ppm, lipid weight) in the southern Baltic proper (*left*) and percentage of successful breeding of sea eagles in Sweden (*right*). (Data from the Swedish National Contaminant Monitoring Program, Swedish Museum of Natural History; see http://www.nrm.se/mg)

the atmosphere is probably now a major contribution to the pollution of the Baltic, now that local or regional contributions have diminished due to regional bans. However, former usage of some substances is still reflected in the fact that soils and water may act as sources, leaking DDTs and highly chlorinated PCBs into the environment.

Although the effects of some toxic substances are now well documented, there is still a danger that new contaminants may be introduced. The M74 syndrome, which causes high mortality in newly hatched salmon fry, was initially suspected to be caused by toxic substances of unknown origin. Recent studies (Bengtsson et al. 1999) show a more complex picture, where the drastic changes in the Baltic food web and the diet of salmon are shown as explanations of the syndrome.

1.3 Critical Processes

An understanding of forcing functions and sources, controlling the inputs and physical distribution of nutrients and toxic substances, should be supplemented by an understanding of the biogeochemical processes that are responsible for redistributions, internal transfers, sources and sinks in the water column and sediments. A challenge to the understanding of Baltic biogeochemistry is that this sea experiences large temporal and spatial variations in physical forcing which, in turn, contribute to spatial and temporal variability in the biogeochemical processes. This is illustrated in Fig. 1.2, which shows typical surface concentrations of phosphate and nitrate in early spring (April 1997). The spring phytoplankton bloom has already started in February in the southern parts, which is seen by the depletion of nutrients in Kattegat and southern Baltic proper. Further into the Baltic, high winter concentrations remain, except in some coastal areas, particularly around the large Rivers Oder and Vistula. Here, the bloom depletes phosphate, resulting in accumulating amounts of unused nitrate. Further into the Baltic, winter concentrations of both nutrients remain and their distributions are still reflections of physical transports, mixing and the distribution of the sources along the coasts. High phosphate concentrations in the Gulf of Finland reflect the large source in the St. Petersburg region and from the River Neva (the largest river flowing into the Baltic Sea). The low concentration further north, particularly in the Bothnian Bay, is partly due to low inputs but also to efficient internal phosphorus sinks. The corresponding high nitrate concentrations cannot be explained by large inputs to this sparsely populated region. An understanding of the observed pattern must be sought in the relative importance of internal biogeochemical processes, discussed later in this chapter. The different regimes are illustrated by large variations in N/P ratios. They are typically 2–4 in the offshore areas in the southern basins, far below the Redfield 16:1 N:P ratio for plankton uptake (Redfield et al. 1963) and far above (40–60) in the northernmost basin.

Thus the nutrient distributions illustrated here are also the result of various internal biogeochemical processes. Plankton production is con-

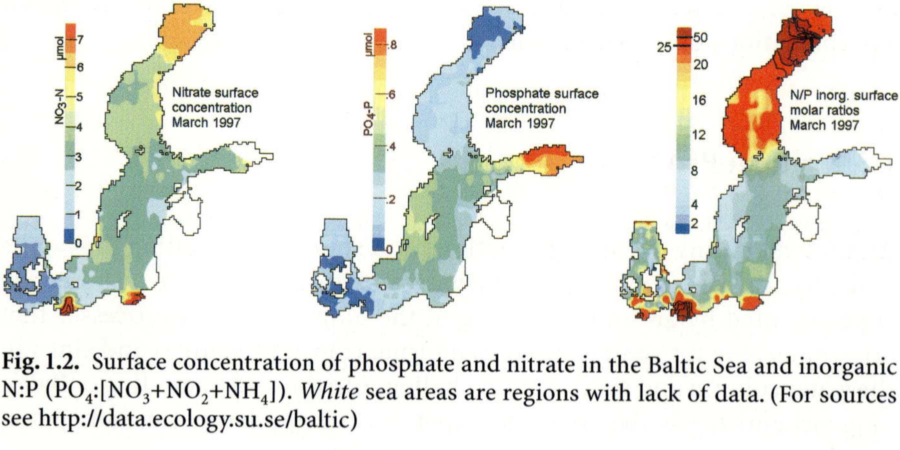

Fig. 1.2. Surface concentration of phosphate and nitrate in the Baltic Sea and inorganic N:P (PO_4:[NO_3+NO_2+NH_4]). *White* sea areas are regions with lack of data. (For sources see http://data.ecology.su.se/baltic)

sidered to be nitrogen limited in the southern basins and phosphorus limited in the north, with a mosaic of alternating P or N limitations in various coastal regions, depending on the nutrient ratios in local inputs and exchange with the sea. This is discussed by Hagström et al. in Chapter 7, while Blomqvist and Heiskanen in the next chapter focus on sedimentation, the particle exchange between the water column and the sediments and some biogeochemical processes in the latter. Jansson follows up by describing the role of sediments in the distribution of nutrients in the Baltic in Chapter 9. Axelman et al. describe the fate of hydrophobic organic compounds in the Baltic and how these distributions can be related to various biogeochemical processes.

Hagström et al. (Chap. 7) review our current understanding of the biogeochemical cycling of carbon and nutrients, with emphasis on microbial processes in the water column. They show how the relative importance of the food web shifts during the year and in different regions, depending on physical forcing, weather and availability of limiting nutrients and dissolved organic matter (DOM). The pool of DOM in the Baltic is about three to four times higher than that in the oceans and it is an important source of both energy and nutrients, channeled into the food webs by microbial processes. Nutrient limitation is discussed in relation to primary production but also in relation to utilization of organic matter and competition between autotrophs and heterotrophs. It has become apparent that also bacteria may experience nutrient limitation and compete with autotrophs; for N in the Baltic proper and P in the Bothnian Bay. Nitrogen-fixing cyanobacteria are common in eutrophic lakes, rare in open oceans and nearly absent in estuaries and coastal seas. The Baltic Sea is one of the few exceptional estuaries where cyanobacterial blooms are common. An increase in these blooms in relation to eutrophication is usually assumed, but it has recently been shown that cyanobacterial blooms have occurred in the Baltic at least during the last 7000 years (Bianchi et al. 2000). The mechanisms behind the blooms are currently under heated debate (Howarth et al. 1999), since the significance of a substantial internal nitrogen source will have implications on possible nutrient reduction policies to counteract coastal eutrophication. In fact, estimates of the net total nitrogen fixation have been carried out by Rahm et al. (2000) using empirical models based on monitoring data. The annual variation is large, but net fixation is in the range of $30–260 \times 10^3$ ton N/year, i.e. in the range originally estimated by Brattberg (1980). Hagström et al. (Chap. 7) review the current understanding of these processes in the Baltic. A paradoxical situation may occur, i.e. while the primary production in the Baltic proper is nitrogen limited (Granéli et al. 1990), the system is phosphorus limited in the longer perspective, demonstrated in model studies of both the Baltic (Savchuk and Wulff 1999) and globally (Tyrrell 1999).

Sedimentation is one of the key processes in the nutrient and pollution dynamics of the Baltic system. Though turbulent mixing may be responsible for certain mass fluxes, the prevailing estuarine circulation coupled with a more or less pronounced stratification will not favor a downward mass transport of dissolved substances towards the sediments. However, particle sedimentation and deposition override these obstacles, making up an efficient mass transport mechanism, especially as the Baltic is relatively shallow compared to oceanic conditions. Further, the Baltic Sea basins usually have their halocline below the entrance sill, making each basin a more or less 'efficient' nutrient and pollutant trap in comparison with for example the Mediterranean (Wulff et al. 1990). In Chapter 8, Blomqvist and Heiskanen thoroughly review the state of the art with regard to the processes that govern the sedimentation of particulate matter in the Baltic. Organic matter represents only a minor part of the depositing matter. A strong seasonal variation characterizes the deposition process. While diatoms and dinoflagellates dominate the spring bloom, only diatoms deposit on the bottoms, supplying the benthic heterotrophic community with organic matter. During other seasons the deposition originating from primary production is more or less insignificant. Plankton community structure, remineralization in the water column and physical conditions, among other factors, control the mass fluxes. Resuspension and lateral transports in the benthic boundary layers are other important factors, although still poorly quantified.

The nitrogen and carbon that are deposited as particulate matter are of organic origin and mineralized by heterotrophs, while phosphorus (usually as orthophosphate) is generally linked to iron and manganese compounds or clay minerals. They are all redox dependent but in different ways. Only a small fraction of these nutrients is buried in the sediments; the bulk is either recycled as dissolved inorganic substances or leaves the system as N_2 or N_2O gas. In Chapter 9, Jansson gives a thorough review of those processes and conditions that control the nutrient remineralization, recycling and burial processes.

The cycling behavior of persistent organic pollutants (POP) in the aquatic environment is summarized in Axelman et al. (Chap. 10). River and atmospheric inputs of PCBs and PAH are estimated, and related to sedimentation and retention in the sediment. Once deposited there, POP may recycle back to the overlying water or be buried and incorporated into the bottoms, depending on chemical properties, sediment characteristics and activities of the sediment-dwelling fauna. Some of the pollutants are transported in the food web, while some remain in the dissolved state and volatilize across the surface microlayer to the atmosphere. The pollutants, exemplified by PCB and PAH, have different source distributions, which affect their fate in the Baltic Sea.

Thus, the sediment pool plays an important role in the nutrient dynamics of the shallow Baltic Sea, but still the uncertainties in sediment nutrient distribution and total amounts are considerable. To improve this situation, the large-scale nutrient distributions in the sediments and of the benthic macrofauna have been studied by Carman and Cederwall (Chap. 11). They showed some typical features of the surficial sediments in the Baltic. Generally, there are low concentrations of inorganic carbon because carbonates are thermodynamically unstable in these environments (Carman and Rahm 1997). Most of the carbon and nitrogen in the sediments are in organic form in contrast to phosphorus, which is mainly associated with mineral particles. The redox-dependent phosphorus behaves qualitatively different in the Gulf of Bothnia (high concentrations in well-oxidized sediments) compared to the Baltic proper. This condition is attributed to the high concentrations of redox metals. Carman and Cederwall (Chap. 11) estimated 'a burial' in the deposition bottoms of the Baltic proper to be about 170,000 ton N/year. Rönner (1985; see also Chap. 9) estimated a typical loss due to denitrification of about 570,000 ton N/year. Hence, the budget estimates presented later in this volume by Wulff et al. (Chap. 13) of a total sink (burial plus denitrification) of more than 600,000 ton N/year corroborate these independent estimates.

The reduced conditions in the Baltic proper deep sediments have increased considerably during the last decades, as reported by Jonsson et al. (1990). They studied the distribution of the laminated sediments, which are a result of anoxia that has eliminated benthic animals and thus bioturbation. Carman and Cederwall (Chap. 11) show considerable increases in benthic biomass in the more shallow areas, compared with earlier studies (e.g. Elmgren 1978). If one compares these studies with ones from the 1920s (Hessle 1924) the increase is roughly of an order of magnitude. At the same time, the abundance of key species in the littoral zone, e.g. bladder wrack, has decreased due to the increase in green algae which are growing on them and hamper their growth (Kautsky et al. 1986). The changes in littoral algal community composition have also affected the food web and, ultimately, the fish communities.

1.4 Budgets and Models

Any budget or dynamical model means a simplified representation of the real ecosystem and involves aggregation of variables into fewer spatial or temporal units. In Chapter 12, Rahm and Danielsson describe and analyze variations in hydrochemical properties of the Baltic within and between sub-basins and in time. For this purpose the large databases available of the Baltic are used and a series of statistical analyses are applied to discern significant variations and trends. Significant differences were found between sub-basins and between seasons. However, it was difficult to find significant spatial variations within the sub-basins, with the exception of the central Baltic proper through the Gulf of Finland. Thus, few stations are usually sufficient to describe the conditions in each sub-basin. Although there have been clear increases in nutrient concentrations in the Baltic in response to increased inputs during the last 50 years, there is no correlation between nutrient loads and concentrations on an annual or decade time-scale. The nutrients show different trends in different sub-basins. While the nitrogen and phosphorus trends seem to increase in consort, the silica to nitrogen ratio indicates that Si may become a limiting nutrient in the near future in the southern regions. While Schelske et al. (1983) suggest that the silicate decrease is an effect of eutrophication, Humborg et al. (1997) came up with a radically different explanation for a similar evolution in the Black Sea. The drastic increase in regulation of the Danube should be the prime cause. Present studies in the Baltic partly seem to support the latter reasoning (Humborg et al. 2000). Rahm and Danielsson also describe large-scale and regional variations in salinity, vertical mixing and oxygen conditions that also must be taken into account in explaining the observed variations in nutrients.

The complex interplay between physical and biogeochemical processes, nutrient inputs and climatic variations can be understood by using models. Such models are also necessary in order to develop a management strategy on how, where and how much the nutrient inputs should be reduced. Two different modeling approaches are applied in the final chapters of this book, i.e. the 'empirical' and the 'mechanistic'. Empirical models, built by using the large data sets from this region (described in previous chapters) to develop nutrient budgets, are described in Chapter 13 by Wulff et al.

Budgets of total and inorganic fractions of nitrogen and phosphorus, and of silica, were set up, showing the relative importance of inputs, advective and diffusive fluxes as well as of internal biogeochemical processes (sources and sinks) in controlling nutrient concentrations in the major sub-basins of the Baltic. Although these models do not contain any process description, they show large regional variations that can be related to variations in hydrographic and biogeochemical conditions in the Baltic sub-basins. Wulff et al. (Chap. 13) show how the high efficiency of the internal nitrogen sink (probably denitrification) almost equals the external nutrient load. Phosphorus and silica, on the other hand, are retained in the water column to a large degree, due to less efficient internal sinks. However, there is a gradient, from the oligotrophic Bothnian Bay to the more eutrophic Baltic proper. While the efficiency of the phosphorus sink is decreasing with increasing eutrophication, the opposite is shown for nitrogen. It is also demonstrated that a nutrient (N or P) that is considered limiting for primary production is also the one with the shortest residence time in a particular basin.

Thus, these budgets yield a system-level principal view on how nutrient concentrations and ratios will change in relation to inputs. This is illustrated in Fig. 1.3, and, consistent with our understanding of key processes, discussed in Chapters 7, 8, 9 and 11. With increasing inputs, primary production and sedimentation will increase, as illustrated in the three panels from left to right in Fig. 1.3. Low sedimentation means that a limited amount of organic matter will be decomposed in the sediments. This means that the sediment surface will remain oxidized and function as an efficient P sink. Most of the organic matter will be decomposed aerobically and only a small amount will be buried deeper into the anoxic part of the sediment. Thus, denitrification, i.e. anaerobic decomposition of organic matter using nitrate, is probably of little importance here. More N compared to P will then be recycled and the primary production will be P limited. With increasing production and sedimentation of organic matter, oxygen consumption in the sediments will increase. If the sediment surface is still oxic, P adsorption will occur but to a lesser extent and more is recycled into the water column. On the other hand, a larger fraction of organic matter is buried and decomposed anaerobically. Denitrification is probably highly efficient in such an environment where high concentration of nitrate from the surface layer can diffuse over a short distance and supply the decomposing bacteria with electron acceptors. Thus, it is likely that the efficiency of the P sink will decrease and the N sink will increase with increasing organic load to the sediment. In such a system, excess P, in comparison to N, will be recycled into the water and primary production is likely to be N limited. In Fig. 1.3, an additional N source, N_2 fixation by cyanobacteria, is favored under such conditions, i.e. with high P but low N concentrations in the water.

Finally, the load of organic matter has increased to such a degree that the increased decomposition has caused the redoxcline to move into the water

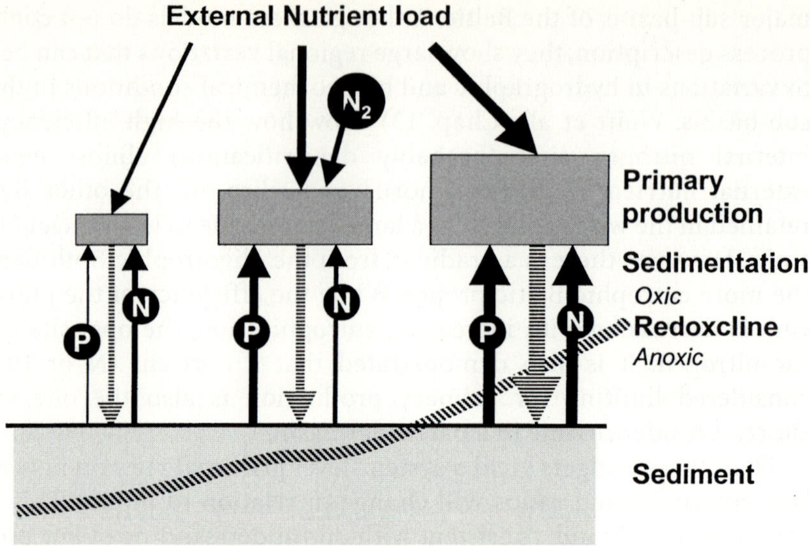

P sink	*High*	*Moderate*	*Poor*
N sink	*Poor*	*High*	*Poor*
Residence times			
P	*short*	*longer*	*long*
N	*long*	*short*	*long*
Production limitation	*P*	*N*	*(light)*

Fig. 1.3. Schematic illustration of how the cycles of nitrogen and phosphorus will change in a shallow marine ecosystem in response to increase in external loads (*left* to *right*)

column. The efficiency of both the P and N sinks of the sediments will decrease drastically. In this case, both nutrients will be recycled into the water column, supplying primary production and accelerating eutrophication. Naturally, this is a highly idealized scenario, which could be modified by a number of additional factors, but it fits actually quite well to the situation depicted in the budget calculations presented here by Wulff et al. (Chap. 13).

Mechanistic models describe various physical and biogeochemical processes controlling nutrient levels and productivity of the sea. These models have been useful tools to test various hypotheses about the relative importance of different processes. In Chapter 14, earlier versions, dealing with only nitrogen, are expanded to also include phosphorus. We take advantage of an improved understanding of nutrient inputs to the Baltic Sea by rivers (Chap. 4) and by the atmosphere (Chap. 5). Our understanding of critical processes, i.e. the pelagic community (Chap. 7), sedimentation (Chap. 8) and processes in the sediments (Chaps. 9 and 11), as well as of physical processes (Chaps. 2 and 3) has also improved and could be used to develop and verify this model. Hence the model has served as a "workbench" where formulations interact with empirical knowledge to test and improve our understanding.

A comparison of simulation results with available data shows that the model is actually capable of reproducing a number of key features of the Baltic proper. For instance, the model generates a nearly quantitative description of the seasonal, interannual, and long-term dynamics of nutrients and oxygen, a plausible description of the seasonal dynamics of plankton variables and a reasonable imitation of the depth distribution of sediment variables, compared to observations. Even more important, the modeled pelagic and benthic biogeochemical fluxes are rather realistic when compared with the corresponding estimates compiled from the literature.

If the verification of a model is successful, it could then be used to forecast possible responses of the Baltic Sea to different scenarios of future variations in climate (see, e.g., Barrow 1993) and different management strategies to reduce pollution (HELCOM 1993). Here, Savchuk and Wulff (1999) model the effects of drastic reduction of inputs of N, P and of both nutrients. The simulations show that a reduction in only N, maintaining the high P concentrations now prevailing in the Baltic, would be compensated by increased nitrogen fixation. In the model, a reduction in large-scale eutrophication is only attainable by reduced inputs of phosphorus. The simulations also show that the effects of altered inputs are seen only after several decades, due to the inertia of the system. This is primarily caused by the long residence time of water and of nutrients. It is also pointed out that management strategies optimal for the overall Baltic might not be optimal seen from a local or regional perspective. Various coastal bays and archipelagos with much shorter water residence times and varying P or N limitation might call for other management strategies, as demonstrated by, for example, Elmgren and Larsson (1997) and Savchuk (2000).

The fate of PCB in the Baltic Sea is modeled by Wania et al. (Chap. 15), using the fugacity approach, which is developed to a steady-state mass balance model. The wide range of climatic, hydrological and ecological variables of the Baltic Sea results in differences in POP behavior in the different sub-basins. Rates of chemical degradation, transports within the water and through the air, and across the interfaces of sediment/water and water/air are calculated. The mass inputs by atmospheric deposition and by rivers are based on results from other studies in this volume (Chaps. 6 and 10). The model outcome is compared with empirical data, and shows good agreement; it shows that the overall pattern of PCB transport in the Baltic Sea environment is downward, from the atmosphere to the water column and further down to surface sediments, to finally being buried in deep sediments. This is in contrast to the Great Lakes of North America, where recent studies suggest the lakes act as sources for PCBs. The major downward transport of PCBs in the Baltic basin is sedimentation and resuspension/focusing of particulate organic matter, to which PCBs are associated. The results further indicate that the 'export/sinks' by volatilization, outflow to the North Sea, sediment burial and degradation are too slow to effectively reduce PCB levels in the surface sediment (and, thereby, transport to water and further to the surface microlayer and air) of the Baltic Sea. Consequently, water concentration will be maintained in the water column for many years, despite reduced loads.

1.5 Conclusions

In summary, the data on climate, hydrology and nutrient inputs and on variations in the Sea indicate that, for the last decades, it is the variations in climate and various internal processes that are responsible for the observed nutrient trends, rather than changes in anthropogenic loads. While the loads have remained rather stable during the last decades, the Baltic is far from in a steady state due to its long residence. Our nutrient budgets confirm the assumed latitudinal trend in Redfield N:P ratio, nitrogen-limited primary production in the southern basins and of phosphorus up north. However, the specific properties of the Baltic with extensive nitrogen fixation relax this nitrogen constraint, compared to most marine coastal regions. On longer (decade) time scales all the ecosystems of the Baltic proper and Bothnian Sea are likely to be phosphorus limited, which is demonstrated in the mechanistic modeling. On an even longer perspective, characteristic properties like extensive phosphorus and silicate recycling (inefficient net sinks) and short nitrogen residence time (efficient net sinks) are responses of the internal biogeochemical processes due to eutrophication, rather than inherent features of the Baltic Sea itself.

Further, similarly to the Great Lakes of North America, the now observed decrease in, for example, PCBs in biota will cease, and stabilize at a level depending on internal sources and recycling in the Baltic Sea. The reduced loads, due to regulation of the use of persistent pollutants, will no longer affect concentration in water and biota. Instead, sediment redistribution and volatilization from the catchment area (i.e. from historically contaminated soil), as well as leakage from PCB dumps (as from contaminated building material), will affect organisms in this semi-enclosed sea. Sources that today dominate the inflow to the Baltic Sea, such as atmospheric deposition and river input, will do so for many years to come, albeit at reduced levels that slowly decrease over a time-scale of several decades. The key processes for persistent pollutant fate in the Baltic Sea are connected to transport over the interfaces of air/water (volatilization/gas exchange) and sediment/water (sediment to water transport/ sediment burial), as presented in the model work in this volume. The story of persistent pollutants in the Baltic Sea, for PCB once started in this region according to studies by Sören Jensen, has unfortunately not been completed as yet.

References

Barret K, Tsyro S, Erdman L, Gusev A, Dutchak S, Pekar M, Lükewille A, Krognes T (1998) Atmospheric supply of nitrogen, lead, cadmium, mercury and lindane to the Baltic Sea. EMEP/MSC-W note 3/98. Research Rep no 70. Meteorological Synthesizing Centre–West Norwegian Meteorological Institute, Oslo, Norway

Barrow EM (1993) Scenarios of climate change for the European Community. Eur J Agron 2:247–260

Bengtsson B-E, Hill C, Bergman Å, Brandt I, Johansson N, Magnhagen C, Södergren A, Thulin J (1999) Reproductive disturbances in Baltic fish: a synopsis of the FIRe project. Ambio 28(1):2–8

Bianchi, T, Westman P, Rolff C, Engelhaupt E, Andrén T, Elmgren R (2000) Cyanobacterial blooms in the Baltic Sea: natural or human induced? Limnol Oceanogr 45(3):716–726

Bignert A, Olsson M, Persson W, Jensen S, Zakrisson S, Litzen K, Eriksson U, Häggberg L, Alsberg T (1998) Temporal trends of organochlorines in northern Europe 1967–1995. Relation to global fractionation, leakage from sediments and international measures. Environ Pollut 99:177–198

Brattberg G (1980) Nitrogen fixation in the marine environment – the Baltic. In: Rosswall T (ed) Processes in the nitrogen cycle (in Swedish). PM 1213, Swedish Environmental Protection Agency, Stockholm, pp 95–103

Carman R, Rahm L (1997) Interstitial and bottom water composition of some Baltic proper deep stations. J Sea Res 37:25–47

Carman R, Aigars J, Danielsson Å, Rahm L (2000) Recent paleoecological sediment records of biogenic silica as an indication of the eutrophication in the Baltic Sea (submitted)

Elmgren R (1978) Structure and dynamics of Baltic benthos communities, with particular reference to the relationship between macro- and meiofauna. Kieler Meeresforsch Sonderh 4:1–22

Elmgren R, Larsson U (1997) Himmerfjärden. Changes in the nutrient loaded coastal ecosystem of the Baltic Sea (in Swedish). Rep 4565, Swedish Environmental Protection Agency, Stockholm, 197 pp

Granéli E, Wallström K, Larsson U, Granéli W, Elmgren R (1990) Nutrient limitation of primary production in the Baltic Sea. Ambio 19:142–151

Gren I-M, Söderqvist T, Wulff F (1997) Nutrient reductions to the Baltic Sea: ecology, costs and benefits. J Environ Manage 51:123–143

HELCOM (1991) The Baltic Sea Joint Comprehensive Programme, interim report of the HELCOM ad hoc High Level Task Force, Helsinki, August 1991

HELCOM (1993) The Baltic Sea Joint Comprehensive Environmental Action Programme. Baltic Sea Environ Proc 48, 150 pp

HELCOM (1996) Third periodic assessment of the state of the marine environment of the Baltic Sea: 1989–1993; background document. Baltic Sea Environ Proc 64b, 252 pp

Hessle C (1924) Bottenboniteringar i inre Östersjön (in Swedish). Medd Kgl Lantbruksstyr 250:1–52

Horstmann U (1975) Eutrophication and mass occurrence of blue-green algae in the Baltic. Merentutkimuslait. Julk/Havsforskningsinst Skr 239:83–90

Howarth RW, Chan F, Marino R (1999) Do top-down and bottom-up controls interact to exclude nitrogen-fixing cyanobacteria from the plankton of estuaries? An exploration with a simulation model. Biogeochemistry 46(1):203–231

Humborg C, Ittekkot V, Cociasu A, von Bodungen B (1997) Effect of Danube River dam on Black Sea biogeochemistry and ecosystem structure. Nature 386:385–388

Humborg C, Conley D, Rahm L, Wulff F, Cociasu A, Ittekkot V (2000) Silica retention in river basins: far-reaching effects on biogeochemistry and aquatic food webs in coastal marine environments. Ambio 29(1):45–50

Jonsson P, Carman R, Wulff F (1990) Laminated sediments in the Baltic Sea – a tool for evaluating nutrient mass balances. Ambio 19:152–158

Kahru M, Horstmann U, Rud O (1994) Satellite detection of increased cyanobacteria blooms in the Baltic Sea: natural fluctuations or ecosystem change? Ambio 23:469–472

Kautsky N, Kautsky H, Kautsky U, Waern M (1986) Decreased depth penetration of *Fucus vesiculosus* (L.) since the 1940s indicates eutrophication of the Baltic Sea. Mar Ecol Prog Res 28(1):1–2

Löfgren S, Gustafson A, Steineck S, Stålnacke P (1999) Agricultural development and nutrient flows in the Baltic states and Sweden after 1988. Ambio 28(4):320–327

Rahm L, Jönsson A, Wulff F (2000) Nitrogen fixation in the Baltic proper: an empirical study. J Mar Syst 25:239–248

Redfield AC, Ketchum BH, Richards FA (1963) The influence of organisms on the composition of sea-water. In: Hill N (ed) The sea, vol 2. Interscience, New York, pp 26–77

Rosenberg, R, Elmgren R, Fleischer S, Jonsson P, Persson G, Dahlin H (1990) Marine eutrophication case studies in Sweden. Ambio 19:102–108

Rönner U (1985) Nitrogen transformations in the Baltic proper: denitrification counteracts eutrophication. Ambio 14:135–148

Savchuk O (2000) Studies of the assimilation capacity and effects of coastal load reductions in the eastern Gulf of Finland with a biogeochemical model. Boreal Environ Res 5:147–163

Savchuk O, Wulff F (1999) Modeling regional and large-scale responses of the Baltic Sea ecosystems to nutrient reductions. Hydrobiologia 393:35–43

Schelske CL, Stoermer EF, Conley DJ, Robbins JA, Glover R (1983) Early eutrophication in the lower Great Lakes: new evidence from biogenic silica in sediments. Science 222:320–322

Stålnacke P (1996) Nutrient loads to the Baltic Sea. Thesis, Linköping University. Linköping Studies Art Sci 146:78

Sweitzer J, Langaas C, Folke C (1996) Land cover and population density in the Baltic Sea drainage basin: a GIS database. Ambio 25(3):191–198

Tyrrell T (1999) The relative influences of nitrogen and phosphorus on oceanic primary production. Nature 400:525–531

Wikner J, Tamminen T, Rahm T, Rahm L (1996) Background document for the expert meeting on sensitive/nonsensitive areas for eutrophication in the Baltic Sea caused by nitrogen. EC BETA 4/96, 20/1 annex 3, Helsinki Commission–Baltic Environment Protection Commission, Helsinki, pp 15–40

Wulff F, Niemi Å (1992) Priorities for the restoration of the Baltic Sea – a scientific perspective. Ambio 21(2):193–195

Wulff F, Stigebrandt A, Rahm L (1990) Nutrient dynamics of the Baltic Sea. Ambio 29(3):126–133

2 Physical Oceanography of the Baltic Sea

A. STIGEBRANDT

2.1 Objectives of this Chapter

The physical transport system of the Baltic Sea is composed of currents and mixing processes. It may in principle explain the distribution of sea salt if the freshwater supply is known. To explain the large-scale ecology and biogeochemistry of the Baltic Sea, however, is a more difficult task since it requires understanding of not only the physical transport system but also important ecological and biogeochemical processes. The main objectives of this chapter are to describe the large-scale physical circulation of the Baltic Sea and the major processes of importance to this.

The large-scale stratification of the Baltic Sea is essentially controlled by the freshwater supply and the supply of seawater. To understand the latter, the roles played by the Belt Sea and Kattegat in the mouth have to be understood because both salinity and flow rate of the 'seawater' entering the Baltic proper are strongly dependent on processes operating in these seas. Furthermore, the vertical stratification and turnover of the deep waters in the different Baltic Sea basins are controlled by interplay between the supply of new deep water and small-scale diapycnal mixing processes operating in the basins.

The distribution of sea salt is governed only by physical processes and external forcing, albeit under strong topographical influence. It is then natural to discuss the large-scale circulation and major physical processes, taking the actual distribution of salinity as a starting point. Among important processes in the Baltic Sea to be discussed in this chapter are: different kinds of strait flow; gravity-forced dense bottom currents; dynamic control of pools of light and dense water at the sea surface and bottom respectively; mixed-layer dynamics; diapycnal mixing in the deep water; and dispersive processes in the surface layer. To complete the picture of the Baltic Sea, obtained from the salt field, observed distributions and variabilities of temperature and sea level variability will be discussed. The

Ecological Studies, Vol. 148
F. Wulff et al. (eds.)
A Systems Analysis of the Baltic Sea

external forcing, which is vital to the state of the Baltic, is described in Bergström et al. (Chap. 3).

2.2 Introduction

To help readers unfamiliar with the physical oceanography of the Baltic Sea to construct a qualitative model of the functioning of the Baltic Sea in their own minds, this introductory section provides a summary of the basic large-scale features of the physical transport system of the Baltic Sea. Later in this chapter the different dynamic processes will be described and discussed in more detail.

The Baltic Sea is composed of a series of connected huge basins with a vast, shallow and partly narrow connection to the ocean. The large-scale vertical circulation in all the basins is of the type encountered in diffusive, stratified filling boxes. This means that waters of different densities are supplied to strata at different levels of a stratified basin where only gentle mixing occurs. Freshwater and water of low salinity from neighboring landward basins are supplied at the sea surface and waters of higher salinity, from the neighboring seaward basin, are supplied at greater depths. A rather homogeneous mixture of these waters occupies the upper layers from which water exits to the neighboring seaward basin (Fig. 2.1).

The Baltic proper, by far the greatest of the filling boxes, is used here to exemplify the features of the large-scale circulation. This filling box is complicated by a number of factors. Firstly, the water exchange across the shallow entrance sills is mainly forced by weather-related sea level variations outside the entrance sills, giving the water exchange an event-like character and making greater inflows, lasting several weeks, infrequent and unpredictable. Secondly, the 'sea waters' flowing into the Baltic proper are mixtures formed outside the entrance sills of low-saline Baltic proper surface water and salty seawater. They span a wide range in salinity depending on the varying degree of recirculation of Baltic proper surface water in the entrance area. Thirdly, due to basin topography the inflowing waters form a suite of stratified pools of deep water in the Baltic proper, connected by gravity-forced, dense bottom currents. Fourthly, dense bottom currents entrain ambient water, by which the volume flow increases and salinity decreases. Interleaving in the downstream deep-water pool occurs at the depth where the density of the flow equals the density of the ambient water. Fifthly, diapycnic or vertical mixing (i.e. mixing across density surfaces) in the deep-water pools, transporting sea

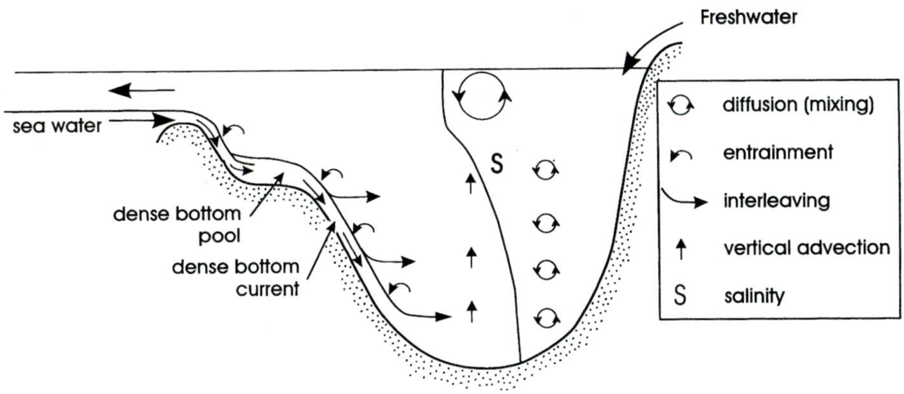

Fig. 2.1. Important dynamic elements of a stratified filling box of the type encountered in the Baltic Sea

salt upwards, is known only quantitatively, but in the virtually tide-less Baltic it should, by different mechanisms, ultimately be forced by the wind.

The Baltic Sea is one of the major estuaries on earth that has the character of a multi-basin fjord, with a broad-crested sill located in the mouth. The long-term net freshwater supply amounts to about 16,000 $m^3 s^{-1}$, as discussed in Bergström et al. (Chap. 3) and the surface area inside the entrance sills is about 370,000 km^2. The Belt Sea occupies the crest of the entrance sill. On the oceanic side of the Belt Sea is the shallow Kattegat that in the eastern part deepens towards the Skagerrak. On the Baltic side is the Arkona Sea, the westernmost part of the Baltic proper. In oceanographic literature the Belt Sea and Kattegat, with a total surface area of ca. 42,000 km^2, are often considered to belong to the Baltic Sea. The Sjælland Island splits the flow between Kattegat and the Arkona Sea into two branches, one through the relatively short and partly quite shallow Öresund and the other through a longer channel system constituted by the parallel Great and Little Belts, separated by the Fyn Island, coupled in series with the Fehmarn Belt (Fig. 2.2). The sill depths of the Fehmarn Belt (Darss Sill) and the Öresund are only 18 and 8 m, respectively.

In the vast shallow area in the mouth, outflowing Baltic surface water (~8 psu) is mixed with quite saline Skagerrak water (~33 psu) which fills up Kattegat beneath the surface layer. The mixtures form surface layers in the Belt Sea and Kattegat, typically 5–15 m in thickness and of relatively low salinity. However, the salinity of the surface water varies greatly in time in response to the recent history of in- and outflows and wind-induced mixing and there is a north-south gradient with increasing salinity towards the Skagerrak. Thus, the Kattegat and Belt Sea usually store large

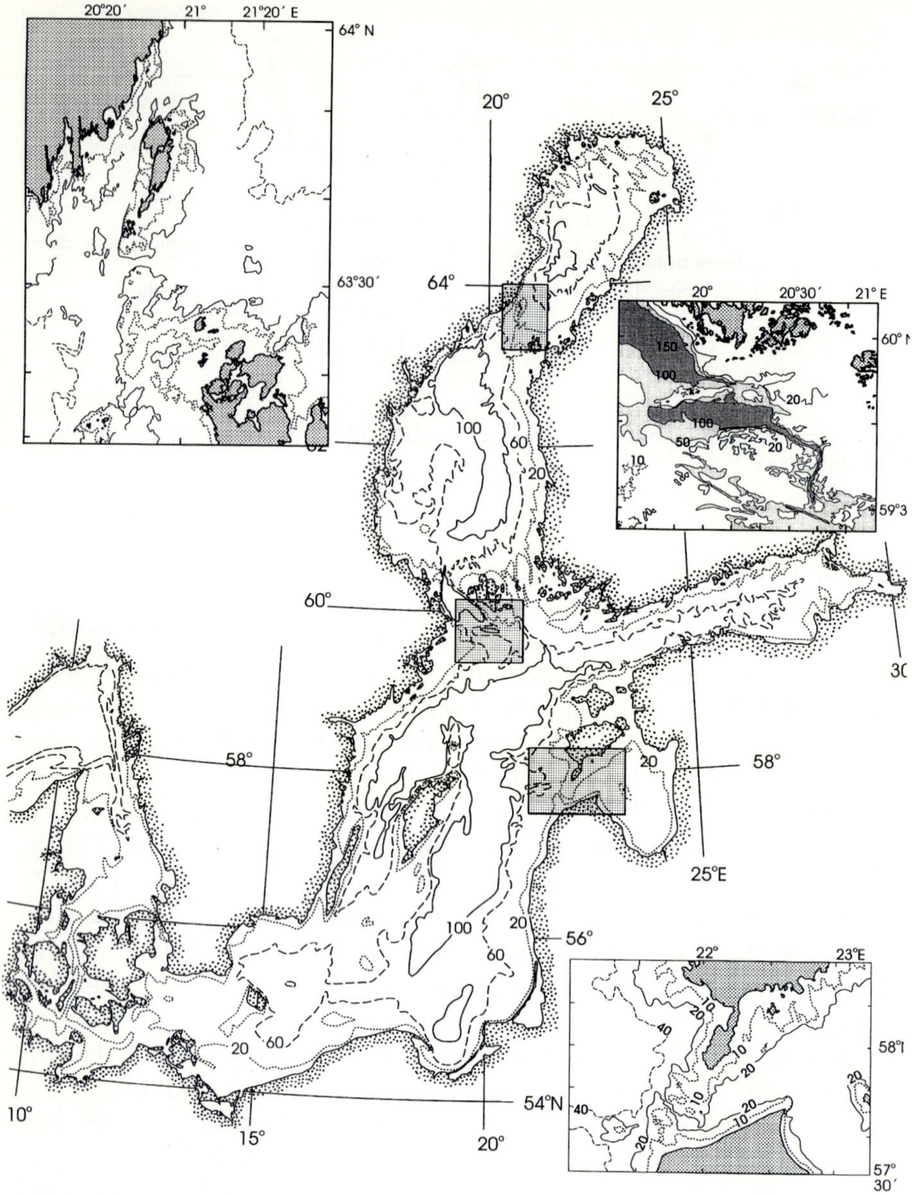

Fig. 2.2. Bathymetric map of the Baltic Sea. Topography of the major straits is shown in the *enlargements*

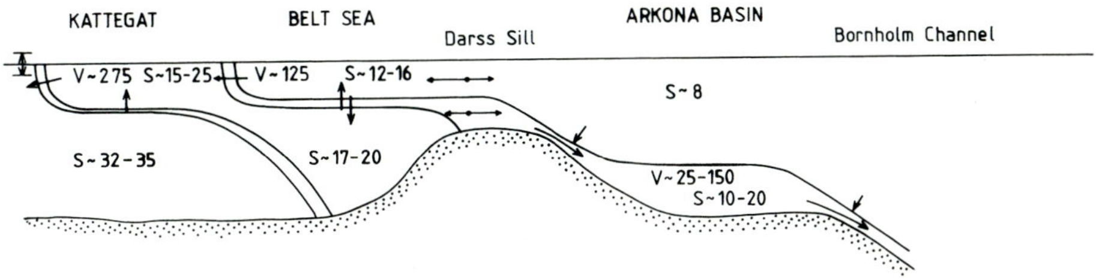

Fig. 2.3. Typical flows (Q; km^3 day^{-1}), volumes (V; km^3) and salinities (S; psu) of water masses in the Baltic entrance area

amounts of low-saline water in their surface layers (Fig. 2.3). The circumstance that the thickness of these layers is usually approximately equal to the sill depths of the entrance straits implies an efficient blocking of direct inflow of high-saline Kattegat deep water to the Baltic Sea. This is one of the major reasons why the salinity of the Baltic Sea is comparatively low.

To drive a current with the capacity to transport the freshwater surplus received from the Baltic Sea into Skagerrak, for transport further into the ocean, the surface layer in Kattegat has to have excess pressure relative to Skagerrak. Since northern Kattegat is rather deep, the excess pressure has to be baroclinic, i.e. primarily caused by the density difference between the surface waters in Kattegat and Skagerrak. Thus, the low-saline water in Kattegat has to form a surface pool of such a thickness that the resulting baroclinic excess pressure is sufficient to force the freshwater obtained from the Baltic into Skagerrak. The front, i.e. the border at the sea surface between the Kattegat surface water pool and Skagerrak surface water, is often quite sharp and oriented roughly west–east, as seen in satellite imagery taken under cloudless conditions (e.g. Aarup 1994). However, close to the Swedish coast the front bends northwards and the flow of surface water from Kattegat becomes parallel to the coast (Fig. 2.4). Because of the relatively large width of the Kattegat, the rotation of the earth influences both the shape and the baroclinic transport capacity of the front (see Stigebrandt 1983; Jacobsen 1997). Several self-controlling pools of dense deep water, dynamically similar to the Kattegat surface pool have been identified both in the entrance area and within the Baltic Sea, as further discussed in Section 2.8.2 below.

In the shallowest parts of the straits in the Belt Sea and Öresund the flow is mainly barotropic, driven by the momentary sea level difference between the southern Kattegat and the western Arkona Sea. Since the sea level difference is quite variable, the flow also varies quite a lot. Instantaneous

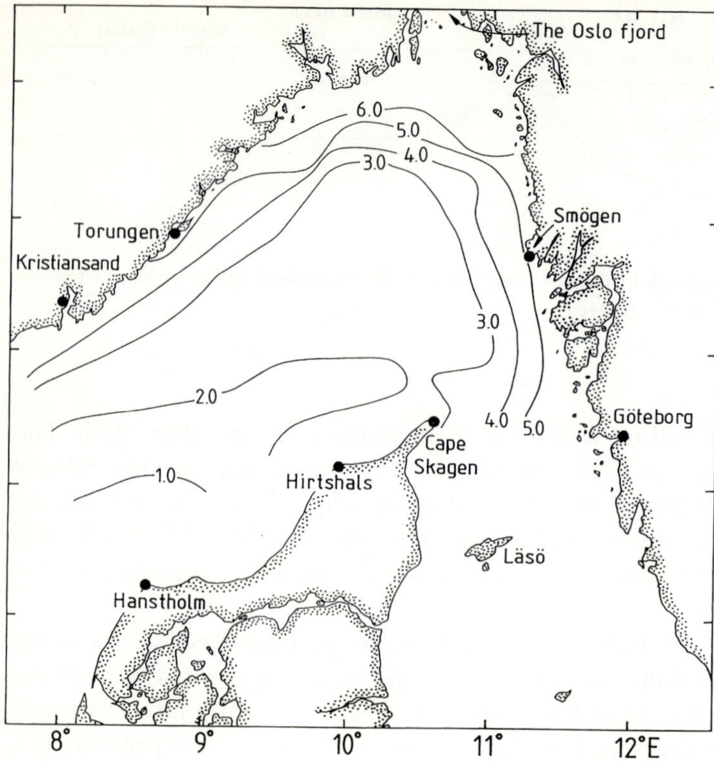

Fig. 2.4. Mean distribution of integrated freshwater content (m) in Skagerrak suggests that the freshwater comes essentially from Kattegat. (Gustafsson and Stigebrandt 1996)

inflows and outflows are in the range 0–300,000 m^3 s^{-1}, with salinity varying between 8 and 28 psu. According to Jacobsen (1980) the partition of the transport between the two routes is about 73 and 27 % through the Fehmarn Belt and Öresund, respectively, but Mattsson (1995) found that the partition is rather 80/20. The salinity of the water transported in the two routes may, however, differ because the relative importance of the routes with respect to salt transports may be different from that of volume transports, as discussed by Mattsson (1996a); see Section 2.6 below.

The flows of sea salt and water into the Baltic proper are evidently determined by the distributions of salinity and current at the boundary between the Baltic proper and the Belt Sea/Öresund. The salinity of the water forced by barotropic currents into the Baltic proper varies quite a lot, depending on the previous history of outflows of low-saline Baltic surface water and diapycnic mixing in the Belt Sea and Kattegat with underlying saltier water, as described above. To compute the volume flow and salinity

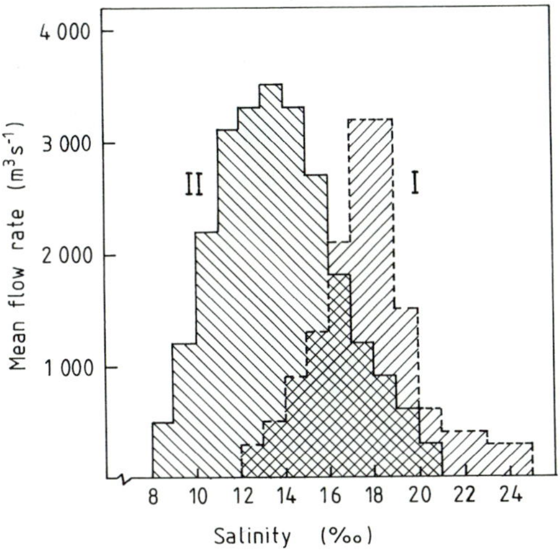

Fig. 2.5. Distribution of flows into the Baltic Sea among different salinities at entrance sills (*I*) and in the Arkona Sea (*II*), respectively. Distribution *I* is a smoothed version of model computations (Stigebrandt 1983) and *II* is a slightly smoothed version of a distribution computed from historical hydrographical data (Stigebrandt 1987b), as discussed further in Section 2.2.8

of the flow into the Baltic Sea one obviously needs a model that is capable of computing not only volume flows but also the vertical and horizontal salinity distributions in the Belt Sea and Kattegat. Already, relatively simple baroclinic models, using a quite rough partition of water masses into a few model boxes, combined with a barotropic model, are able to compute inflows in a reasonably realistic way (cf. Stigebrandt 1983; Gustafsson 1997). A smoothed version of the mean distribution of inflows among different salinities computed by such a model is shown in Fig. 2.5. The most salient feature is that the inflows are distributed over a wide range of salinities. Flows contributing to the high salinity end of the distribution occur only very seldom. The classical description of the mean inflow to the Baltic Sea, due to Knudsen (1900), is a constant inflow of water (15,000 m^3 s^{-1}) of constant salinity (17.4 psu). The fact that the flow rate and salinity of water flowing into the Baltic Sea vary quite a lot is the reason why the deep water in the Baltic proper is strongly stratified, as further discussed in Section 2.8. The strong stratification implies that only waters having relatively high salinity are dense enough to penetrate down into and replace the water in the deepest portions of the Baltic proper. Since such so-called major inflows are only rarely occurring, the deepest deep water in the Baltic Sea may become stagnant for long periods. Then anoxic conditions develop (cf. Fig. 2.6), as described by Fonselius (1969), Fonselius et al. (1984) and others.

From the circumstance that the salinity of inflowing water is due to the salinity in the Kattegat and Belt Sea, it follows that the salinity of the surface layer of the Baltic Sea (S_B) is controlled by the factors controlling the salinity in the Kattegat and the Belt Sea. The steady state response of S_B

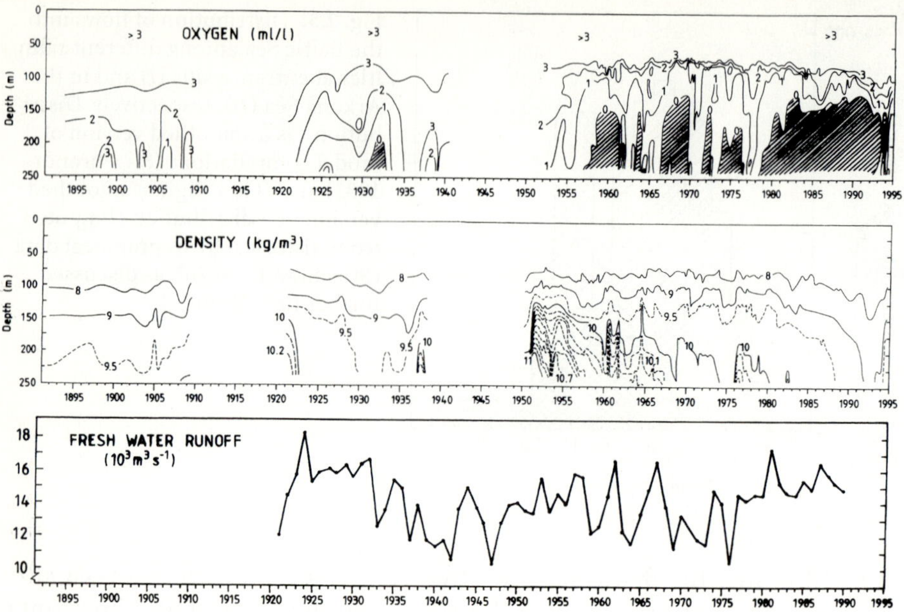

Fig. 2.6. Development of density (expressed as 1000 kg m^{-3}) and oxygen in the Gotland Basin in the period 1895–1994 and runoff of freshwater from land. Runoff data for the period 1921–1949 are from Mikulski (1982) and for the period 1950–1990 from Bergström and Carlsson (1994)

to permanent changes in the forcing controlling the salinity in the Kattegat/Belt Sea, as computed by a model, is shown in Fig. 2.7. It appears that S_B is quite sensitive to changes in the net freshwater supply Q_F to the Baltic Sea and also to changes in the barotropic flow. The sensitivity to changes of the regional wind, over the entrance area, and the salinity of the Skagerrak water, respectively, seems to be much less (Stigebrandt 1983). The response time of the salinity of the Baltic Sea to permanent changes of the inflow is given by the flushing time of the Baltic (~25–30 years). However, changes are seen early in the permanent halocline (cf. Fig. 2.8), where most of the new deep water is supplied (Fig. 2.9).

The water flowing in from the Belt Sea and Kattegat is denser than the Baltic surface water, which is why the flow occurs along the seabed. In the Bornholm Channel the thickness of the bottom current is typically 5–10 m, as observed on numerous occasions by Petrén and Walin (1976) who were the first to estimate the distribution of the deep-water flow with respect to salinity. The bottom current entrains ambient water by which it becomes less saline at the same time as the volume flow increases. The flow distribution with respect to salinity therefore changes with the distance

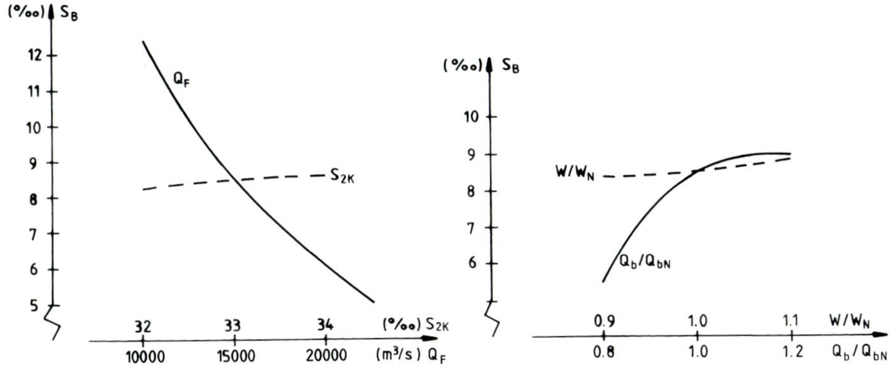

Fig. 2.7. Steady-state response of surface salinity (S_B) in the Baltic proper to permanent changes in total freshwater supply to the Baltic (Q_F), salinity (S_{2K}) of Skagerrak water (*left*), normalized mixing wind (W/W_N) in the Belt Sea/Kattegat and normalized barotropic flow (Q_b/Q_{bN}) through Öresund and the Fehmarn Belt (*right*). (Redrawn from Stigebrandt 1983)

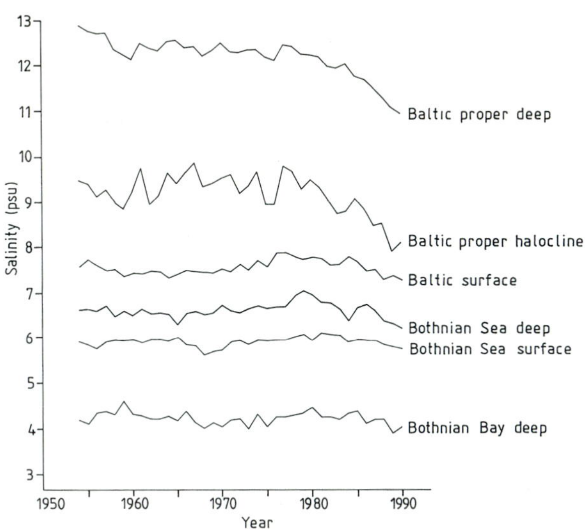

Fig. 2.8. Annual mean salinity in the period 1954–1990 in surface, halocline and bottom waters in the Baltic proper, Bothnian Sea and Bothnian Bay. (Redrawn from Samuelsson 1996)

from the entrance sills, cf. the difference between the distributions I and II in Fig. 2.5. Owing to topographical reasons, however, there is no coherent bottom current carrying the inflowing, 'new' deep water all the way from the entrance sills to the central parts of the Baltic proper. Actually, most of the transportation of the new deep water further into the Baltic takes place in dense bottom pools. Dense bottom currents only transport the new deep water from one pool to the following, like a river transporting water bet-

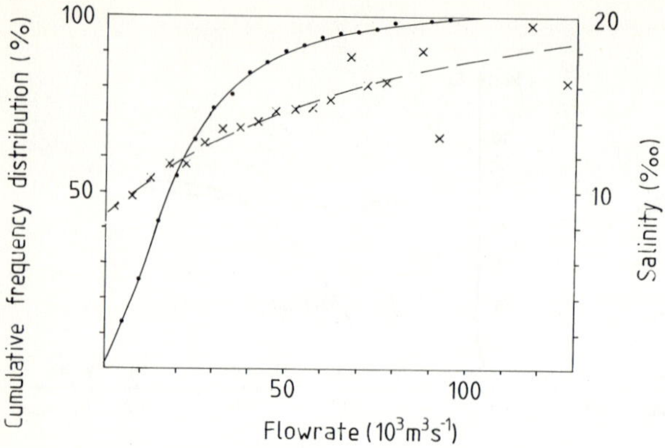

Fig. 2.9. Cumulative frequency distribution of volume flows (*solid line*) and observed mean salinity of inflows (*crosses – broken lined* fitted visually) for different classes of volume flows. (Stigebrandt 1987b)

ween subsequent river-lakes. The first in the suite of pools encountered by new deep water is situated in the Arkona Sea. It has been suggested that the flow of deep water out of a pool is self-controlled. The flow may then be computed from just the vertical stratification of the pool (Stigebrandt 1987b). In the case of the Arkona Sea, the deep-water flow should take place along the northern rim of the pool where a baroclinic-geostrophic force balance is assumed. This is supported by observations of the flow in the Arkona pool as described in Liljebladh and Stigebrandt (1996; see Sect. 2.8 below; Fig. 2.11).

Superimposed upon the basic filling-box circulation, driven by advective inflows and diapycnal mixing, is a circulation driven by wind stress and modified by buoyancy fluxes through the air–sea interface. This has very strong influence on the vertical stratification in the upper layers, as further described in Section 2.6. During seasons with weak stratification and negative buoyancy fluxes through the sea surface (increasing the density of the surface water), wind-driven turbulence manages to homogenize an up to about 60-m-thick surface layer in the Baltic proper. Below this, in the deep water, seasonal variations are much weaker and, as already mentioned, the water is strongly salt-stratified down to greatest depths, starting with very strong stratification (a halocline) in the uppermost about 10 m of the deep water.

In the deep water of the Baltic proper filling box there is an approximate, long-term balance between downward advective salt supply, by inflowing new deep water, and upward diffusive salt loss, by diapycnal mixing. Thus, if one of these flows is known, then also the other is known. The rate of diapycnal mixing has been estimated from the observed inflow, using different kinds of one-dimensional advective-diffusive filling-box models (see Sect. 2.8). Accordingly, diapycnal mixing is known to about the same accuracy as the inflow of deep water. However, from this approach it may be hard to separate the contribution to the total diapycnal mixing by mixing in dense bottom currents and by mixing in pools, respectively, although the observed maximal salinities in the different pools provide constraints on mixing performed by the dense bottom currents. From observed salinity changes during stagnation periods, i.e. periods without vertical advection, it is possible to obtain estimates of the mixing in the pools that are independent of both inflow estimates and mixing in dense bottom currents. The detailed mechanics of diapycnal mixing processes are poorly known. This is particularly true for mixing processes in the deep-water pools, which in some way or another should be forced by the wind since astronomical tides are exceedingly small in the Baltic Sea (see Sect. 2.8).

Although details may differ, the brief sketch of the dynamics and circulation of the Baltic proper given above also applies to the other more remote basins in the Baltic Sea that all have positive freshwater balances and obtain deep water from their seaward, neighboring basin. This is also true for the large number of small inshore basins along the rims of the major sub-sea of the Baltic. As a result, the salinity of the deep waters decreases in the direction away from the Baltic proper (Fig. 2.10). However, compared to the Baltic proper much less is usually known about both diapycnal mixing in the basins and the mean distributions of inflows among different salinities, although the salinity ranges of the distributions certainly are much narrower than the range for the inflow to the Baltic proper.

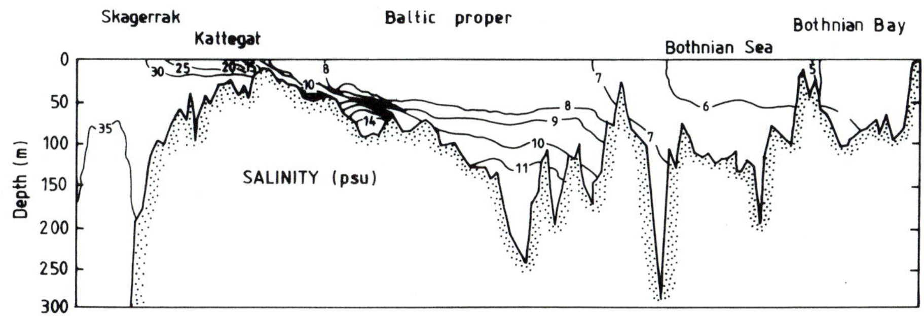

Fig. 2.10. Distribution of salinity in 1988 in a vertical cross section from Skagerrak to the Bothnian Bay. (Andersson et al. 1992)

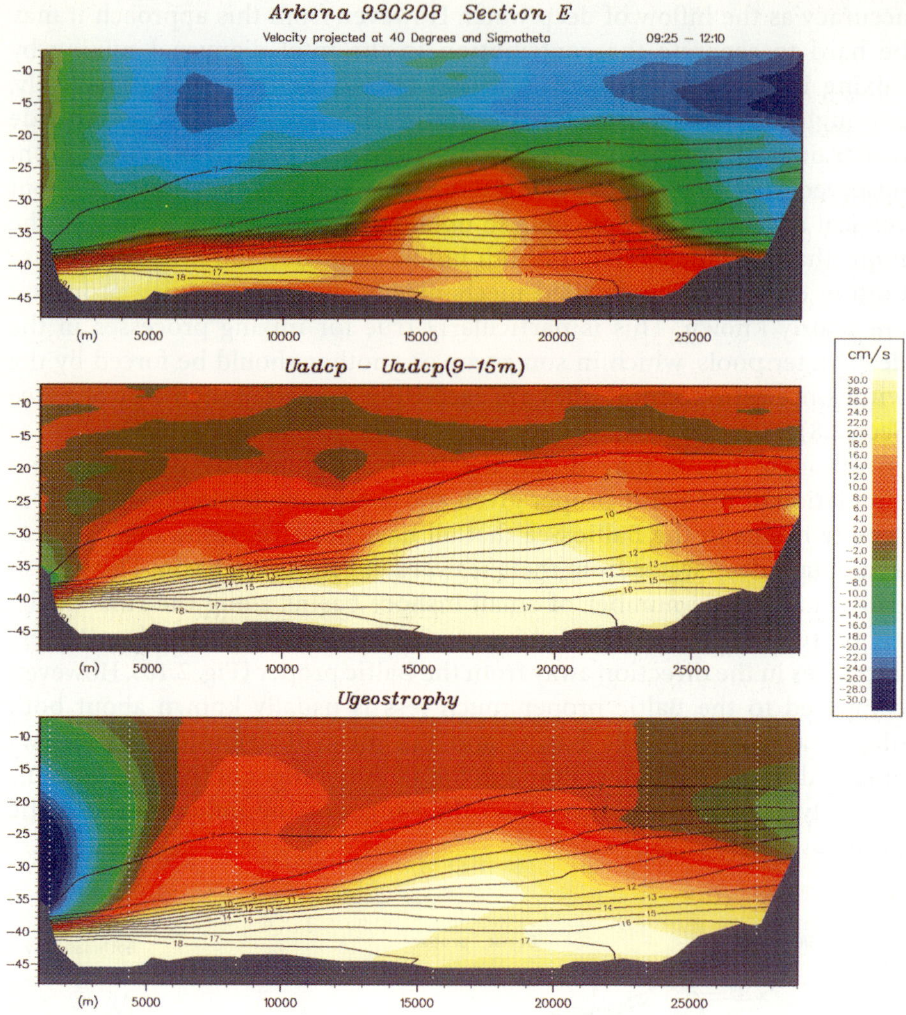

Fig. 2.11. Currents and density (expressed as 1000 kg m^{-3}) in a vertical cross section of the Bornholm Strait, 7 February 1993. Measurements were taken using a ship-mounted ADCP and dense CTD casts. *Above* Total current field; *middle* baroclinic current, obtained by subtraction of barotropic current field as determined in the pycnocline; *below* baroclinic geostrophic current as computed from density field (CTD casts, *dotted lines*). The Swedish coast (Svartgrundet) is to the *left* and Bornholm (Rönne) to the *right*. (Liljebladh and Stigebrandt 1996)

2.3 Overall Description of the Physical State of the Baltic Sea

The overall distributions of sea salt, temperature and currents in the Baltic Sea are described in this section. Also, seasonal and long-term variations are shown. The sea level and its variability are discussed separately in Section 2.5. Overviews of the physical state of the Baltic Sea are given in Kullenberg (1981) and in the nicely illustrated *Swedish National Atlas* (Sjöberg 1992) from which several of the Figures in the present chapter have been adopted. The Gulf of Bothnia is described by Håkansson et al. (1996). The physical state of Skagerrak is well described in a special volume of *Journal of Sea Research*, e.g. Rodhe (1996), and by Gustafsson (1997). The conditions in Kattegat are described in, for example, Svansson (1984), Rydberg (1987) and Gustafsson (1997).

2.3.1 Salinity

The distribution of salinity in a vertical cross section from Skagerrak to the Bothnian Bay (Fig. 2.10) shows that the surface salinity in Kattegat and the Belt Sea is rather low. This is essentially due to the outflow of large amounts of Baltic proper surface water of low salinity (7–8 psu) while local freshwater supplies are only of minor importance. It is mainly water from these surface layers that flows into the Baltic Sea in periods when the sea level stands higher in Kattegat than in the southwestern Baltic Sea. The Baltic proper is strongly stratified below the wintertime homosaline layer extending down to the halocline at about 60 m depth. As explained in Section 2.2, both the volume flow and the salinity of the water entering through the Belts and Öresund vary strongly. The water is saline enough to penetrate down into the deepest parts of the basins of the Baltic proper only during so-called major inflows. In the time interval between such events the water is said to be stagnant and vertical diffusion transports salt upward, whereby the salinity (and density) in the deep water decreases. During stagnation periods, the oxygen content decreases due to consumption connected with the decomposition of organic matter, and may become exhausted below about 130 m in the Baltic proper (Fig. 2.6). As demonstrated in Section 2.8.3, the deep water at higher levels is always well flushed, preventing anoxia from spreading higher up in the water column (cf. Fig. 2.21).

As a consequence of the freshwater supply to the Bothnian Sea and Bay the salinities in these seas are lower than in the Baltic proper. The deep waters are rather strongly stratified also in these basins (Fig. 2.10). The

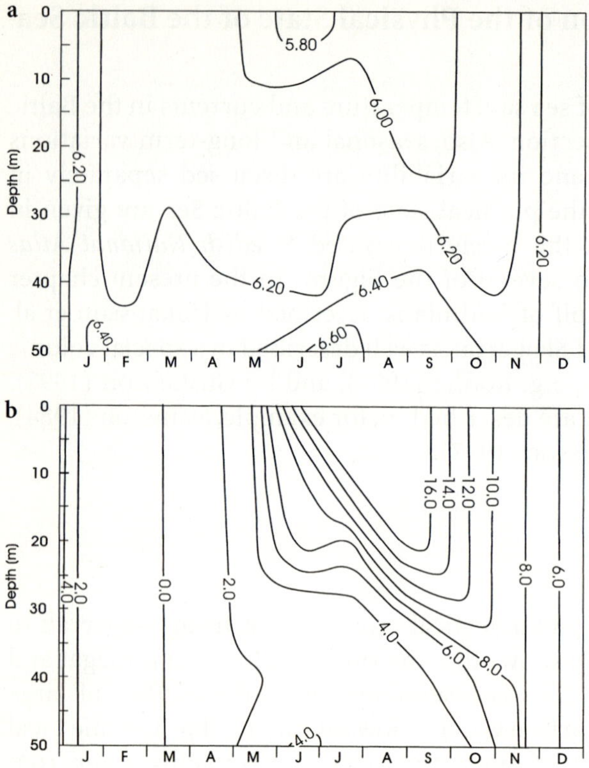

Fig. 2.12. Annual cycles of **a** salinity and **b** temperature in the central part of the Gulf of Riga in 1974

annual cycles of salinity and temperature in the Gulf of Riga are shown in Fig. 2.12. It can be seen that thermal stratification during summer to a high degree protects the deep water from wind-forced turbulent mixing, thus saltier water flowing in from the Baltic proper may build up a salt stratification below the thermocline.

A rough, diagnostic estimate of the long-term net flow through the straits connecting the different sub-basins may be obtained using conservation of salt and volume under stationary conditions. For the calculations it is assumed that the mean salinity of inflowing (Q_0) and outflowing (Q_1) waters is S_0 and S_1, respectively. Under stationary conditions the following equations apply:

$$Q_1=Q_0+Q_F \tag{2.1}$$

$$Q_0S_0=Q_1S_1 \tag{2.2}$$

Here, Q_F is the long-term mean freshwater supply landward of the strait.

Equations (2.1) and (2.2), known as Knudsen's relationships, constitute a budget model that often gives useful first-order estimates of the mean flows of volume and salt through straits. The Knudsen volume exchange between the major sub-basins in the Baltic has been computed by, for example, Wulff and Stigebrandt (1989). Yurkovskis et al. (1993) computed the exchange with the Gulf of Riga. As described in Section 2.4.1 below, estimates of the net freshwater supply to the Baltic have increased in the last few years, mainly due to decreased estimates of the evaporation from the Baltic. Recomputed Knudsen volume exchanges, based on the new figures for the net freshwater supply but using the salinities from Wulff and Stigebrandt (1989), are given in Fig. 2.13. The water exchange with the Gulf of Riga is taken from a recent estimate presented in Eilola and Stigebrandt (1998). Dynamic calculations of water exchange between the major sub-basins are discussed in Section 2.6.

Long-term salinity variations in the Baltic proper and the Bothnian Sea and Bay are evident from Fig. 2.8. Most conspicuous is the large salinity decrease in the halocline and deep water of the Baltic proper during the 1980s. The salinity decrease starts, and has greatest amplitude, in the halocline, indicating that it is caused by a decreasing salinity of the inflowing 'seawater'. It can be seen that the salinity variations in the three basins are fairly parallel, suggesting a close coupling of the basins.

The seasonal variation in salinity in the Baltic is particularly evident in the surface layers. In spring and summer 'juvenile freshwater', from rivers and low-saline water from adjacent seas, is trapped in the surface layer by the seasonal thermal stratification, leading to a minimum salinity at the end of the summer (Fig. 2.14b). The salinity is restored by entrainment of

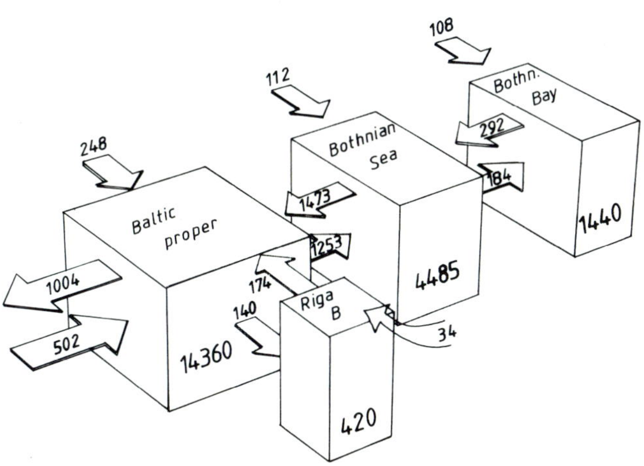

Fig. 2.13. Mean water exchange (km^3 year^{-1}) through major straits of the Baltic Sea. Also shown is net freshwater supply by runoff from land to the different subseas. *Numbers* near base of boxes show volumes of basins (km^3)

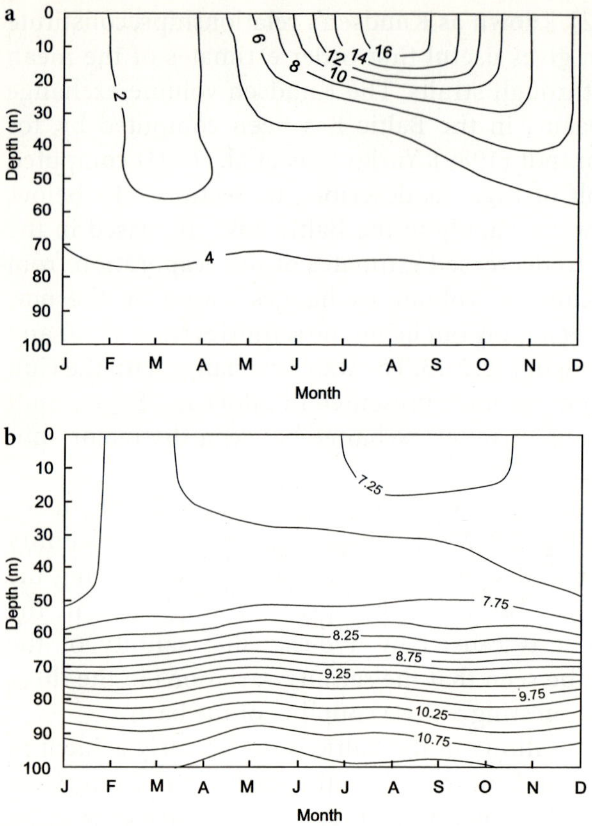

Fig. 2.14. Mean seasonal cycles of **a** temperature and **b** salinity in the upper Baltic proper for the period 1959–1975. Data from the East Gotland Basin

saltier water at the permanent halocline in late autumn and winter, as further discussed in Section 2.7 below.

The residence time of water in the Baltic Sea, as determined by the ratio between the total basin volume (21×10^3 km^3) and the mean rate of total inflow (freshwater: ~16,000 m^3 s^{-1} + seawater (35 psu): ~4000 m^3 s^{-1}), is about 32 years. This gives the hydrographical state of the Baltic large inertia to short-term variations in the external forcing. The physical state of the Baltic has been recorded during a century, which thus is about four times longer than the residence time for water and conservative substances. The most important forcing functions have been reasonably well known by measurements since about 1920, although the meteorological forcing is not yet available in a useful format for the whole period. Thus, using historical data it should be possible to learn quite a lot about the response of the Baltic to observed long-term variations in the forcing functions. Indeed, there have been many attempts to explain observed long-term hydrographical variations in terms of changed forcing. However, many of these seem to be purely qualitative, trying to explain observed changes in the state focusing on the variation of one single component of the forcing. Since the system is quite complex with intricate non-linear feedback loops, analyses of changes in the state of the Baltic Sea should be based upon at least a good conceptual understanding or, preferably, upon an appropriate quantitative model.

2.3.2 Temperature

Solar radiation drives many important processes in the sea. Almost all the heat received by solar radiation is sooner or later submitted to the atmosphere by long-wave radiation, evaporation and conduction. The accompanying energy fluxes involve important short-term changes in the temperature and its vertical distribution in the sea. A small fraction of the solar radiation fuels phytoplankton and other plants and by that the ecological systems, and even this part of the radiation ultimately ends up as heat when organic matter is respired. The biological energy loop is coupled to the loops of carbon, plant nutrients and oxygen. The solar radiation and its variability over the southern Baltic Sea are measured and discussed by Dera (1983), Dera and Rozwadowska (1991) and Bergström et al. (Chap. 3) who also describe the seasonal ice cover of the Baltic Sea. The heat budget of the Baltic Sea is discussed in Section 2.4 and the dynamics of the surface layers in Section 2.7. In the present section only the distribution of temperature and its variability are described.

The distributions of temperature in summer and winter in a vertical section from Skagerrak to the Bothnian Bay are shown in Fig. 2.15. The surface layer shows a large annual temperature cycle (see also Fig. 2.14a), while deep water in the Baltic Sea basins, below the perennial haloclines, shows almost no seasonal variation.

The seasonal temperature cycle is typical of waters of relatively high latitudes, with strong seasonal variation in solar radiation, although the amplitude is greater in the Baltic Sea than in the open Atlantic Ocean (Fig. 2.14a). This is a result of the great freshwater supply and the accompanying strong salt stratification in the Baltic, limiting the vertical extension of wind-forced and thermohaline convection in winter to the permanent halocline. A permanent seasonal feature of the vertical temperature distribution is the cold winter water residing below the seasonal thermocline during the whole summer.

In cold winters the surface water in the Baltic proper is cooled to temperatures below the temperature of maximum density (T_{max})

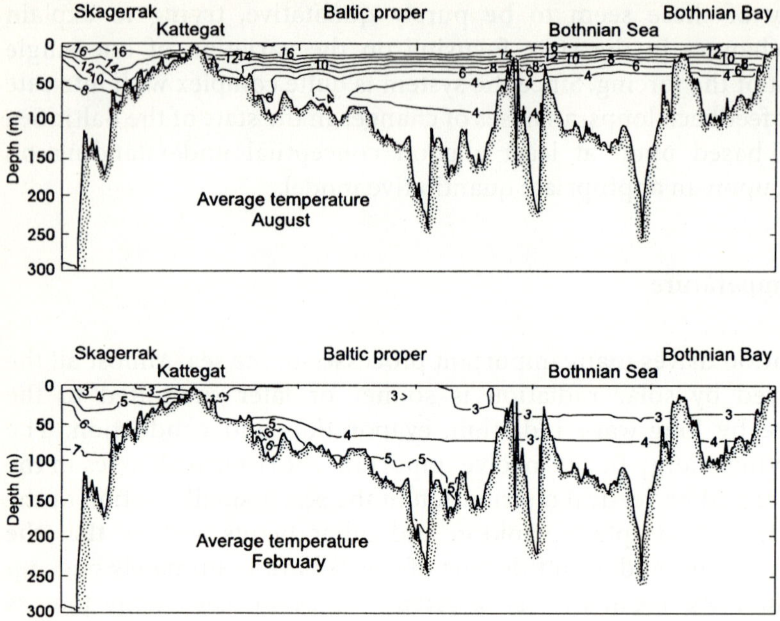

Fig. 2.15. Distribution of temperature in a vertical cross section from Skagerrak to the Bothnian Bay in winter and summer. (Andersson et al. 1992)

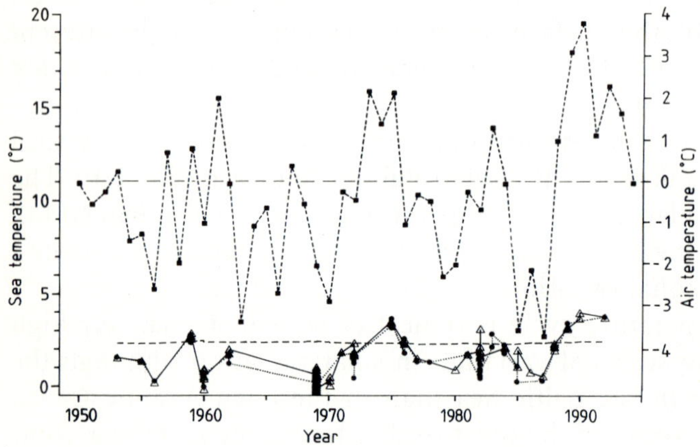

Fig. 2.16. Temperature at 50-m depth in June in the Baltic proper in the period 1950–1990 (upper line). Also shown is winter (January to March) air temperature over the western Baltic proper as observed at Hoburgen in the southern part of Gotland (lower line). (Eilola 1997)

(T_{max}=4–0.22 S, where S is the salinity in psu). It is observed that the development of the seasonal pycnocline in springs following cold winters occurs before the surface water has been heated to T_{max} (Fig. 2.16). This is due to the stabilizing effect of the freshwater supply by rivers, as explained by Eilola (1997; see also Sect. 2.7).

Upwelling at coasts occurs when the wind-driven so-called Ekman transport (directed to the right of the wind direction) is directed seawards (e.g. Gill 1982). In the warm seasons, large horizontal temperature gradients may develop close to coasts due to upwelling of water of lower temperature, as described by Walin (1972). Gidhagen (1987) made a thorough analysis of the occurrence of upwelling in the Baltic using satellite imagery. Kahru et al. (1995) analyzed thermal fronts in the Baltic from satellite imagery. They found particularly persistent and coherent fronts along the eastern coast of the Bothnian Sea where bottom topography is relatively smooth. The irregular topography on the western coast seems to cause more irregular and less persistent frontal systems.

2.3.3 Currents

Several mechanisms operating on wide ranges in time and space force the current fields in the Baltic Sea. On the longest time-scale one finds baroclinic currents related to the freshwater supply. This estuarine baroclinic current system has a filling-box character, with denser water flowing into the deeper parts and fresher water from the surface layers flowing out of the basins. New deep water enters the Baltic proper through a system of dense bottom pools connected by dense bottom currents (Sects. 2.2 and 2.8). Current speeds in the dense bottom currents in the Bornholm Channel rather frequently exceed 25 cm s^{-1} (cf. Petrén and Walin 1976; Fig. 2.11). Baroclinic velocities in the Arkona dense bottom pool have been observed to exceed 40 cm s^{-1} and superposed on these there may be equally strong but rather narrow short-period barotropic currents (Liljebladh and Stigebrandt 1996). East of Bornholm, currents in the deep-water pools of the Baltic proper seem to be generally rather slow but inhomogeneous. Stronger currents (15 cm s^{-1}) occur in anticyclonic, slowly drifting, lens-like eddies in and beneath the halocline (e.g. Zhurbas and Paka 1996).

The long-term mean horizontal current fields in the surface layers of the different basins in the Baltic Sea seem to be quite weak and cyclonic (anticlockwise), with currents from the north along the western and from the south along the eastern margins, respectively. This circulation pattern is consistent with the distribution of surface salinity. There is a wide and slow southward current along the western coast of the Bothnian Sea and the characteristics of this have been described by Omstedt et al. (1993). The synoptic scale wind field is typical of the latitude range of the Baltic Sea with rapidly shifting wind direction and speed as atmospheric highs and lows are passing (see Chap. 3). The accompanying current field of course encompass unsteady (inertial) and steady Ekman currents and different kinds of wave motions forced by convergence/divergence of Ekman transports. As mentioned above, satellite imagery of surface temperature shows strong gradients and fronts, but the accompanying current field may be hard to deduce. Håkansson et al. (1996) present some Lagrangian current measurements with drifters in the Bothnian Bay.

The mean circulation of the surface layers in the Baltic proper in spring and summer may be studied by making use of 'juvenile freshwater' caught by the seasonal thermal stratification. Juvenile freshwater, from river runoff, precipitation minus evaporation at the sea surface and freshwater contained in the water of low salinity coming from adjacent basins, then merely acts like a non-buoyant tracer. The buoyancy of freshwater in the Baltic (S ~7) is only about 20 % of what it is in the ocean (S ~35). Eilola and

a

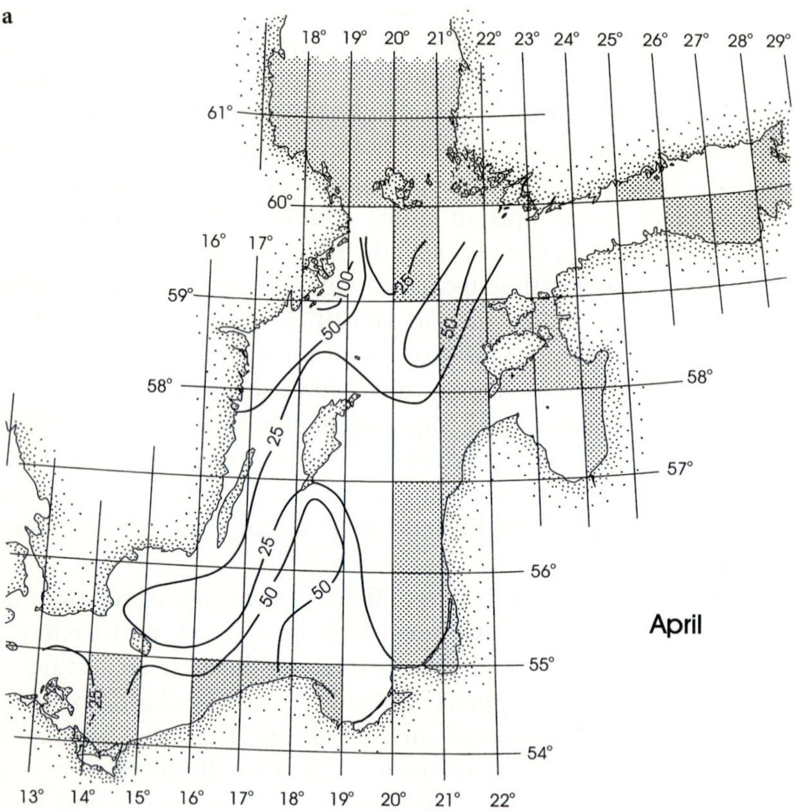

Fig. 2.17. Mean distribution of juvenile freshwater (m) in surface layers of the Baltic proper in **a** April and **b** July. There are insufficient amounts of data from the *shadowed squares*. (Eilola and Stigebrandt 1998)

Stigebrandt (1998) studied the spreading of juvenile freshwater in the Baltic proper. The amount of juvenile freshwater was computed on a monthly basis using historic hydrographic data from squares with a latitudinal-longitudinal resolution of 1×1°. The main routes for the spreading of juvenile freshwater from the Bothnian Gulf and the Gulf of Finland seem to be along the Swedish coast and along the longitudinal central axis east of Gotland. Figure 2.17 shows the distributions of juvenile freshwater in April and July. Only a smaller part of the spreading appears to be due to the weak mean circulation (1–2 cm s^{-1}), so most of the spreading (~5 cm s^{-1}) seems to be due to the action of fluctuating winds, as further discussed in Section 2.7.2.

Relatively strong currents occur due to topographical reasons in the straits connecting the major basins. Strait flows are discussed in

b

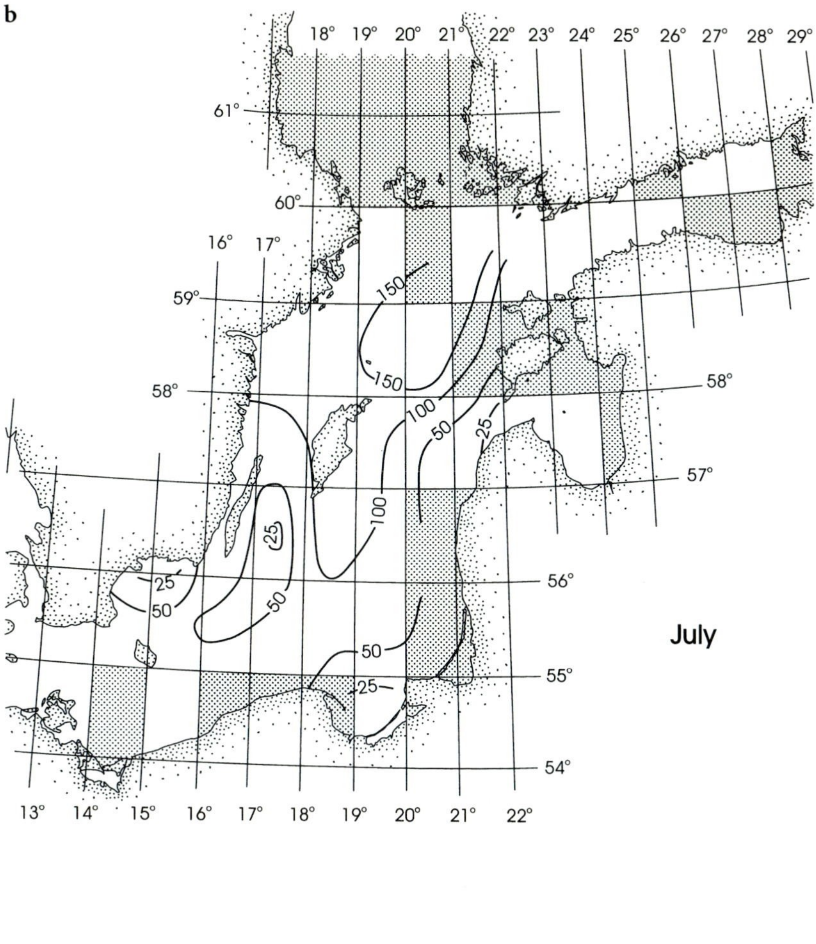

Section 2.6. Currents at the seabed are important to the oxygen supply to the benthic community and to erosion and transport of bottom sediments. It seems that wind waves during strong wind conditions are essential for erosion and sediment deposits. The frequency of resuspension can apparently be predicted from a simple wind-wave model if the topography is realistically described (see Brydsten 1993).

2.4 Water and Heat Budgets

The supplies of heat, freshwater and saline water represent, together with the wind forcing, the major physical forcing of the Baltic Sea. These boundary fluxes also constitute an interface between different academic disciplines because meteorologists need to compute the water and heat exchange through the air-sea interface of the Baltic Sea for weather predictions and hydrologists may offer observations or computations on the freshwater runoff from land. The interdisciplinary character motivates specific sections in this chapter on water and heat balances.

2.4.1 Water Budget

In Section 2.2, it was described how the distribution of sea salt in the Baltic Sea in a rather complicated way is determined by the water exchange with the ocean, the supply of freshwater, mixing processes and water exchange between the sub-basins. Knowledge of the different terms in the water balance is fundamental to an understanding of many aspects of both the physical oceanography and the ecology and biogeochemistry of the Baltic Sea. The short-term water balance may be discussed starting from the continuity equation for volume that reads

$$\frac{dV}{dt} = Q_F + P - E + Q_0 - Q_1 + T_E + S_E \tag{2.3}$$

Here, V is the volume of the Baltic Sea, t time, Q_F the freshwater supply by runoff from land, P the freshwater supply by precipitation on the sea surface, E evaporation from the sea surface, Q_1 (Q_0) outflow (inflow) across the entrance sills and, finally, T_E and S_E volume changes due to thermal expansion and salt contraction, respectively. There are also small volume changes in the Baltic Sea due to the postglacial crustal uplift and sedimentation of particles of continental origin. These contributions are <50 and <5 $m^3\ s^{-1}$, respectively (cf. Winterhalter et al. 1981), and will be neglected in the following. Accumulation of organic matter in the bottom sediments in the Baltic Sea should contribute with appreciably less than 1 $m^3\ s^{-1}$. This estimate is based on permanent burial of less than 15 % of the net production (about 50 g C m^{-2} $year^{-1}$).

The terms on the right-hand side of Eq. (2.3) span at least two orders of magnitude. The largest terms are the flows over the entrance sills, i.e. Q_I and Q_0. These are typically either 0 or 100,000 $m^3\ s^{-1}$, indicating that the flow momentarily is essentially unidirectional but with varying direction. However, in the short term maximum transports may be as great as 300,000 $m^3\ s^{-1}$ (see e.g. Jacobsen 1980). The long-term average of the runoff Q_F inside the entrance sills is about 14,150 $m^3\ s^{-1}$ (1950–1990), but the mean monthly runoff may be in the interval 7000–32,000 $m^3\ s^{-1}$ (see Chap. 3). There are large interannual and seasonal variations in P and E. The long-term annual mean of P is ca. 7300 $m^3\ s^{-1}$ (about 620 mm $year^{-1}$; Dahlström 1986) and E is ca. 5800 $m^3\ s^{-1}$ (493 mm $year^{-1}$; Henning 1988). These values of P and E are, however, generally considered rather uncertain. As argued in Section 2.4.2 below, the rate of evaporation estimated by Henning seems to be overestimated and should probably be reduced by some 20 %. A higher P-E is supported by the results of Omstedt et al. (1997) who also accounted for effects upon evaporation of an ice cover. The long-term net supply of freshwater to the Baltic Sea inside the entrance sills thus seems to be about 16,000 $m^3\ s^{-1}$. The amplitude of the seasonal cycle of T_E is about 1400 $m^3\ s^{-1}$, based upon a sinusoidal steric sea level cycle of amplitude 2 cm due to local heating/cooling.

Jacobsen (1980) discusses in detail the different terms in Eq. (2.3), paying particular attention to measurements and computations of the flow through the Öresund and the Danish Belts. He concluded due to the large superposed fluctuating flow component that it is almost impossible to determine the long-term discharge of the Baltic Sea (i.e. Q_F+P-E) from measurements in the Belts and Öresund. Mattsson (1995) arrived at a similar conclusion from considerations of the precision of current measurements in these straits. Jacobsen (1980) also made a thorough review of earlier attempts to estimate the different elements of the water balance of the Baltic Sea (see also Ehlin 1981; Mikulski and Falkenmark 1986). The latter paper appeared in a special volume from the Helsinki Commission on the water balance of the Baltic Sea.

The terms P-E, T_E and S_E are quite small compared to the other terms in Eq. (2.3) and, besides, they are hard to estimate. Neglecting these terms, this equation may be used to compute short-term volume changes in the Baltic Sea forced by changes in the sea level in Kattegat (h_k) and the freshwater runoff to the Baltic Sea (Q_F). For these computations the term dV/dt is replaced by Adh/dt, where A is the surface area and h the height of the horizontally averaged sea level of the Baltic Sea. Q_1 and Q_2 are taken as proportional to $|h_k-h|$ (Wyrtki 1954) or $|h_k-h|^{1/2}$ (Stigebrandt 1980; see also Sect. 2.6 below). The computations in these papers, and early computations by Bergsten (1933) and Hela (1944), demonstrate very clearly that the water exchange, in terms of volume flows between Kattegat and the Baltic Sea is mainly driven by the instantaneous sea level difference between these seas.

Model computations (Stigebrandt 1983) show that the surface salinity of the Baltic Sea is sensitive to long-term changes in the freshwater supply, as discussed in connection with Fig. 2.6 above. Thus, it should in principle be possible to use salinity changes in the Baltic Sea and high precision estimates of the inflow of saltier water from the Belt Sea and Kattegat to close the long-term water budget.

2.4.2 Heat Budget

The local net heat flux H through the sea surface obtains contributions from several processes and may be computed from

$$H=H_s-H_b-H_c-L\cdot E \tag{2.4}$$

Here, H_s is the absorbed solar radiation, H_b is the effective (net) long-wave back radiation, H_c is the sensible heat flux, $L\cdot E$ is the heat loss due to evaporation E and L is the latent heat of vaporization of water. A positive (negative) value of H leads to heating (cooling) of the surface water. This implies that time series of observed ocean temperatures (heat content) may be used to check estimates of the net heat flux. Monthly means of H_s for the Baltic proper at 54, 57 and 60°N are given by Dera (1983). The fluxes H_b, H_c and E during ice-free conditions may be computed from expressions given by, for example, Gill (1982) provided that air and sea temperatures, air humidity, cloudiness and wind speed are known. Evaporation (E), for instance, is computed from

$$E=\rho_a C_e W\,(q_s-q_a) \tag{2.5}$$

where ρ_a is the density of air, W the wind speed, C_e an empirical coefficient, q_a the specific humidity of air and q_s the specific humidity at the air–sea interface (assumed to be the saturation value at the temperature T of the sea surface). Exchange coefficients for heat and water vapor are given by

Smith (1980), for example. The coefficients used by Henning (1988) are appreciably greater than the coefficients given by Smith and this may be the reason why Henning probably overestimated evaporation from the Baltic Sea, as argued below. There seems, however, to be great uncertainty about the exchange coefficients under high-latitude conditions, as shown by Simonsen and Haugan (1996).

If the surface temperature is at the freezing temperature ($T_f = -k_f S$ where $k_f = 0.054\,°C\ psu^{-1}$ and S is the salinity in psu) and the water is cooled ($H<0$), sea ice forms whereby heat is released. During melting, heat has to be supplied, either from the atmosphere or from underlying water at temperatures above T_f. When the water is ice-covered, heat conduction and evaporation practically vanish (cf. Omstedt et al. 1997). The distribution of sea ice in the Baltic is described in Bergström et al. (Chap. 3).

From a heat budget including his own evaporation estimates, Henning (1988) found that the Baltic Sea on an annual basis must import by advection about 6.7×10^{12} W. This should imply, however, that the inflowing water (~15,000 $m^3\,s^{-1}$) has to be cooled by about 100 K during the stay in the Baltic Sea! The inevitable conclusion is that Henning's heat budget is in serious error and if the whole error is ascribed to the evaporation estimate this should be reduced by some 20 %.

The dynamics of the surface layer of the sea are intimately coupled to buoyancy fluxes due to heat fluxes through the air–sea interface, including evaporation, and to precipitation (and runoff), and there is a mutual feedback between such fluxes and the temperature and salinity of the surface layer. Thus, computations of the rate of evaporation and the other terms in the heat budget should not be done as an isolated task, as in, for example, Henning (1988), but using a seasonal pycnocline model that computes the heat storage and gives the correct feedback due to changes in the surface temperature. From such a model, with negligible advective import of heat from the North Sea, the present author obtained a long-term mean evaporation of about 430 mm $year^{-1}$ for the Baltic proper (cf. Fig. 2.6 in Stigebrandt 1985), which is only about 78 % of the evaporation obtained by Henning (1988) for the same area. A similar computation was done by Omstedt et al. (1997) for the whole Baltic Sea. The dynamics of the surface layer are further discussed in Section 2.7 below.

If the sea surface is at the freezing temperature and the heat flux H is negative [cf. Eq. (2.4)], ice is formed. When an ice cover has been established, the sea becomes thermally insulated and evaporative and other buoyancy fluxes are radically changed, as demonstrated by Omstedt et al. (1997). An ice-cover accumulates precipitation in the form of snow which further affects heat fluxes and short-wave radiation from the sun. Fluxes of momentum and mechanical energy from the wind to the water are effectively stopped by land-fast ice. Correct descriptions of the thickness and the physical properties of sea ice and a possible snow cover are necessary for computations of the penetration of solar radiation and the heat flux through the sea surface. The dynamics and thermodynamics of sea ice are discussed in Omstedt et al. (1994), Omstedt and Nyberg (1995, 1996) and Haapala and Leppäranta (1996). These papers also contain extensive lists of references (see also Bergström et al., Chap. 3).

2.5 Sea Level and Its Variability

The mean sea level drops from the Bothnian Bay to Skagerrak by about 35–40 cm (see Ekman and Mäkinen 1996; Carlsson 1998b). Superposed on this long-term feature, the sea level varies in time both in response to external sea level changes, i.e. in the Kattegat, transmitted by flow through the Danish Sounds, and by direct meteorological and hydrological forcing of the Baltic Sea. As shown in Section 2.6 below, the sea-level variations in Kattegat and the different Baltic subseas are quite important for the transports through the Baltic Sea straits and by that for the water exchange between the subseas and, in particular, between the Baltic Sea and Kattegat.

Since the last glaciation of Scandinavia, the Baltic Sea has undergone a remarkable evolution with respect to salinity. As a result of variations in the depth of the outlets relative to the oceanic sea level, the Baltic has, since the final stages of the last glaciation, switched between states characteristic of a freshwater lake and a semi-enclosed part of the ocean, respectively. The Baltic Ice Lake, which existed when the Gulf of Bothnia was still filled with inland ice, lasted to about 10,000 B.P. Then a sea strait across southern Sweden, connecting the Baltic Ice Lake and the Skagerrak, was established whereby the sea level in the Baltic sank by about 26 m in a very short time and the Baltic soon became a marine environment called the Yoldia Sea. Due to rapid land rise the strait was closed at about 9000 B.P. and an inland lake, the Ancylus Lake, resulted. This state ended at about 8000 B.P. when the present connection with Kattegat was opened and the Litorina Sea was established. Reviews of the post-glacial hydrological history of the Baltic can be found, for example, in Svensson (1989) and Björck (1995).

The contemporary apparent land rise is 8–9 mm year^{-1} in the Bothnian Bay, while it is 0–1 mm year^{-1} in the southern Baltic proper (e.g. Ekman 1992). The world's longest continued sea level record, from Stockholm starting in 1774, was put together by Ekman (1988). The apparent land rise was about 1 mm year^{-1} less in the second than in the first half of the record.

Ekman attributed this to a rising sea level in the ocean in the second half. A possibly continued rising sea level in the ocean of course affects the sea level in the Baltic Sea. Topographical changes in the sea straits in the Baltic Sea because of the apparent land rise occur, however, on very long time scales compared to the residence time of water (~30 years). Therefore, the topography of the Baltic straits can be considered constant when discussing the oceanography of the contemporary Baltic Sea. This is true also for the Great Belt and Öresund where recently built bridges (2000) are designed not to change transports through the straits (cf. Sect. 2.6.1).

2.5.1 Externally and Internally Forced Sea Level Variability

Due to the limited transport capacity of the Danish Straits, the response (co-oscillation) of the sea level in the Baltic Sea to the varying sea level in Kattegat is strongly damped for higher frequencies. The frequency response of the Baltic Sea to externally forced sea level variability shows that only external signals of periods from about 1 month and longer may penetrate into the Baltic without noticeable damping (see Sect. 2.6; Stigebrandt 1984). The damping of shorter periods is clearly demonstrated by the semidiurnal tidal component M_2 having an amplitude of about 10 cm in the southwestern Kattegat but only about 1 cm in the Baltic proper. Due to the strong choking effect exerted by the Belts and Öresund, tides are generally quite small in the Baltic (see Defant 1960 for a compilation of tidal constants).

An example of the sea-level variability in the Baltic Sea area is given in Fig. 2.18. Samuelsson and Stigebrandt (1996) studied sea-level variability in the Baltic Sea for periods from 2 days to several years. The sea-level variance

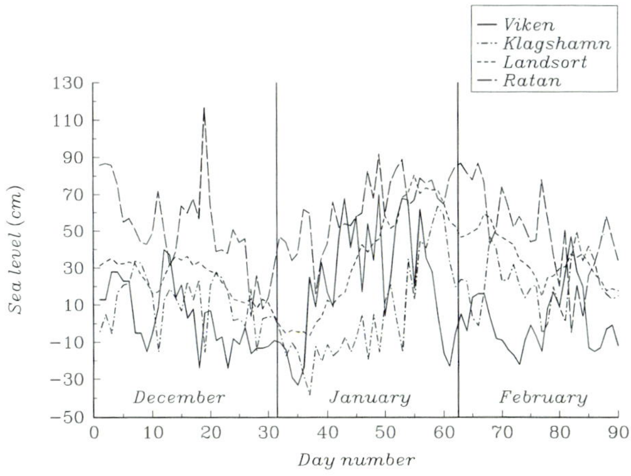

Fig. 2.18. Daily mean sea level (in the NH60 height system) in Kattegat (Viken), the southwestern Baltic proper (Klagshamn), the northern Baltic proper (Landsort) and Bothnian Bay (Ratan) for the period 1 December 1992 to 28 February 1993. Note the major inflow in January, raising sea level in the Baltic Sea by about 0.8 m in 3 weeks

Table 2.1. Contributions from different period bands to sea-level variance (cm²); *Var* is variance around mean sea level for the whole period. (Samuelsson and Stigebrandt 1996)

Station	Var	Period band 2–9 years	8 months –2 years	2–8 months	2 weeks– 2 months	2 days– 2 weeks
Viken	365.5	13.5	79.5	66.4	79.6	126.5
Klagshamn	303.7	16.1	57.2	77.4	74.2	78.9
Stockholm	434.1	28.9	136.8	160.1	84.1	24.2
Forsmark	482.6	29.8	153.0	171.0	89.8	39.1
Spikarna	500.2	28.6	158.8	164.6	97.0	51.1
Ratan	630.8	35.4	189.6	181.2	122.7	104.6

at different coastal stations is shown in Table 2.1, where the variance has also been partitioned into different period bands. These authors also show that the annual variance has rather large interannual variations, in particular in the Bothnian Bay. This gives the transports through the Baltic straits an evident annual cycle, in particular the transports through the Belts and Öresund (cf. Carlsson 1998a).

Samuelsson and Stigebrandt (1996) subtract the externally forced variability computed from the barotropic choking model described in Section 2.6.1 below. The internally forced variability found in this way shows that the Baltic Sea may be regarded as a closed system for periods of less than about 1 month but as an open system for longer periods. Thus, for periods longer than 1 month the amplitudes near the mouth of the Baltic proper are the same as in Kattegat. However, the amplitudes increase with the distance from the mouth, with maximum amplitudes in the Bothnian Bay, as is clearly seen in Fig. 2.19. This is typical of an open system. For periods shorter than 1 month, however, the variability is greatest at the ends (i.e. in the southern Baltic proper and the Bothnian Bay), with a minimum in between as expected in a closed system (Fig. 2.19). Much of the internally forced variability is apparently due to varying winds along the main axis of the Baltic Sea (Carlsson 1998a).

The natural modes of basin-wide oscillations (seiches) of the Baltic Sea have been investigated by several authors (see Kullenberg 1981 for a review). Because of the topographical complexity of the system, with coupled basins of complicated topography, one may arrive at different periods of oscillation depending on the assumptions used in the analysis. Krauss (1974) found eigen-oscillations of 32 h in a two-dimensional model of the Baltic Sea, a period that generally occurs in sea level spectra of the Baltic proper. The fundamental seiche in the Baltic proper may have large

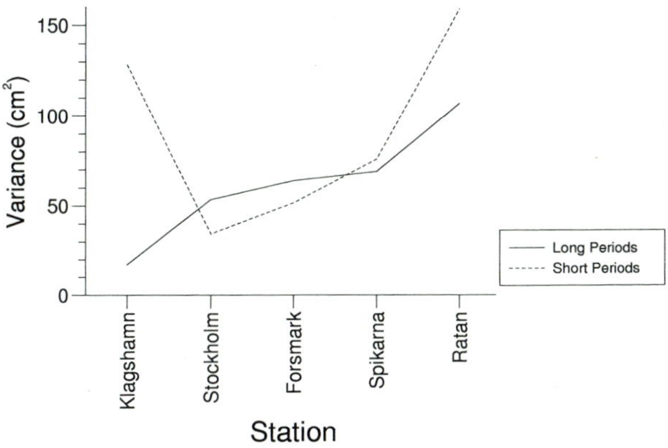

Fig. 2.19. Variability of internally forced sea level variations in the Baltic Sea of periods less and greater than 1 month, respectively. (Redrawn from Samuelsson and Stigebrandt 1996)

amplitude, particularly in the innermost part of the Gulf of Finland. Here it may contribute to sea levels about 4 m above the mean sea level, leading to severe flooding of St. Petersburg (cf. Fonselius 1996). To prevent future flooding a wall with gates is under construction west of St. Petersburg. In the Bothnian Bay an oscillation with a period of ~2 days often occurs (cf. Lisitzin 1974).

2.5.2 Distribution of Sea Level Variability

Because of the closed character of the Baltic Sea, for sea level oscillations of periods less than about 1 month, as discussed in the previous section, one finds a minimum of the sea level variability at about the nodal line for the first fundamental seiche mode of the Baltic Sea and increasing variability with increasing distance away from this (see Fig. 2.19). This property has often been used, letting sea level observations from Stockholm (or Landsort nearby) represent the mean sea level of the Baltic Sea. The open character of the Baltic Sea for long-period sea-level oscillations means, as mentioned above, that amplitudes increase from the mouth towards the extreme inner parts of the Baltic Sea (Fig. 2.19). This property has been demonstrated for the annual and other long periods by Ekman (1996).

Secular changes in some long-period sea-level components were studied by Ekman and Stigebrandt (1990) using the world's longest continued sea-level record from Stockholm mentioned above. They found that compared to the previous century significant changes have occurred in the present century in the annual, semi-annual and Chandler (or pole tide, ~430 days) periods. The most conspicuous change is a large increase in the annual period. The authors concluded that due to the open character of the Baltic Sea, for long period oscillations the changes most certainly are imported from the North Sea via Kattegat.

The coupling between anomalous winter climate and extreme annual means in the Baltic sea level was studied by Ekman (1997) who showed that extremely high annual mean sea levels occur when the winter is extremely mild due to strong and persistent wind from the sector south to west. Extremely low annual mean sea levels occur in years with extremely cold weather in winter due to persistent winds from the sector from north to east.

2.6 Water Exchange Between Basins – Strait Dynamics

The Baltic Sea is composed of a number of major basins connected by sea straits. In addition to these, there are also hundreds of minor basins along the rims of the major basins. The basins obtain their deep waters from the seaward, neighboring basin. A close connection between the salinity of the deep water in one basin and the surface salinity in the neighboring seaward basin is suggested by the observations compiled in Fig. 2.8. Since all major basins have positive freshwater balances, the surface salinity decreases from basin to basin in the direction away from the mouth (Kattegat), with apparent salinity jumps at the straits and usually rather small horizontal salinity gradients within the basins (cf. Fig. 2.10; Wulff and Rahm 1988). The long-term mean exchange of water and salt through the different straits is reasonably well known from calculations based on conservation of water and sea salt (see Fig. 2.13). However, such descriptions do not explain how the flow rate and density (salinity, temperature) of water flowing through a specific strait vary with respect to time which is of utmost importance to the maintenance of the vertical stratification in the deep water of the 'up-system' basin. For computations of the time-dependent exchange through a strait one needs a model that includes the dominating dynamic strait processes and, besides, the forcing (boundary conditions) must be known.

Water transports through a strait connecting two basins are performed by two dynamically different modes. The barotropic mode is forced by the barotropic (surface) pressure gradient along the strait, caused by the sea-level difference between the adjacent basins. The baroclinic mode is forced by the baroclinic (internal) pressure gradient along the strait, caused by different vertical density distributions near the strait in the two basins. Barotropic and baroclinic flows often occur simultaneously, but the baroclinic mode may be insignificant when barotropic velocities are high (see

Sect. 2.6.2). The amplitudes of currents in a strait may be modified by different kinds of flow resistance and by the rotation of the Earth through the Coriolis force, both depending on the topography of the strait. Transports through straits are usually negligibly influenced by direct (local) wind stress on the strait. The main effect of winds upon strait flows is through the generation of large-scale barotropic and baroclinic horizontal pressure gradients in the adjacent basins, which in turn cause barotropic and baroclinic pressure gradients along the strait.

The topographies of the major straits in the Baltic Sea are shown in Fig. 2.2. Öresund and the Belts connect the Baltic proper and Kattegat. The sill depths for the two routes are only 8 m (between Malmö and Copenhagen) and 18 m (the Darss sill), respectively. The Southern Kvark Strait connects the Baltic proper and the Bothnian Sea. However, south of the Åland Sea there is a wide sill (~40 m) intersected by a narrow channel with a sill depth of about 70 m. The Northern Kvark Strait, connecting the Bothnian Sea and Bay, is quite shallow but has two narrow channels, each with a sill depth of ca. 25 m. The outer part of the Irbe Strait, the major connection between the Gulf of Riga and the Baltic proper, is quite shallow but has a narrow channel with a sill depth of about 20 m.

Several authors have described different aspects of the flow through Öresund and the Belts, as discussed in Section 2.6.1 below. The water exchange through the Åland Sea was measured and analyzed by Ehlin and Ambjörn (1977) and Eilola and Stigebrandt (1998) made geostrophic computations. Marmefelt and Omstedt (1993) discussed the water exchange of the Gulf of Bothnia. Theoretical discussions of the flows through these straits are given in Section 2.6.2 below. The flow in the Irbe Strait is described by Lilover et al. (1997) and geostrophic computations were made by Eilola and Stigebrandt (1998). It appears that descriptions of the flows of new deep water through the straits as functions of time and salinity, indispensable in explaining the deep-water stratification of the basins (cf. Sects. 2.2 and 2.8), have so far only been derived for the flows coming through the Belts and Öresund.

2.6.1 Barotropic Strait Flows

Large volume changes occur in the Baltic Sea as a result of inflows and outflows through the Belt Sea and Öresund, as previously discussed in connection with the water budget of the Baltic (Sect. 2.4.1). An inflow of $37\,km^3$ will raise the mean sea level in the Baltic Sea by 0.1 m. During extreme inflow events (so-called major inflows), the sea level in the Baltic Sea may rise by as much as 1 m (e.g. Matthäus and Franck 1992). An example of a large inflow is shown in Fig. 2.18 . Barotropic flows between the Baltic proper and Kattegat are essentially forced by the changing sea level in Kattegat (cf. Sect. 2.5). Using the simple model described below, where the Baltic Sea is connected to the Kattegat by narrow and shallow channels with frictionally balanced barotropic flow, the flow through the Belt Sea and Öresund and the horizontal mean sea level in the Baltic Sea may be computed astonishingly well when the model is driven exclusively by the observed sea level in Kattegat. If the freshwater supply to the Baltic Sea is accounted for in the model, the results change slightly but systematically such that flows out of the Baltic are prolonged and inflows are shortened.

The sea-level difference Δh between two basins connected by a strait forces a barotropic flow Q through the strait. Flow resistance due to both bed friction and large-scale topographic form-drag, caused by acceleration–deceleration at the ends of the strait, is proportional to the flow speed squared, thus:

$$Q^2 = \frac{1}{\phi}\Delta h \tag{2.6}$$

Here, ϕ [s^2m^{-5}] is a strait-specific coefficient of flow resistance which may be determined using simultaneous time series of observed sea levels in the two basins and current speed in the strait, respectively (e.g. Stigebrandt 1992; Mattsson 1995; Jacobsen et al. 1997).

The horizontal mean sea level h_B in the Baltic Sea may be computed from the known sea level h_K in Kattegat ($\Delta h = h_K - h_B$) using the formula above, with a value of ϕ valid for the coupled channel system in the mouth, and the following volume budget for the Baltic Sea:

$$A\frac{dh_B}{dt} = Q + Q_F \tag{2.7}$$

Here, A is the horizontal surface area of the Baltic Sea and Q_F is the net freshwater supply. Note that Eq. (2.7) is a simplified version of Eq. (2.3) from which small terms have been removed. By integrating this model with respect to time, using the observed sea level in Kattegat $h_K(t)$ and the net freshwater supply $Q_F(t)$ to the Baltic Sea, both the barotropic flow through the mouth and $h_B(t)$ may be computed. The model shows that instantaneous in- and outflows to the Baltic Sea may amount to 300,000 $m^3\ s^{-1}$ which is a factor of 20 greater than the net outflow from the Baltic Sea. This simple model, developed by Stigebrandt (1980), has been adopted in several other models (e.g. Omstedt 1987). Carlsson (1998a) showed that the model is only marginally improved if the horizontal mean sea level is replaced by the local sea level in the southwestern Baltic proper.

The model shows that sea level oscillations in the Kattegat of periods shorter than about 1 month are strongly damped by the Danish Straits while oscillations of longer periods are transmitted with most of their amplitudes preserved (cf. Stigebrandt 1984). As discussed in Section 2.5.2 above, this implies that for internally forced periods shorter than 1 month the Baltic Sea may be regarded as a closed system while for longer periods it may be regarded as open, as explained by Samuelsson and Stigebrandt (1996). A clear verification of this is presented in Fig. 2.19 above, showing that sea-level oscillations forced within the Baltic Sea have a node at the Danish Straits for periods of longer than 1 month but a belly for shorter periods. Carlsson (1998a) computed from a sea-level model for the Baltic Sea, which includes forced longitudinal seiche modes in the three major basins, daily mean barotropic transports through the major straits.

2.6.2 Baroclinic Strait Flows

2.6.2.1 Narrow Straits

Baroclinic flows in straits are forced by internal horizontal pressure gradients, caused by different vertical density stratification in the two water bodies connected by the strait. There is a vast literature on steady two-layer flows through narrow straits. The most important contribution to the dynamics of steady two-layer flows was given by Stommel and Farmer (1953) who introduced the concept of baroclinic hydraulic control governed by the following equation that applies in a so-called critical section:

$$\frac{u_1^2}{g'H_1} + \frac{u_0^2}{g'H_0} = 1 \tag{2.8}$$

Here, $u_{1(0)}$ is the velocity and $H_{1(0)}$ is the thickness of the upper (lower) layer and $g'=g\Delta\varrho/\varrho_0$ is the so-called reduced gravity, where g is the acceleration of gravity, $\Delta\varrho$ the density difference between the two layers and ϱ_0 a reference density.

For a steady-state situation, one requires conservation of salt and volume as expressed by Eqs. (2.1) and (2.2). Furthermore, since horizontal density differences in the Baltic Sea and in estuaries in general are mainly due to salinity differences, one may use the following equation of state for brackish water:

$$\varrho=\varrho_0(1+\beta S) \tag{2.9}$$

where $\beta \approx 8\times10^{-4}$ psu^{-1} is the salt contraction coefficient [cf. Eq. (2.16) for the definition of β].

From Eqs. (2.1), (2.2), (2.8), and (2.9) one may derive the following equation (e.g. Stigebrandt 1975, 1981)

$$P^3(1+\frac{\eta^3}{(1-\eta)^3})-2P^2+P=\frac{\eta^3}{F_e^2} \tag{2.10}$$

Here, $P=Q_1/Q_F$ is a mixing parameter, $\eta=H_0/H$, $H=H_0+H_1$, W is the width of the strait and the estuarine Froude number F_e is defined by

$$F_e^2=\frac{Q_F^2}{g\beta S_0 H^3 W^2} \tag{2.11}$$

Equation (2.10) has (two) real roots only when $F_e \leq 1$. The baroclinic flow is maximum when the two roots become equal (double root). Whether or not this state is attained depends on the rate of mixing in the estuary and also on the topography of the strait (e.g. Stigebrandt 1975, 1981; Armi and Farmer 1986; Farmer and Armi 1986). If the maximal baroclinic flow rate is attained, the estuary is said to be overmixed (Stommel and Farmer 1953). The reason for this is that the maximal baroclinic transport capacity of the strait is utilized. The baroclinic flow in straits to semi-enclosed basins may temporarily be superseded by strong barotropic flows ($>u_{1,0}$) whereby the baroclinic control of the flow is eliminated and this may considerably increase the capacity of the strait with respect to transports of water of different densities, as demonstrated in Stigebrandt (1977). The vast shallow sill area in the southern Öresund is a very good example of a case where the baroclinic two-layer flow is usually superseded by strong barotropic currents.

2.6.2.2 Application to the Northern Kvark Strait

In the Northern Kvark Strait there are two deep channels on each side of a larger island (Fig. 2.2). Both channels have sill depths of slightly more than 20 m and widths of less than 2 km. For the application of the theory above, we assumed that H=20 m, the salinity of the Bothnian Sea S_0=5.5 psu, the freshwater supply to the Bothnian Bay Q_F=3425 m^3 s^{-1} and the composite width of the channels W=3300 m. This gives F_e=0.055 and the condition for overmixing (maximal baroclinic transport) gives $P\approx3.1$, implying that

Q_1=10600 and Q_0=7125 $m^3 s^{-1}$ and $S_1 \approx 3.7$ psu. This model thus gives a salinity difference, $S_0 - S_1$, between the Bothnian Sea and Bothnia Bay of the observed magnitude (~2 psu). The composite width of the two channels used here is admittedly rather arbitrary and it is of course possible to tune it so that the computed and observed salinity differences conform. The same is true for the depths of the channels. To verify the applicability of this model to the strait of the Northern Kvark, vertical current and salinity profiles should be obtained from the two channels for a sufficiently long period covering both winter and summer conditions. If the model turns out to be good, it follows that the Bothnian Bay is an overmixed estuary in the sense of Stommel and Farmer (1953).

2.6.2.3 Wide Straits

If a strait is wider than the so-called internal Rossby radius $r_{Roi}=c_i/f$, where c_i is the speed of internal waves and $f=2\Omega\sin\varphi$ the Coriolis parameter (here, Ω is the rate of angular rotation of the earth and φ the latitude), effects of the rotation of the earth may possibly hinder a two-layer flow from occupying the whole width of the strait by limiting the flow to a width in the order of r_{Roi}. This situation has been observed to occur in, for example, Irbe Strait where relatively fresh water flowing out of the Gulf of Riga may be kept along the northern part of the strait like a narrow coastal current (see Lilover et al. 1998).

Assuming vertical side-walls in the strait, the baroclinic geostrophic transport Q in cases like this may be obtained from integration of the so-called thermal wind equation:

$$Q_1 = \frac{g' H_1^2}{2f} \tag{2.12}$$

Here, H_1 is the thickness of the flowing layer at the wall.

The present author modeled the stratification in Kattegat and the Belt Sea using a few coupled two-layered boxes where a barotropic transport, computed as described in Section 2.6.1, is superposed on baroclinic geostrophic transports computed from Eq. (2.12) (Stigebrandt 1983). The model simulated the development of the stratification in these two seas during 18 months reasonably well and the resulting distribution of inflows to the Baltic Sea is shown in Fig. 2.5. Equation (2.12) has also been used to compute the flow from the dense bottom pool in the Arkona Sea (see Sect. 2.8.2 below).

2.6.2.4 Application to the Southern Kvark Strait

The sill section between the Bothnian Sea and the Baltic proper is much wider than the internal Rossby radius. It is therefore likely that Eq. (2.12) may describe the transport of less dense water from the Bothnian Sea to the Baltic proper. Since horizontal density differences are mainly due to salinity differences, one may use the equation of state for brackish water, Eq. (2.9), to rewrite Eq. (2.12) in the following way:

$$Q_1 = \frac{g\beta\Delta S H_1^2}{2f} \tag{2.13}$$

Here, $\Delta S = S_0 - S_1$ and $S_1(S_0)$ is the salinity of the outflowing (inflowing) water.

For a steady-state situation one requires conservation of salt and volume as expressed by Eqs. (2.1) and (2.2). Combining these with Eq. (2.13), one obtains

$$\Delta S = \sqrt{\frac{2Q_F f S_0}{g\beta H_1^2}} \tag{2.14}$$

With H_1=40 m (the depth of the western part of the sill), $f=1.25\times10^{-4}\,s^{-1}$, S_0=7 psu and the freshwater supply to the Gulf of Bothnia (i.e. Bothnian Sea and Bothnian Bay) $Q_F=7000\,m^3\,s^{-1}$, one obtains $\Delta S\approx1$ psu and $Q_1\approx50{,}000\,m^3\,s^{-1}$ ($\approx1575\,km^3\,year^{-1}$). This is quite close to the transport estimated by Wulff and Stigebrandt (1989) using only conservation of salt and volume and a value of ΔS estimated from observations. Thus it seems that an assumption of geostrophically controlled baroclinic outflow limited vertically by the local depth of the sill may explain the salinity difference between the Bothnian Sea and the Baltic proper.

2.6.3 Water Exchange of Fjords, Bays and Archipelagoes

The water exchange in semi-enclosed water bodies along the rim of the Baltic Sea is also forced by barotropic and baroclinic forces. It seems that a varying density field in the coastal water provides the major forcing of water exchange in small semi-enclosed water bodies with not too shallow mouths. The water exchange (remotely) driven by a varying offshore density field is a general case of time-dependent baroclinic flows. It has been termed intermediary water exchange and was modeled in a pragmatic way in Stigebrandt (1990). This model was applied by Stigebrandt and Aure (1990) to a wide spectrum of fjords (bays) with various combinations of mouth topography (mean width B_m and sill depth H_t) and horizontal surface area A_f. The following formula for the mean intermediary circulation was fitted to the model results:

$$Q_i = \beta\sqrt{B_m H_t A_f \frac{g\Delta M}{\varrho_0}} \qquad (2.15)$$

The forcing of the intermediary circulation is here given by the standard deviation ΔM of the weight M (kg m^{-2}) of a water column from the mean sea level down to the sill depth and the value of the empirical non-dimensional constant β was estimated to 17×10^{-4}. The forcing ΔM for some parts of the Baltic Sea area is given in Table 2.2.

From Table 2.2 it can be seen that the variability in the density field is about a factor of 5 greater in northern Kattegat than in the Baltic proper, while it is still slightly smaller in the Bothnian Sea and Bay. Thus, for topographically similar bays and fjords, the water exchange forced by the variability of the density field in the coastal water should be more than a factor of 2 ($\sim5^{1/2}$) larger in Kattegat than in the Baltic proper.

Aure et al. (1997) took flow measurements in the mouth of the Topdal Fjord at the Norwegian Skagerrak coast that largely verify Eq. (2.15) above. Intermediary circulation as modeled in Stigebrandt (1990) is included in, for example, Engqvist and Omstedt (1992), Engqvist (1993) and Mattsson (1996b) who computed the water exchange in Himmerfjärden, Örefjärden and Öresund, respectively. Engqvist (1997) enlightens the complicated mechanics for quasi-steady multi-layer flow in straits.

Table 2.2. Standard deviation ΔM of weight of a water column from mean sea level down to different depths (kg m^{-2}) at some locations in the Baltic Sea area. Data from the Northern Kattegat are from Gustafsson and Stigebrandt (1996) and data from the Baltic were supported by Anders Engvist

Depth (m)	N. Kattegat	S. Baltic proper	N. Baltic proper	Bothnian Sea	Bothnian Bay
5	18	3.1	3.4	2.9	2.7
10	32	5.7	5.9	4.9	4.8
20	53	10.5	9.7	7.4	7.5
30	67	14.9	13.3	9.3	9.6
40	77	19.3	17.9	11.2	11.6
50	87	No data	23.0	13.2	13.9

2.7 Dynamics of the Surface Layer

The surface layer of the sea is exposed to exchange of energy and matter with the atmosphere. This makes the surface layer the most dynamic part of the sea, with ever-changing wave and current conditions, temperature, salinity and production of organic matter. The dynamics regulating the annual variations in temperature, vertical stratification and nutrient concentrations are maybe the most interesting from a biological point of view. One of the themes of the present section is therefore the seasonal dynamics of the surface layer. Another theme is horizontal diffusion, which may be of interest because of the spreading in the sea of different substances such as freshwater, nutrients and planktonic living matter including, for instance, genetic material.

Most of the solar radiation reaching the sea surface is absorbed by the sea – only some few percent (varying with the height of the sun above the horizon, the state of the sea surface with respect to wind-waves and the cloudiness) are reflected (cf. Jerlov 1968). Due to high concentrations of dissolved organic humic (yellow) substances in the Baltic, most of the radiation is absorbed in the upper few meters and the euphotic zone is only some 15 m thick. Because the Baltic Sea is situated at relatively high latitudes, the received solar radiation has a strong seasonal variability. The temperature of the surface layer therefore shows a large seasonal variation, from 15 to 20 °C in summer to the freezing temperature in cold winters. The ice production varies strongly with the coldness of the winter. Wind speeds show an annual cycle, with a maximum in December–January and minimum in summer (see Bergström et al., Chap. 3) for a thorough discussion of atmospheric and ice conditions.

The response of the Baltic Sea to the varying wind stress spans wide ranges in space and time. The direct response is a turbulent surface layer with steady and transient Ekman currents. However, currents may be organized on larger scales due to convergence/divergence of Ekman transports, in particular at the coasts where coastal currents and wave motions may be generated. Unsteady Ekman currents, so-called inertial currents of period 2 π/f (~13.9 h at latitude 60°), were observed in the surface layer of the Baltic proper by Gustafsson and Kullenberg as early as 1935 (Gustafsson and Kullenberg 1936; see also Fig. 2.20). Observations of inertial currents are quite common in the open Baltic.

Fluxes of heat and freshwater through the sea surface may change temperature and salinity of the surface layer. The dynamic effects of heat and freshwater fluxes through the sea surface are best described in terms of local buoyancy fluxes (B) given by

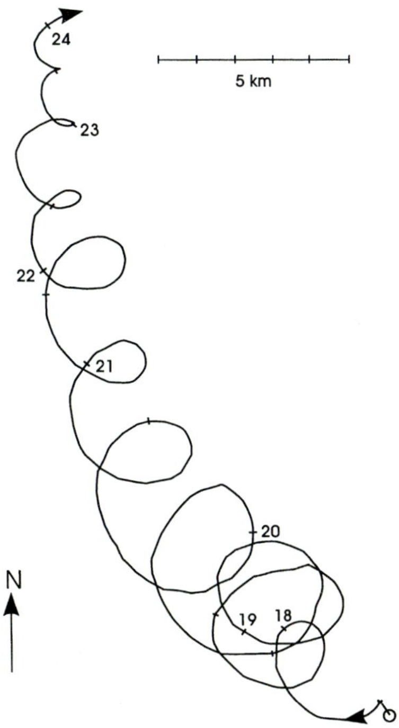

Fig. 2.20. Inertial currents in surface layer of the Baltic proper. (Redrawn from Gustafsson and Kullenberg 1936)

$$B = g(\frac{\alpha}{\varrho C_p} H + \beta F S) \qquad \text{where } \alpha = -\frac{1}{\varrho}\frac{\partial \varrho}{\partial T}, \beta = \frac{1}{\varrho}\frac{\partial \varrho}{\partial S} \tag{2.16}$$

Here, H is the local heat flux through the sea surface as described in Section 2.4 above, F is the local freshwater flux, C_p the specific heat and T, S and ϱ are temperature, salinity and density, respectively, of the surface water. The freshwater flux F equals q_F+p-e, where q_F is the contribution by runoff (Q_F) per unit surface area A, thus $q_F = Q_F/A$, and p and e are local values of precipitation and evaporation. The rationale for inclusion of Q_F in the expression for F is given below. The heat expansion coefficient α defined in Eq. (2.16) changes sign at $T_{\varrho max}$. Below this temperature, cooling thus produces lighter and heating produces denser water. However, the numerical value of α is quite small in an interval around the temperature of maximal density. It appears that freshwater supplies to the Baltic proper contribute more than heating to the buoyancy fluxes in this interval.

2.7.1 Annual Cycles

The response of the Baltic proper to the annual cycles of heating/cooling and freshwater supply is shown in Fig. 2.14. The thickness of the surface layer is dependent not only on the magnitude and sign of B but also on the intensity of the turbulent, mainly wind-forced, mixing. The turbulent state of the surface layer is often described by the so-called friction velocity u_* which is calculated from the wind speed by

$$u_*^2 = C_D \frac{\varrho_a}{\varrho} W^2 \tag{2.17}$$

Here, W is the wind speed, C_D the drag coefficient given by, for example, Smith (1980), and ϱ_a the density of air. Much of the uncertainty of the corresponding coefficient C_e for evaporation estimates discussed in Section 2.4.2 is valid also for C_D. Carlsson (1998a) found that C_D may increase by as much as a factor of 2 compared to the value given by Smith over open water during winter.

If the wind speed is increasing and/or the buoyancy flux B decreasing, the well-mixed surface layer may expand vertically by erosion of underlying water which thus becomes entrained (included) into the surface layer. The entrainment velocity w_e is given by (Stigebrandt 1985):

$$w_e = \frac{2m_0 u_*^3}{g'h} - \varepsilon \frac{B}{g'} \tag{2.18}$$

Here, m_0 ($\approx$0.6) is an efficiency factor, h the thickness of the surface layer, $g' = g\Delta\varrho/\varrho$ the reduced gravity and $\Delta\varrho$ is the density difference between the surface water and the underlying water. The buoyancy flux B contributes to the entrainment when B is negative (penetrative thermohaline convection), but the efficiency of this process is low and ε=0.05 if B<0. Otherwise ε=1 (B>0). The entrainment velocity vanishes if

$$h = \frac{2m_0 u_*^3}{B} \tag{2.19}$$

This depth of the surface layer is the so-called Monin-Obukhov depth; the wind-driven turbulence is not able to keep well mixed a surface layer thicker than this (B>0). If w_e computed from Eq. (2.18) formally becomes negative, the surface layer retreats (becomes thinner). The basic physics of surface layers are discussed in Stigebrandt (1985), for example.

For a large part of the year, heating and cooling processes contribute much more than the supply of freshwater to the buoyancy flux at the sea surface in the Baltic proper which then is dynamically similar to a freshwater lake (e.g. Stigebrandt 1985). In winter and early spring, however, the temperature of the surface water is close to the temperature of maximal density $T_{\varrho max}$ (~2.5 °C) and α is close to zero. It is readily seen from Eq. (2.16) that freshwater supply then may have a decisive influence upon the buoyancy flux at the sea surface. Evidence for this is that the seasonal pycnocline after cold winters becomes established before the surface layer has been heated to T_{max}, as shown by Eilola (1997; cf. Sect. 2.2).

It is obvious that buoyancy fluxes due to local evaporation and precipitation affect the sea surface instantaneously. A fundamental question for the Baltic and other estuaries is how the freshwater Q_F from river runoff becomes distributed over the sea surface. Eilola (1997) found that freshwater buoyancy came in pulses to the PEX area (the Patchiness EXperiment, carried out in the open Baltic proper in the spring of 1986). The local precipitation was negligible during the experiment. The freshwater pulses might have been connected to passages of fronts. Thus, the freshwater buoyancy from runoff certainly spreads over the whole surface of the Baltic proper, as evidenced by the horizontally rather homogeneous salinity distribution. However, the spreading pattern is complicated as described by Eilola and Stigebrandt (1998) for the summer period (cf. Fig. 2.17). The observed spreading pattern of juvenile freshwater fits quite well with the spatial progress of the start of the phytoplankton spring bloom described by Kahru and Nommann (1990). These authors attributed the start of the bloom in the open sea to the establishment of salt stratification by advection of less saline coastal water. They found that the spring bloom starts in the northern Baltic proper from where it progresses roughly in the southerly direction.

At present, it is not possible to formulate a simple relationship between the supply of freshwater buoyancy to a certain point in the sea and the freshwater supply by runoff to the Baltic. However, horizontally integrated seasonal pycnocline models, see below, seem to compute in a quite realistic way the evolution of the salinity field in the Baltic proper using the freshwater flux by runoff from land without any time lag (e.g. Stigebrandt 1985). The reasons for the apparently good performance of the models may perhaps be found in the weak seasonal variation of Q_F and the large time constant of the Baltic proper surface layer.

The dynamics of the mixed layer at the sea surface is of general nature and there are no processes specific to the Baltic. Thus, dynamic models for the well-mixed surface layer of the sea, usually called seasonal pycnocline models, should apply equally well to the Baltic as to other parts of the ocean. There are a variety of globally applicable seasonal pycnocline models, forced by buoyancy fluxes through and wind stress on the sea surface. The perhaps most commonly used model is based on the so-called K-ε model which has a rather detailed turbulence parameterization (see, e.g. Mellor and Durbin 1975; Svensson 1979; Omstedt et al. 1983). Simpler integrated models of the Kraus-Turner type, assuming a well-mixed surface layer, are also in common use (e.g. Gill and Turner 1976). The present author added to the development of this type of model by the inclusion of rotational constraints upon the vertical penetration of turbulence during periods with neutral or weakly positive buoyancy fluxes through the sea surface and applied the model to the Baltic proper (Stigebrandt 1985). With the same atmospheric forcing the two kinds of seasonal pycnocline model compute almost identical seasonal evolution of the surface mixed layer.

Some observations of general interest for mixed layer dynamics have been carried out in the Baltic, e.g. observations of the erosion of a thermocline by strong shear of inertial waves made during BOSEX-77 (Baltic Opens Sea EXperiment in September 1977) and described by Krauss (1981). The efficiency of the wind in generating vertical mixing during the same experiment was treated by Jensen and Kullenberg (1981). Some of the energy supplied by the wind to the surface layer apparently leaks down into and below the seasonal pycnocline to eventually become the major energy source for turbulence and mixing in these stratified layers, as further discussed in Section 2.8 below.

2.7.2 Horizontal Diffusion

The spreading of juvenile freshwater in the seasonal thermal stratification in the Baltic proper (cf. Fig. 2.17) cannot be explained as due to the slow mean circulation (see Sect. 2.3 for a short description of currents in the Baltic proper). Since juvenile freshwater acts as an almost buoyancy-free tracer, the processes dispersing juvenile freshwater should also disperse dissolved and suspended substances released at the coasts. Large-scale patterns of distributions of such substances may therefore, possibly, be explained by the spreading patterns of juvenile freshwater. Earlier studies of dispersion of tracer clouds in oceanic environments show that the dispersion may be computed using a coefficient of horizontal diffusion that increases with the size of the cloud (e.g. Okubo 1974). This suggests that larger and larger scales of motion contribute to the dispersion process as the tracer cloud grows.

So-called shear dispersion is a well-known mechanism which has been studied using small tracer clouds created by instantaneous injection of tracer in a stratified water mass (e.g. Kullenberg 1972). In the simplest case, the vertical gradient of a large-scale horizontal current is responsible for both the stirring and the small-scale mixing, the latter due to turbulence derived from small-scale dynamic instabilities growing on the vertical velocity gradient. Peeters et al. (1996) describe small-scale tracer experiments, of length scale in the order of thousands of meters, i.e. O (10^3 m) and time scale O (1 day) in some Swiss lakes. They also give a good review of models for horizontal dispersion in aquatic environments. In an idealized numerical experiment Rahm and Svensson (1993) found that the dispersion in a non-stratified ocean surface layer with Ekman dynamics follows the empirical relationship by Okubo (1974) for time scales up to a few days. For longer time scales they found that dispersion is slower than in field experiments, indicating that motion on larger scales, not present in their experiment, is of crucial importance.

The spreading of juvenile freshwater in the seasonal thermal stratification of the Baltic proper is a large-scale natural experiment repeated each year, with length scale O (10^5 m) and time scale O (10^2 days). The pattern of spreading suggests that the juvenile freshwater from the Gulfs of Bothnia and Finland passes the northwestern corner of the Baltic proper from where it spreads southwards, partly along the Swedish coast and partly along the east coast of Gotland in the middle of the Baltic proper. It takes a few months for juvenile freshwater to traverse the Baltic proper from north to south, implying an effective dispersion velocity in the order of 0.05 m s^{-1} (see Eilola and Stigebrandt 1998). The analysis in that paper suggests that the juvenile freshwater is spread mainly by unsteady stirring motions induced by fluctuating winds, but geostrophic baroclinic currents due to long-term horizontal salinity gradients have some importance, especially in the northwestern Baltic proper. The spreading is well described by Okubo's (1974) empirical relationship between the horizontal variance of a tracer cloud and the time scale of the experiment. The use of juvenile freshwater as a tracer should help in the interpretation of spreading of tracers (e.g. pollutants) from local sources as well as it may help to verify recipes for calculating the spread of tracer clouds by different mechanisms.

2.8 Dynamics of the Deep Water

The overall description of the Baltic Sea in Section 2.2 stressed that the vertical circulation may be described as occurring in a kind of filling box (Fig. 2.1). Several processes such as advection and diapycnal mixing in dense bottom currents and pools are important in the deep-water domain. This section describes essential facts and features of the dynamics of the deep water of the Baltic Sea.

2.8.1 Inflow of Saline Water to the Baltic Sea

Flow rates and salinity of the dense water entering the Baltic proper through the Fehmarn Belt and Öresund span wide ranges. From estimates based on different kinds of measurements in the southwestern Baltic proper we know reasonably well how contemporary inflows are distributed with respect to salinity and time. Petrén and Walin (1976) and Walin (1981) described high-resolution transport measurements from about 30 occasions of the dense bottom current in the Bornholm Strait. The rationale for this measurement program was to obtain the advective forcing of a one-dimensional advection-diffusion model with salinity as space variable, developed by Walin (1977) for the description of the large-scale circulation of estuaries. Figure 2.11 shows a well-developed bottom current in the Bornholm Channel about 10 days after the end of the major inflow in January 1993.

A different approach to estimate the distribution of inflows with respect to flow rate and salinity was taken by Stigebrandt (1987b) who assumed that the inflowing water should form a pool of dense water at the bottom of the Arkona Basin before reaching the Bornholm Strait. He also assumed that the flow of dense bottom water from the pool and further into the Bornholm Channel is hydraulically controlled by the vertical stratification of the pool itself. Using a tuned version of Eq. (2.12), the inflow of dense water to the Baltic Sea may be computed using standard hydrographic observations from one single station in an area where barotropic flows do not give a net contribution to the transport of new deep water. If applicable, this method should make expensive current measurements in vertical cross sections superfluous. Statistics of flows into the Bornholm Channel, computed from almost 200 historical vertical density profiles from the western Arkona Sea, are presented in Stigebrandt (1987b). A slightly smoothed version of this inflow statistics with respect to salinity is shown in Fig. 2.5 (distribution II). Quite similar statistics were also obtained by Kõuts and Omstedt (1993) using almost 300 profiles from the eastern part of the Arkona Sea.

The inflow statistics based upon historical vertical hydrographic profiles from the Arkona Sea are probably the most reliable inflow estimates available at present. The Arkona inflow statistics were used for long-term simulations (20 years) with a one-dimensional advection–diffusion model for the vertical circulation of the Baltic proper (Stigebrandt 1987a). Coupled to this diffusive filling-box model is a sub-model for an entraining dense bottom current, carrying the new deep water from the Arkona Sea deep-water pool to the Baltic proper deep-water pool. The model reproduces the main observed large-scale properties of the time-dependent vertical stratification in the whole water column. This can be achieved only if the inflow is well described with respect to the flow distributions in the time and salinity domains respectively. In an experiment, the model was run with stationary inflow of water of constant salinity ('Knudsen inflow; cf. Sect. 2.3.1). In this experiment the deep water became almost homogeneous below the halocline. Thus it may be concluded that the circumstance that the dense waters entering the Baltic Sea span wide ranges in flow rate and salinity is the ultimate reason why the Baltic Sea is strongly stratified also below the halocline.

The inflow statistics determined from the historical hydrographical data from the Arkona Sea show an additional interesting feature, namely that large flow rates are positively correlated with higher salinities (Fig. 2.9). The explanation for this is as follows. About 37 km^3 of water has to flow into the Baltic to raise the sea level by 0.1 m. However, most inflow events comprise less than 100 km^3 of water. From the sketch in Fig. 2.3, showing typical volumes and salinities of water masses in the entrance area, it follows that the water forced into the Baltic usually has rather low salinity. Inflowing waters attain high salinities only when inflow rates are large and persistent (so-called major inflows) and preceded or accompanied by strong wind-driven vertical mixing in Kattegat and the Belt Sea, raising the surface salinity in these seas. Due to the small volume of the surface layer in the Öresund, inflows taking this route may rather quickly attain the salinity of the Kattegat surface layer (cf. Mattsson 1996b). Major inflows of water of high salinity occur irregularly and seldom, usually with intervals of several years (cf. Fig. 2.6; Matthäus and Franck 1992 for a statistical analysis of inflows).

It is recalled that the distributions of inflows in the domains of time and salinity actually are determined by advective and diffusive processes in Kattegat and the Belt Sea. Thus, empirical inflow distributions, like those referred to above, may not be used to predict the response of the Baltic Sea to long-term changes in the external forcing, for instance in the net freshwater supply to the Baltic Sea. This emphasizes what has already been discussed above, namely that the prediction of the long-term response of the Baltic Sea to changes in the external forcing requires a model for the stratification in the Belt Sea and Kattegat (cf. Fig. 2.7).

A major inflow of highly saline water occurred in January 1993 when the sea level in the Baltic Sea rose by about 0.8 m, implying an inflow of about 300 km^3 (cf. Fig. 2.18). The inflow through the Öresund is documented by Håkansson et al. (1993) and the flow through the Great Belt is shown in Liljebladh and Stigebrandt (1996). The latter authors also investigated the Arkona dense water pool shortly after the inflow, as described in the following section. Matthäus and Lass (1995) present measurements from the Darss Sill during the inflow period and give a good overview of the inflow including effects of the inflow in the Baltic proper. This major inflow in 1993 was followed by additional major inflows in 1994 and the changes of the distributions of density and oxygen in the Baltic proper were quite drastic (cf. Fig. 2.6).

2.8.2 Dense Bottom Pools and Bottom Currents in the Baltic Sea

The dynamics of the flow of new deep water in the Baltic Sea have been studied by several authors. The stratification and flow in the Arkona Basin were observed and discussed by Liljebladh and Stigebrandt (1996) who in particular focused on the question of possible self-control of the deep-water flow by the stratification in the basin, as suggested in Stigebrandt (1987b). The idea behind self-control of the flow is that the outflow from the pool equals the anticyclonic (clockwise) flow along the rim of the pool which is geostrophically balanced and given by Eq. (2.12), with H_l equal to the pool height. The existence of such a flow control thus makes it possible to compute the flow using hydrographical observations from pools of dense water as mentioned above in connection with computations of inflow statistics. Several field expeditions showed that the predicted pool structure in the Arkona Sea really exists, with a deepening of isohaline surfaces on the Swedish side and an accompanying strong eastward baroclinic current along the northern boundary of the pool. Liljebladh and Stigebrandt (1996) found that barotropic currents modulated the baroclinic flow but did not destroy the baroclinic structure of the pool.

Several authors have discussed and modeled the dynamics of the dense bottom currents in the Bornholm and Stolpe Channels, e.g. Pedersen (1977) and Lundberg (1983). Rydberg (1980) considered rotational critical flow and hydraulic control in the Stolpe Channel. Gidhagen and Håkansson (1992) suggested that there is upstream influence from the Bornholm

Channel upon the Arkona deep-water pool. This was questioned, however, by Liljebladh and Stigebrandt (1996) who argued that such an influence should be seen in a varying depth of the northern flank of the Arkona pool (i.e. where the deep-water front hits the bottom). Their measurements indicate that the flank is quite stable at a water depth of about 41 m. Results by Krauss and Brügge (1991), using a 3-D numerical circulation model, suggest that wind-induced barotropic flow modulates the deep-water flow in the Stolpe Channel.

It is generally believed that the inflowing dense water entrains ambient water. By this, the volume flow increases and the distribution of flows among different salinities shifts towards lower salinities. In a simple model for an entraining dense bottom current developed by Stigebrandt (1987a) the rate of entrainment was tuned to give a realistic dilution of the inflowing deep water. The model shows that much of the diapycnal mixing in the Baltic proper actually is done before the new deep water is incorporated into the deep-water pool of the Baltic proper. Figure 2.5 shows how the distribution among salinities of the inflowing deep water changes from the entrance sills to the Arkona Basin. Estimates of diapycnal mixing performed by dense bottom currents have been made by, for example, Pedersen (1977), Rydberg (1987), Stigebrandt (1987a,b) and Kõuts and Omstedt (1993), but there seem to be no direct field observations of the mixing process. It might be that dense bottom currents are not always vertically well mixed. If so, the concept of entrainment does not hold. In this case one would expect a dominance of shear-induced mixing with overlying waters and more or less continuous detachment of the more diluted upper parts of the current. Clearly, a field program aimed at studies of the dynamics of dense bottom currents under different conditions is badly needed.

2.8.3 Interleaving of Inflows into the Baltic Proper – Halocline Ventilation

In the Arkona Sea some of the new deep water may be entrained into the surface layer when the surface layer is deep and the wind strong, as discussed by, for example, Omstedt (1987) and Liljebladh and Stigebrandt (1996). However, most of the inflowing dense water seems to be interleaved into the Baltic deep-water pool. Most of the interleaving occurs in the halocline region, which by that is well ventilated (Fig. 2.21). This is also clearly illustrated by the observed oxygen development in the East Gotland Basin (Fig. 2.6), showing that anoxic conditions in this basin do not expand higher in the water column than about 120 m below the sea surface.

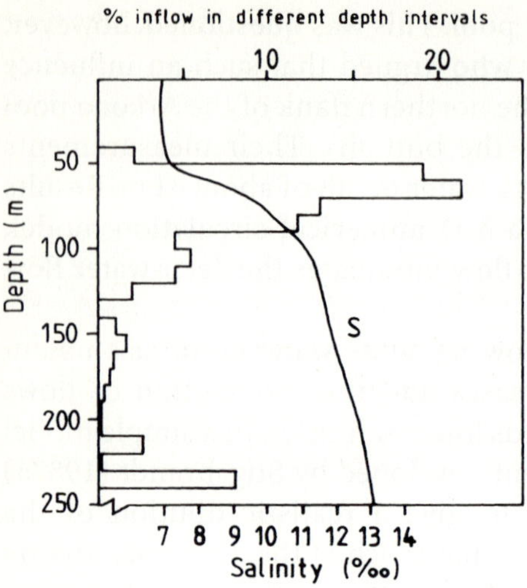

Fig. 2.21. Computed mean interleaving of new deep water into different depths of the Baltic proper. Also shown is a typical vertical salinity profile (*S*). (Stigebrandt 1987a)

Replenishment of the deeper parts of the deep water occurs more seldom and mostly in connection with so-called major inflows. The final incorporation of recently interleaved parcels of dense water into the Baltic proper deep water has been studied by Zhurbas and Paka (1996) using high-resolution field observations. Elken et al. (1988) and Kõuts et al. (1990) have found lens-shaped meso-scale eddies in and below the halocline, similar to the lenses of Mediterranean water (Meddies) in the Atlantic. Some of the eddies appear to have their origin in pulsating inflows from the Bornholm Basin, but also other places and ways of generation might be possible.

2.8.4 Diapycnal Mixing Beneath the Seasonal Pycnocline

In the long term, there is an approximate advective–diffusive balance in the deep water. However, in the shorter term, there are salinity variations because of the discontinuous character of the advective supply of deep water. Advective supplies of new deep water tend to increase and diffusive flows tend to decrease the salinity. There was an extremely rapid decrease in the salinity in the Baltic proper deep water during the 1980s and at the beginning of the 1990s, in particular in and below the halocline (cf. Fig. 2.8). A comparison between the two high-runoff periods 1922–1932 and 1977–1990 suggests that high freshwater supply by runoff is not the only external factor of importance for rapid salinity decreases in the deep water (Fig. 2.6). Samuelsson (1996) concludes that the salinity decrease during the 1980s was due to both an increased mixing in the deep water and a strongly decreased inflow of water of high salinity, the latter probably caused by the unusually high freshwater runoff from land during this period.

Because astronomical tides are extremely small in the Baltic Sea, it seems quite certain that most of the energy sustaining turbulence in the deep-water pools must be provided by the wind. From long-term modeling (decades) of the large-scale vertical circulation in the Baltic proper we know that under contemporary conditions the basin-wide vertical (diapycnal) diffusivity in the deep-water pools is reasonably well described by

$$\kappa = a_0 N^{-1} \tag{2.20}$$

where $a_0 \approx 2.0 \times 10^{-7}$ (m/s)^2 (Stigebrandt 1987a; Omstedt 1990). This is also consistent with estimates using observational data from shorter periods of stagnation (e.g. Svansson 1980; Matthäus 1990). A recent analysis of a_0 applying the budget method to historical data obtained from the deep water in the East Gotland Basin in the Baltic proper confirms that a_0 has the mean value given above. However, it also shows that a_0 has an evident seasonal variation in phase with the variation of the wind speed (Axell 1998).

Rahm (1985) used the one-dimensional advection–diffusion model by Walin (1977) with salinity as space variable. In the salinity interval 8–12 (psu) his κ values are well described as proportional to N^{-1} but with a_0 in the range $3.5–5 \times 10^{-7}$ (m/s)^2. The difference between the results of Rahm and those presented above may be explained by the circumstance that Rahm's model also includes mixing performed by dense bottom currents, whereas this mixing is treated separately in the papers mentioned above.

A layered TS structure has been observed in the deep water of the Baltic (e.g. Prandke and Stips 1990). However, both heat and salt stratification are usually stable, so a significant contribution of double diffusion processes to vertical fluxes seems unlikely (Prandke and Stips 1992).

The small-scale turbulence and the accompanying diapycnal mixing in a stratified sea are not directly coupled to the shear of the large-scale flow field. At the present level of understanding of diapycnal mixing below the seasonal pycnocline in the Baltic Sea, the main question is to find the sources and paths for the energy sustaining the turbulence. It appears almost certain that the wind is the ultimate energy source for turbulence in the deep water, away from the dense bottom currents formed by inflowing new deep water. It is also very likely that internal waves play a key role in the transfer of energy down into the deep water. Several mechanisms may generate internal waves in the sea. Near-inertial internal gravity waves are excited by inertial currents in the surface layer (cf. Krauss 1972; Mälkki 1975). Since inertial currents were first observed (cf. Fig. 2.20), it has become clear that they are prominent and energetic features of the surface layer in the Baltic Sea (e.g. Kullenberg 1981). Internal waves may also be generated by resonant interaction with wind-waves on the sea surface, by interaction with long barotropic waves and inertial currents at sloping bottoms (topographical generation), by advection of wind stress fluctuations along the sea surface and by extraction of mean current shear.

Because of convergence and divergence of wind-driven flows, the coastal boundary generates more intense vertical motions than the open sea. The coastal boundary is therefore expected to be more active than the open sea with respect to internal wave generation and diapycnic mixing in the deep water (e.g. Shaffer 1979). However, the baroclinic coastal boundary layer is only 5–10 km wide in the Baltic proper (Walin 1972). Only along a small fraction of the coasts of the Baltic proper is the deep water, delimited upwards by the permanent halocline, that close to the shore (cf. the 50-m isobath in Fig. 2.2). Thus, only in quite limited parts of the Baltic proper should wind-forced coastal dynamics be able to directly excite mixing in the deep water. Most of the mixing in the deep water of the Baltic proper is therefore probably forced and performed in the open sea, which of course includes the benthic (bottom) boundary layer. It is observed in the open ocean that vertical mixing is intensified in areas with rough bottom topography (ridges, etc.) compared to areas with smoother topography (e.g. Polzin et al. 1997). This indicates that much of the mixing energy actually is being fed to the turbulence by tides at sloping bottoms, maybe in a way similar to that suggested by Sjöberg and Stigebrandt (1992). It is not known if this is true also for the Baltic proper where tidally forced mixing should be negligible.

Acknowledgments. The help with preparation of drawings by Agneta Malm and Kari Eilola is greatly acknowledged. Anders Engqvist kindly supplied data on the variability of the density field along the Swedish coast of the Baltic Sea.

References

Aarup T (1994) Satellite imagery of Danish and neighbouring waters. Interpretation of satellite ocean colour data of the transition zone between the North Sea and the Baltic Sea. PhD Diss, University of Copenhagen, Havforskning fra Miljøstyrelsen, 52, 154 pp

Andersson L, Håkansson B, Lundqvist J-E, Omstedt A, Rahm L, Sjöberg B, Svensson J (1992) Vattnet i väster och öster. In: Sjöberg B (ed) Hav och Kust. Sver Nationalatlas 7:56–72

Armi L, Farmer DM (1986) Maximal two-layer exchange through a contraction with barotropic net flow. J Fluid Mech 164:27–51

Aure J, Molvær J, Stigebrandt A (1997) Observations of inshore water exchange forced by a fluctuating offshore density field. Mar Pollut Bull 33:112–119

Axell L (1998) On the variability of the Baltic Sea deepwater mixing. J Geophys Res 103:21667–21682

BALTEX (1995) Baltic Sea Experiment, BALTEX. Initial implementation plan. International BALTEX Secretariat Publ no 2, GKSS Research Center, Geesthacht, Germany, 84 pp

Bergsten F (1933) Wasserstandsvariationen des Baltischen Meeres. 4. Hydr Konf Balt Staten, Leningrad, 21 pp

Bergström S, Carlsson B (1994) River runoff to the Baltic Sea: 1950–1990. Ambio 23:280–287

Björck S (1995) A review of the history of the Baltic Sea, 13.0–8.0 ka B.P. Quat Int 27:19–40

Brydsten L (1993) Characterization of transport bottoms in the Gulf of Bothnia – a model approach. Aqua Fenn 23(2):153–164

Carlsson M (1998a) A coupled three-basin sea level model for the Baltic Sea. Cont Shelf Res 18:1015–1038

Carlsson M (1998b) The mean sea level topography in the Baltic Sea determined by oceanographic methods. Mar Geod 21:203–217

Dahlström B (1986) Determination of the areal precipitation for the Baltic Sea. Meteorology and climatology. Rep 54. SMHI, Norrköping, 72 pp

Defant A (1960) Physical oceanography, vol 2. Pergamon Press, Oxford, 598 pp

Dera J (1983) Fizyka morza. Panstwowe Wydawnictwo Naukowe, Warszawa, 432 pp

Dera J, Rozwadowska A (1991) Solar radiation variability over the Baltic Sea due to weather conditions. Oceanologia 30:5–36

Ehlin U (1981) Hydrology of the Baltic Sea. In: Voipio A (ed) The Baltic Sea. Elsevier Oceanogr Ser 30. Elsevier, Amsterdam, pp 123–134

Ehlin U, Ambjörn C (1977) Water transports through the Åland Sea. Ambio Spec Rep 5:117–125

Eilola K (1997) The development of a spring thermocline in low-saline seawater at temperatures below the temperature of maximum density with application to the Baltic Sea. J Geophys Res 102:8657–8662

Eilola K, Stigebrandt A (1998) Spreading of juvenile freshwater in the Baltic proper. J Geophys Res 103:27795–27807

Ekman M (1988) The world's longest continued series of sea level observations. Pure Appl Geophys 127:73–77

Ekman M (1992) Postglacial rebound and sea level phenomena, with special reference to Fennoscandia and the Baltic Sea. In: Kakkuri J (ed) Geodesy and geophysics. Publ Finn Geod Inst 115:7–70

Ekman M (1996) A common pattern for interannual and periodical sea level variations in the Baltic Sea and adjacent waters. Geophysica 32:261–272

Ekman M (1997) Anomalous winter climate coupled to extreme annual means in the Baltic sea level during the last 200 years. Small publications in historical geophysics no 3, 14 pp. (Can be ordered from M. Ekman, National Land Survey, 80182 Gävle, Sweden)

Ekman M, Mäkinen J (1996) Mean sea surface topography in the Baltic Sea and its transition area to the North Sea: a geodetic solution and comparisons with oceanographic models. J Geophys Res 101:11993–11999

Ekman M, Stigebrandt A (1990) Secular change of the seasonal variation in sea level and of the pole tide in the Baltic Sea. J Geophys Res 95C:5379–5388

Elken J, Pajuste M, Kõuts T (1988) On intrusive lenses and their role in mixing in the Baltic deep layers. In: Proc 16th Conf of the Baltic Oceanographers, Kiel, pp 367–376

Engqvist A (1993) Water turnover in Örefjärden. Aqua Fenn 23:17–27

Engqvist A (1997) Self-similar multi-layer exchange flow through a contraction. J Fluid Mech 328:49–66

Engqvist A, Omstedt A (1992) Water exchange and density structure in a multi basin estuary. Cont Shelf Res 12:1003–1026

Falkenmark M (1986) Hydrology of the Baltic Sea area: temporal fluctuations in water balance. Ambio 15:97–102

Farmer DM, Armi L (1986) Maximal two-layer exchange over a sill and through the combination of a sill and contraction with barotropic flow. J Fluid Mech 164:53–76

Fonselius SH (1969) Hydrography of the Baltic deep basins III. Fish Board of Sweden, Ser Hydrogr Rep 23, Göteborg, 97 pp

Fonselius S (1996) Västerhavets och Östersjöns oceanografi. SMHI, Norrköping, 200 pp

Fonselius S, Szaron J, Öström B (1984) Long-term salinity variations in the Baltic Sea deep water. Papp Réun Cons Int Explor Mer 185:140–149

Gidhagen L (1987) Coastal upwelling in the Baltic Sea – satellite and in situ measurements of sea-surface temperatures indicating coastal upwelling. Estuarine Coastal Shelf Sci 24:449–462

Gidhagen L, Håkansson B (1992) A model of the deep water flow into the Baltic Sea. Tellus 44A:414–424

Gill AE (1982) Atmosphere–ocean dynamics. Int Geophys Ser 30. Academic Press, New York, 662 pp

Gill A, Turner JS (1976) A comparison of seasonal thermocline models with observation. Deep Sea Res 23:391–401

Gustafsson B (1997) Interaction between the Baltic Sea and the North Sea. Dtsch Hydrogr Z 49:163–181

Gustafsson B, Stigebrandt A (1996) Dynamics of the freshwater-influenced surface layers in the Skagerrak. J Sea Res 35:39–53

Gustafsson T, Kullenberg B (1936) Untersuchungen von Trägheitsströmungen in der Ostsee. Svenska Hydrogr-Biol Komm Skr Ny Ser Hydrogr 13, Lund, pp 1–28

Håkansson B, Broman B, Dahlin H (1993) The flow of water and salt in the Sound during the Baltic major inflow event in January 1993. ICES Statutory Meeting, Dublin. International Council for the Exploration of the seas, Copenhagen, C.M. 1933/C:57

Håkansson B, Alenius P, Brydsten L (1996) Physical environment in the Gulf of Bothnia. Ambio Spec Rep 8:5–12

Haapala J, Leppäranta M (1996) Simulating the Baltic Sea ice season with a coupled ice–ocean model. Tellus 48A:622–643

Hela I (1944) Über die Schwankungen des Wasserstandes in der Ostsee mit besonderer Berücksichtigung des Wasseraustausches durch die dänischen Gewässer. Meerentutkimuslait. Julk./Havsforskningsinst Skr 134,1–108. Helsinki, Finland

Henning D (1988) Evaporation, water and heat balance of the Baltic Sea. Estimates of short- and long-term monthly totals. Meteorol Rdsch 41:33–53

ICES (1989) Baltic Sea Patchiness Experiment (PEX-86), part I and II, general report. ICES Cooperative Research Rep no 163. ICES, Copenhagen

Jacobsen F (1997) Hydrographic investigations of the northern Kattegat front. Cont Shelf Res 17:533–554

Jacobsen F, Lintrup MJ, Steen Møller J (1997) Observations of the specific resistance in the Öresund. Nordic Hydrol 28(3):217–232

Jacobsen TS (1980) The Belt Project. Sea water exchange of the Baltic – measurements and methods. National Agency of Environmental Protection, Denmark, 106 pp

Jensen TG, Kullenberg G (1981) On the efficiency of the wind to generate vertical mixing. Geophysica 17:47–61

Jerlov N (1968) Optical oceanography. Elsevier Oceanogr Ser 5. Elsevier, Amsterdam, 194 pp

Kahru M, Nommann S (1990) The phytoplankton spring bloom in the Baltic Sea in 1985, 1986: multitude of spatio-temporal scales. Cont Shelf Res 10:329–354

Kahru M, Håkansson B, Rud O (1995) Distributions of the sea-surface temperature fronts in the Baltic Sea as derived from satellite imagery. Cont Shelf Res 15:663–679

Knudsen M (1900) Ein hydrographischer Lehrsats. Ann Hydr 28:316–320

Kõuts T, Omstedt A (1993) Deepwater exchange in the Baltic proper. Tellus 45A:311–324

Kõuts T, Elken J, Lips U (1990) Late autumn intensification of deep thermohaline anomalies and formation of lenses in the Gotland Deep. In: Proc 17th Conf of the Baltic Oceanographers, Norrköping, pp 280–293

Krauss W (1972) Wind-generated internal waves and inertial-period motions. Dtsch Hydrogr Z 25:241–250

Krauss W (1974) Two-dimensional seiches and stationary drift currents in the Baltic Sea. ICES special meeting on models and water circulation in the Baltic. Pap 10, ICES, Copenhagen, 32 pp (mimeogr)

Krauss W (1981) The erosion of a thermocline. J Phys Oceanogr 11:415–433

Krauss W, Brügge B (1991) Wind-produced water exchange between the deep basins of the Baltic Sea. J Phys Oceanogr 21:373–384

Kullenberg G (1972) Apparent horizontal diffusion in stratified vertical shear flow. Tellus 24:17–28

Kullenberg G (1981) Physical oceanography. In: Voipio A (ed) The Baltic Sea. Elsevier Oceanogr Ser 30. Elsevier, Amsterdam, pp 135–218

Liljebladh B, Stigebrandt A (1996) Observations of the deepwater flow into the Baltic Sea. J Geophys Res 101:8895–8912

Lilover M, Lips U, Laanearu J, Liljebladh B (1998) Flow regime in the Irbe Strait. Aquat Sci 60:253–265

Lisitzin E (1974) Sea level changes. Elsevier, Amsterdam, 286 pp

Lundberg P (1983) On the mechanics of the deep-water flow in the Bornholm Channel. Tellus 35A:149–158

Mälkki P (1975) On the variability of currents in a coastal region of the Baltic Sea. Merentutkimuslaitoksen Julk. Havforskningsinst Skr 240:3–56

Marmefelt E, Omstedt A (1993) Deepwater properties in the Gulf of Bothnia. Cont Shelf Res 13:169–187

Matthäus W (1990) Mixing across the primary Baltic halocline. Beitr Meereskd (Berlin) 61:21–31

Matthäus W, Franck H (1992) Characteristics of major Baltic inflows – a statistical analysis. Cont Shelf Res 12:1375–1400

Matthäus W, Lass HU (1995) The recent salt inflow into the Baltic Sea. J Phys Oceanogr 25:280–286

Mattsson J (1995) Observed linear flow resistance in the Öresund due to rotation. J Geophys Res 100(C10):20779–20791

Mattsson J (1996a) Some comments on the barotropic flow through the Danish Straits and the division of the flow between the Belt Sea and the Öresund. Tellus 48A:456–464

Mattsson J (1996b) Analysis of the exchange of salt between the Baltic and the Kattegat through the Öresund using a three layer model. J Geophys Res 101:16571–16584

Mellor GI, Durbin PA (1975) The structure and dynamics of the ocean surface mixed layer. J Phys Oceanogr 5:718–728

Mikulski Z (1982) River inflow to the Baltic Sea. Polish Academy of Sciences, Polish National Committee of the IHP. Faculty of Geography and Regional Studies University of Warsaw

Mikulski Z, Falkenmark M (1986) Calculated freshwater budget of the Baltic as a system. In: Water balance of the Baltic Sea, chapter 10. Proc Symp on the Baltic Sea Environment, no 16. Baltic Marine Environment Protection Commission–Helsinki Commission, 174 pp

Okubo A (1974) Some speculations on oceanic diffusion diagrams. Rapp P-v Réun Cons Int Explor Mer 167:77–85

Omstedt A (1987) Watercooling in the entrance of the Baltic Sea. Tellus 39A:254–265

Omstedt A (1990) Modelling the Baltic Sea as thirteen sub-basins with vertical resolution. Tellus 42A:286–301

Omstedt A, Nyberg L (1995) A coupled ice-ocean model supporting winter navigation in the Baltic Sea, part 2. Thermodynamics and meteorological coupling. Rep RO 21. SMHI, Norrköping, 30 pp + appendices

Omstedt A, Nyberg L (1996) Response of the Baltic Sea ice to seasonal, interannual forcing and climate change. Tellus 48A:644–662

Omstedt A, Sahlberg J, Svensson U (1983) Measured and numerically simulated autumn cooling in the Bay of Bothnia. Tellus 35A:231–240

Omstedt A, Marmefeldt E, Murphy R (1993) Some flow characteristics of the coastal boundary layer in the Bothnian Sea. Aqua Fenn 23:5–16

Omstedt A, Nyberg L, Leppäranta M (1994) A coupled ice-ocean model supporting winter navigation in the Baltic Sea. Part 1: ice dynamics and water levels. Rep RO17. SMHI, Norrköping, 17 pp

Omstedt A, Mueller L, Nyberg L (1997) Interannual, seasonal and regional variations of precipitation and evaporation over the Baltic Sea. Ambio 26(8):484–492

Pedersen FB (1977) On dense bottom currents in the Baltic deep water. Nord Hydrol 8:297–316

Pedersen FB (1993) Fronts in the Kattegat: the hydrodynamic regulating factor for biology. Estuaries 16:104–112

Peeters F, Wüest A, Piepke G, Imboden DM (1996) Horizontal mixing in lakes. J Geophys Res 101:18361–18375

Petrén O, Walin G (1976) Some observations of the deep flow in the Bornholm Strait during the period June 1973–December 1974. Tellus 28:74–87

Polzin KL, Toole JM, Ledwell JR, Schmitt RW (1997) Spatial variation of turbulent mixing in the abyssal ocean. Science 276:93–96

Prandke HA, Stips A (1990) Statistische Analyse lokaler Gradienten in den Distesprungschichten der Ostsee. Beitr Meereskd 61:93–102

Prandke HA, Stips A (1992) A model of the Baltic thermocline structure patches, deduced from experimental investigations. Cont Shelf Res 12:643–659

Rahm L (1985) On the diffusive salt flux of the Baltic proper. Tellus 37A:87–97

Rahm L, Svensson U (1993) Note on dispersion in an ocean surface Ekman layer due to variable wind forcing. Dtsch Hydrogr Z 45(1):43–54

Rodhe J (1996) On the dynamics of the large-scale circulation of the Skagerrak. J Sea Res 35:9–21

Rydberg L (1980) Rotating hydraulics in deep-water channel flow. Tellus 32:77–89

Rydberg L (1987) Hydrography, oxygen and nutrient balance of the Kattegat and related phenomena. Doctoral Diss, Department of Oceanography, Göteborg University, 40 pp

Samuelsson M (1996) Interannual salinity variations in the Baltic Sea during the period 1954–1990. Cont Shelf Res 16:1463–1477

Samuelsson M, Stigebrandt A (1996) Main characteristics of the long-term sea level variability in the Baltic Sea. Tellus 48A:672–683

Shaffer G (1979) Conservation calculations in natural coordinates (with an example from the Baltic). J Phys Oceanogr 9:847–855

Simonsen K, Haugan PM (1996) Heat budgets of the Arctic Mediterranean and the sea surface heat flux parameterizations for the Nordic Seas. J Geophys Res 101:6553–6576

Sjöberg B (ed) (1992) Kust och Hav. Sveriges Nationalatlas, vol 7

Sjöberg B, Stigebrandt A (1992) Computations of the geographical distribution of the energy flux to mixing processes via internal tides: its horizontal distribution and the associated vertical circulation in the ocean. Deep Sea Res 39:269–291

Smith SD (1980) Wind stress and heat flux over the ocean in gale force winds. J Phys Oceanogr 10:709–726

Stigebrandt A (1975) Stationär tvålagerströmning i estuarier. Vassdrags-og Havnelaboratoriet. Rep STF60 A75120. SINTEF, Trondheim, 68 pp

Stigebrandt A (1977) On the effect of barotropic current fluctuations on the two-layer transport capacity of a constriction. J Phys Oceanogr 7:118–122

Stigebrandt A (1980) Barotropic and baroclinic response of a semi-enclosed basin to barotropic forcing from the sea. In: Freeland HJ, Farmer DM, Levings CD (eds) Fjord oceanography. Plenum, New York, pp 151–164

Stigebrandt A (1981) A mechanism governing the estuarine circulation in deep, strongly stratified fjords. Estuarine Coast Shelf Res 13:197–211

Stigebrandt A (1983) A model for the exchange of salt and water between the Baltic and the Skagerrak. J Phys Oceanogr 13:411–427

Stigebrandt A (1984) Analysis of an 89-year-long sea level record from the Kattegat with special reference to the barotropically driven water exchange between the Baltic and the sea. Tellus 36A:401–408

Stigebrandt A (1985) A model for the seasonal pycnocline in rotating systems with application to the Baltic proper. J Phys Oceanogr 15:1392–1404

Stigebrandt A (1987a) A model for the vertical circulation of the Baltic deep water. J Phys Oceanogr 17:1772–1785

Stigebrandt A (1987b) Computations of the flow of dense water into the Baltic from hydrographical measurements in the Arkona Basin. Tellus 39A:170–177

Stigebrandt A (1990) On the response of the horizontal mean vertical density distribution in a fjord to low-frequency fluctuations in the coastal water. Tellus 42A:605–614

Stigebrandt A (1992) Bridge-induced flow reduction in sea straits with reference to effects of a planned bridge across Öresund. Ambio 21:130–134

Stigebrandt A, Aure J (1990) De ytre drivkreftenes betydning for vannutskiftningen i fjordene fra Skagerrak til Finnmark. Rapport FO 9003. Institute for Marine Research, Bergen, 29 pp

Stommel H, Farmer HG (1953) Control of salinity in an estuary by a transition. J Mar Res 12:13–20

Svansson A (1980) Exchange of water and salt in the Baltic and adjacent seas. Oceanol Acta 3:431–440

Svansson A (1984) Hydrographic features of the Kattegat. Reports et Proces-Verbaux des Reunions. Cons Int Explor Mer 185:79–90

Svensson N-O (1989) Late Weichselian and early Holocene shore displacement in the central Baltic, based on stratigraphical and morphological records from eastern Småland and Gotland, Sweden. Lundqua Thesis 25, Diss, University of Lund, Lund, Sweden, 195 pp

Svensson U (1979) The structure of the turbulent Ekman layer. Tellus 31:340–350

Walin G (1972) On the hydrographic response to transient meteorological disturbances. Tellus 24:169–186

Walin G (1977) A theoretical framework for the description of estuaries. Tellus 29:128–136

Walin G (1981) On the deep water flow into the Baltic. Geophysica 17:75–93

Winterhalter B, Flodén T, Ignatius H, Axberg S, Niemistö L (1981) Geology of the Baltic Sea. In: Voipio A (ed) The Baltic Sea. Elsevier Oceanogr Ser 30. Elsevier, Amsterdam, pp 1–121

Wübber C, Krauss W (1979) The two-dimensional seiches of the Baltic Sea. Oceanol Acta 2:435–446

Wulff F, Rahm L (1988) Long-term, seasonal and spatial variations of nitrogen, phosphorus and silicate in the Baltic: an overview. Mar Environ Res 26:19–37

Wulff F, Stigebrandt A (1989) A time-dependent budget model for nutrients in the Baltic Sea. Global Biogeochem Cycles 3:63–78

Wyrtki K (1954) Schwankungen im Wasserhaushalt der Ostsee. Dtsch Hydrogr Z 7:91–129

Yurkovskis A, Wulff F, Rahm L, Andruzaitis A, Rodriguez-Medina M (1993) A nutrient budget of the Gulf of Riga; Baltic Sea. Estuarine Coast Shelf Sci 37:113–127

Zhurbas VM, Paka VT (1996) Observations of meso-scale eddy-like structures and thermohaline intrusions in the Gotland Basin after the 1993 major Baltic inflow. In: Hagen E (ed) GOBEX summary report. Marine Sci Rep 19. Baltic Sea Research Institute, Warnemünde, pp 38–59

3 Climate and Hydrology of the Baltic Basin

S. Bergström, H. Alexandersson, B. Carlsson,
W. Josefsson, K.-G. Karlsson, and G. Westring

Due to its semi-enclosed character the Baltic Sea is very vulnerable to pollution and its environmental status is a major concern for the northern European countries. The water body is affected by contributions of freshwater and nutrients from rivers, pollution from industries, municipalities and shipping and by direct atmospheric deposition. The inflow of water of high salinity and high oxygen concentrations via the Danish Sounds is another critical factor for the ecosystem of the Baltic Sea. Many of these factors are dependent on the climate and the hydrology of the Baltic drainage basin.

3.1 Climate of the Baltic Basin

3.1.1 General Overview

The whole Baltic basin is well within the west wind zone where cyclones coming from the west or south-west dominate the weather scene. Many of these become weaker after they have passed the Scandinavian fells. However, also cyclones from a more southerly direction enter the region. These are of great importance as they quite often give large amounts of rain to the coastal areas of the Baltic Sea.

Two maps (Figs. 3.1 and 3.2) illustrating the average pressure distribution and winds over Europe in January and July, respectively, show that the average air flow is strong and from south to west in winter, weak and more westerly in summer. The individual cyclones are much more intense in winter but often loose much energy as they penetrate inland. In summer small cyclones may become very intense inland and yield heavy precipitation in fairly concentrated bands.

So-called blocking episodes, with an extensive anticyclone often centered near southern Scandinavia, can for several weeks steer the cyclones towards

Ecological Studies, Vol. 148
F. Wulff et al. (eds.)
A Systems Analysis of the Baltic Sea

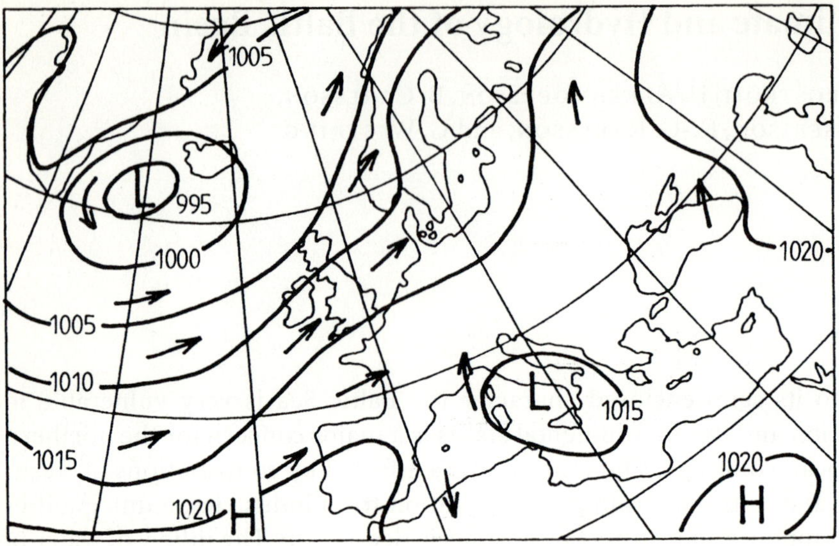

Fig. 3.1. Average air pressure and air flow over Europe in January. (Redrawn from Liljequist 1970)

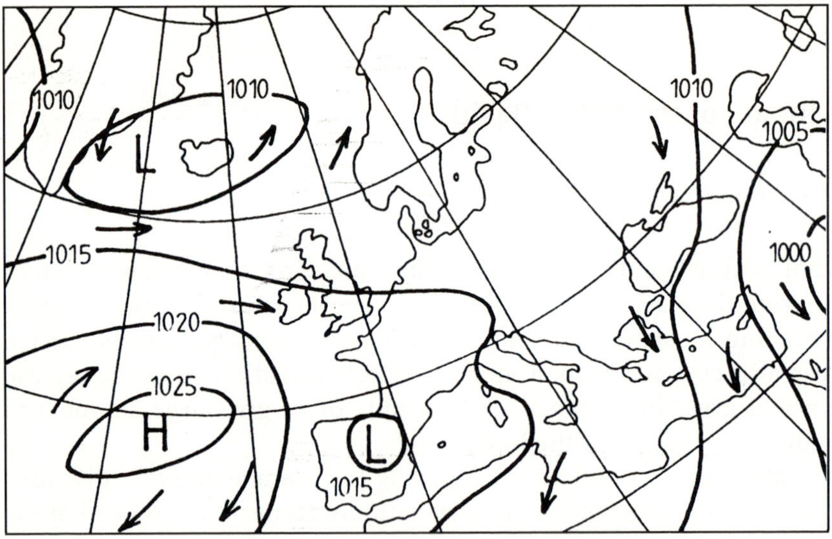

Fig. 3.2. Average air pressure and air flow over Europe in July. (Redrawn from Liljequist 1970)

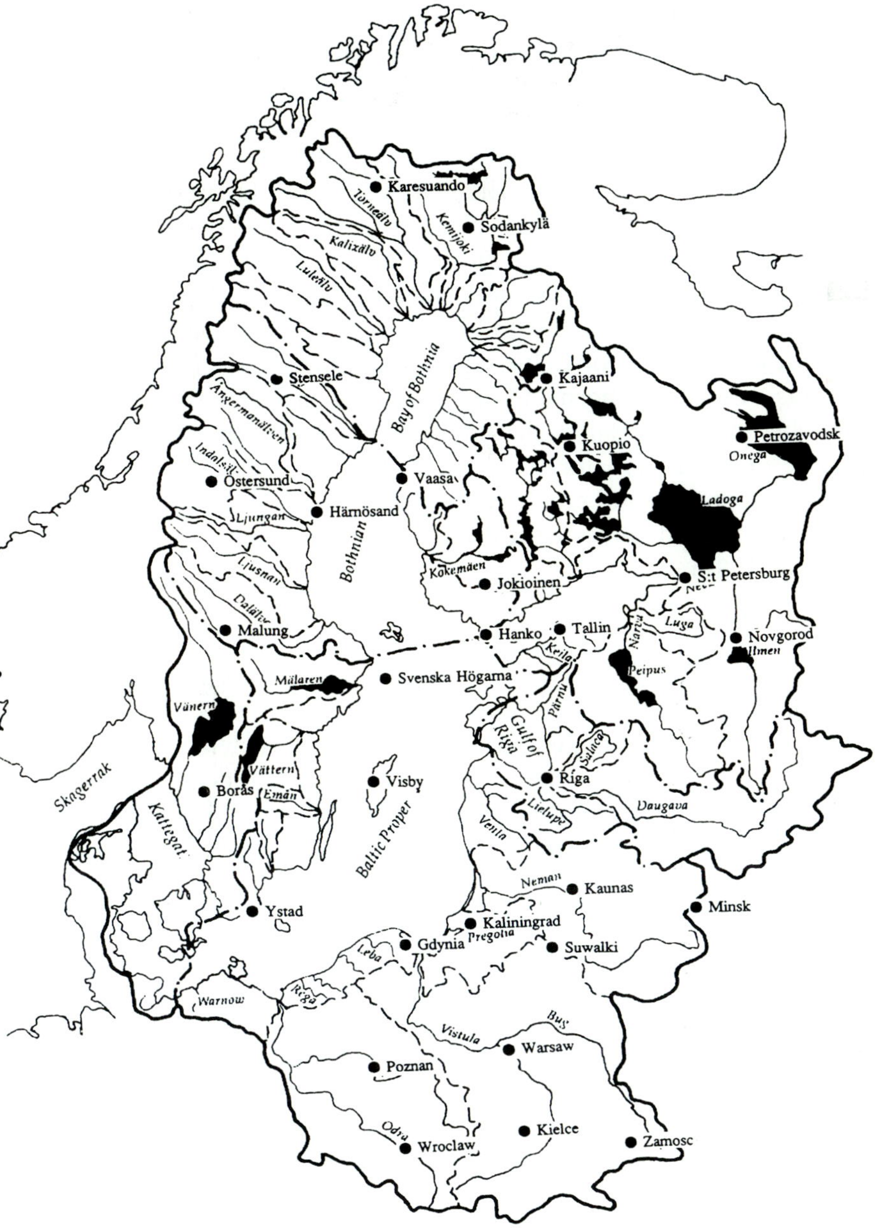

Fig. 3.3. Map of the Baltic basin and location of the 30 stations used to illustrate air temperature and precipitation

the northern Norwegian Sea or towards the Bay of Biscay and southern Europe. The northern cyclones can give some rain over the north-western slopes of the Scandinavian fells and later also over the eastern fringes of the region as some of these cyclones penetrate towards the south-east into Russia. However, during these blocking episodes it is very dry throughout almost the whole region. In winter it is cold, but in summer the daytime temperatures are high.

A recent perspective of the European climate can be found in the work by the European Climate Support Network (1995). In Fig. 3.3 the locations of 30 stations used in the following sections to give a more detailed climatic overview of the Baltic region are shown. Two groups of additional stations, one for northern and one for southern Sweden, have been used to describe the temporal variation of the temperature and precipitation climate of the region for the period 1860–1995. Each group consists of seven temperature and precipitation stations which all have been tested for homogeneity and corrected if it was found necessary (Alexandersson and Eriksson 1989).

3.1.2 Air Temperature

The climate of the Baltic region depends to a large extent upon the position of the polar front which separates cold arctic air from warmer and moister air to the south. Most of the cyclones develop along this front. The cloud cover also plays an important role in surface temperatures. In winter more continental parts of the basin can suffer from intense cold waves during clear and anticyclonic weather.

The climate has its two extremes in the south-west over the water, with maritime conditions and modest daily and annual variations, and in the valleys up in the northern outskirts of the basin with much larger amplitudes, both annually and daily. In January the warmest parts in the south-west have monthly mean temperatures of around +2 °C while some cold valleys in Swedish Lapland have −17 °C on average. In July the corresponding values are around 16 and 12 °C, respectively (Alexandersson et al. 1991).

Figures 3.4 and 3.5 illustrate, in a simplified manner, the average temperature distribution over Europe in winter and summer, respectively. More details are shown in Table 3.1, which contains monthly mean temperatures for the 30 stations shown in Fig. 3.3. As no more recent normal values could be obtained (except from Sweden and Finland), these data are for the standard period 1931–1960. The data are also presented graphically in the map in Fig. 3.6.

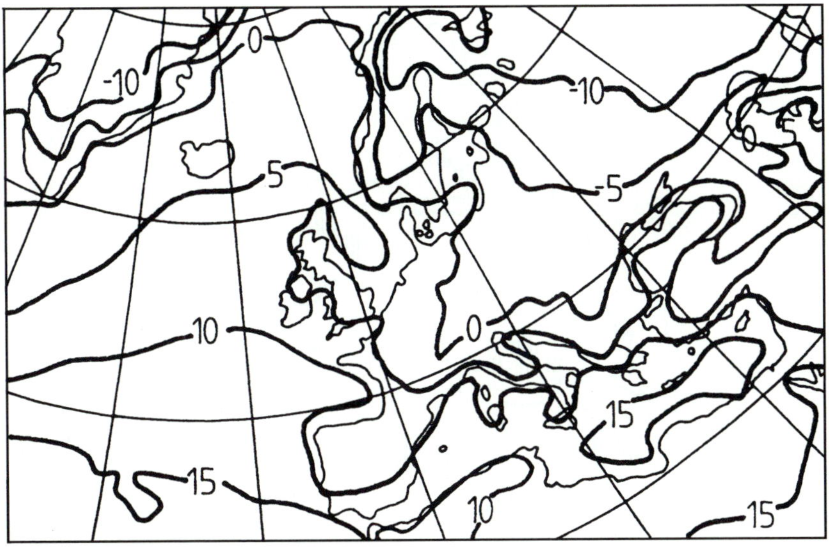

Fig. 3.4. Simplified, average temperature conditions over Europe in winter. (European Climate Support Network 1995)

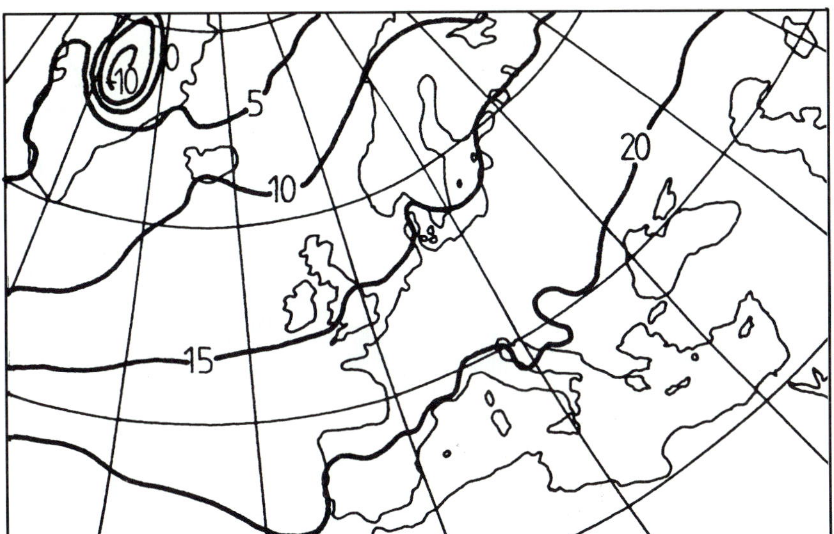

Fig. 3.5. Simplified, average temperature conditions over Europe in summer. (European Climate Support Network 1995)

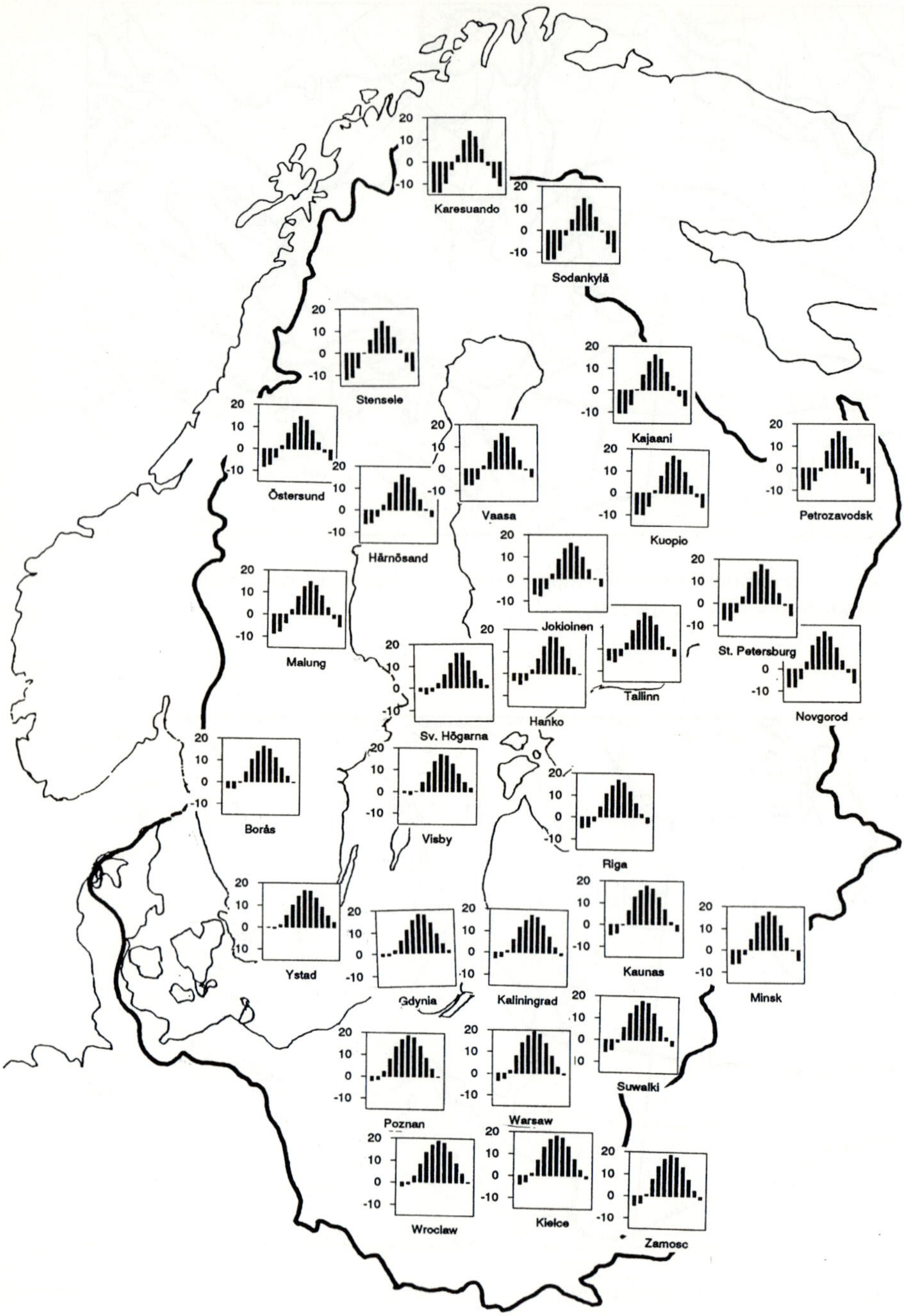

Fig. 3.6. Mean monthly temperatures at 30 stations in the Baltic basin for the period 1931–1960. Data from Landsberg (1985) and official Finnish and Swedish climatological yearbooks

Table 3.1. Mean monthly and annual temperatures at 30 stations in the Baltic basin for the period 1931–1960. Data from Landsberg (1985) and official Finnish and Swedish climatological yearbooks. Locations of stations are shown in Fig. 3.1; data are presented graphically in Fig. 3.6

Station	Jan	Feb	Mar	Apr	May	Jun	Jul	Aug	Sep	Oct	Nov	Dec	Year
Ystad	−0.2	−0.6	1.2	5.3	10	14.1	16.7	16.4	13.4	9.2	5.3	2.4	7.8
Borås	−2.9	−3	−0.4	4.7	10.5	14.2	16.5	15.4	11.4	6.7	2.7	−0.1	6.3
Visby	−0.7	−1.4	0	4.3	9	13.9	17.1	16.6	12.9	8.3	4.4	1.8	7.2
Sv. Högarna	−1.5	−2.7	−1.5	2	6.1	11.4	15.9	16	12.4	7.7	3.9	1.2	5.9
Malung	−8.8	−7.8	−4	2	8.2	12.5	15	13.2	8.6	3.2	−1.7	−5.4	2.9
Härnösand	−6.2	−5.8	−2.8	2.2	7.8	12.7	16.3	15	10.4	4.9	0.6	−2.7	4.4
Östersund	−8.4	−7.3	−4	1.5	7.2	11.7	14.8	13.2	8.6	3.2	−1.2	−4.7	2.9
Stensele	−12.2	−11.1	−6.8	−0.2	5.9	11.1	14.3	12.2	7.1	1	−4.2	−8.4	0.7
Karesuando	−13.8	−13.9	−9.9	−3.6	3	9.8	13.7	11.2	5.4	−1.6	−7.3	−11.2	−1.5
Sodankylä	−13.5	−13	−9	−2.1	4.9	11.3	14.7	12	6.2	−0.5	−5.8	−9.8	−0.4
Kajaani	−10.6	−10.5	−6.7	0.3	6.9	12.9	16.1	14	8.3	2.1	−2.6	−7	1.9
Vaasa	−7.3	−7.5	−4.7	1.3	7.5	12.8	16.2	14.6	9.6	3.8	−0.5	−3.7	3.5
Kuopio	−9.8	−10	−6.1	1.1	7.9	14	17.1	15.3	9.7	3.6	−1.4	−6.1	2.9
Jokioinen	−7.2	−7.8	−4.6	2.2	8.8	13.7	16.2	14.7	9.7	4.3	−0.1	−3.5	3.9
Hanko	−3.5	−5.1	−3.1	1.7	7	12.4	16.8	16.5	12.4	7.3	3.2	−0.1	5.5

Table 3.1. (*continued*)

Station	Jan	Feb	Mar	Apr	May	Jun	Jul	Aug	Sep	Oct	Nov	Dec	Year
Petrozavodsk	-9.7	-9.9	-5.8	-1.3	7.7	13.3	16.6	14.3	9.1	3.2	-2.2	-7	2.4
St. Petersburg	-7.5	-7.9	-4.1	2.9	9.6	14.5	17.7	15.7	10.7	4.7	-0.6	-5.3	4.2
Tallinn	-5	-5.8	-3	2.6	8.4	13.1	16.4	15	11	5.6	0.8	-3	4.7
Novgorod	-8.4	-8.4	-4.5	3.4	10.6	14.6	17	14.7	9.8	4	-1.4	-6.3	3.8
Riga	-5	-4.8	-2	4.6	10.7	14.3	17.1	15.7	11.7	6.2	1.5	-2.6	5.6
Kaunas	-4.7	-4	-0.2	6.3	12.7	15.6	17.8	16.4	12.6	7	1.2	-2.8	6.5
Kaliningrad	-2.7	-2.1	0.8	6.2	11.8	15	17.3	16.4	12.9	7.7	2.6	-1	7.1
Minsk	-6.6	-6.2	-2.1	5.1	12.5	15.6	17.6	15.9	11.4	5.7	-0.2	-4.7	5.3
Gdynia	-1.3	-0.9	1.5	6.1	11	15.3	17.9	17.6	14	9.1	4.3	1.1	8.0
Suwalki	-5.6	-4.8	-1.3	5.7	12.2	15.8	17.7	16.8	12.3	6.5	1.5	-2.4	6.2
Poznan	-2.2	-1.6	2.3	8	13.6	17	18.8	17.9	14	8.5	3.7	0.2	8.4
Warsaw	-3.5	-2.5	1.4	8	14	17.5	19.2	18.2	13.9	8.1	3	-0.6	8.1
Wroclaw	-2	-1.1	2.8	8.3	13.6	17	18.8	17.9	14.1	8.7	4.1	0.3	8.5
Kielce	-4.1	-3	1	7.1	13.1	16.5	18.2	17.3	13.3	7.7	2.7	-1.1	7.4
Zamosc	-4.4	-3.3	0.7	7.6	13.5	16.9	18.6	17.6	13.3	7.6	2.6	-1.3	7.5

In the coastal areas, the influence from the Baltic Sea on air temperatures is very clear, with cold springs (especially April–June) and much milder autumn and winter conditions than at inland stations. This can, for example, be seen if observations from Svenska Högarna are compared with those from Malung or if Hanko is compared with Jokioinen. It is also noted that the easternmost stations, Petrozavodsk and Novgorod, have moderately cold winters. The European part of Russia is fairly flat, giving more even temperatures than in northern Scandinavia. However, further eastwards close to the Ural Mountains and Siberia the temperatures drop to much lower values.

Comparisons of monthly averages for the two normal periods 1961–1990 and 1931–1960 for Swedish and Finnish stations show that May and June have become slightly warmer, and July, August, September, November and December have all become clearly colder. A more detailed time evolu-

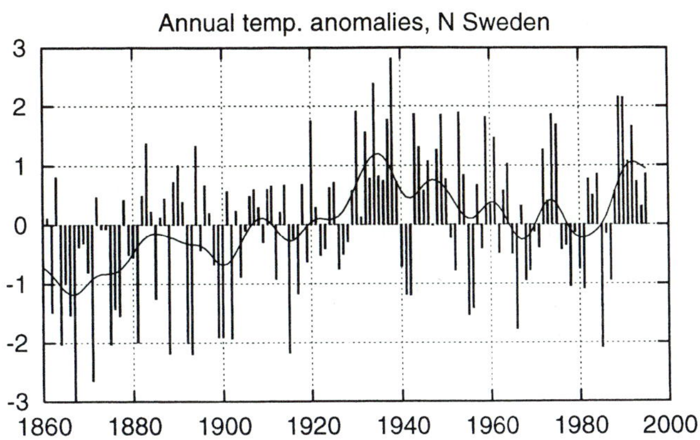

Fig. 3.7. Annual air temperature anomalies in northern Sweden, 1860–1995, determined from data from seven stations (homogenised). Also included is a Gaussian low pass filter curve corresponding to a 10-year moving average

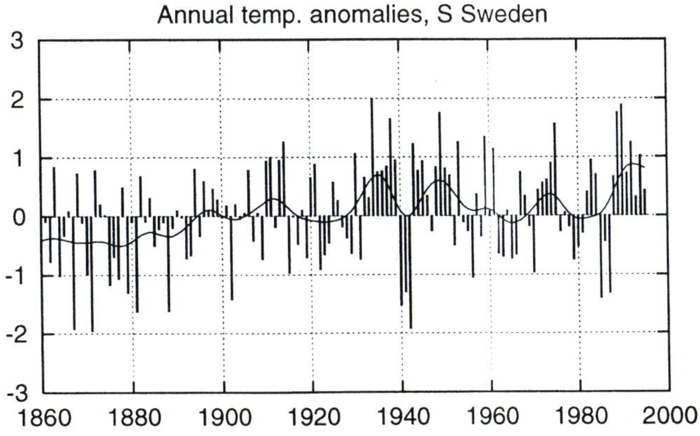

Fig. 3.8. Annual air temperature anomalies in southern Sweden, 1860–1995, determined from data from seven stations (homogenised). Also included is a Gaussian low pass filter curve corresponding to a 10-year moving average

tion, based on the two groups of stations from northern and southern Sweden, is shown in Figs. 3.7 and 3.8. These show a clear trend to a maximum during the warm 1930s and then a decrease, especially during the cold 1960s and 1980s. After that, the recent warm years and especially the mild winters have caused a rapid and strong recovery. The general features are equal in the southern and northern region but stronger in absolute values in the north.

Time series of seasonal temperatures are shown in Figs. 3.9–3.12. The winter temperatures show stronger interannual variations than the summers and thus dominate the annual variabilities. It is therefore not surprising to find winter maxima and minima in the same years as the annual extremes. For the northern region the smoothed value rises from about −11 °C around 1870 to −7 °C in the 1930s, and after the colder 1960s and 1980s is again up around −7 °C. In southern Sweden as well as in

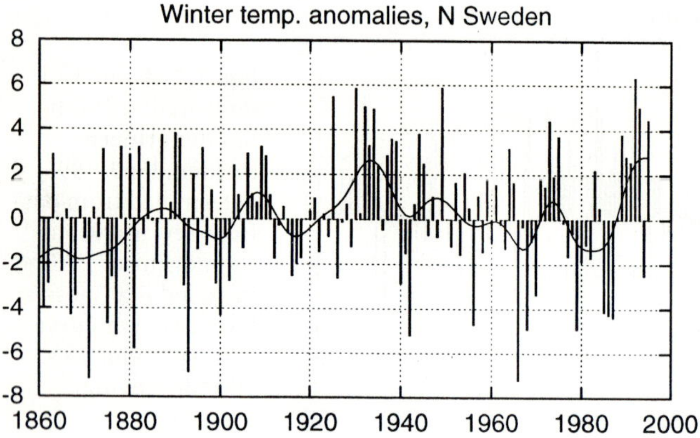

Fig. 3.9. Winter (December–February) air temperature anomalies in northern Sweden, 1860 1995, determined from data from seven stations (homogenised). Also included is a Gaussian low pass filter curve corresponding to a 10-year moving average

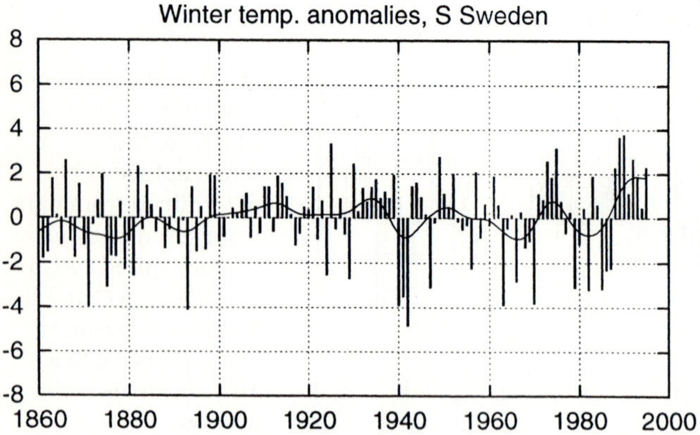

Fig. 3.10. Winter (December–February) air temperature anomalies in southern Sweden, 1860–1995, determined from data from seven stations (homogenised). Also included is a Gaussian low pass filter curve corresponding to a 10-year moving average

central and south-western Europe, the recent mild winters have climbed to new record levels, in Scandinavia especially in 1989 (i.e. December 1988 to February 1989) and in 1990.

The summer temperatures (June–August) also had their maxima in the 1930s or early 1940s, but in other respects they differ much from the annual and the winter curves (Figs. 3.11 and 3.12). In the last 15 years, the summer mean temperatures have approached the long-term average for 1860–1993 in the whole region, while several winters have been very mild. Spring temperatures have shown a stable increase during this whole period, whereas the autumn temperatures were similar to those between summer and winter.

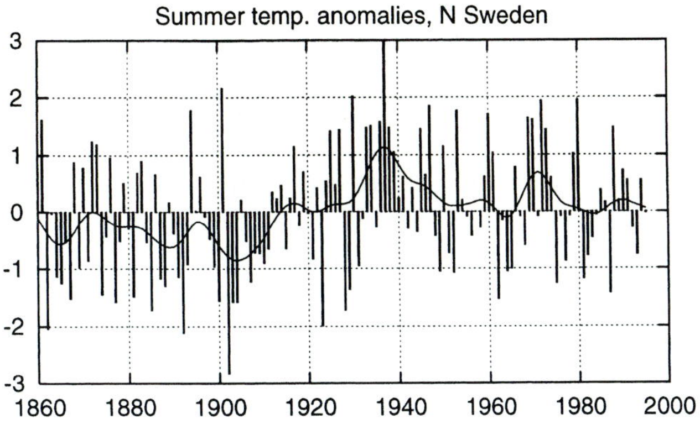

Fig. 3.11. Summer (June–August) air temperature anomalies in northern Sweden, 1860–1995, determined from data from seven stations (homogenised). Also included is a Gaussian low pass filter curve corresponding to a 10-year moving average

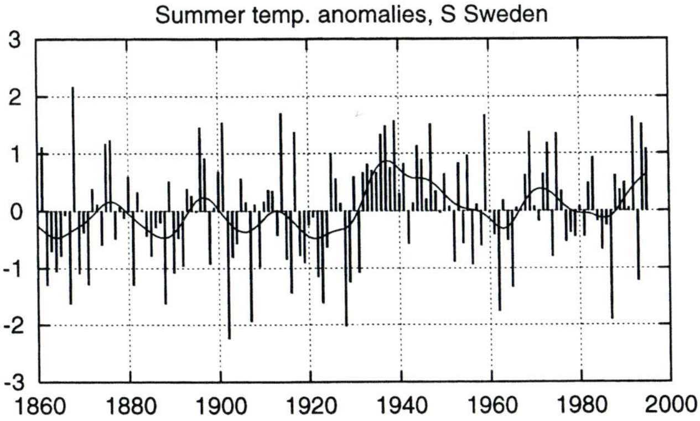

Fig. 3.12. Summer (June–August) air temperatures in southern Sweden, 1860–1995, determined from data from seven stations (homogenised). Also included is a Gaussian low pass filter curve corresponding to a 10-year moving average

3.1.3 Precipitation

In winter most of the precipitation in the Baltic basin is of the frontal type connected to extensive cyclones, especially inland. In summer around half of the precipitation is of the convective, showery type and more abundant inland than over the water. Especially during spring and early summer, the cold water of the Baltic Sea moderates convective precipitation.

The spatial precipitation pattern is to a high degree related to the topography, but there is also a general trend towards a wetter climate in the south and west and drier in the north and east. The largest annual amounts within the basin are estimated to be 1500–2000 mm in some parts of the Scandinavian fells. In a separate pronounced maximum in south-western Sweden the annual mean totals range from 1000 to 1200 mm. Further eastward in southern Sweden it is much lower with around 500 mm on Öland, an island in the western part of the Baltic Sea.

Further east the precipitation increases again in Latvia with up to 700–800 mm in a maximum zone some 20–40 km from the coast. East of the Baltic States in Russia the amounts drop, but fairly slowly as the topography is smooth. Another maximum can be found along the southern limit in the Polish parts of the Carpathian Mountains with annual totals of up to around 1000 mm.

In Finland the overall precipitation pattern has similarities with the Swedish one with a maximum in the south-west (locally slightly above 700 mm) and lower values up in the north (close to 400 mm in some regions). In offshore areas of the Baltic Sea the average annual amount is estimated to be 450–500 mm, as observed at the best sheltered island stations. The estimate for the open sea is, naturally, uncertain. As for air temperatures, the lowest precipitation values are found in valleys penetrating into the Scandinavian fells. Thus, for example, the measured mean values are only slightly above 300 mm close to the lake Torneträsk in the northernmost part of Sweden.

Table 3.2 shows monthly and annual mean precipitation for 30 stations in the area for the period 1931–1960. These data are also presented graphically in the map in Fig. 3.13. As can be seen, the minimum values occur mostly in March, while maximum precipitation is found either in July or August for almost all stations. There is a clear tendency towards later minima and maxima at more maritime sites, just as is the case with air temperatures (cf. Table 3.1).

A comparison between the two normal periods 1931–1960 and 1961–1990 for Sweden shows that there has been a pronounced increase in precipitation in most of the western and inner parts of Sweden, while some parts close to the Baltic Sea have experienced stable conditions or small

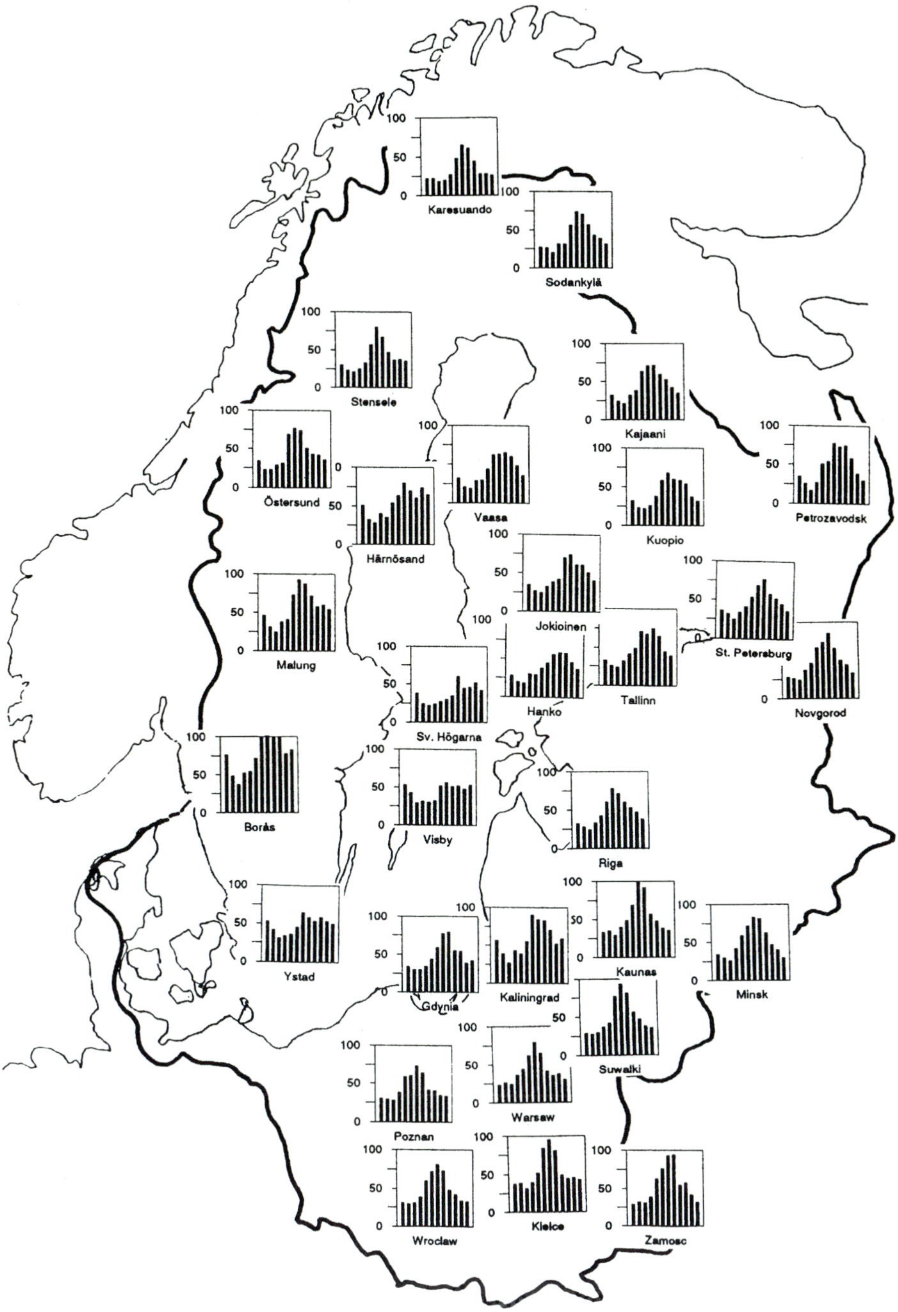

Fig. 3.13. Mean monthly precipitation at 30 stations in the Baltic basin for the period 1931–1960. Data from Landsberg (1985) and official Finnish and Swedish climatological yearbooks

Table 3.2. Mean monthly and annual precipitation amounts at 30 stations in the Baltic basin for the period 1931–1960. Data from Landsberg (1985) and official Finnish and Swedish climatological yearbooks. Locations of stations are shown in Fig. 3.1; data are presented graphically in Fig. 3.13 (values are not corrected for undercatch due to aerodynamic effects, wetting or other error sources)

Station	Jan	Feb	Mar	Apr	May	Jun	Jul	Aug	Sep	Oct	Nov	Dec	Year
Ystad	53	42	31	34	36	46	65	59	54	59	54	50	583
Borås	76	48	37	52	54	72	100	101	98	99	78	83	898
Visby	53	42	29	31	30	32	52	56	51	52	48	53	529
Sv. Högarna	38	24	22	24	27	30	35	60	45	46	52	42	445
Malung	46	31	25	38	41	73	93	88	72	58	60	54	679
Härnösand	51	32	28	40	35	54	64	80	70	61	74	65	654
Östersund	34	23	23	29	31	69	77	74	51	43	42	36	532
Stensele	30	23	21	25	33	57	80	67	47	37	38	36	494
Karesuando	22	22	18	20	27	48	65	61	44	28	28	26	409
Sodankylä	27	26	20	31	31	56	74	71	57	43	39	31	508
Kajaani	32	24	21	32	38	64	71	72	60	53	42	35	544
Vaasa	32	21	19	29	30	44	62	63	65	60	48	36	519
Kuopio	32	23	22	26	38	59	68	60	59	54	37	32	510
Jokioinen	35	27	25	33	39	42	70	74	61	61	51	41	559
Hanko	28	20	18	30	29	38	45	56	59	58	46	37	464
Petrozavodsk	35	26	17	27	51	54	77	73	74	58	39	30	559
St. Petersburg	36	32	25	34	41	54	69	77	58	52	45	36	559
Tallinn	33	26	24	32	41	49	71	68	75	65	45	39	568
Novgorod	28	26	24	37	46	67	74	86	66	50	44	33	581
Riga	32	28	24	33	42	60	78	71	60	53	47	38	566
Kaunas	33	35	29	40	49	69	98	92	57	48	39	36	625
Kaliningrad	56	39	27	43	39	55	90	84	83	70	52	59	698
Minsk	34	30	26	42	59	72	83	81	62	47	40	30	606
Gdynia	34	30	30	34	44	59	77	79	55	54	39	42	576
Suwalki	29	28	30	37	42	77	93	81	56	47	38	36	594
Poznan	30	29	28	38	59	60	73	64	41	40	34	33	528
Warsaw	23	26	24	36	44	62	79	65	41	35	37	30	502
Wroclaw	31	30	31	39	60	72	81	73	48	42	34	33	574
Kielce	38	39	32	40	52	84	95	81	49	45	46	43	644
Zamosc	29	32	31	39	63	76	93	94	55	58	42	32	644

decreases. This increase is mainly concentrated in the winter (Alexandersson et al. 1991). The month of March was generally much wetter during 1961–1990 than 1931–1960 so the driest month has instead become February (or sometimes April or May).

A detailed study of the changes in annual precipitation in Finland (Heino 1994) shows that precipitation has increased over the whole of Finland, although only marginally in the middle part of the country.

Precipitation records for the two regions in northern and southern Sweden are shown in Figs. 3.14 and 3.15. They are not at all representative for the whole region but may serve as an indication of the time evolution and variability. Although these data have been tested for homogeneity, it cannot be excluded that older data suffer from larger measuring losses than recent data. This might explain the lower values during the previous century. There may also be some reason to believe that it was slightly drier because of the colder winters and springs.

The two curves show that the 1980s was a wet decade. Other less pronounced wet periods were centered around 1930, 1950 and 1965. The

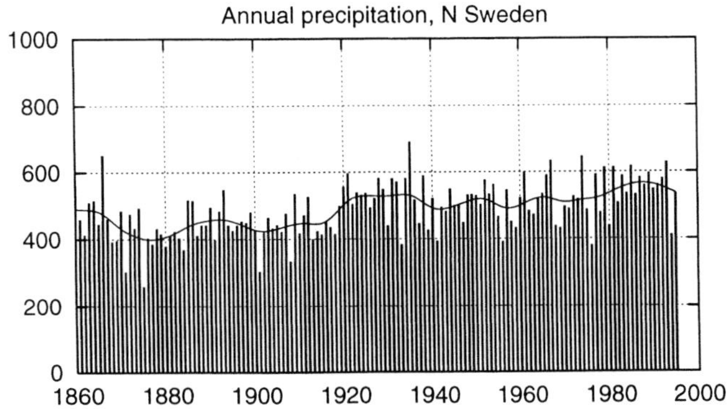

Fig. 3.14. Annual precipitation in northern Sweden, 1860–1995, determined from data from seven stations (homogenised). Also included is a Gaussian low pass filter curve corresponding to a 10-year moving average

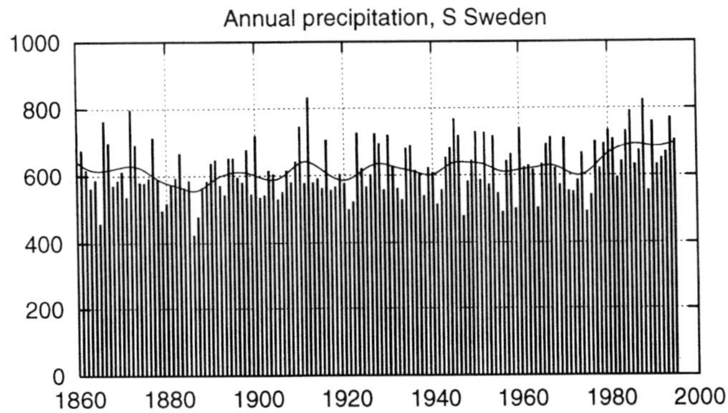

Fig. 3.15. Annual precipitation in southern Sweden, 1860–1995, determined from data from seven stations (homogenised). Also included is a Gaussian low pass filter curve corresponding to a 10-year moving average

recent mild winters with predominant westerly winds have had a strong effect on the winter precipitation in the Scandinavian mountains. Extreme accumulation of snow has been reported.

3.1.4 Winds

Winds are closely related to the cyclones and the pressure gradients they create. Winds of storm force, i.e. at least 25 m/s, are almost exclusively formed by deep cyclones that originated to the west of Scandinavia and occur mainly from September to March. The strongest winds over the Baltic Sea have, as a rule, a westerly to northerly direction behind these deep cyclones some 100 km from the storm centers.

The long-term trends in the wind climate can be explored by deriving geostrophic winds from air pressure data. Pressure data are far more exact

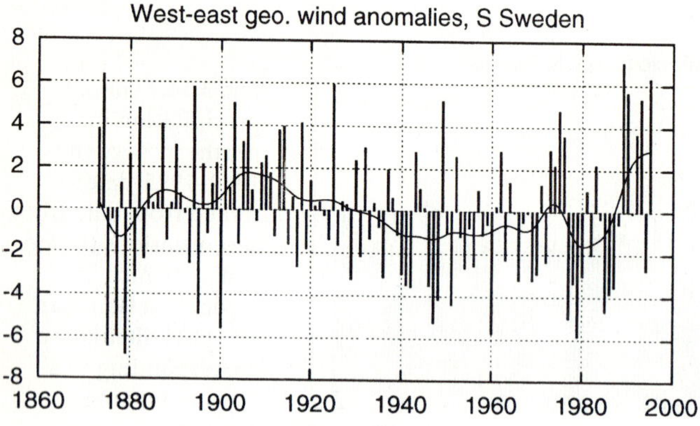

Fig. 3.16. Zonal (westerly) geostrophic wind in southern Sweden during winter, 1873–1995. Also included is a Gaussian low pass filter value corresponding to a 10-year moving average

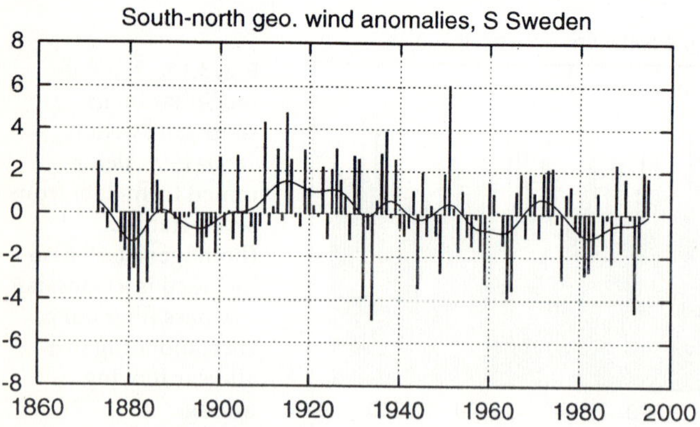

Fig. 3.17. Meridional (southerly) geostrophic wind in southern Sweden during winter, 1873–1995. Also included is a Gaussian low pass filter curve corresponding to a 10-year moving average

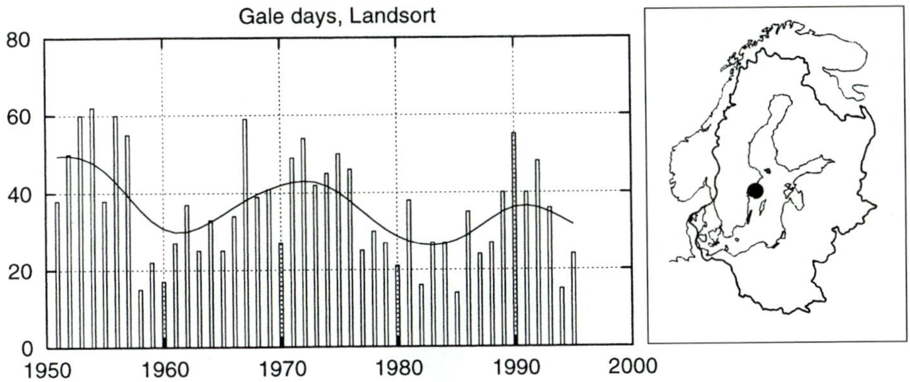

Fig. 3.18. Days with gale force winds (≥15 m/s) at Landsort, 1951–1993

and homogeneous than wind speed data, which suffer from homogeneity problems and where existing objectively measured records are lacking. In Figs. 3.16 and 3.17 long records of the zonal (westerly) and the meridional (southerly) geostrophic winds for winter conditions in southern Sweden are presented. As air pressure, especially on a longer time scale such as months, is smoothly varying, the results are believed to be representative at least of the southern half of the Baltic basin.

The zonal wind has apparently increased in recent winters (Fig. 3.16). This is coupled to the mild winters from 1987/1988 to 1994/1995. The winter of 1995/1996 (not included in the figures) broke this series of mild winters in southern Scandinavia. The meridional component is practically uncorrelated with temperature in winter. It has generally negative values during the latest decade (Fig. 3.17). This means that several of these winters have had a west-north-westerly mean flow.

Direct wind observations are needed to illustrate more extreme wind conditions. Figure 3.18 shows days per year with at least gale-force winds (≥15 m/s at one or more of the three main observations per day) obtained from direct wind observations at Landsort (58°44'N, 17°52'E) on the Baltic coast south of Stockholm. This record indicates calmer periods at the beginning of the 1960s and 1980s over the northern parts of the Baltic proper.

3.1.5 Cloudiness

The spatial distribution of clouds can be obtained with high resolution from satellite imagery. As it is a newly developed technique, no long-term trends or representative time averages can yet be derived from these types

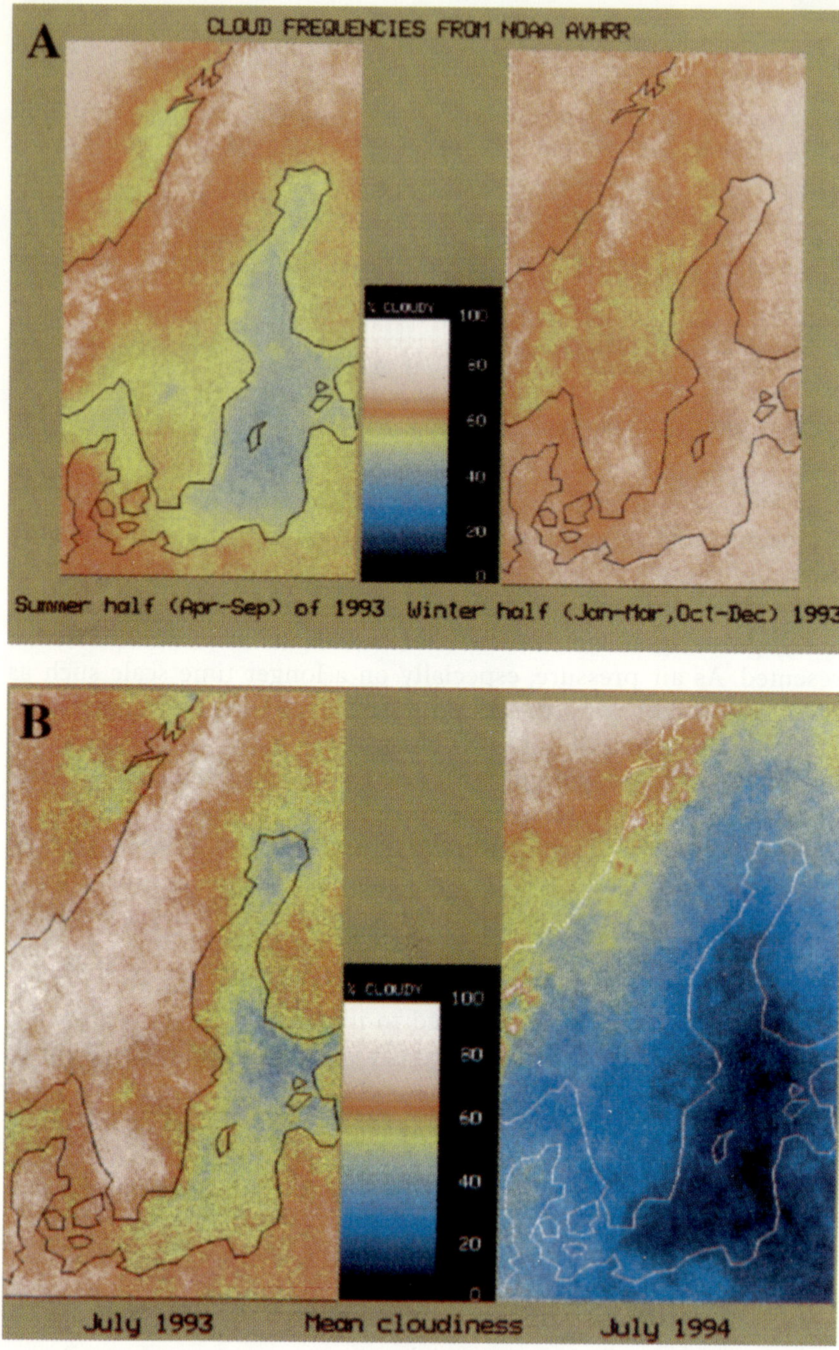

Fig. 3.19. Analysis of spatial distribution of clouds derived from NOAA AVHRR data. **A** Seasonal variation in cloudiness (%) in 1993; **B** comparison of mean cloudiness (%) in July in 1993 and 1994

of data. The Nordic cloud climate of the entire 1993 has been investigated by using AVHRR measurements (Karlsson 1994, 1995). First daily means of cloud cover were computed by using four satellite measurements with the approximate valid times 04:00, 7:30, 14:00 and 17:30 UTC (universal time coordinated). Then averages for longer periods were computed.

The seasonal variation in cloudiness in 1993 is demonstrated in Fig. 3.19A, where the mean cloudiness (in percent) is shown for the summer and winter halves of 1993. The summer half is defined as the period April to September, while the winter half is defined as the period January to March plus October to December. As can be seen, the winter half of the year is in general more cloudy (55–80 % mean cloudiness) than the summer half (40–80 % mean cloudiness). Large differences are found between land and sea. The suppression of cloudiness in the summer seems to be remarkably effective over the Baltic Sea and over the large lakes. On the contrary, mountain and inland areas show high cloudiness throughout the year.

The interannual variations in the cloud climate are substantial. This is exemplified in Fig. 3.19B, where satellite-estimated cloudiness for July is

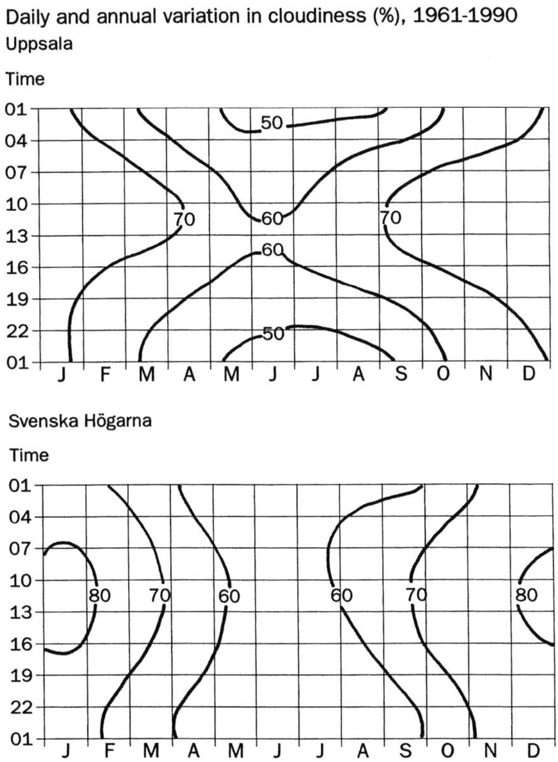

Fig. 3.20. Daily and annual variation in cloudiness (%) for the period 1961–1990 at Uppsala and Svenska Högarna. Uppsala is located some 80 km inland from the Baltic Sea coastline and Svenska Högarna is situated on an island some 40 km out at sea. (National Atlas of Sweden 1995)

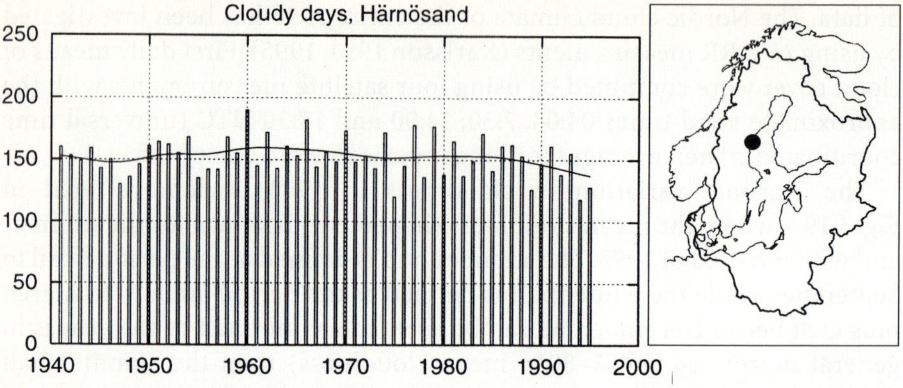

Fig. 3.21. Number of cloudy days in Härnösand on an annual basis for the period 1940–1995. Also included is a Gaussian low pass filter value corresponding to a 10-year moving average

analysed for the 2 years 1993 and 1994. July 1993 was dominated by a cyclonic weather type giving high mean cloudiness in the Nordic area, while July 1994 was dominated by a strong anticyclone (a blocking weather type) giving extremely low mean cloudiness in most places. Mean cloudiness as low as 5 % was found in the Baltic Sea east of Gotland. Note the suppression of cloudiness over sea areas under both cyclonic (July 1993) and anticyclonic (July 1994) conditions.

Figure 3.20 illustrates the daily and seasonal course of cloudiness at two nearby stations not far from Stockholm. Uppsala is situated some 80 km inland from the Baltic Sea coastline and Svenska Högarna is situated on an island some 40 km from the coast. The annual cycle is clearly shown as is the difference in cloudiness during daytime in summer.

The long-term variation pattern of cloudiness is exemplified by the number of cloudy days at Härnösand (62°37'N, 17°56'E) in Fig. 3.21 (a cloudy day is defined as one with the average cloud cover at 06, 12 and 18 UTC exceeding 75 %). Some particularly cloudy years are recognised, for example, 1960, a famous year for its extremely wet and cloudy summer and autumn in the coastal areas of the Baltic Sea. Data from Härnösand may be considered as fairly homogeneous from 1940 and onwards.

3.1.6 Radiation

The global radiation is the integrated energy of the solar radiation on a horizontal surface. More than 99 % of this lies within the spectral range 0.29–3.0 μm at the surface of the Earth. The variation in the global radiation primarily depends on the solar elevation, which causes the yearly and daily cycles. The yearly cycle dominates at the latitudes of the Baltic basin. The northern parts are located north of the Arctic Circle and thus the winter season has very low incoming radiation.

The main meteorological factor affecting radiation is the cloudiness. However, the reflectance of the surface also has an impact, both on a local scale by reflections on nearby slopes and on a larger scale where the multiple reflections between the ground and the atmosphere increase the global radiation. These factors can be highly variable due to variations in the snow cover.

The amount of aerosols affects the radiation. It is connected to the type of air mass and therefore its effect on the radiation is regional. There might also be large local sources of aerosols in some industrial areas with local impact. After the eruptions of the El Chichon (May 1982) and the Pinatubo (June 1991) volcanoes there was a global increase in the aerosols in the lower stratosphere. This affected the input of solar radiation especially during the first 2 years after the eruption. The effect was most easily detected in observations of the direct solar radiation on cloud-free days.

The data coverage of radiation measurements is relatively poor over the Baltic basin and non-existent over the sea. Traditionally, only observations of global radiation and sunshine duration are available in records long enough for an analysis of the climate. There are some shorter series

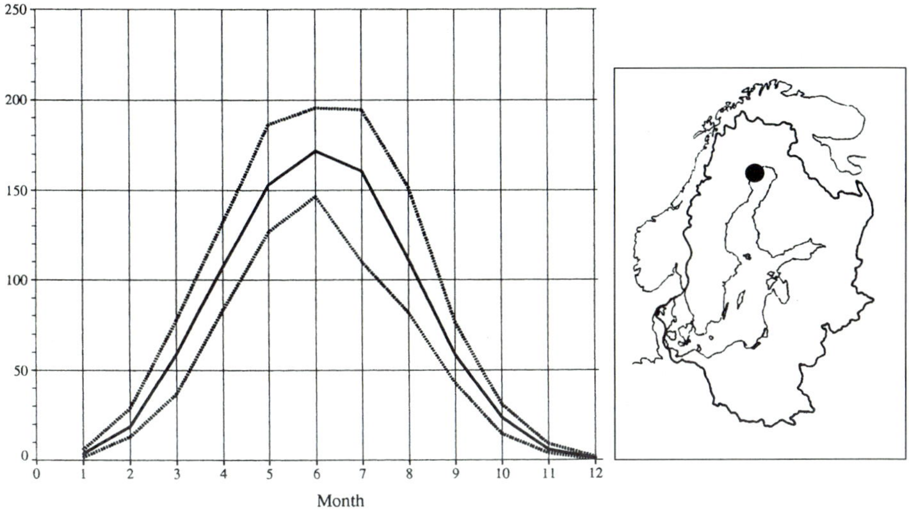

Fig. 3.22. Annual course of global radiation at Luleå based on monthly values 1961–1990. *Solid curve* is the mean and *hatched curves* are maximum and minimum monthly values (kW h m^{-2})

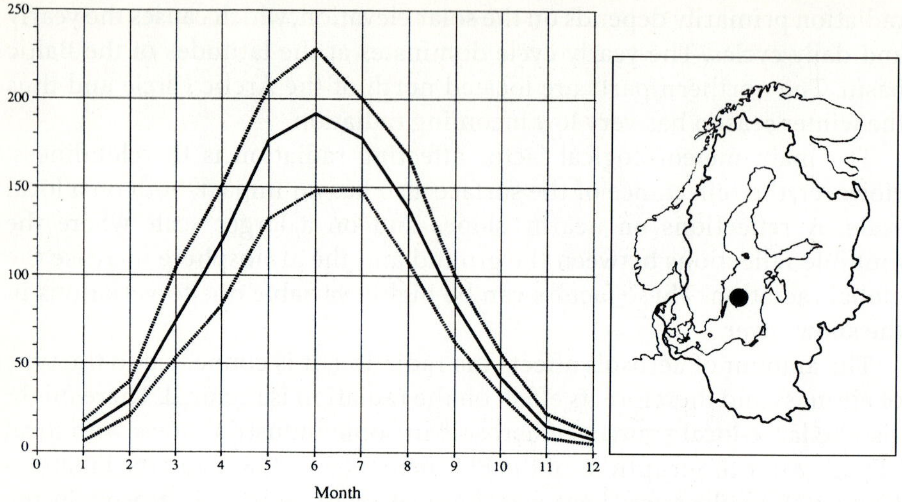

Fig. 3.23. Annual course of global radiation at Visby based on monthly values 1961–1990. *Solid curve* is the mean and *hatched curves* are maximum and minimum monthly values (kW h m^{-2})

available of direct solar radiation and longwave downward radiation. Examples of the yearly and seasonal variation in monthly values of global radiation from Luleå (65°37'N, 22°08'E) and Visby (57°38'N, 18°20'E) are given in Figs. 3.22 and 3.23. The data are from the reference period 1961–1990.

The long-term variation and seasonal distribution of the global radiation are also represented by data from Luleå (1962–1993) and Visby (1957–1993) (Figs. 3.24 and 3.25). Most of the solar radiation in the Baltic region is found during spring and summer. In Luleå only the year 1976 exceeds the lowest values recorded at Visby (for the period 1962–1993).

Both Luleå and Visby are located in a coastal area, i.e. in a zone with a large radiation gradient. Therefore, on a yearly average higher values are expected over the open sea and lower values inland, as illustrated in the cloud analyses in Figs. 3.19 and 3.20. Estimates of typical average yearly global radiation for some areas of the Baltic basin for the period 1961–1990 are summarised in Table 3.3.

The direct solar radiation is more variable than is global radiation, mostly due to the variations in cloudiness. Data on direct solar radiation recorded at Visby for the period 1983–1993 are presented in Fig. 3.26 for each season. It is notable that there are spring values as large as many of the values recorded in summer.

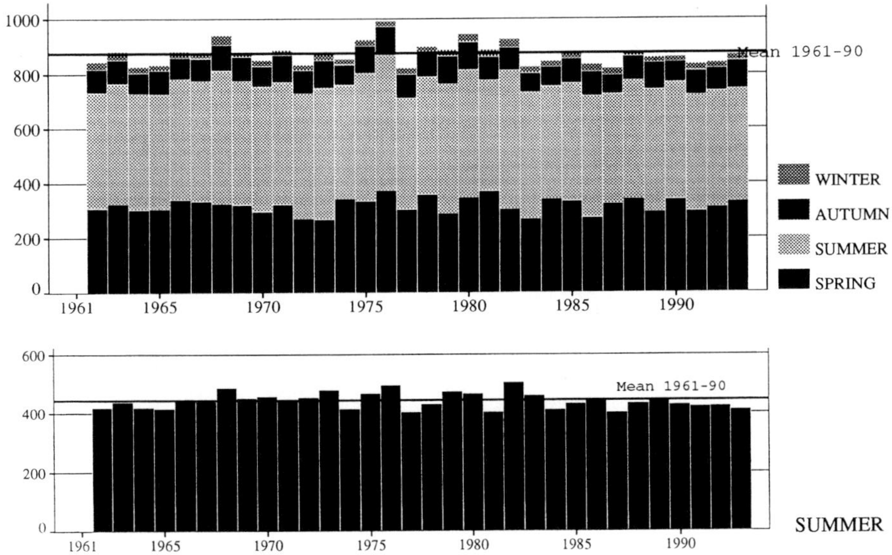

Fig. 3.24. Long-term variation in global radiation in Luleå, 1962–1993. *Bars* also show seasonal variation, with summer values shown separately (kW h m^{-2})

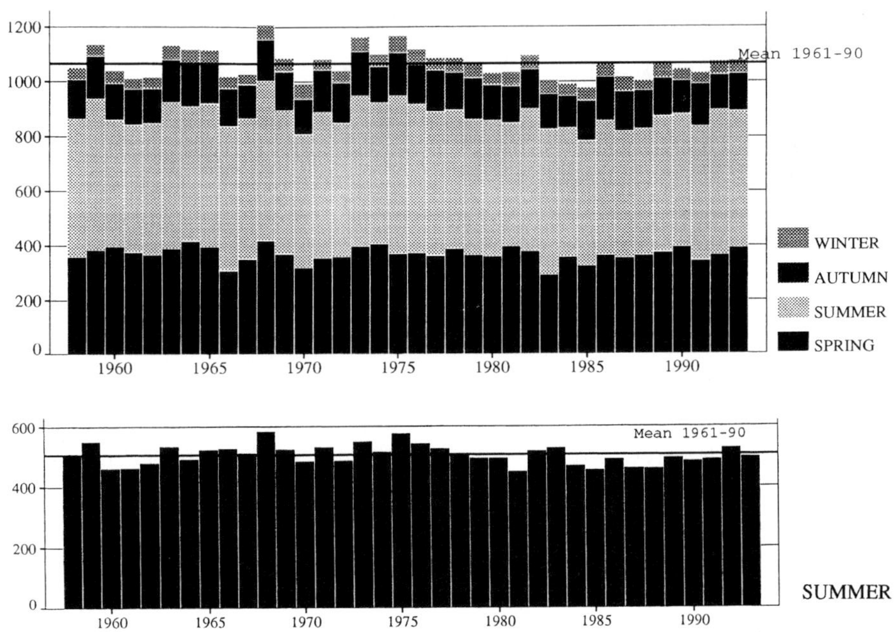

Fig. 3.25. Long-term variation in global radiation in Visby, 1958–1993. *Bars* also show seasonal variation, with summer values shown separately (kW h m^{-2})

Table 3.3. Typical average yearly global radiation (kW h m^{-2}) for some areas of the Baltic basin for the period 1961–1990. Values for positions at sea are estimates from National Atlas of Sweden (1995) based on a simplified model. Standard deviation of yearly values measured in Luleå and Visby is about ±5 %

Site	Yearly global radiation (kW h m^{-2})
Luleå	876
Bothnian Bay	910
Bothnian Sea	975
Visby	1067
c proper	1075

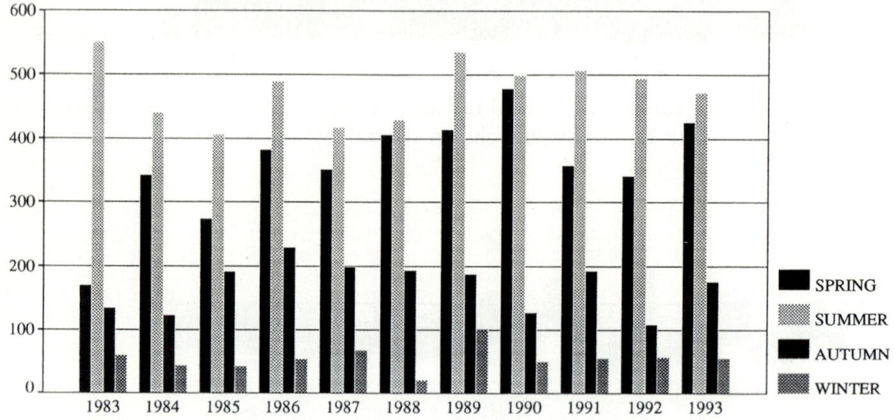

Fig. 3.26. Long-term variation in direct solar radiation in Visby, 1983–1993, for the four seasons (kW h m^{-2})

3.1.7 Ice Conditions

Ice occurs in all parts of the Baltic Sea. Certain areas will during some winters remain ice free, but ice will always occur somewhere each winter and all areas have at some time experienced ice. Ice conditions vary widely – locally, regionally, in- and offshore, from day to day, from year to year, and over longer time scales. The most severe winters, which occur only a few times per century, will cover the whole Baltic Sea with ice, although only for limited periods. During very mild winters the sea ice cover will not extend beyond the Bothnian Bay, with a little coastal ice further south in inner parts of bays, gulfs and archipelagos (see Figs. 3.27 and 3.28).

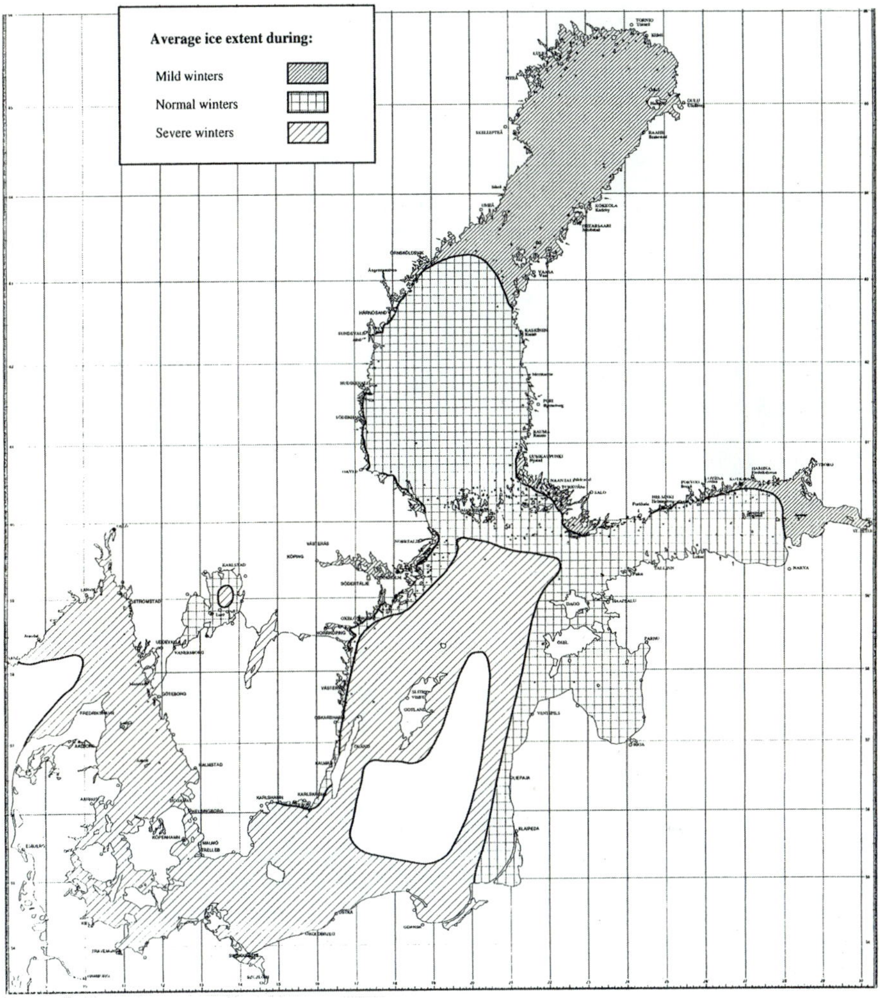

Fig. 3.27. Average ice cover for typical mild, normal and severe winters of the Baltic Sea, Danish Sounds, Kattegat and parts of Skagerrak. (Swedish Meteorological Institute and Finnish Institute of Marine Research 1982; Håkansson et al. 1995)

Observations of ice conditions in Swedish waters have been carried out since 1926. Statistics based on the observations have been compiled for two normal periods 1930/1931–1959/1960 and 1960/1961–1989/1990 (Thorslund 1966; Westring 1993). Despite large variations in ice winter severity and recent mild winters (Fig. 3.29) an analysis of the two normal periods suggested that long-term average ice conditions have grown worse (Westring 1995). Initial results indicated an overall increased severity in the order of 10 %, but its statistical significance was low.

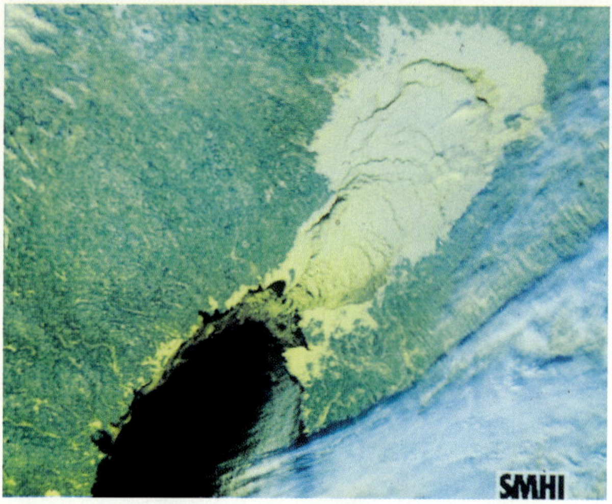

Fig. 3.28. Mild ice winter conditions in the Bothnian Bay, NOAA satellite image of 15 March 1991

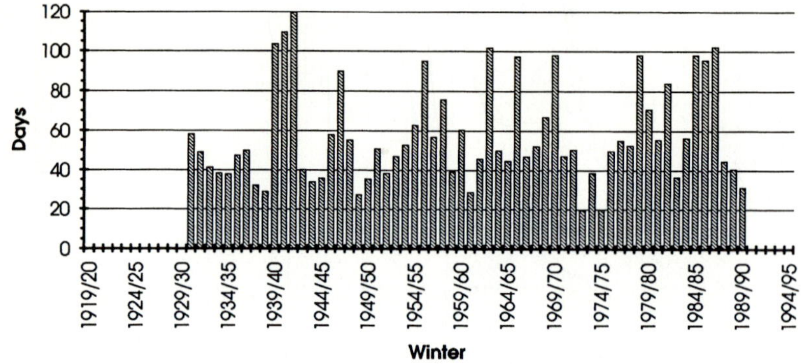

Fig. 3.29. Ice winter severity index 1930/1931–1989/1990, expressed as annual mean number of ice days for all observations in Swedish coastal waters. (Westring 1995)

The ice winter severity, expressed in terms of annual mean number of ice days, is well correlated with air temperatures. As shown by Alexandersson et al. (1991) the air temperatures for the latest normal period (1961–1990) are about 0.5 °C lower than for the earlier normal period (1931–1960) in northern Sweden. For southern Sweden this difference is less than 0.2 °C.

The severity of ice conditions can also be described by the annual maximum ice coverage in the Baltic Sea as shown in Fig. 3.30. This pattern shows great similarities with that of the index of severity in Fig. 3.29.

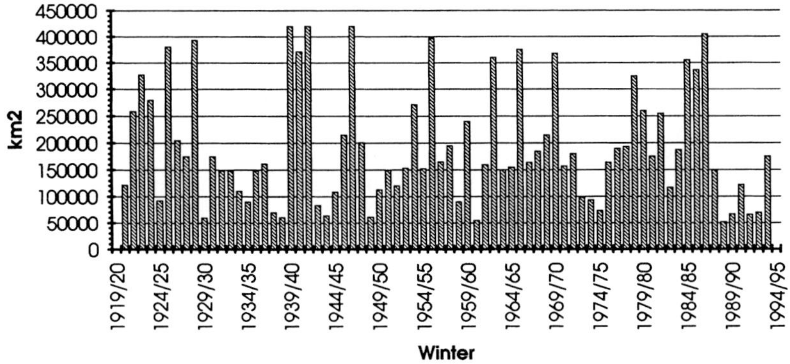

Fig. 3.30. Ice winter severity 1920/1921–1993/1994, expressed as annual maximum ice coverage in the Baltic Sea, Danish Sounds, Kattegat and parts of Skagerrak (Seinä and Palosuo 1993). A. Seinä, Finnish Institute of Marine Research, Helsinki, provided data for 1993/1994–1994/1995

3.2 River Runoff to the Baltic Sea

Nine countries share the shorelines of the Baltic Sea and its drainage basin includes territories from altogether fourteen countries. The size of the total area is approximately 1,729,000 km^2 with a distribution over sub-basins according to Table 3,4.

Several large rivers discharge into the Baltic Sea, the largest being Neva River with a mean annual flow of approximately 2500 m^3/s. Other large rivers are shown in Table 3.5. As can be seen from the table, the specific runoff is highly variable for the main river basins.

There are several large lakes in the area, the largest being Lake Ladoga in Russia. The lakes have a strong influence on the dynamics of the rivers as they damp the fluctuations in runoff. They also affect the water balance of catchment as they cause increased evaporation.

Many of the rivers in the Baltic basin are affected by regulation. This is particularly pronounced in Sweden where hydroelectric power covers about half of the demand for electricity. The development of hydropower leads to a more artificial hydrograph, where most of the flow-peaks (but not all!) are stored in the reservoirs and the winter base-flow is increased. Maybe the best illustration of this is the record from River Luleälven in northern Sweden (Fig. 3.31). The short-term variability is increased to meet the fluctuations in power demand.

The water balance of the Baltic Sea and its sub-basins has attracted the interest of many scientists since the beginning of the previous century

Table 3.4. Approximate sizes of land areas draining into different sub-basins of the Baltic Sea

Sub-basin	Size (km^2)
Bothnian Bay	261,000
Bothnian Sea	230,000
Gulf of Finland	421,000
Gulf of Riga	132,000
Baltic proper	584,000
Danish Sounds and Kattegat	101,000
Total area	1,729,000

Table 3.5. The ten largest rivers of the Baltic Sea system, Danish Sounds and Kattegat, their approximate drainage area, mean annual and specific runoff during 1950–1990

River	Drainage area (km^2)	Mean annual runoff (m^3/s)	Specific runoff [$l/(s \cdot km^2)$]
Neva	281,000	2460	8.8
Vistula	194,400	1065	5.5
Daugava	87,900	659	7.5
Neman	98,200	632	6.4
Odra	118,900	573	4.8
Kemijoki	51,400	562	11.0
Göta älv	50,100	574	11.5
Ångermanälven	31,900	489	15.0
Luleälven	25,200	486	19.0
Indalsälven	26,700	443	16.5

(Falkenmark and Mikulski 1975). In recent times the compilation of river inflow by Mikulski has played a key role in this respect (Mikulski 1970, 1982). Mikulski's work has been used for water balance estimates presented by, for example, Falkenmark and Mikulski (1975, 1988), Ehlin (1981), by the Helsinki Commission (HELCOM 1986) and Henning (1988).

Lately, a database on inflow from the drainage basin to the Baltic Sea system was established within the Swedish research program „Large-scale Environmental Effects and Ecological Processes in the Baltic Sea" (Bergström and Carlsson 1993, 1994). The database covers the time period 1950–1990 and consists of data from altogether some 200 river flow stations, most of them near the mouth of the rivers. The catchment areas of these stations represent 86 % of the total drainage basin. Runoff from areas not covered by measurements, mostly coastal areas located between major

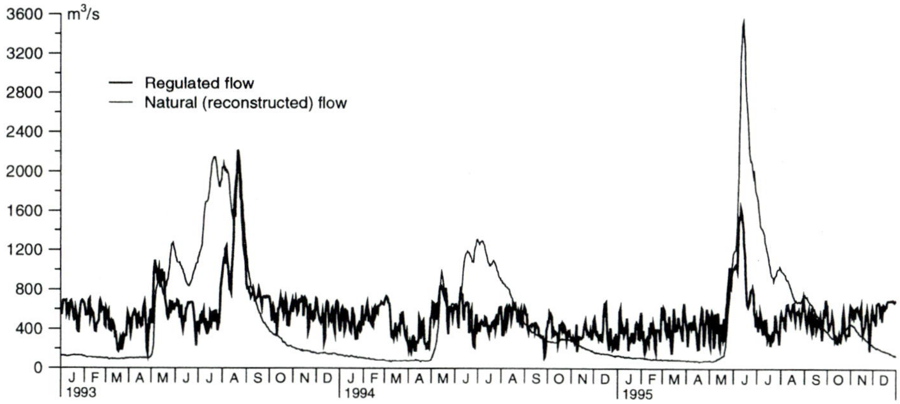

Fig. 3.31. Example of influence of river regulation on flow in River Luleälven, near outlet into Bothnian Bay. Peak in August 1993 occurred when reservoirs were full and therefore regulated flow came close to natural flow. In absence of regulation, peak in June 1995 would have become the highest in the century

rivers, was estimated by the use of calculations of specific runoff for neighbouring stations which were considered representative.

The database was recently updated to include the period 1991–1993. In this updating process a few gaps have been filled in by comparisons with nearby stations. This is not considered to affect the results significantly.

3.2.1 Annual River Runoff to the Baltic Sea

The total contribution of river runoff to the system and individual contributions to the sub-basins for the period 1950–1990 are shown in Table 3.6. The mean annual contribution to the Baltic Sea system, excluding the Danish Sounds and Kattegat, is 14,151 m^3/s or 446 km^3/year for the period 1950–1990. The latter figure corresponds to a water body of 1.18 m covering the Baltic Sea and means that the total volume of water, 21,200 km^3, would be replaced in 48 years, if no mixing and other sources are considered.

A total flow of 14,151 m^3/s is further comparable to that of a mighty river like the Paraná, St. Lawrence or Irrawaddy and slightly less than the average flow of the Mississippi.

The estimates of freshwater inflow to the different basins of the Baltic Sea made by Mikulski (1982) offer an opportunity for comparison and extension of the records as they overlap the present database. Mikulski's estimates are based on a sample of 17 rivers and the time period is 1921–

Table 3.6. Mean annual river runoff to the Baltic Sea system and its sub-basins during the period 1950–1990

Sub-basin	Mean annual runoff (m^3/s)	Mean annual runoff (km^3/year)
Bothnian Bay	3104	98
Bothnian Sea	2860	91
Gulf of Finland	3556	112
Gulf of Riga	1020	32
Baltic proper	3610	114
Danish Sounds and Kattegat	1159	37
Total area	15,310	483

1975. The areal representation of the catchments of these rivers is 63 % of the total drainage area. Figure 3.32 shows a composite presentation of the annual data from the Mikulski and Bergström and Carlsson projects covering the 73 years from 1921 to 1993. Corresponding data for the entire system per decade are shown in Table 3.7; Mikulski's data represent the period 1921–1950.

Figure 3.32 shows that the estimates made by Mikulski are well in agreement with those made by Bergström and Carlsson (1993) for the whole system and for some of the sub-basins. The greatest differences are found in the estimates of contributions to the Bothnian Sea and the Baltic proper. These differences are due to different data coverage in the two studies.

The long-term record of runoff in the total Baltic basin shows a remarkable annual variability, with high values at the beginning and end of the record. The last decade is the wettest of the whole period. 1924 was an outstandingly wet year with a mean annual inflow of 19,500 m^3/s, while the inflow during the driest year, 1976, was as low as 11,100 m^3/s.

Since 1981 a marked drop in runoff to the Baltic proper is observed while the other basins mostly have kept their high values. As can be seen in Fig. 3.33, the recent drop in freshwater contribution to the Baltic proper is strongly associated with dry conditions in the Vistula basin in Poland.

The data can also be used to estimate the relative contributions from each one of the countries around the Baltic Sea. It is important to note, however, that the water that discharges into the sea from one country very well may have its origin in another country further upstream. It is thus more appropriate to say that the water reaches the Baltic sea *via* a particular country. Rivers floating along the borderlines between two countries, like the Rivers Torneälven, Narva and Neman, cause another problem. In the following presentation, the river flows of these are shared equally between

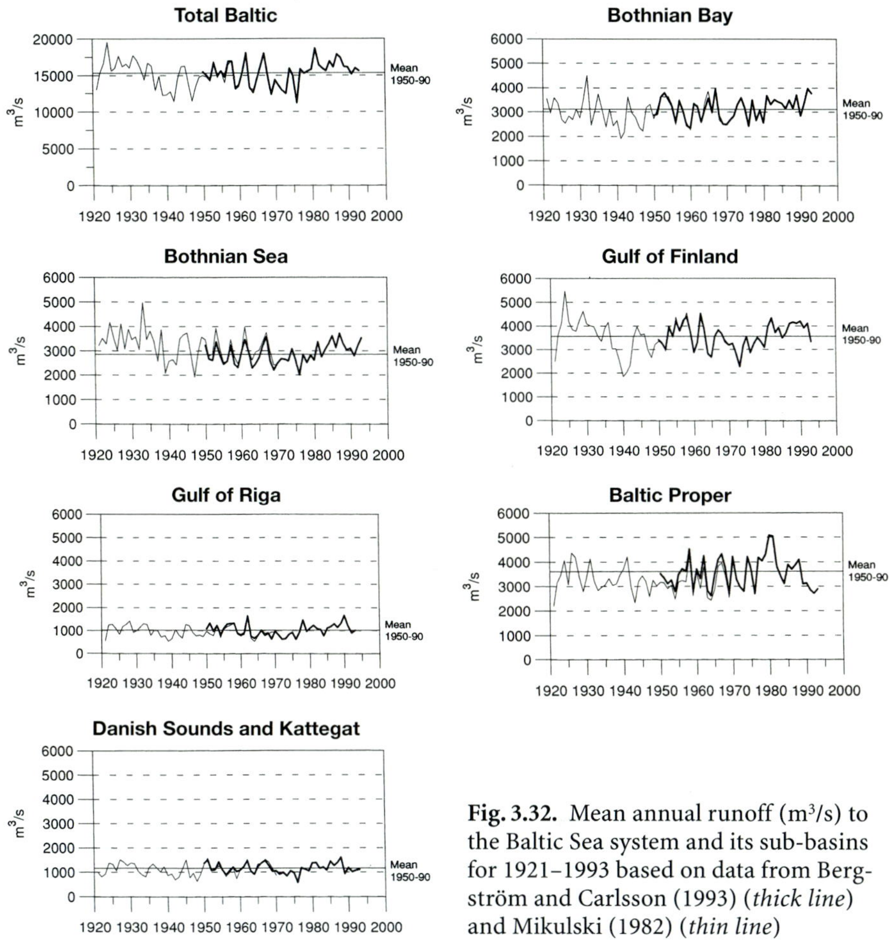

Fig. 3.32. Mean annual runoff (m³/s) to the Baltic Sea system and its sub-basins for 1921–1993 based on data from Bergström and Carlsson (1993) (*thick line*) and Mikulski (1982) (*thin line*)

Table 3.7. Average runoff and total volumes of freshwater to the Baltic Sea system and Danish Sounds and Kattegat for each decade during the period 1921–1990. The period 1921–1950 is based on data from Mikulski (1982) and the period 1951–1990 on data from Bergström and Carlsson (1993)

Decade	Average inflow (m³/s)	Total volume (km³)
1921–1930	16,246	5120
1931–1940	15,039	4740
1941–1950	13,940	4400
1951–1960	15,217	4800
1961–1970	14,997	4730
1971–1980	14,318	4520
1981–1990	16,692	5260

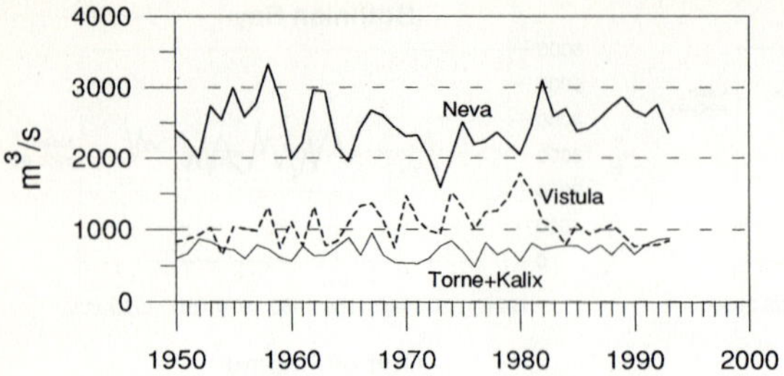

Fig. 3.33. Records of annual runoff for Rivers Torne and Kalix, Neva and Vistula for the period 1950–1993 (m^3/s)

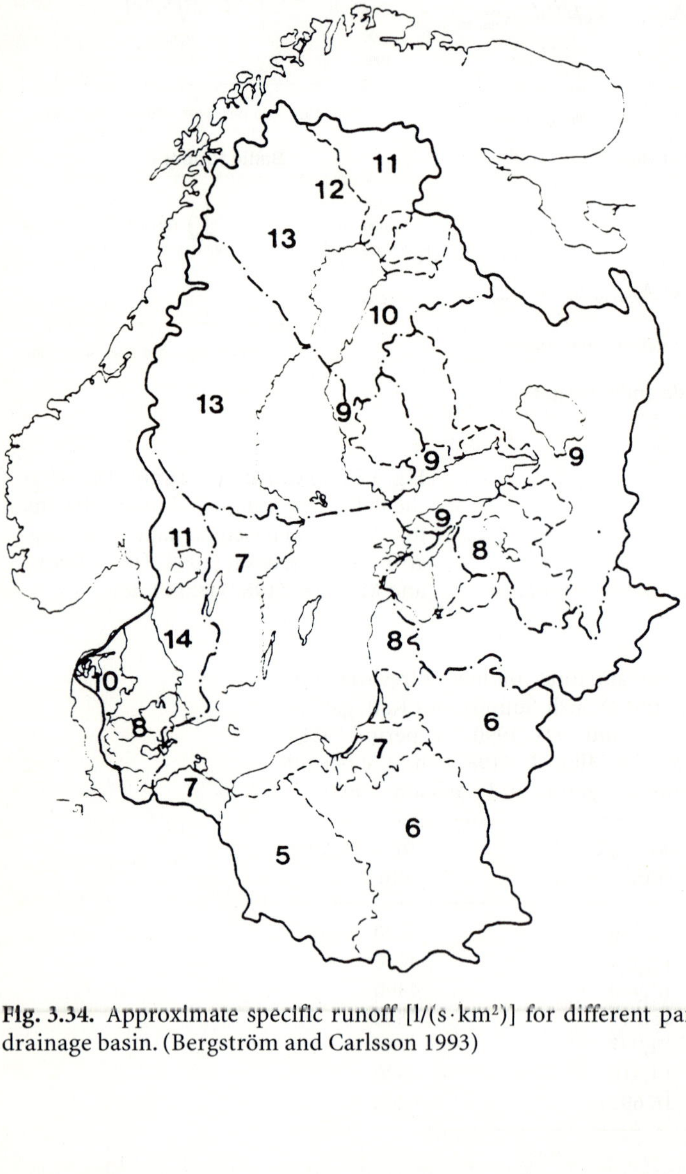

Fig. 3.34. Approximate specific runoff [$l/(s\cdot km^2)$] for different parts of the Baltic Sea drainage basin. (Bergström and Carlsson 1993)

Table 3.8. Approximate relative contributions (%) to the Baltic Sea via different countries and distribution over each sub-basin. Mean values for the period 1950–1990. (Bergström and Carlsson 1993)

Country	Total basin	Bothnian Bay	Bothnian Sea	Gulf of Finland	Gulf of Riga	Baltic proper	Danish Sounds and Kattegat
Sweden	35	10	16	–	–	3	6
Finland	16	10	3	3	–	–	–
Russia	22	–	–	19	–	3	–
Estonia	3	–	–	2	0.5	0.5	–
Latvia	7	–	–	–	6	1	–
Lithuania	2	–	–	–	–	2	–
Poland	12	–	–	–	–	12	–
Germany	1	–	–	–	–	1	–
Denmark	2	–	–	–	–	<0.5	2
Total	100	20	19	24	6.5	22.5	8

the two countries. A summary of relative contributions via each country, under the assumptions given above, is given in Table 3.8.

Due to the differences in specific runoff from different regions, which is strongly related to climatic conditions, the relative contributions in Table 3.8 do not fully reflect the areal conditions. As can be seen in Fig. 3.34, the specific runoff in Poland is only about half of that of northern Sweden. The highest specific runoff is found in the north-west of Sweden and in Norway where precipitation is high and evapotranspiration is low.

3.2.2 Seasonal Contributions from Rivers

Seasonal runoff is of interest in environmental studies as the concentrations of nutrients in the rivers may vary considerably over the year due to biological activities and other conditions (see, for example, Taylor 1987; Brandt 1990; Stålnacke 1996). In particular, the nitrate concentrations in lake-rich systems are low in summer. The dynamics in concentrations interact with river flow and have to be considered when estimating the total load to the coastal waters.

The seasonality of the river flow to the Baltic Sea system is illustrated as regimes in Fig. 3.35 for the Baltic sub-basins. Time series of inflow to the total system for winter, spring, summer and autumn respectively are shown in Fig. 3.36.

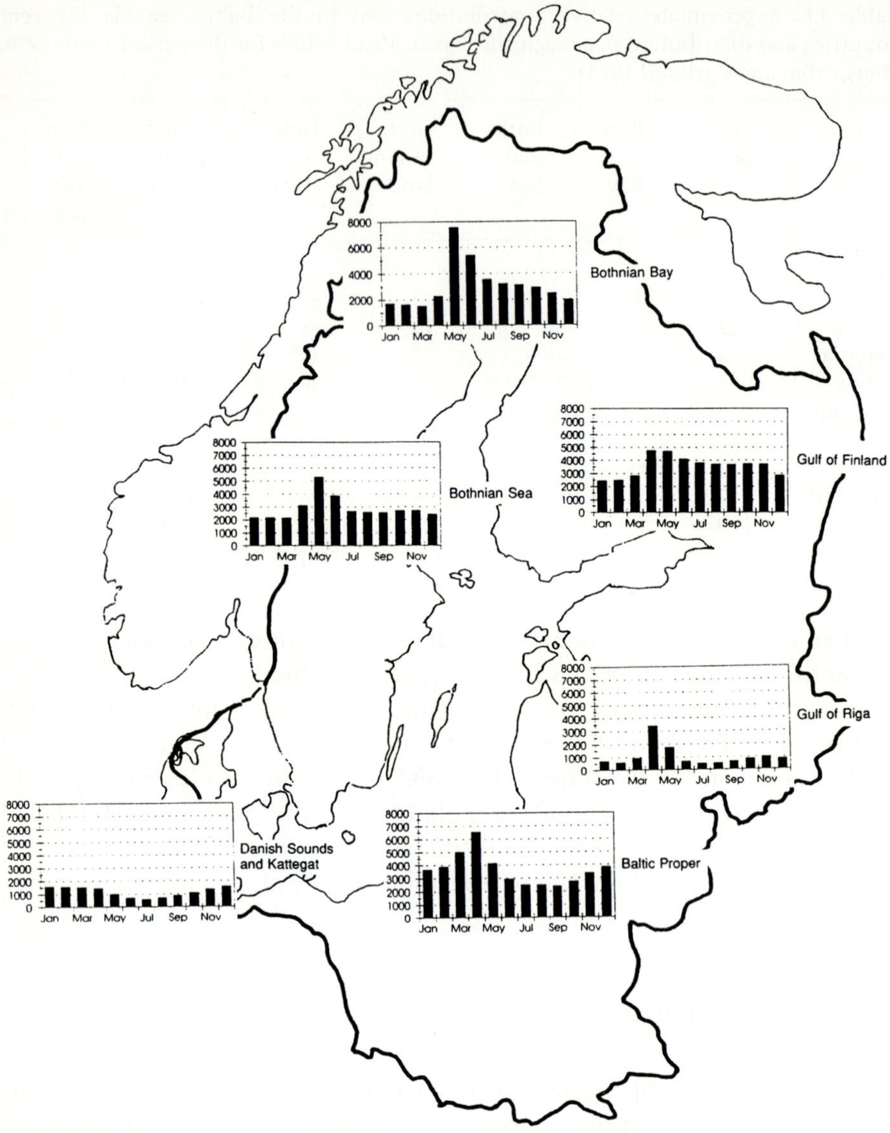

Fig. 3.35. Monthly means of runoff, showing hydrological regime of sub-basins of the Baltic Sea for the period 1950–1990

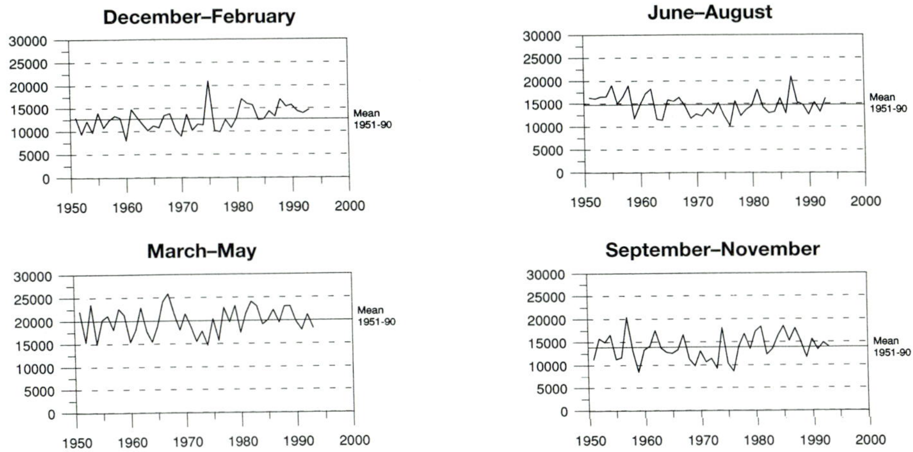

Fig. 3.36. Annual means of seasonal runoff from land to the Baltic Sea for the period 1951–1993

Figure 3.35 shows that the hydrological regimes are highly variable for the different sub-basins. The northern sub-basins have a pronounced spring flood due to snowmelt and low winter runoff, while winter runoff is gradually becoming more important further south in the drainage area and is dominating the contribution to the Danish Sounds and Kattegat.

From Fig. 3.36 it can be seen that the pronounced increase in runoff during the late 1980s is dominated by increasing winter flows, with some influence from increased flows in spring and autumn. A closer analysis reveals that the increase in winter runoff is a general tendency for all basins, with the possible exception of the Gulf of Finland which shows a more irregular pattern (Bergström and Carlsson 1993). Increasing flows in February for the whole of Finland have previously been reported by Kuusisto (1992).

The increase in winter runoff is partly an effect of the recent mild winters, with high precipitation and with little or no accumulation of snow in the south. For the northern basins, the Bothnian Bay and the Bothnian Sea, there is also a gradual effect of hydropower developments that result in increasing winter discharges and decreasing discharge other times of the year. This effect has stabilised since very few new hydropower projects have been completed since 1980.

3.2.3 Monthly Values and Extremes

Monthly values is the highest resolution which can be analysed with the present database. This is a relevant resolution if extreme values from larger areas are of interest and can also, to some extent, help detect the effects of river regulation. Monthly and annual extremes are summarised in Table 3.9.

The maximum recorded monthly average contribution of freshwater to the system occurred in May 1966 and was 32,411 m^3/s, which corresponds to a volume of 87 km^3. It would cause a rise in the sea level of 23 cm if other factors were disregarded.

A closer look at the runoff values gives no indication of a change in the pattern of extremes. They seem to be fairly randomly distributed over the analysed period. There are, however, clear signs of effects of river regulation in the monthly data for the Bothnian Bay and the Bothnian Sea. The effects show up as increasing base flow levels over time.

Table 3.9. Annual and monthly extreme mean inflows to the Baltic Sea and its subbasins during 1950–1990 (m^3/s)

	Highest annual	Year	Lowest annual	Year	Highest monthly	Date	Lowest monthly	Date
Total system	18,660	1981	11,132	1976	32,411	May 1966	7,635	Dec 1959
Bothnian Bay	3940	1967	2326	1960	11,380	May 1953	910	Jan 1957
Bothnian Sea	3711	1987	1992	1976	9391	May 1966	1100	Oct 1959
Gulf of Finland	4521	1962	2267	1973	7194	May 1955	1405	Jan 1973
Gulf of Riga	1654	1962	629	1976	5593	Apr 1951	162	Feb 1960
Baltic Proper	5072	1981	2602	1964	12,167	Apr 1979	1276	Oct 1959
Danish Sounds and Kattegat	1607	1988	571	1976	2841	Mar 1951	309	Jul 1957

References

Alexandersson H, Eriksson B (1989) Climate fluctuations in Sweden 1860–1987. Rep Meteorol Climatol RMK 58. SMHI, Norrköping, 55 pp

Alexandersson H, Karlström C, Larsson-McCann S (1991) Temperature and precipitation in Sweden 1961–90. Reference normals (in Swedish with an English summary). Ser Meteorol 81. SMHI, Norrköping, 87 pp

Bergström S, Carlsson B (1993) Hydrology of the Baltic basin – inflow of fresh water from rivers and land for the period 1950–1990. Rep RH 7. SMHI, Norrköping

Bergström S, Carlsson B (1994) River runoff to the Baltic Sea: 1950–1990. Ambio 23(4/5):280–287

Brandt M (1990) Simulation of runoff and nitrate transport from mixed basins in Sweden. Nordic Hydrol 21:13–34

European Climate Support Network (1995) Climate of Europe – recent variation, present state and future prospects. ECSN, KNMI, De Bilt

Ehlin U (1981) Hydrology of the Baltic Sea. In: Voipio A (ed) The Baltic Sea. Elsevier Oceanogr Ser 30. Elsevier, Amsterdam, pp 123–134

Falkenmark M, Mikulski Z (1975) The Baltic Sea – a semi-enclosed sea, as seen by the hydrologist. Nordic Hydrol 6:115–136

Falkenmark M, Mikulski Z (1988) The Baltic Sea as an example: hydrological conditions as a determinant of the ecosystem in a semi-enclosed sea. Nature Resour 24(1):14–25

Heino R (1994) Climate in Finland during the period of meteorological observations. Contrib 12. Finnish Meteorological Institute, Helsinki, 209 pp

HELCOM (1986) Water balance of the Baltic Sea. Baltic Sea Environ Proc 16, Helsinki

Henning D (1988) Evaporation, water and heat balance of the Baltic Sea. Estimates of short- and long-term totals. Meteorol Rundsch 41:33–53

Håkansson B, Moberg M, Thompson T (1995) Real-time use of ERS-1 SAR imagery for ice service and icebreaking operation in the Baltic Sea. Int J Remote Sensing 16(17):3441–3458

Höybye J (1991) Ferskvandstilströmning til danske farvande 1990. Hedeselskabet, Fagdatacenter for hydrometriske data, Slagelse, Denmark

Karlsson K-G (1994) Satellite-estimated cloudiness from NOAA AVHRR data in the Nordic area during 1993. Rep Meteorol Climatol RMK 66. SMHI, Norrköping, 51 pp

Karlsson K-G (1995) Estimation of cloudiness at high latitudes from multispectral satellite measurements. Ambio 24(1):33–40

Kuusisto E (1992) Runoff from Finland in the period of 1931–1990. Aqua Fenn 22(1):9–22

Landsberg HE (1985) World survey of climatology, vols 1–15. Elsevier, Amsterdam

Liljequist G-H (1970) Climatology (in Swedish). Generalstabens Litografiska Anstalt, Stockholm, 527 pp

Lindström G (1993) Floods in Sweden – trends and occurrence. Rep RH6. SMHI, Norrköping

Mikulski Z (1970) Inflow of river waters to the Baltic Sea in 1961–1970. Nordic Hydrol 4:216–227

Mikulski Z (1982) River inflow to the Baltic Sea. Polish Academy of Sciences, Polish National Committee of the IHP, University of Warsaw, Faculty of Geography and Regional Studies

Seinä A, Palosuo E (1993) Itämeren suurimpien vuotuisten jääpeitteen laajuuksien luokittelu 1720–1992 [The classification of maximum annual extent of ice cover in the Baltic sea 1720–1992] (in Finnish with English Abstract). MERI 20. Finnish Institute of Marine Research, Helsinki

Swedish Meteorological Institute and Finnish Institute of Marine Research (1982) Climatological ice atlas for the Baltic Sea, Kattegat, Skagerrak and Lake Vänern (1963–1979). SMHI, Norrköping, and FMI, Helsinki

National Atlas of Sweden (1995) Climate, lakes and rivers. National Atlas of Sweden, Stockholm

Stigebrandt A (1983) A model for the exchange of water and salt between the Baltic and the Skagerrak. J Phys Oceanogr 13:411–427

Stålnacke P (1996) Nutrient loads to the Baltic Sea. Linköping Stud Arts Sci 146. Linköping University

Taylor R (1987) Factors forming the level and dynamics of nutrient runoff from an agricultural watershed. Pol Arch Hydrobiol 34(3):285–295

Thorslund B (1966) Isförhållanden i svenska farvatten under normal perioden 1931–1960 [Ice conditions in Swedish fairways during the normal period 1931–1960] (in Swedish). Ser Meteorol 13. SMHI, Norrköping

Westring G (1993) Isförhållanden i svenska farvatten under normal perioden 1961–1990 [Ice conditions in Swedish fairways during the normal period 1931–1960] (in Swedish). Oceanogr 59. SMHI, Norrköping

Westring G (1995) Isförhållanden utmed Sveriges kust [Ice conditions along the Swedish coast] (in Swedish). Oceanogr 63. SMHI, Norrköping

4 Riverine Inputs of Nutrients to the Baltic Sea

A. Grimvall and P. Stålnacke

4.1 Introduction

Rivers play a crucial role in the total input of nutrients to the Baltic Sea. Several studies have indicated that nutrient delivery along this route may exceed the combined contribution from atmospheric deposition on the sea, point emissions along the coast, and nitrogen fixation by marine organisms (Larsson et al. 1985; Rosenberg et al. 1990; HELCOM 1993; Enell and Fejes 1995; Stålnacke et al. 1999a). There are also strong indications that the riverine loads of nutrients have increased dramatically over the past century (Larsson et al. 1985). However, data regarding the historical development of nutrient loading of the Baltic Sea are uncertain, and even loading during the past few years has been a matter of controversy. According to the second pollution load compilation undertaken by HELCOM (1993), rivers were responsible for an annual input of 550,000 tons of nitrogen, whereas Enell and Fejes (1995) estimated that this route, including some point emissions along the coast, contributed about 980,000 tons year^{-1}.

The purpose of this chapter is to elucidate riverine inputs of nutrients to the Baltic Sea by examining whole time series of water quality and discharge data. After a brief review of databases and background documents, we present estimates of the average annual inputs and consider reasons for the substantial discrepancies between the estimates published thus far. Thereafter, the discussion is devoted to the spatial and temporal variation of nutrient inputs, in particular considering which river basins make the largest contributions to the total input and the question of when the major increases in the input occurred. Lastly, we scrutinize the controllability of riverine loads, weighing the theoretical and practical feasibility of achieving a rapid reduction in the amount of nutrients carried in rivers to the Baltic Sea.

Ecological Studies, Vol. 148
F. Wulff et al. (eds.)
A Systems Analysis of the Baltic Sea

4.2 Databases and Background Documents

The Baltic Sea and its tributaries form one of the most intensively monitored aquatic systems in the world. Water discharge or runoff data have long been available for the major rivers (Mikulski 1970, 1982) and, more recently, Bergström and Carlsson (1994) compiled time series of monthly runoff values for about 200 sites in the Baltic Sea drainage basin. The investigated catchments cover about 86% of the total drainage area, and the study period ranged from 1950 to 1993. There is also a relatively long tradition of monitoring nutrient concentrations. Systematic measurements of inorganic nitrogen and phosphorus were begun as early as the 1950s at some sites in Latvia and Estonia (Tsirkunov et al. 1992), and the Swedish and Finnish national monitoring programs have their roots in measurements initiated in the mid 1960s. In 1977, the former Soviet Union introduced an extended program for water quality monitoring, and figures for concentrations of inorganic nitrogen and phosphorus are available from that year forward for most of the large and medium-sized rivers in the Baltic Sea drainage basin. Data regarding organic forms of nitrogen and phosphorus are less complete. Long time series of such data exist primarily for rivers in Sweden, Finland, Denmark and Poland, and only during the past few years have large parts of the Baltic States and Russia been systematically monitored in this respect. The frequency of the sampling that has been carried out has varied from twice a week in parts of the Polish program to once a quarter in parts of the Estonian program.

By combining available water quality and runoff data or by compiling previously published load estimates, several investigators have calculated the total riverine input of nutrients to the Baltic Sea for single years or for periods of a few years (Table 4.1). The time series of riverine loads reported here are based on calculations performed by Stålnacke et al. (1999a). Runoff data were taken from the database established by Bergström and Carlsson (1994), and concentration data for almost 150 rivers and streams were obtained directly from the agencies responsible for the national monitoring programs of the Baltic Sea countries. In addition, our research team at Linköping University organized a more than 2-year-long sampling program in the Vistula River (Sundblad et al. 1994; Tonderski et al. 1995; Tonderski 1997) and undertook three, seven, eight, and six sampling tours to the Neva, Narva, Daugava, and Neman Rivers, respectively, in 1992–1994. Data from these sampling tours were used to cross-check data from the national monitoring programs and to provide rough estimates of the concentrations of total nitrogen and phosphorus in some rivers, e.g. the Neman, for which such data were missing.

Table 4.1. Reported estimates of total riverine loads of nitrogen and phosphorus to the Baltic Sea

Reference	Study period	Nitrogen (tons year^{-1})	Phosphorus (tons year^{-1})
Pawlak (1980)	1972–1977	265,225[a,b]	22,112[a,b]
Maksimova (1982)	1968–1978	508,000	45,400
Larsson et al. (1985)	1977–1983	640,500[b]	50,170[b]
Bernes (1988)	1977–1988	831,245[c]	74,560[c]
Rosenberg et al. (1990)	1982–1987	535,400[c]	47,290[c]
Rybinski (1989)	1985–1989	692,894	38,119
HELCOM (1993)	1990	567,739	33,360
Fotyma and Terelak (1994)	1990	733,000	53,100
Swedish Ministry of Environment and Natural Resources (1990); Enell and Fejes (1995)	1990–1993	979,000[c]	58,860[c]
Stålnacke et al. (1999a)	1980–1993	825,000	41,000
Total range		265,000–979,000	22,000–59,000

[a] Excluding loads from the former Soviet Union (Russia and the Baltic States) to the Gulf of Riga and Baltic proper
[b] Excluding loads to the Kattegat
[c] Including coastal point source emissions

The distribution of nutrient loads among major river basins has also been compiled (Stålnacke et al. 1999a), and more detailed information about the spatial distribution of nutrient loads and the role of various point and non-point sources of nutrients in different parts of the Baltic Sea drainage area can be found in a number of other articles, e.g. Löfgren and Olsson (1990); Kronvang et al. (1996); Arheimer et al. (1997); Tonderski (1997); Laznik et al. (2000). The existence of long-term water quality trends has been addressed mainly in studies of specific rivers (Hagebro et al. 1983; Kauppi 1984; Grimvall et al. 1991; Tsirkunov et al. 1992; Kauppi et al. 1993; Kronvang et al. 1993; Stålnacke et al. 1999b).

4.3 Data Quality and Statistical Methods

The quality of available nutrient concentration and runoff data has a significant impact on calculations of riverine loads of nutrients. Our inspection of time series of runoff data did not reveal any serious data quality problems, whereas examination of concentration data produced more alarming results. In fact, some of the collected time series of NO_3-N and PO_4-P data exhibited a temporal variation that was difficult to explain. We have previously (Stålnacke et al. 1999a) attempted to localize and quantify potential data quality problems by computing coefficients of variation for time series of flow-weighted mean annual concentrations. The results of that study show that data from the Neman River produced the largest coefficients of variation. Moreover, a remarkable interannual variation was observed for NO_3-N in the Vistula and Narva Rivers and PO_4-P in the Neva, Narva, and Göta Rivers. In general, however, the flow-weighted mean annual concentrations of NO_3-N and PO_4-P varied in a fashion that was consistent with mechanisms known to operate in the loss of nutrients from soil to water and in freshwater environments.

We made the assumption that the suspected data quality problems were restricted to a relatively small number of rivers or laboratories, and closer examination of collected data indicated that the problems were also limited to specific time periods. Therefore, we decided to reject only a rather small fraction of all collected concentration data. Whole sequences of data were removed only if collected data were clearly contradicted by the results of our own sampling campaigns or by theoretical calculations based on emission coefficients for urban areas and agricultural land. A complete list of the rejected data and detailed information regarding methods for identification of outliers can be found in the above-mentioned article by Stålnacke et al. (1999a). That report also provides information about the extrapolation and interpolation procedures used to extend available time series of concentration data to complete series of monthly data.

Nutrient loads from catchments not included in the national monitoring programs were estimated by extrapolating runoff and concentration data from adjacent basins. This is, of course, another source of uncertainty. However, in a Baltic Sea perspective, the errors introduced by such extrapolations are fairly small. Due to the dominating role of large rivers, such as the Neva, Daugava, Neman, Oder, and Vistula, it is also possible to produce accurate estimates of annual nutrient loads, even if samples for water quality measurements are collected only once a month (Tonderski et al. 1995; Grimvall 1996). Higher sampling frequency is needed primarily in small catchments exhibiting rapid variation in nutrient concentrations

(Rekolainen et al. 1991; Kronvang et al. 1996), and some of the national monitoring programs have been adapted to such requirements.

When temporal trends in time series of data were analyzed, observed concentrations or loads were flow-normalized to display more clearly possible anthropogenic trends. Unless otherwise stated, flow normalization was carried out according to a semiparametric procedure developed by Stålnacke and Grimvall (2000).

4.4 Total Input of Nutrients to the Baltic Sea

In the introduction it was mentioned that the size of the total input of nutrients to the Baltic Sea has been a matter of controversy. We found that the most complete sets of concentration data were available for the years 1980–1993, and, for that period, we estimated the average input of total nitrogen to be 825,000 tons year^{-1} and the average input of total phosphorus to be 41,000 tons year^{-1}. Our estimates and values published by other investigators are listed in Table 4.1.

Two of the studies cited in Table 4.1, a report by the Ministry of the Environment and Natural Resources (MENR) (1990) and a follow-up article by Enell and Fejes (1995), presented emission-based estimates of the input of nutrients to the Baltic Sea; to obtain those values, emission coefficients were assigned to different types of point and non-point sources and then combined with source-to mouth retention in the rivers. All other study teams have based their estimates on water discharge and nutrient concentration data from national monitoring programs, which is essentially the same source of information as we used in the present study. Therefore, it is noteworthy that our estimate of the total riverine input of nitrogen to the Baltic Sea was similar to that of Enell and Fejes (1995) but substantially higher than the estimates presented by most other investigators. There are also marked differences in regard to estimates of phosphorus loads.

Closer examination of the discrepancies between our estimates and the other values listed in Table 4.1 provided two major explanations for the differences in nitrogen loads. First, we tried to estimate missing values of organic nitrogen, whereas other investigators, e. g. HELCOM (1993), simply omitted such forms from their calculations when no real data were available. Second, we rejected some nitrate concentration data that we considered to be unrealistically low (Stålnacke et al. 1999a). Considering phosphorus loads, the discrepancies were attributed primarily to differences in study periods and large interannual variation in reported loads.

Table 4.2. Estimated annual input of nutrients to the Baltic Sea along four major pathways

Pathway	Study period	Nitrogen load (tons year^{-1})	Phosphorus load (tons year^{-1})
Riverine load	1980–1993	830,000	41,000
Coastal point sources	1990	100,000	13,000
Atmospheric deposition	1985–1989	300,000	5,500
N_2 fixation	1980	130,000	–
All pathways combined		1360,000	59,500

Differences in the definition of the Baltic Sea drainage area and the handling of contributions from small streams (<5 m^3/s) and coastal areas located between the monitored catchments should also be taken into account when comparing our estimates with the other values presented in Table 4.1.

The careful examination of reported data that was done in the study by Stålnacke et al. (1999a) has hopefully removed most of the uncertainty regarding the total riverine load of nutrients delivered to the Baltic Sea. In any case, there is now convincing evidence that the rivers are responsible for the major fraction of the total input of both nitrogen and phosphorus (see Table 4.2). The sum of all direct emissions from industries on the Baltic Sea coast and from cities such as St. Petersburg, Tallinn, Riga, Gdynia-Gdansk, Szczecin, Kiel, Copenhagen, Stockholm, Helsinki, Kaliningrad, and Gothenburg is much smaller (HELCOM 1993). Moreover, it is generally accepted that the input of nitrogen with atmospheric deposition is about 300,000 tons year^{-1}, i.e. less than 40 % of our estimate of the riverine contribution (Areskoug 1993; Chap. 5). Data regarding atmospheric deposition of phosphorus and nitrogen fixation by cyanobacteria (blue-green algae) indicate that these two sources represent 10 % or less of the total input of nutrients (Table 4.2).

4.5 Spatial Distribution of Riverine Loads

The Neva, Narva, Daugava, Neman, Vistula, and Oder Rivers are all located east or south of the Baltic proper. These rivers are the largest in the Baltic Sea drainage area, and they are also responsible for the largest input of nutrients to the sea. In 1980–1993, the Vistula accounted for more than 14 % (119,000 tons year^{-1}) of the total riverine input of nitrogen, and the Oder River was responsible for 16 % (6000 tons year^{-1}) of the total riverine input of phosphorus. The catchment upstream of St. Petersburg in the Neva River basin delivered the greatest amount of water (20 % of the total runoff) but a considerably smaller amount of nitrogen (7 %) and phosphorus (8 %). The diagrams in Fig. 4.1 further illustrate the dominating role of the eleven major Baltic rivers, which together contributed approximately 48 % (401,000 tons year^{-1}) of the nitrogen and 56 % (23,000 tons year^{-1}) of the phosphorus carried in rivers to the Baltic Sea.

Normalization of the riverine loads with respect to the size of the drainage area revealed several spatial patterns. As was expected, the lowest values were obtained for almost pristine areas, such as the Torne River

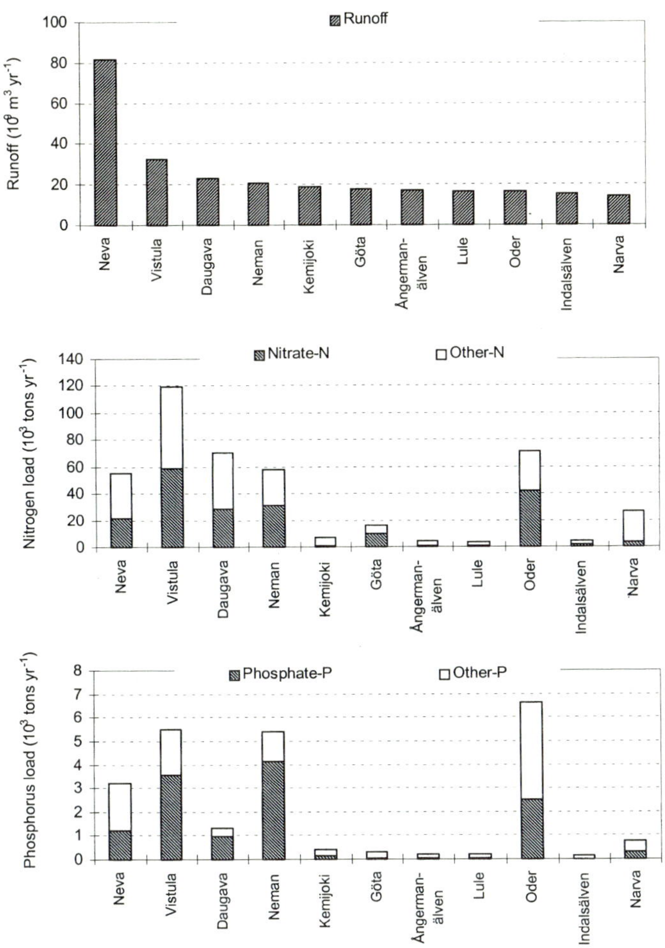

Fig. 4.1. Time-averaged runoff and riverine loads of NO_3-N, total-N, PO_4-P and total-P for the eleven largest rivers in the Baltic Sea drainage basin. (Stålnacke et al. 1999a)

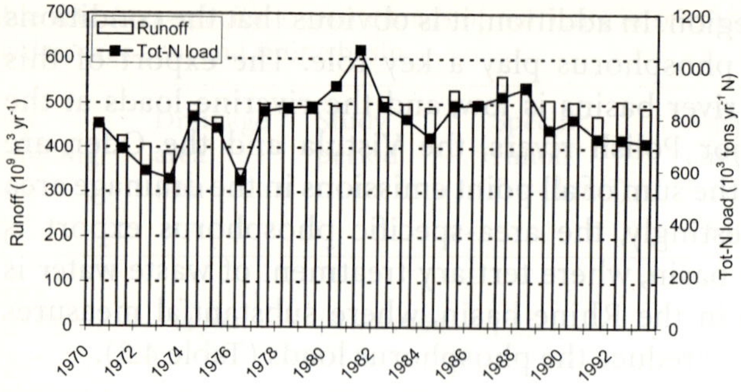

Fig. 4.2. Annual riverine input of water, nitrogen and phosphorus to the Baltic Sea. (Stålnacke et al. 1999a)

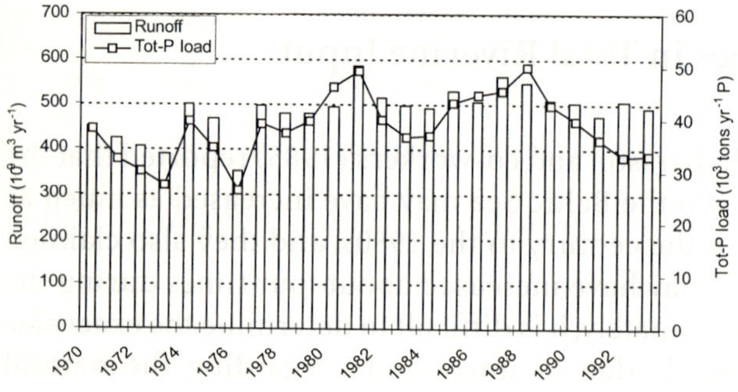

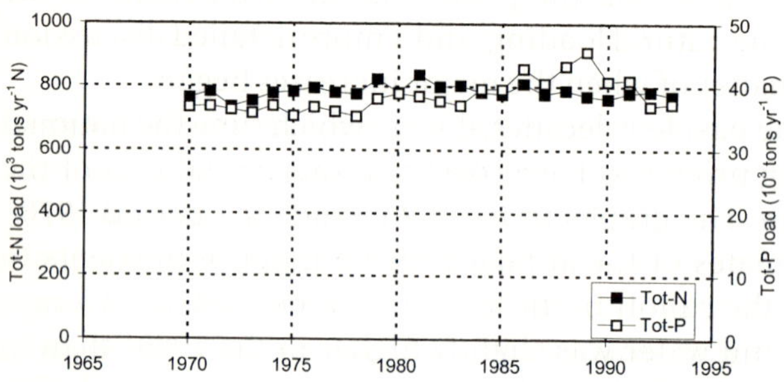

Fig. 4.3. Flow-normalized input of nitrogen and phosphorus from rivers to the Baltic Sea. (Stålnacke et al. 1999a)

increased due to urbanization, and expanded tilling and ditching favored erosion and leaching processes. Studies of lake sediments have demonstrated that the loss of phosphorus increased long before commercial fertilizers became widely used (Dearing et al. 1987). Furthermore, both model calculations (Hoffman 1999) and the classical Rothamsted experiments (Lawes et al. 1881) show that older agricultural practices also resulted in substantial export of nitrogen to water. Notwithstanding, it is completely clear that the turnover of nutrients in the environment has accelerated since World War II, in particular due to increased use of commercial fertilizers, changes in the agricultural landscape and the formation of NO_x in combustion processes. In the following section, we will examine the combined impact of these factors on the concentrations of nutrients in some rivers that have been monitored for an exceptionally long period of time.

4.7 Temporal Trends in Specific Rivers

The previously mentioned measurements in the Daugava River (Tsirkunov et al. 1992) appear to represent the longest series of nutrient concentrations data available for the Baltic Sea drainage basin. As can be seen in Fig. 4.4, the nitrate concentration in this river increased during the post-war period, and the critical decade was the 1960s. Data from other Eastern European rivers, such as the Danube and its Hungarian tributaries (Oláh and Oláh 1996), further indicate that nitrate levels increased significantly in that part of Europe in the 1960s and, to some extent, also in the 1970s. Strongly elevated nitrate concentrations seem to have occurred somewhat earlier in Western Europe. In the Lower Rhine River, for example, an upward trend in nitrate began in the early1950s, and from that time up to 1980 the nitrate concentration approximately tripled (Behrendt and Böhme 1993; van Dijk et al. 1996). A similar long-term increase has been reported by Hagebro et al. (1983), who found that in two Danish streams the nitrate level doubled from the late 1960s to the early 1980s. Other studies performed in the Nordic countries have also revealed upward nitrogen trends (Grimvall et al. 1991; Kauppi et al. 1993), although the scientists responsible for those investigations emphasized that the mild and rainy winters in the 1980s may have been a main reason for the relatively high nutrient concentrations observed at that time.

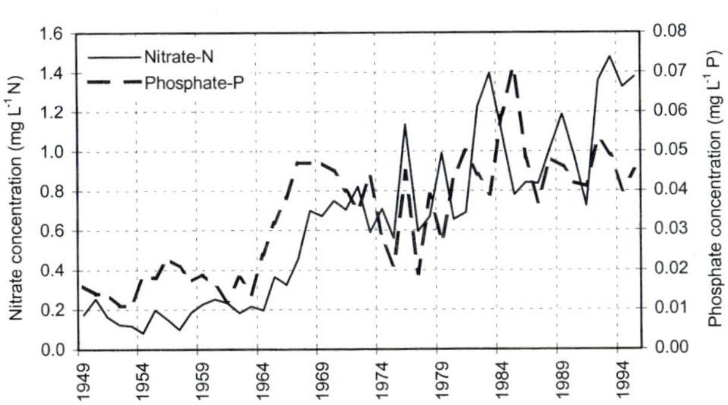

Fig. 4.4. Flow-weighted annual concentrations of nitrate-N and phosphate-P at Daugavpils on the Daugava River 1949–1995. (Latvian Hydrometeorological Agency)

In contrast to the concentration of nitrate (and total nitrogen), which obviously increased dramatically after World War II, it is rather uncertain when the greatest rise in phosphate (and total phosphorus) concentrations took place. Tsirkunov and colleagues (1992) reported a clear increase in this ion from the 1970s to the 1980s, and the previously mentioned studies of the Lower Rhine (Behrendt and Böhme 1993; van Dijk et al. 1996) revealed an almost fivefold increase in phosphate in that river from the late 1950s to 1980. However, it is difficult to know whether these trends are representative of the Baltic Sea drainage basin, hence we can only conclude that the major increases in phosphate and total phosphorus probably occurred before the onset of water-quality monitoring in most of the Baltic Sea nations.

4.8 Controllability of Riverine Loads of Nutrients

There is a widespread conception that it is relatively easy to control riverine loads of phosphorus. Point emissions often play a major role in the most severe cases of river pollution, and efficient techniques for removal of phosphorus can be introduced in existing wastewater treatment systems. The rehabilitation of the Rhine River provides a striking example of how successful such measures can be. When several major point emissions of phosphorus to this river were practically eliminated and phosphorus-containing detergents were phased out in the 1980s, it took less than a decade for the phosphate concentration in the Lower Rhine to reach the much lower level that had prevailed in the 1950s (Behrendt and Böhme 1993; van Dijk et al. 1996).

The results of almost daily measurements in the Odense River in Denmark illustrate more moderate decreases in phosphorus that have been observed in some rivers leading to the Baltic Sea. As can be seen in Fig. 4.5, a downward trend prevailed in 1987–1991, and thereafter the phosphate concentration leveled out. Furthermore, the decrease was accompanied by a significant shift in the relationship between concentration and flow: the concentration peaks that previously appeared during low-flow conditions have been practically absent since 1992, which indicates that the present riverine loads are dominated by non-point sources. It is also noteworthy that the amounts of phosphate leveled out at values exceeding 0.05 mg l^{-1}, which is much higher than the natural background concentration. Hence, it seems to be comparatively easy to reduce phosphorus from high to moderate concentrations, whereas further reduction is a more difficult and time-consuming task.

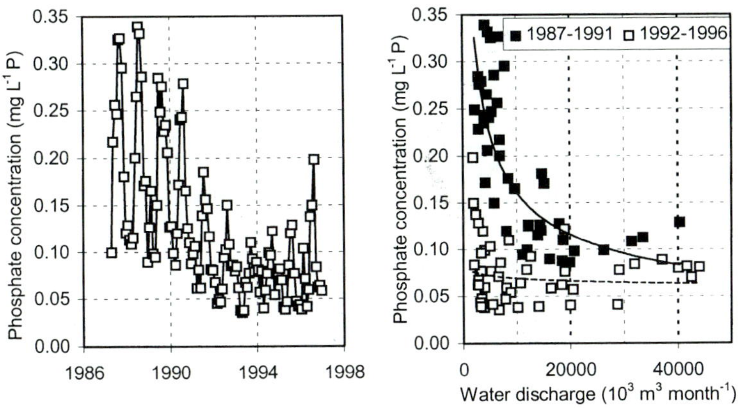

Fig. 4.5. Monthly flow-weighted mean concentrations of PO_4-P at Kratholm on the Odense River. *Left* Time series plot; *right* scatter chart of concentration vs. discharge (County Board of Funen, Denmark; Grimvall et al. 2000)

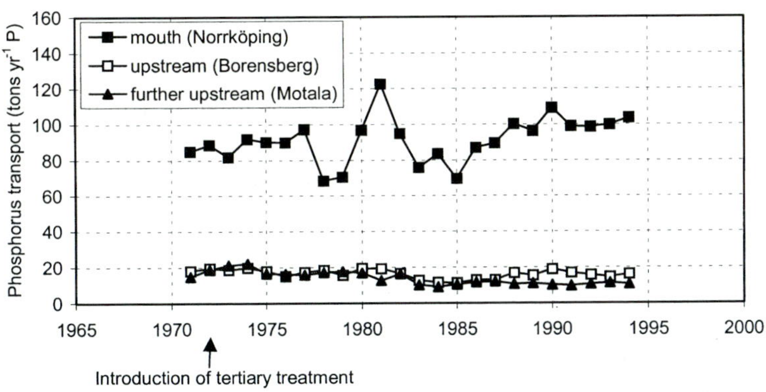

Fig. 4.6. Annual phosphorus transports at three sites in the Motala River basin in Sweden. Tertiary treatment was introduced around 1975 at all major municipal treatment plants located between sampling sites at Motala and Norrköping. (Grimvall et al. 2000)

Observations in the lake-rich Motala River basin in Sweden (Fig. 4.6) provide a thought-provoking example of the inertia of the systems that control the transport of phosphorus with rivers. In the mid 1970s, tertiary treatment was introduced at all major wastewater treatment plants in this basin, and the total point emissions of phosphorus in the study area dropped by almost 200 tons year^{-1} to less than 10 % of the previous level (Karlsson 1989). In spite of that, the flow-normalized transport of phosphorus at the river mouth has remained at about 100 tons year^{-1} for at least 20 years.

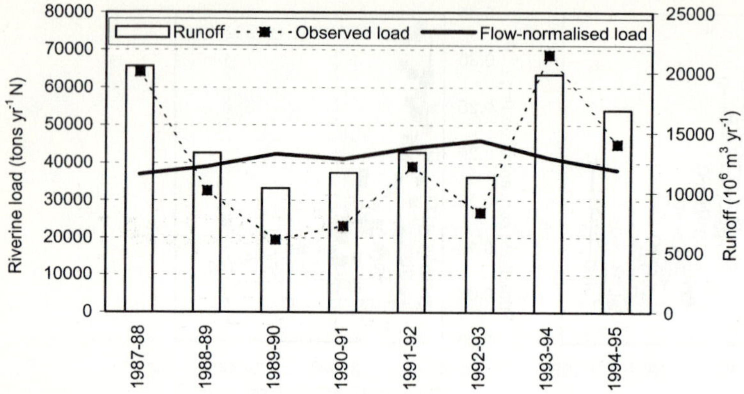

Fig. 4.7. Annual values of runoff and observed and flow-normalized riverine loads of NO_3-N at Krajnik on the Oder River, from November to October. (Tonderski 1997)

The recent dramatic decrease in the use of mineral fertilizers in Eastern Europe has created new possibilities to test hypotheses regarding large-scale fluxes of nitrogen in the Baltic Sea drainage basin. Thus far, studies of small catchments in Estonia (Loigu and Vasilyev 1997; Mander and Kull 1997) have provided clear evidence of downward trends in the output of nitrate with water. In contrast, it is remarkable that no trends have appeared in several major rivers entering the Baltic Sea. Statistical trend assessments of nutrient loads carried by the Daugava and other Latvian rivers showed that most of the interannual variation could be attributed to natural changes in runoff (Laznik et al. 2000). Moreover, a similar study of the Oder and Vistula Rivers in Poland revealed that the flow-normalized loads remained practically constant from 1989 to 1995 (Fig. 4.7).

There are several possible reasons for the lack of downward trends in the mentioned rivers. Organic nitrogen accumulated in soil during periods of more extensive use of fertilizers may continue to be an important source of nitrate in agricultural runoff for quite some time (Jenkinson 1991). Another plausible explanation is the retention of nitrogen in large groundwater aquifers. In regard to the Oder and Vistula Rivers, the following should also be taken into account: (1) there was little loss of nitrogen from cultivated land in Poland even before recent changes in fertilizer application; and (2) fluctuations in sources other than agricultural runoff may have concealed a moderate decrease in leaching from cultivated land (Tonderski 1997; Grimvall et al. 2000).

A recent study of data on nitrogen in Swedish rivers in 1971–1994 further illustrated the inertia and complexity of nitrogen fluxes from land to sea (Stålnacke et al. 1999b). That investigation showed that all major

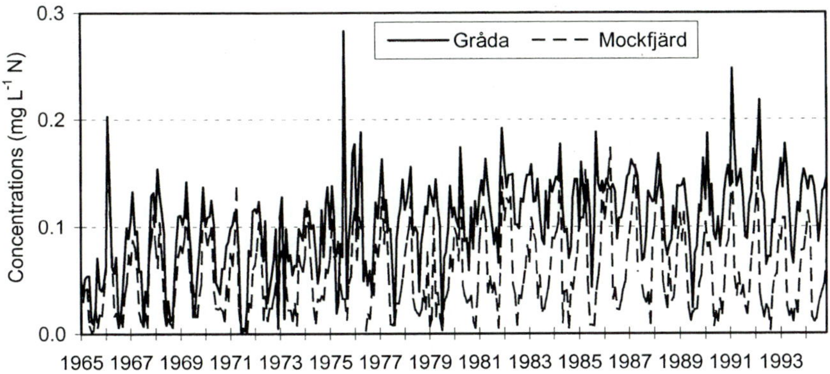

Fig. 4.8. Nitrate concentrations at Mockfjärd on western branch of the Dalälven River and at Gråda on eastern branch of same river. Gråda is located downstream of Lake Siljan. (Stålnacke et al. 1999b)

changes in nitrate and total nitrogen had occurred in the 1970s, and the most pronounced upward trends were observed at sites located downstream of major lakes. Figure 4.8 depicts the development of nitrate concentrations in the eastern and western branches of the Dalälven River. At Gråda, downstream of Lake Siljan on the eastern branch of the river, there was a significant increase in nitrate in the late 1970s, whereas at Mockfjärd on the western branch of the Dalälven, time series of nitrate concentrations do not exhibit any significant trend. Further examination of the upward nitrogen trends indicated that the increase in nitrate coincided in both time and space with decreased emissions of phosphorus to lakes located upstream of the sampling sites under consideration. In a Baltic Sea perspective, the described trends in Swedish rivers are of little importance, but it is possible that such trends will have a significant impact when phosphorus removal is introduced at wastewater treatment plants in most of the major cities in the Baltic Sea drainage basin.

Together, the examples used to illustrate the controllability of riverine loads of nutrients have drawn attention to the role of long-term processes. Although it is often easy to reduce very high concentrations and loads to more moderate levels, it may take decades to achieve further reduction. In addition, successful abatement of riverine nutrient pollution of the Baltic Sea will probably require coordinated control of both phosphorus and nitrogen emissions.

4.9 Conclusions

- The riverine input of nutrients to the Baltic Sea was found to be larger than the input along all other routes combined.
- Our estimates of the average annual riverine loads of nitrogen and phosphorus during the time period 1980–1993 (825,000 and 41,000 tons year^{-1}, respectively) are higher than those presented by HELCOM and several other investigators.
- The discrepancies between different estimates of riverine loads were due primarily to the handling of both missing data on total-N concentrations in the rivers of the former Soviet Union and nutrient concentration data that were obviously inconsistent or incorrect.
- The appearance of strongly elevated concentrations of nitrate (and total nitrogen) in major rivers in the Baltic Sea drainage basin is primarily a post-war phenomenon, whereas it is more uncertain as to when the major increases in phosphorus occurred.
- Despite considerable changes in land use, point source emissions and atmospheric deposition in certain parts of the Baltic Sea drainage basin, the total annual riverine loads of nitrogen and phosphorus transported to the sea have been fairly constant from 1970 to 1993.
- The inertia of the systems that control the loss of nutrients from land to sea was underestimated when the present goal of a 50 % reduction was adopted.

Acknowledgements. The authors are grateful to many organizations and scientists for supplying data that enabled us to conduct this study. Nutrient concentration data were obtained from the following organizations or agencies: the Swedish University of Agricultural Sciences in Uppsala, the National Board of Waters and the Environment in Helsinki (Finland), the State Hydrological Institute in St. Petersburg (Russia), Tallinn Technical University (Estonia), the Latvian Hydrometeorological Agency in Riga, the Institute of Meteorology and Water Management in Wroclaw and Gdansk (Poland), and the National Environmental Research Institute in Silkeborg and the County Board of Funen (Denmark). Water discharge data for the entire Baltic Sea drainage area were obtained from the Swedish Hydrological and Meteorological Institute in Norrköping.

References

Areskoug H (1993) Nedfall av kväve och fosfor till Sverige, Östersjön och Västerhavet (in Swedish). Rep 4148. Swedish Environmental Protection Agency, Solna, 41 pp

Arheimer B, Brandt M, Grahn G, Roos E, Sjöö A (1997) Modellerad kvävetransport, retention och källfördelning för södra Sverige (in Swedish). Rep Hydrol 13. SMHI, Norrköping, 78 pp

Behrendt H (1996) Inventories of point and diffuse sources and estimated nutrient loads – a comparison of different river basins in central Europe. Water Sci Technol 33: 99–107

Behrendt H, Böhme M (1993) Point and diffuse loads of selected pollutants in the River Rhine and its main tributaries. International Institute for Applied Systems Analysis (IIASA), Res Rep RR-1–93, Laxenburg, 84 pp

Bergström S, Carlsson B (1994) River runoff to the Baltic Sea: 1950–1990. Ambio 23:280–287

Bernes C (ed) (1988) Monitor 1988. Sweden's marine environment-ecosystem under pressure. Swedish Environmental Protection Agency, Solna, 207 pp

Dearing JA., Håkansson H, Liedberg-Jönsson B, Persson A, Skansjö S, Widholm D, El-Daoushy F (1987) Lake sediments used to quantify the erosional response to land use change in southern Sweden. Oikos 50:60–78

Enell M, Fejes J (1995) The nitrogen load to the Baltic Sea – present situation, acceptable future load and suggested source reduction. Water Air Soil Pollut 85:877–882

Fotyma M, Terelak H (1994) Environmental challenges in Polish agriculture covering a main area of the Baltic Sea basin. Mar Pollut Bull 29:455–463

Grimvall A (1996) Uncertainty of estimates of riverine loads of pollutants. In: Riverine input to coastal areas. Tema Nord 1996:529, Nordic Council of Ministers, Copenhagen, pp 74–82

Grimvall A, Stålnacke P (1996) Statistical methods for source apportionment of riverine loads of pollutants. J Environ 7:201–213

Grimvall A, Stålnacke P, Sundblad K, Niemirycz E, Pitkänen H, Bruhn A (1991) Trend analysis of nutrient concentrations in Baltic Sea rivers. In: Fenger J, Halsnaes K, Larsen H, Schroll H, Vidal V (eds) Environment, energy and natural resource management in the Baltic region. In: Proc 3rd Int Conf Systems Analysis. Nord:48, Nordic Council of Ministers, Copenhagen

Grimvall A, Stålnacke P, Tonderski A (2000) Time scales of nutrient losses from land to sea – a European perspective. Ecol Eng 14(2000):363–371

Hagebro C, Bang S, Somer E (1983) Nitrate load/discharge relationships and nitrate load trends in Danish rivers. In: Dissolved loads of rivers and surface water quantity/ quality relationships. Proc of the Hamburg Symp, August. IAHS Publ 141, pp 377–386

Hoffmann M (1999) Assessment of leaching loss estimates and gross load of nitrogen from arable land in Sweden. PhD Thesis, Department of Soil Sciences, Swedish University of Agricultural Sciences, Uppsala

HELCOM (1993) Second Baltic Sea pollution load compilation. Baltic Sea Environ Proc 45. Helsinki Commission, Helsinki, 161 pp

Jenkinson DS (1991) The Rothamsted long-term experiments: are they still of use? Agron J 83:2–10

Karlsson G (1989) Dynamics of nutrient mass transport – a river basin evaluation. PhD Thesis, Department of Water and Environmental Studies, Linköping University

Kauppi L (1984) Nitrate in runoff and river waters in Finland in the 1960s and 1970s. Publ Water Res Inst 57. National Board of Waters, Helsinki

Kauppi L, Pietiläinen O-P, Knuuttila S (1993) Impacts of agricultural nutrient loading on Finnish watercourses. Water Sci Technol 28:461–471

Kronvang B, Aertebjerg G, Grant R, Kristensen P, Hovmand M, Kirkegaard J (1993) Nationwide monitoring of nutrients and their ecological effects: state of the Danish aquatic environment. Ambio 22:176–187

Kronvang B, Grœsbøll P, Larsen SE., Svendsen LM, Andersen HE (1996) Diffuse nutrient losses in Denmark. Water Sci Technol 33:81–88

Larsson U, Elmgren R, Wulff F (1985) Eutrophication and the Baltic Sea: causes and consequences. Ambio 14:9–14

Laznik M, Stålnacke P, Grimvall A, Wittgren HB (2000) Riverine input of nutrients to the Gulf of Riga – temporal and spatial variation. J Mar Syst 23(1999):11–25

Lawes JB, Gilbert JH, Warrington R (1881) On the amount and composition of rain and drainage waters collected at Rothamsted. J R Agric Soc Engl (Ser 2) 17:241–279; 311–350; 18:1–71

Loigu E, Vasilyev A (1997) Water quality response to sudden changes in the amounts of fertilizers used in Estonia. In: Proc Int Conf on Hydrology and Environment, Kaunas, 22–24 May

Löfgren S, Olsson H (1990) Nitrogen and phosphorus entering water courses in inland Sweden (in Swedish). Rep 3692, Swedish Environmental Protection Agency, Uppsala, 100 pp

Maksimova M-P (1982) The balance of nutrients and organic matter in the Baltic Sea during the period of intensive anthropogenic action. Oceanology 22:555–559

Mander Ü, Kull A (1997) Influence of land use change and climatic fluctuations on runoff and nutrient fluxes in an agricultural watershed. In: Proc 7th Stockholm Water Symp and 3rd Int Conf on the Environmental Management of Enclosed Coastal Seas, Stockholm, 10–15 Aug

Swedish Ministry of Environment and Natural Resources (1990) Vem förorenar Sverige. Underlagsrapport till utredningen om Sveriges internationella miljösamarbete (in Swedish). Swedish Ministry of Environment and Natural Resources, Stockholm, 105 pp

Mikulski Z (1970) Inflow of waters to the Baltic Sea in 1961–1970. Nordic Hydrol 4:216–227

Mikulski Z (1982) River inflow to the Baltic Sea 1921–1975. Polish Academy of Sciences, Polish National Committee of the IHP, University of Warsaw

Oláh J, Oláh M (1996) Improving landscape nitrogen metabolism in the Hungarian lowlands. Ambio 25:331–335

Pawlak J (1980) Land-based inputs of some major pollutants to the Baltic Sea. Ambio 9:163–167

Rekolainen S, Posch M, Kämäri J, Ekholm P (1991) Evaluation of the accuracy and precision of annual phosphorus load estimates from two agricultural basins in Finland. J Hydrol 128:237–255

Rosenberg R, Elmgren R, Fleischer S, Jonsson P, Persson G, Dahlin H (1990) Marine eutrophication case studies in Sweden. Ambio 19:102–108

Rybinski J (1989) The Baltic Sea. Nutrient loads – their sources and consequences. In: Proc Conf ELMIA – Air and Water, May 1989, Jönköping, 13 pp

Stålnacke P, Grimvall A (2000) Semiparametric approaches to flow-normalisation and source apportionment of substance transport in rivers. J Environ (in press)

Stålnacke P, Grimvall A, Sundblad K, Tonderski A (1999a) Estimation of riverine loads of nitrogen and phosphorus to the Baltic Sea 1970–1993. Environ Monit Assess 58:173–200

Stålnacke P, Grimvall A, Sundblad K, Wilander A (1999b) Trends in nitrogen transport in Swedish rivers. Environ Monit Assess 59:47–72

Sundblad K, Tonderski A, Rulewski J (1994) Nitrogen and phosphorus in the Vistula River, Poland – changes from source to mouth. Water Sci Technol 30:177–186

Tonderski A (1997) Control of nutrient fluxes in large river basins. PhD Thesis, Department of Water Environmental Studies, Linköping University

Tonderski A, Grimvall A, Sundblad K, Stålnacke P (1994) An East–West perspective on riverine loads of nutrients in the Vistula and Rhine basins. Water Sci Technol 30:121–130

Tonderski A, Grimvall A, Dojlido J, van Dijk G (1995) Monitoring nutrient transport in large rivers. Environ Monit Assess 34:245–269

Tsirkunov VV, Nikanorov AM, Laznik MM, Dongwei Z (1992) Analysis of long-term and seasonal river water quality changes in Latvia. Water Res 26:1203–1216

van Dijk GM, Stålnacke P, Grimvall A, Tonderski A, Sundblad K, Schäfer K (1996) Long-term trends in nitrogen and phosphorus concentrations in the Lower River Rhine. Arch Hydrobiol Suppl 113:99–109

5 Deposition of Nitrate and Ammonium from the Atmosphere to the Baltic Sea

L. GRANAT

5.1 Introduction

The wet deposition over the Baltic Sea of inorganic nitrogen compounds (NO_3 and NH_4) has previously been estimated from precipitation sampling (e.g. Rodhe et al. 1980; HELCOM 1991). The deposition of organic nitrogen is less well known. Recently, Rahm et al. (1995) measured organic and inorganic nitrogen compounds accumulated on the ice in Bothnian Bay. Dry deposition has been estimated from concentrations in the air by the dry deposition velocity concept (Lindfors et al. 1991, 1993). The wet and dry deposition have also been calculated from emission data and long-range transport and from deposition models (e.g. Barrett and Berge 1996). This model included a number of source areas for deposition of the compounds in the Baltic. The present investigation shows that present deposition estimates differ from previous ones and this is caused mainly by a higher spatial ($1\times1°$) and temporal (monthly) resolution. Further, a longer time period (1971–1991) has been used, including periods when few sampling stations were in use in the region. There are also some differences compared to previous estimates with regard to data quality control, interpolation technique and estimation of dry deposition. The main objective of this study was to calculate monthly deposition estimates for inorganic nitrogen species. The study first describes data treatment, then the mean values and trends that characterize the material are shown, and finally comparisons with previous estimates are carried out. The results show, similarly to previous studies, that the cyanobacteria-fixed nitrogen from the atmosphere is an important part of the total nitrogen input to the Baltic Sea region. It is also clear that the atmospheric deposition of nitrogen is a major nutrient source to the Baltic Sea.

Ecological Studies, Vol. 148
F. Wulff et al. (eds.)
A Systems Analysis of the Baltic Sea

5.2 Estimation of Wet Deposition

5.2.1 Data

Data on precipitation chemistry were taken from the stations and time periods shown in Table 5.1 and locations shown in Fig. 5.1. These were the data that were available when the deposition matrix was prepared. Operation has continued at several of the stations and more data are now available both from precipitation and air sampling. These latter data have not been considered here, except for a trend diagram, since the purpose is to highlight the calculation of the deposition matrix provided early in the project.

The European Monitoring and Evaluation Project (EMEP) stations (see, for instance, Schaug et al. 1993) have daily sampling and these data have been converted to monthly weighted mean values for the present work by K. Törseth from the Norwegian Institute of Air Research (NILU). Values from the HELCOM database (K. Törseth, NILU) and from the Swedish National Environmental Supervision Network (Granat 1991; Sjöberg et al. 1992) are based on monthly sampling and were used as such. Monthly data on precipitation amount for the period 1979–1992 were obtained from SMHI (see Chap. 3).

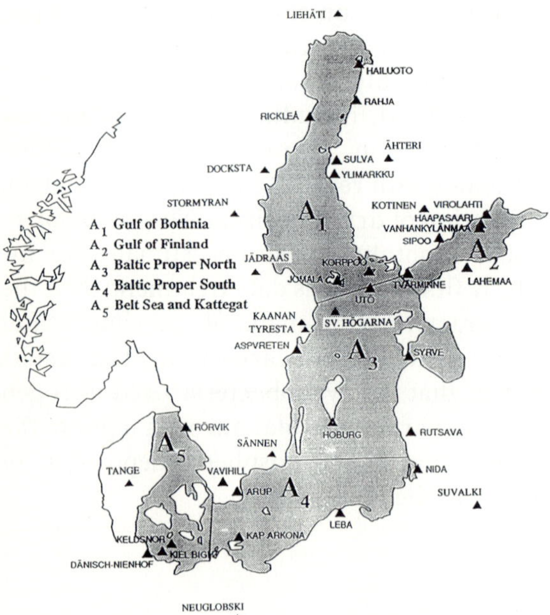

Fig. 5.1. The Baltic, with divisions into subregions and measuring sites. (Modified from HELCOM 1991)

Table 5.1. Available data on precipitation and air chemistry for the present deposition estimate. *E* EMEP network; *H* HELCOM; *P* Swedish PMK network

	E	H	P	Precipitation chemistry From	To	Air data From	To	Arbitrary station no.
Liehäti		×	×	1984	1991			1
Rahju		×		1986	1990			4
Rickleå			×	1983	1993			5
Sulva		×		1986	1990			6
Docksta			×	1983	1993			7
Ähteri	×			1978	1992	1989	1992	10
Stormyran		×		1983	1993			13
Jädraås			×	1985	1993			14
Korpoo		×		1986	1990			16
Sippo		×		1986	1990			18
Virolathi		×		1986	1990	1987	1990	20
Lahem		×		1986	1990	1986	1990	25
Tvärminne		×		1986	1990			26
Utö	×			1990	1992			28
Tyresta			×	1985	1993			32
Aspvreten			×	1988	1993	1986	1992	34
Syrve		×		1982	1991			39
Sännen	×			1985	1993			44
Rucava		×		1986	1990	1989	1990	45
Nida		×		1984	1990			47
Nida		×				1989	1990	46
Arup			×	1983	1993			48
Vavihill	×			1985	1993	1986	1992	49
Suvalki	×			1979	1992			50
Leba		×		1987	1990			51
Kap Arkona	×			1986	1990			55
Neuglobski	×			1981	1990			56
Keldsnor	×			1978	1992	1989	1992	58
Tange	×			1978	1992	1989	1992	59

5.2.2 Data Scrutinization

The quality assurance procedure for EMEP data includes a statistical test to reveal outliers based on data from previous measurements, ion balance and conductivity criteria and deviations from established interdependencies (i.e. between Na and Cl). Analytical methods used in the EMEP study are intercalibrated regularly. Analytical methods and sampling equipment have been compared under field conditions, with participation of several of the investigators who are carrying out measurements in the region (Granat et al. 1992). Generally, the analytical performance was satisfactory with regard to nitrate and ammonium analysis. Comparison between collectors showed that there are some deviations, mainly explained by different designs rather than by collection principles (i.e. wet only vs. bulk collectors) or varying sampling time. The comparisons were, however, carried out at an inland site (Aspvreten) where deviations between wet only and bulk collectors are known to be less than on islands. Stations in the Swedish national network usually have two parallel bulk collectors which facilitates identification (and exclusion) of contaminated samples.

The most severe influence on the data quality seems to come from unidentified contamination of the samples, resulting in high concentrations of NH_4 and in some cases also of NO_3. Another problem is micro-organisms that when abundant may lead to a substantial loss of NH_4 and NO_3. Contaminated samples may be identified by comparison with nearest collector(s). If the distance is 1 km or less, the normal deviation between collectors suggests that a difference greater than 30 % is due to contamination, or some other bias. One (in some cases both) of the samples is therefore not representative and should not be used for further calculations (except for investigations of sampling problems; see for instance Granat 1991). If the comparison has to be made with the nearest station in the sampling network, the distance is usually at least 150 km. A deviation factor of two is then acceptable, and probably due to normal spatial variability for monthly samples. This is the case for samplers in the Swedish network for comparable distances.

Based on these arguments, the following screening was used for the Baltic catchment area: monthly concentration values from one station at a time were compared month by month with those from two to four nearby stations. The value was rejected if it was *both* greater than three times the median value from the other stations and larger than the mean value for the actual station. The latter procedure was considered necessary to give a stable routine (not removing all data in a region). A similar procedure was also carried out in cases when stations had unusually low values (real or due to changes during storage of samples). Any rejected value was replaced by the median for the nearby stations. About 10 % of monthly data were modified in this way. In addition, the data from four stations (Utö, Hoburgen, Ylimarku and Jomala) were excluded due to frequent very high values (compared to neighboring stations) and where modest contamination (undetectable by the screening method) was suspected to occur frequently.

5.2.3 Interpolation

During the period 1986 through 1990, when several stations were operated all around the Baltic, concentration in the grid squares was estimated from actual data from these stations month by month. Interpolation was carried out in six areas. In each area, the monthly concentrations ($c_{i,j}$) in $1 \times 1°$ grids (index i and j, locations given in Fig. 5.2) were calculated as a distance-weighted contribution from several stations in each area as:

$$c_{i,j}=k_A \cdot c_A+k_B \cdot c_B+k_C \cdot c_C+k_D \cdot c_D + \ldots\ldots$$

where $c_{i,j}$ is concentration in a given grid area, c_A, c_B, c_C, etc. are concentrations at stations A, B, C, etc. in the region, and k_A, k_B, k_C, etc. are weighting functions for respective station data defined as

$$k_A=f(1/(s_{i,j,A}+r)^2)$$

and

$$\Sigma k_n=1$$

Here, $s_{i,j,A}$ is the distance from the centre of the grid (i,j) to a station (A), r is an empirical distance to allow for some influence from other stations in the region even close to the particular station (A), and n represents all stations (A, B, C, etc.).

For the period 1970–1985 and for 1991 when only a few stations were operated east and south of the Baltic (none before 1979), the concentration of a substance at a specific station was estimated. The estimation was carried out by using the mean during 1986 through 1990 and with adjustment for seasonal variation and trend thus

$$c_{A,y,m}=c_{A,m} \cdot yrel_{A,y} \cdot seasrel_{A,m}$$

where $c_{A,y,m}$ is the concentration at station A for any particular year and month (y,m), $yrel_{A,y}$ is the relative yearly concentration (relative to years 1986–1990) from a mean of several stations in central Sweden, and $seasrel_{A,m}$ is the seasonal variation estimated from the 1986–1990 period from stations around the Baltic. The interpolation to obtain concentration in grid squares was then carried out as described earlier. This procedure is more robust than using data from the very few stations in operation close to the Baltic during the period before 1986. Some comparisons with actual data are given below.

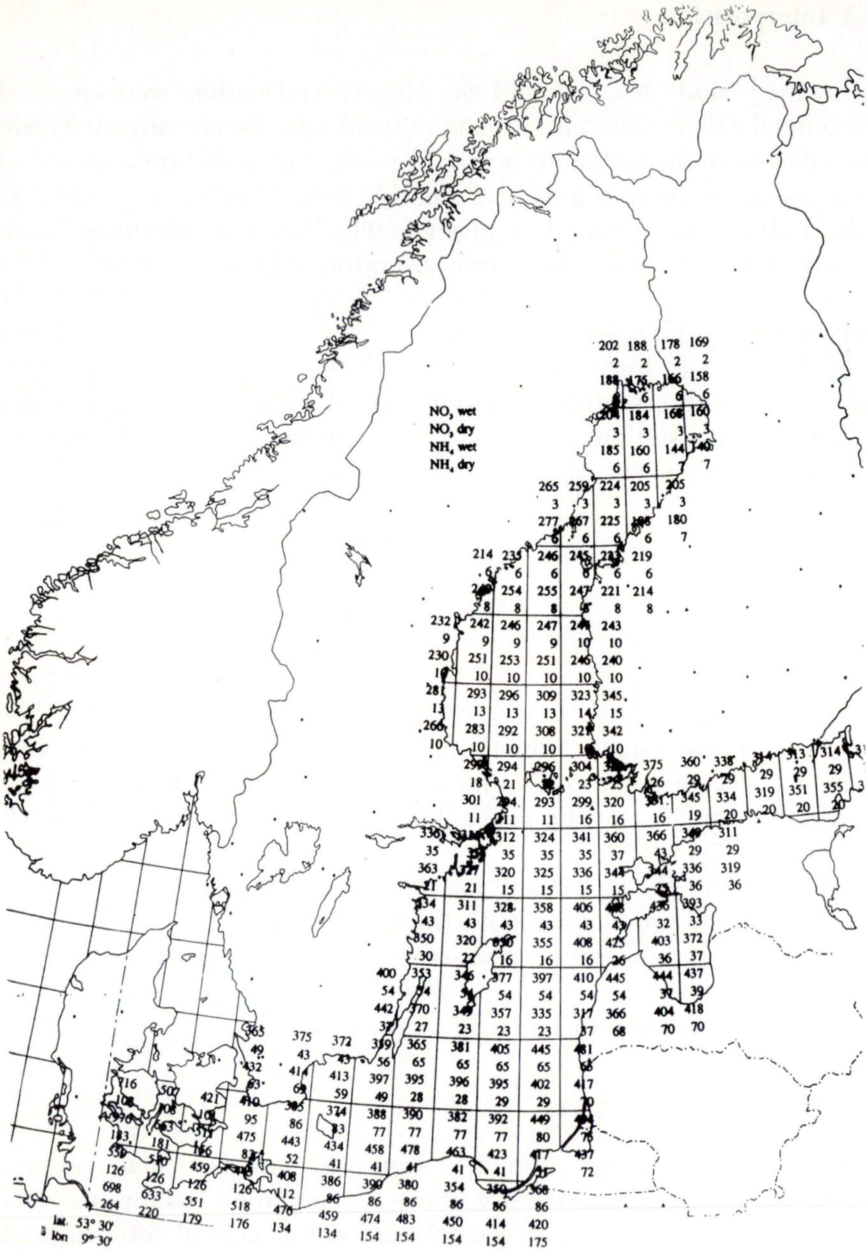

Fig. 5.2. Wet and dry deposition (in mg N m^{-2} year^{-1}). estimated in 1 × 1° grid squares for 1986–1990. In each grid is given nitrate wet, nitrate dry, ammonium wet and ammonium dry deposition. *Bottom left corner* is at 53°30'N and 9°30'E

It is generally considered that a deposition field is best estimated from the product of independently estimated concentration and precipitation fields. There are indications that under normal conditions precipitation and concentration of nitrogen species do not spatially correlate. Precipitation is also more spatially variable than the concentration of substances in the precipitation. Further, the number of precipitation stations are far more numerous than those where analysis of different substances in precipitation also are included. Consequently, when estimating deposition over a large area, precipitation data can be satisfactory described while concentration data do not have similar resolution. The concept is reasonable but has not been extensively verified quantitatively by field measurements (i.e. Granat 1988a,b).

Deposition ($d_{i,j,m,y}$) was thus calculated for each grid (i,j) month and year (m,y) as

$$d_{i,j,m,y}=c_{i,j,m,y}\cdot p_{i,j,m,y}$$

Here, p is the precipitation estimated from meteorological data and available for the period 1979–1994. For the period before 1979, mean values from the period 1986–1990 for each grid square and calendar month were used, assuming no trend in precipitation or change in seasonal variation. Precipitation data are available for the period before 1979 but not over open waters. The 1986–1990 values were then considered to be a reasonable alternative, when one considers the low amount of concentration data during the previous period.

5.3 Estimation of Dry Deposition

5.3.1 Concentration Data

Air concentrations of nitrogen components were obtained from the EMEP database (with monthly averages calculated from the reported daily values), from the HELCOM database and a few stations in national networks in Denmark and Sweden. They are listed in Table 5.1. Air concentration is measured with filter packs, with one filter collecting particulate material, followed by filters impregnated with a sorbing material. With this method both the particulate and gaseous compounds were sampled. Nitrogen dioxide was collected in a device called a „bubbler“. Additional information on partitioning between gaseous and particulate nitrogen compounds was obtained from measurements in Sweden and Denmark by

water/air interface will be upward. The flux will decrease or change direction if concentration in the water is decreasing. In the summer in the southern part of the Baltic, concentration of ammonia in the air is high and dry deposition increases in importance compared to precipitation. As a consequence, dry deposition of ammonia will impact on growth of plankton and biomasses in this part of the aquatic ecosystem. This somewhat speculative reasoning would be strengthened by measurements of daily concentration data in air and water and a better knowledge of the partitioning coefficient, resulting in better understanding of resistance in the surface film.

Deposition of N_2 by nitrogen fixation has not been included in this calculation since this flux is dependent on biological processes rather than transfer processes at or above the surface.

5.4 Results and Discussion

The main result of this study is the calculated wet and dry deposition month by month in 1×1° grid squares for the period 1970–1991 for nitrogen compounds NO_3, NH_3 and NH_4. These data are used for further calculations (see Chap. 13). Some results are given here that reveal general features in the spatio-temporal variability and the uncertainty of the interpolation.

The annual deposition for the period 1986–1990 is given in Fig. 5.2 for wet and dry deposition of nitrate and ammonium, respectively. Mean values for five major basins are given in Table 5.3 and compared with other estimates. Wet deposition of nitrate is about the same as for ammonium in the whole area, except in the most southern part where ammonium deposition is about 20 % higher. Deposition decreases by a factor of about 2.5 from the southern part of the Baltic to the most northern part of the Bothnian Bay. Gradients in the west–east direction are quite weak and irregular.

Long-term trends are estimated by data from an area in central Sweden (Fig. 5.3). The increasing trend of nitrate and ammonium has mainly been attributed to European emissions (rather than from Sweden) and may, therefore, be relevant also for the Baltic region. However, a rather large deviation can be suspected for individual years due to the natural spatial variability. Concentration increased from the mid-1950s to a maximum in the mid-1970s and has since then remained about constant. Recent calculations (Barrett and Berge 1996), which will be discussed further below, result in a different picture, with a constant total (wet and dry) deposition

Table 5.3. Deposition estimates (in mg N m^{-2} $year^{-1}$) for sub-regions obtained from the EMEP model (Barrett and Berghe 1996), this study and the HELCOM (1991) study. *Wet-h* refers to the hybrid method with an alternative estimate of amount of precipitation. Note that Kattegat is only partially covered in the present study

	NO_3-N		NH_4-N		NO_3-N +NH_4-N
EMEP model	Dry	Wet	Dry	Wet	Dry+wet
Gulf of Bothnia	61	170	35	105	371
Gulf of Finland	97	237	95	219	648
Baltic proper, north	117	285	114	214	734
Baltic proper, south	187	399	231	340	1157
Belt Sea and Kattegat	175	406	322	485	1388
This study					
Gulf of Bothnia	9	248	8	243	508
Gulf of Finland	29	326	20	344	719
Baltic proper, north	42	371	30	359	802
Baltic proper, south	78	397	82	438	995
Belt Sea and Kattegat	106	505	178	638	1427
HELCOM, 1991	NO_3-N		NH_4-N		
	Wet	Wet-h	Wet	Wet-h	
Gulf of Bothnia	284	318	345	385	
Gulf of Finland	322	406	357	448	
Baltic proper, north	384	351	361	325	
Baltic proper, south	419	496	432	511	
Belt Sea and Kattegat	489	536	571	628	
	Dry Total N	Wet Total N	Wet+dry Total N		
North Baltic	84	650	734		
Central Baltic	84	910	994		
South Baltic	349	1140	1489		

of nitrogen components from 1985 to 1988 and then a decrease of 30% until 1995.

Seasonal variation in wet deposition of NO_3 and NH_4 is shown in Fig. 5.4. Deposition is generally lower in summer, down to half the winter values. This is not the case in the southern part of the Baltic proper, where wet deposition of nitrate and especially ammonium is markedly higher in summer. Contrary to other areas, concentrations of both nitrate and ammonium in precipitation are in this area higher in summer than in winter. This might be of importance with regard to biological demand for nitrogen in summer.

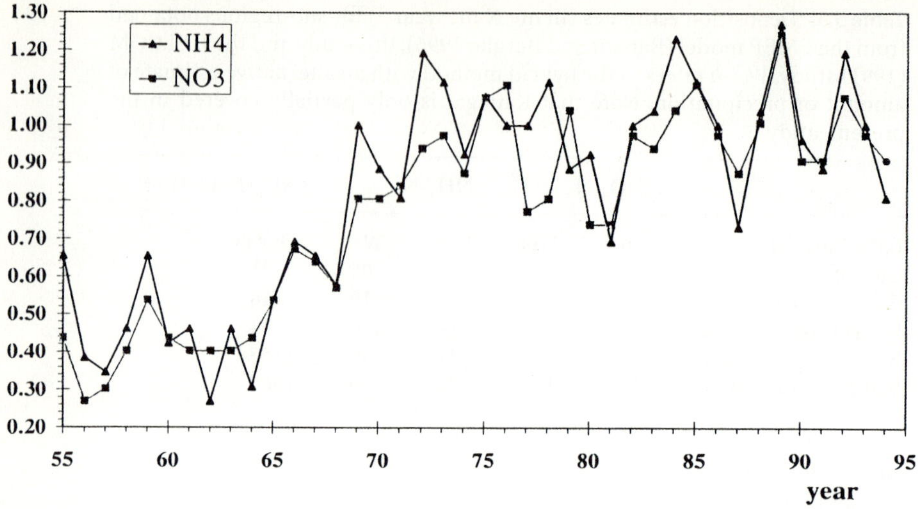

Fig. 5.3. Long-term trends in concentrations of nitrate and ammonium in precipitation. Values are ratios between concentration during a particular year and a mean for the period 1986–1990

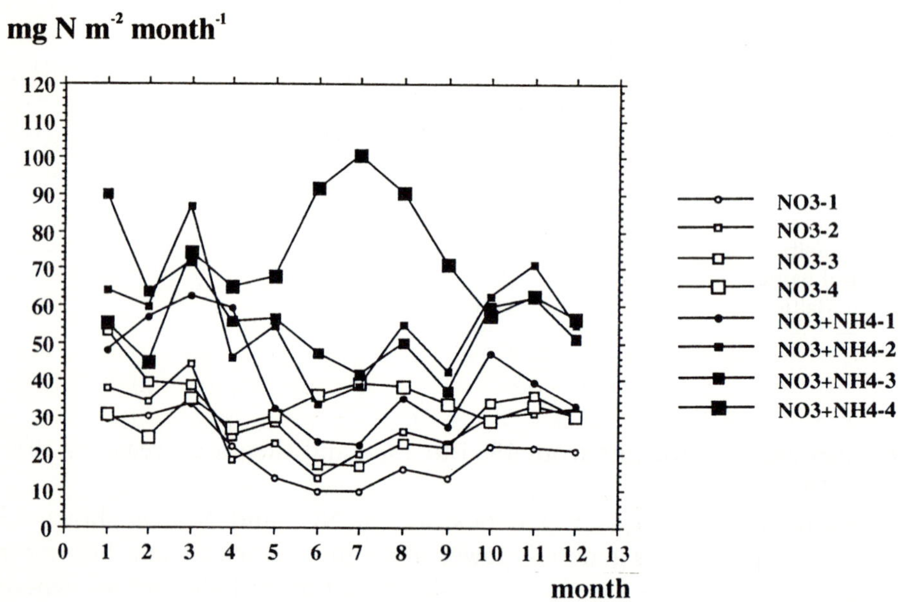

Fig. 5.4. Seasonal variation in wet deposition (in mg N m^{-2} $month^{-1}$) of nitrate-N and nitrate-N+ammonium-N in Bothnian Bay (*1*), Finnish Bay (*2*), Baltic proper (*3*) and southern Baltic (*4*)

When the present deposition estimate was made, organic nitrogen in precipitation was considered to give only a minor contribution, say 10 %, based on occasional measurements in the Swedish rain chemistry network. The study by Rahm et al. (1995) shows that the additional amount could be as much as 36 % of the inorganic nitrogen in winter, over ice and far from land. Another recent study (Cornell et al. 1995) indicates that dissolved organic nitrogen is a ubiquitous and significant component both in polluted and remote areas. It now appears that the deposition of organic nitrogen has been underestimated. Biological nitrogen fixation was not considered in this work. as explained above.

Dry deposition of inorganic nitrogen is, according to the estimates, quite small compared to corresponding wet deposition in the northern part of the Baltic (5–10 %), it increases to 10–15 % in the middle of the Baltic proper and makes a significant contribution in the southern part (around 25 %). The latter part of the gradient is due to the inferred strong concentration gradients supported by model calculations, but not verified by any long-term measurements. If the high deposition (measured at some land stations) should extend further out over the sea area, it will cause an underestimation of the deposition. On the other hand, in areas with low ammonia concentration, deposition may be overestimated, due to the effect of a compensation point.

Deposition of the gaseous components nitric acid and ammonia is important, although the concentrations are much lower than the particulate phase. It should be noted, however, that deposition velocities are deduced from theoretical considerations, with in principal no experimental verification over sea and under the specific conditions in the Baltic Sea. A crucial factor is the size distribution of nitrate particles, which is not based on empirical data from this region. (A bimodal distribution is assumed, making the deposition velocity almost a factor 10 higher than for NH_4-p.) Consequently, the estimated mean depositions have an uncertainty of at least a factor of 2, depending on both concentration and deposition velocity. Further, additional uncertainty arises as a result of the method used to interpolate between stations for individual months. Actual monthly measurements could not be considered owing to the few and widely separated stations.

5.4.1 Uncertainties in Wet Deposition Estimates

The deposition over the sea is estimated from land-based stations. The interpolation has to be made over rather large distances. An estimate of the uncertainty can be made in the following way: concentration of the actual substances in rainwater was measured for monthly periods during 3 years on two islands, Svenska Högarna and Finska Utö. These data were assumed to represent conditions over the open sea due to the rather small size of the islands. Concentrations of the substances were then compared by a linear interpolation with land-based stations some distance from the coast in Sweden and Finland. The agreement between measured and interpolated values was surprisingly good. The deviation was 0.03 mg N l^{-1} (7%) for nitrate and 0.04 mg N l^{-1} (10%) for ammonium. The standard deviation of the difference for individual months was 0.11 mg N l^{-1} (24%) for nitrate and 0.20 mg N l^{-1} (48%) for ammonium. The magnitude of uncertainty probably reflects real conditions when interpolating from land over open water over distances of some 400 km on monthly time scales. Sampling and analytical errors are considered to be minor in this regard, and in the order of 10% for individual measurements. An estimate of the uncertainty in deposition is more difficult to do, since such a calculation also involves the uncertainty in amount of precipitation over the sea

5.4.2 Comparison with Other Estimates

Depositions of nitrogen compounds and trace metals were estimated in a comprehensive study (HELCOM 1991) and based on a similar data set as was used in the present study. The results are given in Table 5.3 with comparisons. This "hybrid estimate" was an attempt to estimate precipitation amount at sea rather than to use land-based measurements. Comparison can also be made with a major modeling effort, based on emission inventories in Europe (Barrett and Berge 1996), which includes a special section regarding deposition over the Baltic. The EMEP data are generally lower then those obtained in the present study with regard to wet deposition (except for nitrate in the southern Baltic where the results are similar), but the dry deposition data are substantially higher. Difference in total (wet and dry, nitrate and ammonium) deposition ranged from 10% higher in the southern Baltic to 26% lower in the Bothnian Bay. Differences in wet deposition seem to be inherent in the model for the northern areas. Dry deposition is higher since the estimated air concentrations are higher. The HELCOM study suggested a higher wet deposition rate for both nitrate and ammonium. The reasons for this discrepancy are not clear but may be due to a more stringent data quality assurance in the present study. The dry deposition in the HELCOM study was adopted from Lindfors et al. (1991), and the data given were not quite compatible with the divisions of the Baltic area used in Table 5.3.

Acknowledgements. The Swedish Environmental Protection Agency supported this work financially.

References

Asman W (1990) Atmosfaerisk ammoniak og ammonium I Danmark (in Danish). NPO-forskning fra Miljöstyrelsen, no A18, Miljostyreisen, Kobenhavn, Denmark

Asman W, Harrision R, Ottley C (1994) The net air–sea flux of ammonia over the southern Bight of the North Sea. In: Borrell PM, Borrell P, Cvitas T, Seiler W (eds) Proc of EUROTRAC Symp 1994. SPB Academic Publishing, The Hague

Barrett K, Berge E (1996) Transboundary air pollution in Europe. EMEP/MSC-W Rep 1/96, Norwegian Meteorological Institute, Oslo, Norway

Cornell S, Rendell A, Jickells T (1995) Atmospheric inputs of dissolved organic nitrogen to the oceans. Nature 376:243–246

Granat L (1988a) Concentration gradients in atmospheric precipitation in areas of high annual precipitation. In: Unsworth MH, Fowler D (eds) Acid deposition at high elevation sites. Kluwer, London, pp 431–441

Granat L (1988b) A comparative study of precipitation chemistry at inland, coastal and island sites in the Bothnian Bay area. Rep CM-73, International Meteorological Institute/Department of Meteorology, Stockholm University

Granat L (1991) Air and precipitation chemistry network. Annual report for 1990 (in Swedish). Naturvårdsverket Rapport 3942. Naturvårdsverket, Informations-avdelningen, Solna, Sweden, 64 pp

Granat L (1993) Particulate and gaseous nitrate and ammonium compounds in air. Annual report (in Swedish). Naturvårdsverket Rapport 4214. Naturvårdsverket, Informationsavdelningen, Solna, Sweden

Granat L, Areskoug H, Hovmand M, Devenish M, Schneider B, Bieber E, Marquardt W, Reissell A, Järvinen O, Hanssen J-E, Sjöberg K (1992) Intercomparison of precipitation collectors for chemical analysis, HELCOM intercalibration – second stage, Aspvreten, Sweden, 1987–1988. Baltic Marine Environment Protection Commission – Helsinki Commission – Mannerheimintie 12A, Helsinki

HELCOM (1991) Airborne pollution load to the Baltic Sea 1986–1990. Baltic Sea Environ Proc 39

HELCOM (1992) Intercalibrations and intercomparisons of measurement methods for airborne pollutants. Baltic Sea Environ Proc 41, 92 pp

Lindfors V, Joffre M, Damski J (1991) Determination of the wet and dry deposition of sulfur and nitrogen compounds over the Baltic Sea area using actual meteorological data. Contribution no 4, Finnish Meteorological Institute, Helsinki, 111 pp

Lindfors V, Joffre M, Damski J (1993) Meteorological variability of the wet and dry deposition of sulfur and nitrogen compounds over the Baltic Sea. Water Air Soil Pollut 66:1–28

Rahm L, Håkansson B, Larsson P, Fogelqvist E, Bremle G, Valderma J (1995) Nutrient and persistent pollutant deposition on the Bothnian Bay ice and snow fields. Water Air Soil Pollut 84:187–201

Rodhe H, Söderlund R, Ekstedt J (1980) Deposition of airborne pollutants on the Baltic. Ambio 9:168–1873

Schaug J, Pedersen U, Skjelmoen JE, Kvalvågnes I (1993) Data report 1991, part 1: annual summaries. EMEP/CCC Rep 4/93, Norwegian Institute for Air Research, Lilleström

Sjöberg K, Persson K, Lövblad G, Granat L (1992) Air and precipitation chemistry network. Annual report for 1991 (in Swedish). Naturvårdsverket Rapport 4090, Naturvårdsverket, Informationsavdelningen, Solna, Sweden

Voldner EC, Barrie LA, Sirois A (1986) A literature review of dry deposition of oxides of sulfur and nitrogen with emphasis on long-range transport modelling in North America. Atmos Environ 20(11):2101–2123

6 Atmospheric and River Input of PCBs, DDTs and HCHs to the Baltic Sea

C. Agrell, P. Larsson, L. Okla, G. Bremle, N. Johansson, M. Klavins, O. Roots, and A. Zelechowska

6.1 Introduction

The Baltic Sea is vulnerable to pollution due to its semi-enclosed character and hydrography. The long residence time of water has led to the accumulation of nutrients as well as of persistent pollutants. At present, the levels of PCBs, HCHs and DDTs are high in the Baltic ecosystem. All three substance groups are persistent, bioaccumulative and cause reproductive disturbances. The effects of PCBs and DDTs include reproduction disturbances in white-tailed eagles (*Haliaeetus albicilla*; Helander et al. 1982) and gray seals (*Halichoerus grypus*; Zakharov and Yablokov 1990).The levels of persistent pollutants in the Baltic biota are comparable with those found in the Great Lakes of North America, e.g. the levels of PCB in pelagic predators like lake trout (*Salvelinus namaycush*) in the Great Lakes are about 2–3 mg/kg (fresh weight; Borgmann and Whittle 1991) compared with 0.5–1 mg/kg for salmon *(Salmo salar)* in the Baltic Sea (Larsson et al. 1996). Recently, an increased risk of low birth weights in infants, especially boys, has been associated with a high consumption of contaminated fish from the Baltic Sea by their mothers (Rylander et al. 1995). Also, breast cancer incidents were higher than expected in women from the Baltic coast. These women consumed locally caught fatty fish at least twice as often as the control group (Rylander and Hagmar 1995).

In the Baltic Sea, levels of PCBs and DDTs in herring (*Clupea harangensis*) and guillemot (*Uria algae*) eggs have been documented since 1969 (Bignert et al. 1995). Levels of PCBs in herring have decreased from 15 mg/kg in 1974 to about 4 mg/kg in 1989. However, the decreasing trend seen in the late 1970s in levels of PCBs has ceased. Similar trends have been registered in Lake Michigan (Stow et al. 1993). No trend studies exist for lindane in the Baltic region. Concentration measurements in animals are, however, not the best way to estimate the load of persistent pollutants to an ecosystem as the concentration is affected by other factors than the load.

Ecological Studies, Vol. 148
F. Wulff et al. (eds.)
A Systems Analysis of the Baltic Sea

No trend studies on the "direct load" to the ecosystem exist for any of the compounds.

Although trends in levels of persistent pollutants in some Baltic biota are well documented, little is known about the origin of these pollutants and the flow between different compartments in the ecosystem. It is unclear whether the atmosphere, the rivers or the sediments constitute the dominant source of pollutants to the biota of the system. Furthermore, many of the fundamental processes that these compounds undergo need to be better understood if we are to quantify and predict their environmental fate and transport in the Baltic ecosystem. These processes include deposition from the atmosphere, transfer across the surface microlayer, partitioning between different compartments in the water phase, interaction with biota, incorporation into the sediment, redistribution from sediment to water and volatilization across the water-air interface.

6.1.1 The Compounds

Within the framework of the project "Organic Pollutants, Load and Critical Processes", the industrial mixture of polychlorinated biphenyls (PCBs), the pesticides hexachlorocyclohexanes (HCHs; γ-HCH or lindane, which is the active ingredient, and the by-product, α-HCH) and the pesticide dichlorodiphenyltrichloroethane (p,p´-DDT), together with its metabolite p,´p-DDE (DDTs), have been used as model compounds to evaluate the contamination situation. The HCHs have been included because they are still in use in proximity to the Baltic Sea (Hilbert and Poulsen 1992; Voldner and Li 1995). DDTs are of interest because time trend data are available (Olsson and Reutergårdh 1986; Larsson and Okla 1989; Bignert et al. 1995) and also because the use of this compound group was restricted in almost all of the Baltic countries at the beginning of the 1970s.

PCBs were first detected in environmental samples in 1966 (Jensen et al. 1969), and at least two important routes of entry into the environment have been documented: losses during the process of manufacture and leakage from products containing PCBs. There have been little or no quantitative emission data from various processes including PCBs. Losses from open-end and nominally closed systems have been large (Anonymous 1979). These systems may still contribute to contamination of soil, water and atmosphere.

DDT has been a widely used insecticide and the total amount used worldwide between 1950 and 1995 is estimated to be 2,600,000 tons (Voldner and Li 1995). The use of DDTs was severely restricted or banned in all countries surrounding the Baltic Sea during the 1970s and 1980s.

Other major consumption areas have been Central America, India, North America and the former Soviet Union. Today, these countries have also banned or severely restricted the use of DDTs. There are only a few tropical countries still permitting use, but for many countries, especially Africa, no data are available (Voldner and Li 1995).

In some countries adjacent to the Baltic Sea, e.g. Denmark, Poland, Germany and Norway, lindane is still used for controlling crop and forest pests and fumigating storage rooms. The technical mixture of HCH consists of about 55–70% α-HCH, 5–14% γ-HCH and 10–18% other isomers (Barrie et al. 1992). In Norway, there are no restrictions against using this product, but it is banned in Poland, Denmark and Germany. In Sweden (1988), Estonia, Latvia and Lithuania, a total ban against use of both technical-HCH and lindane exists. Finland has banned the use of lindane, but the technical mixture of HCH is unregistered (Voldner and Li 1995). The major consumption areas of both the technical mixture of HCHs and lindane are concentrated in the equatorial regions, especially India, Central America and the Euro-Asian continent (Voldner and Li 1995).

6.1.2 Transport Processes

Hydrophobic organic compounds readily partition into biotic compartments, and especially PCBs are very lipophilic and resistant to biological, chemical and physical degradation. These properties are strengthened as the chlorine substituents on the biphenyl rings increase in number. The PCB congeners are semivolatile and the volatility decreases with increasing number of chlorine substituents. Lindane is considerably more water soluble and slightly more volatile than PCBs and DDTs. The lower Henry's law constant for lindane indicates that it partitions more into the waterphase than into the atmosphere compared to PCBs and DDTs.

Semivolatile compounds can be efficiently transported as vapors in the troposphere (Ballschmiter 1990). Apart from occurring and being transported as vapors, persistent organic pollutants (POPs) may be bound to microparticles, and in this phase, they may also be dispersed into the atmosphere. This transport can occur on a regional (10–100 km), meso (100–1000 km) or hemispheric (1000–40,000 km) scale and depends partly on the size of the particles to which the pollutants are bound.

Adsorption of gas-phase molecules on aerosol particles or the vapor-to-particle (V/P) ratio is controlled by the vapor pressure of the compound, the temperature and the available surface of suspended particles per volume unit (Junge 1977). As air moves over the sea, the concentration of particles decreases because of processes such as sedimentation, coagula-

some stations. The study was completed in February 1993. In this chapter, PCBs, DDTs and HCHs in air and precipitation and their calculated deposition will be evaluated with respect to temperature, latitude and sampling site.

We also studied POPs in the Rivers Wistula and Odra in Poland during 1991–1993, in order to calculate the river input from densely populated and industrialized catchment areas. During the autumn of 1993, PCBs, DDTs and lindane were also measured in two smaller Swedish rivers, the southern river Mörrum and the northern river Ume.

6.2 Material and Methods

During the study, a total of 299 air samples, 192 samples of precipitation and 75 river water samples were collected around the Baltic Sea. All three media were sampled with polyurethane foam filters (PUFF), which after extraction were analyzed for PCBs, DDTs and HCHs using gas chromatography and electron capture detection (ECD). The concentration data and additionally obtained meteorological data were analyzed statistically using ANCOVA and regression analysis.

6.2.1 Air and Precipitation Sampling

Sampling of persistent pollutants in the atmosphere was carried out at 16 locations surrounding the Baltic Sea (Fig. 6.1). The sampling stations were preferentially located in the vicinity of meteorological stations and as close to the sea as possible. Nine stations were less than 5 km from the shoreline and five stations were situated on islands. The air and rain samplers were mounted 2 and 1 m, respectively, above ground (Fig. 6.2). As collection media, two PUFFs (27×40 mm of colorless polyester, density 30 kg/m^3), connected in series, were used (Fisher et al. 1983; Braun et al. 1985). The PUFFs were pre-cleaned in a modified Soxhlet apparatus with acetone:n-hexane and kept frozen until use.

About 1000 m^3 of air was drawn through the PUFF by means of an electric pump at a rate of 30–50 l/min (registered by a flowmeter). The PUFFs were changed for new ones on a 14-day basis.

Precipitation was collected by a stainless steel funnel (1 m^2) and percolated through the PUFF. The accumulated amount of water was measured in a calibrated jar. PUFFs were changed monthly or when 40 l had passed the filters. The measured deposition was a sum of wet and dry de-

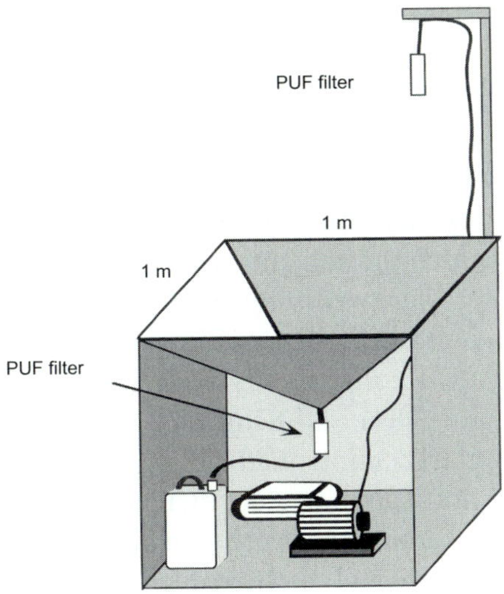

Fig. 6.2. Simplified drawing showing techniques to obtain samples of persistent pollutants in air and precipitation

position as the samplers were open to the atmosphere all the time. A heater was installed inside the sampling unit and switched on at temperatures below 5 °C. This enabled the snow that the sampler collected to melt, and the water to percolate through the PUFF. All samples were stored in a refrigerator or freezer before being sent to our laboratory.

6.2.2 River Sampling

Sampling locations were selected close to the river mouth in two rivers in Sweden, the Mörrum and Ume, and in two rivers in Poland, the Wistula and Odra (Fig. 6.1). Water samples were taken continuously between 1991 and 1993 in Poland and in the autumn of 1993 in Sweden. At each station about 100 l (Rivers Mörrum and Ume) and 300 l (Rivers Wistula and Odra) of water were filtered through PUFFs. The water was filtered at a rate of 10–20 ml/min. Each sampling lasted for 7–14 days after which the filter was changed and stored in a refrigerator or freezer until analysis. In total, 75 samples were collected. It has been shown previously that the PUFFs adsorb both particulate and dissolved POP (Larsson et al. 1990).

6.2.3 Sample Preparation and Analysis

PCBs, DDTs and HCHs adsorbed on the PUFF from air, precipitation and river water were extracted in a modified Soxhlet apparatus with a mixture of acetone and n-hexane (Bremle et al. 1995). Octachloronaphthalene and pentachlorobenzene were used as an internal standard and chromatographic standard, respectively.

The extracts were equilibrated with water to remove the acetone and polar impurities. The hexane fraction was evaporated in a vacuum centrifuge, and the sample redissolved in n-hexane. The samples were cleaned and fractionated in an open silica, after which the solvents were evaporated in a vacuum centrifuge and the samples were redissolved in isooctane. Samples were analyzed for persistent pollutants by capillary-gas chromatography/ECD (Bremle et al. 1995). PCB components were identified and quantified according to Mullin et al. (1984) and Schulz et al. (1988). The analytical performance was regularly controlled with pesticide standards, PCB standards, Aroclor 1242 and Clophen A60. PCB congeners that could not be properly identified or quantified were excluded from analysis. Extraction efficiencies of the internal standard octachloronaphthalene were 91±23 % for air, 93±21 % for rain and 97±27 % for river samples. All samples where corrected individually for recovery. Laboratory blanks were analyzed regularly and found to be satisfactory. Unexposed PUFFs from one station in Sweden and one station in Lithuania were found to be uncontaminated.

6.2.4 Statistical Analysis

We have used "analysis of covariance", i.e. ANCOVA, as the statistical method to validate the results. The ANCOVA tests a dependent variable Y (concentration of PCBs, DDTs and HCHs) for homogeneity among group means (sampling station) in a design similar to "analysis of variance", i.e. ANOVA. Before testing for homogeneity, the means are adjusted for the group differences in the independent variable X (temperature) also known as the "covariate" (Sokal and Rohlf 1995).

The temperature data were taken from the meteorological database created by SMHI within the subproject "Meteorological Forcing Functions". This database consists of fields of meteorological parameters which have been analyzed by one-dimensional univariate objective interpolation (Gandin 1963). This method is used at SMHI for numerical weather prediction and for analysis of surface parameters (Andersson et al. 1986).

Surface weather observations with a 3-hourly time resolution and 12-h accumulated precipitation observation were used as input to the statistical analysis. The analyzed meteorological parameters are represented in a regular horizontal latitude/longitude grid with a resolution of $1 \times 1°$ covering the whole Baltic drainage basin.

To obtain a comparable set of data from each station with a minimum of bias of season, we limited the time period from which data were selected to approximately one consecutive year. Due to lack of data, the two Polish stations Dziwnow and Swibno were omitted from these analyses. For three stations, Lahemaa, Ventes Ragos and Vilsandi, there were periods during which no air samples were taken. These data are thus less reliable. For Lahemaa, no spring samples have been obtained, for Ventes Ragos there are no summer samples and for Vilsandi winter data are lacking. Thus, for pollutants in air, 315±90 days of sampling were included in the statistical analysis and for pollutants in rain and deposition, 357±50 days of sampling. All POP data used in the statistical analysis were log transformed to obtain normally distributed values (Newton 1988).

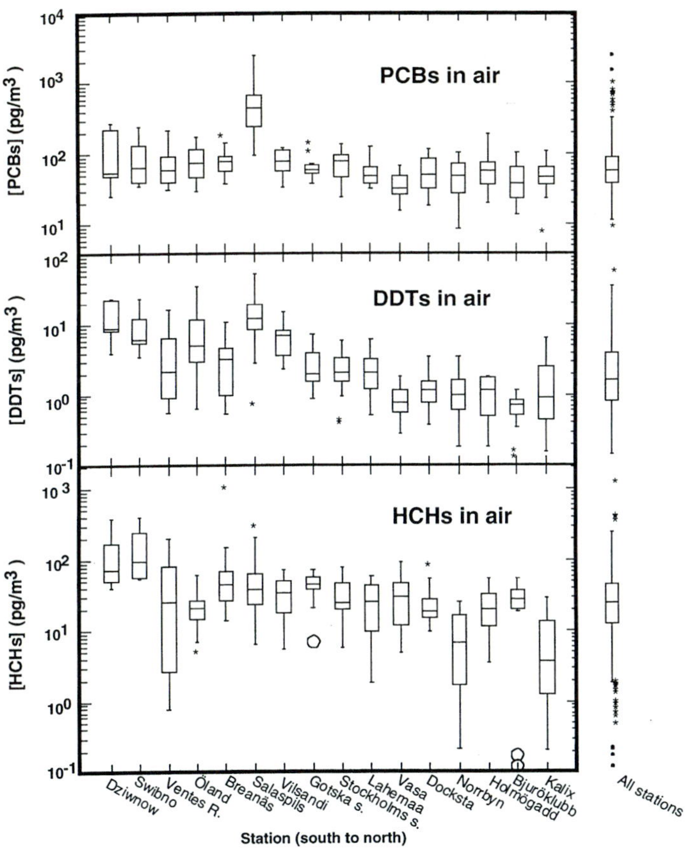

Fig. 6.3. Box plot showing concentration of PCBs, DDTs and HCHs in air (pg/m³) from sampling stations. Median, 25th and 75th percentiles and values that are outliers are plotted

6.3 Results and Discussion

6.3.1 Concentration of Persistent Pollutants in the Atmosphere

The median concentrations in the air samples for all stations were 57 pg/m^3 for PCBs, 1.6 pg/m^3 for DDTs (sum of p,p´-DDT and p,p´-DDE) and 25 pg/m^3 for HCHs (sum of a-HCH and g-HCH; Fig. 6.3, Table 6.1). The ratio between maximum and minimum median concentration between stations was 2.5 for PCBs at 15 of the stations (excluding Salaspils; for reasons see below). The concentrations ranged from 32 to 80 pg/m^3. Due to this homogeneity, this PCB concentration interval could be considered as background levels for the Baltic Sea. The ratio between maximum and minimum in median concentration between stations was larger for DDTs, and especially for HCHs, 13 and 26 times, respectively. One explanation for this discrepancy between the pesticides and PCBs may be that the diffuse sources of PCBs are more evenly distributed over the Baltic Sea area, whereas the source areas of DDTs and HCHs are found mainly in the agricultural, southern parts. Also, the variation in concentration of pollutants within a station was greater for the pesticides than for PCBs. This could be explained by the fact that HCHs and DDTs as pesticides are used seasonally, which is not the case for the industrial chemical PCB. The finding that concentrations in the atmosphere within a station generally show greater variation for pesticides than for PCBs agrees with a similar study on deposition of POPs to the Skagerack area (west of Sweden; Brorström-Lundén 1995).

The station in Latvia (Salaspils) consistently showed the highest values of PCBs and DDTs in air, with a median of 454 pg/m^3 of PCBs and 12 pg/m^3 of DDTs. These levels, which are about eight and six times higher than the median concentrations of all stations, are most likely the result of local sources. This sampling station was situated only 10 km from Riga, an industrial city with 1 million inhabitants. On the other hand, the levels of HCHs in Salaspils were similar to other stations in the vicinity (median of 39 pg/m^3; Fig. 6.3), indicating that Riga does not act as a point source for HCHs. Instead, the highest median concentrations of HCHs were found in Swibno (103 pg/m^3) and Dziwnow (72 pg/m^3), the two Polish stations that also had high median concentrations of DDTs (6 and 9 pg/m^3, respectively). Both stations are situated in agricultural areas and it is possible that the insecticides have been used either recently, as Poland does not have restrictions against usage of g-HCH (Voldner and Li 1995), or previously (HCHs and DDTs) in such large amounts that revolatilization from the ground is reflected in the samples of the present study.

Table 6.1. Summary of the different concentrations and calculated depositions at the stations

Latitude	Station	PCBs Air (pg/m3)	Precipitation (ng/l)	Deposition (ng m^{-2} day)
54°00'	Dziwnow	55 (*n*= 5)	1.4 (*n*= 2)	2.3 (*n*= 2)
54°15'	Swibno	69 (*n*= 6)	4.4 (*n*= 4)	5.0 (*n*= 4)
55°25'	Ventes R.	61 (*n*= 10)	2.0 (*n*= 15)	3.7 (*n*= 15)
56°14'	Öland	76 (*n*= 21)	8.3 (*n*= 15)	3.5 (*n*= 15)
56°17'	Breanäs	79 (*n*= 21)	2.8 (*n*= 12)	2.8 (*n*= 12)
56°50'	Salaspils	454 (*n*= 20)	10.7 (*n*= 15)	17.9 (*n*= 15)
58°20'	Vilsandi	79 (*n*= 9)	1.5 (*n*= 9)	2.2 (*n*= 9)
58°21'	Gotska s.	60 (*n*= 24)	2.0 (*n*= 15)	3.0 (*n*= 15)
59°17'	Stockholms s.	80 (*n*= 21)	1.3 (*n*= 10)	2.4 (*n*= 10)
59°30'	Lahemaa	49 (*n*= 16)	0.8 (*n*= 12)	1.8 (*n*= 12)
63°02'	Vasa	32 (*n*= 27)	0.9 (*n*= 12)	1.2 (*n*= 12)
63°03'	Docksta	50 (*n*= 24)	1.8 (*n*= 15)	2.6 (*n*= 15)
63°32'	Norrbyn	48 (*n*= 24)	1.8 (*n*= 17)	3.2 (*n*= 14)
63°36'	Holmögadd	57 (*n*= 23)	4.9 (*n*= 12)	5.7 (*n*= 12)
64°31'	Bjuröklubb	38 (*n*= 24)	2.9 (*n*= 13)	2.2 (*n*= 13)
65°44'	Kalix	47 (*n*= 24)	2.4 (*n*= 14)	1.5 (*n*= 14)
	All stations	57 (*n*=299)	2.3 (*n*=192)	2.7 (*n*=192)

Latitude	Station	DDTs Air (pg/m^3)	Precipitation (ng/l)	Deposition (ng m^{-2} day)
54°00'	Dziwnow	9.0 (*n*= 5)	0.21 (*n*= 2)	0.30 (*n*= 2)
54°15'	Swibno	6.3 (*n*= 6)	1.24 (*n*= 3)	1.0 (*n*= 3)
55°25'	Ventes R.	2.3 (*n*= 10)	0.18 (*n*= 15)	0.38 (*n*= 15)
56°14'	Öland	5.1 (*n*= 21)	0.71 (*n*= 13)	0.38 (*n*= 13)
56°17'	Breanäs	3.3 (*n*= 20)	0.17 (*n*= 12)	0.19 (*n*= 12)
56°50'	Salaspils	12.4 (*n*= 20)	0.40 (*n*= 15)	0.64 (*n*= 15)
58°20'	Vilsandi	6.9 (*n*= 8)	0.28 (*n*= 5)	0.23 (*n*= 5)
58°21'	Gotska s.	2.0 (*n*= 24)	0.15 (*n*= 15)	0.19 (*n*= 15)
59°17'	Stockholms s.	2.0 (*n*= 21)	0.09 (*n*= 10)	0.12 (*n*= 10)
59°30'	Lahemaa	2.0 (*n*= 16)	0.06 (*n*= 12)	0.1 (*n*= 12)
63°02'	Vasa	0.8 (*n*= 26)	0.03 (*n*= 12)	0.05 (*n*= 12)
63°03'	Docksta	1.2 (*n*= 24)	0.08 (*n*= 15)	0.08 (*n*= 15)
63°32'	Norrbyn	0.9 (*n*= 24)	0.07 (*n*= 16)	0.09 (*n*= 16)
63°36'	Holmögadd	1.2 (*n*= 22)	0.18 (*n*= 8)	0.16 (*n*= 8)
64°31'	Bjuröklubb	0.7 (*n*= 22)	0.04 (*n*= 11)	0.05 (*n*= 11)
65°44'	Kalix	0.9 (*n*= 24)	0.07 (*n*= 14)	0.05 (*n*= 14)
	All stations	1.6 (*n*=281)	0.13 (*n*=178)	0.15 (*n*=178)

Table 6.1. (*continued*)

Latitude	Station	HCHs Air (pg/m³)	Precipitation (ng/l)	Deposition (ng m⁻² day)
54°00'	Dziwnow	72 (*n*= 5)	0.63 (*n*= 2)	1.4 (*n*= 2)
54°15'	Swibno	103 (*n*= 6)	8.65 (*n*= 3)	5.7 (*n*= 3)
55°25'	Ventes R.	26 (*n*= 10)	1.63 (*n*= 15)	3.2 (*n*= 15)
56°14'	Öland	20 (*n*= 21)	2.5 (*n*= 13)	0.98 (*n*= 13)
56°17'	Breanäs	45 (*n*= 21)	1.8 (*n*= 12)	1.9 (*n*= 12)
56°50'	Salaspils	39 (*n*= 20)	1.3 (*n*= 15)	2.5 (*n*= 15)
58°20'	Vilsandi	33 (*n*= 28)	2.1 (*n*= 5)	3.7 (*n*= 5)
58°21'	Gotska s.	45 (*n*= 24)	1.4 (*n*= 15)	2.2 (*n*= 15)
59°17'	Stockholms s.	24 (*n*= 21)	1.0 (*n*= 10)	1.3 (*n*= 10)
59°30'	Lahemaa	26 (*n*= 16)	0.31 (*n*= 12)	0.53 (*n*= 12)
63°02'	Vasa	30 (*n*= 16)	0.38 (*n*= 12)	1.3 (*n*= 12)
63°03'	Docksta	18 (*n*= 24)	0.92 (*n*= 15)	1.7 (*n*= 15)
63°32'	Norrbyn	7 (*n*= 24)	0.16 (*n*= 17)	0.61 (*n*= 17)
63°36'	Holmögadd	20 (*n*= 23)	1.3 (*n*= 8)	0.82 (*n*= 8)
64°31'	Bjuröklubb	28 (*n*= 15)	0.46 (*n*= 10)	0.22 (*n*= 10)
65°44'	Kalix	4 (*n*= 21)	0.33 (*n*= 14)	0.16 (*n*= 14)
	All stations	25 (*n*=275)	1.0 (*n*=178)	1.3 (*n*=178)

In 1984–1985, Larsson and Okla (1989) reported median atmospheric concentrations at five coastal stations in Sweden to be 62 pg/m³ for PCBs and 11 pg/m³ for DDTs. A comparison with the present study indicates that, whereas there has been no major decline in PCB concentration, the levels of DDTs have declined by an order of magnitude. Backe et al. (1994) measured concentrations in air, rain and deposition of PCBs and DDTs at 11 sites in southern Sweden during 1992–1993. The levels were 91 and 11 pg/m³, respectively (geometrical means). For PCBs the levels are similar to our study, but the levels of DDTs were one order of magnitude higher. This could be due to the fact that Backe et al. were measuring in southern Sweden. The highest values of both PCBs and DDTs in the study by Backe et al. were found in the city of Landskrona (224 and 25 pg/m³, respectively), a town with 40,000 habitants and much heavy industry. These values are in parity with those recorded in Salaspils close to Riga. In the Skagerack region, Brorström-Lundén (1995) found concentrations of about 100–200 pg/m³ PCBs and 200 pg/m³ HCHs. These levels are somewhat higher than those found in our study.

We found that concentrations of PCBs, DDTs and HCHs in the atmosphere were dependent on sampling station and generally positively related to temperature. The temperature relationship had different slopes at the

Table 6.2. Effects of sampling station and temperature on concentrations (log-transformed) of PCBs, DDTs and HCHs in the atmosphere. Analysis performed with ANCOVA using temperature as a covariate (n=288 for PCBs, n=270 for DDTs and n=264 for HCH)

Substance	Source	*df*	F-ratio	R^2	*p*
PCBs	Station	13	3.23	0.63	0.000
	Temp. (K)	1	8.85		0.003
	Station * temp.	13	3.32		0.000
DDTs	Station	13	2.40	0.68	0.005
	Temp. (K)	1	36.32		0.000
	Station * temp.	13	2.36		0.006
HCHs	Station	13	2.14	0.43	0.013
	Temp. (K)	1	1.48		0.225
	Station * temp.	13	2.20		0.010

different stations (i.e. significant interaction term; Table 6.2). This, in turn, could be because the sources were different for the stations. Also, the surroundings of the sampling stations were different so local influence (e.g. vegetation type and cover, closeness to sea) may also be important for the slope. At the main part of the stations there was a positive relationship between temperature and levels of PCBs and DDTs. For PCBs 10 and for DDTs 13 out of 14 stations had a positive slope of the regression curve. This positive temperature dependence is in agreement with the findings of several authors (Doskey and Andrén 1981; Swackhammer et al. 1988; Hermanson and Hites 1989; Larsson and Okla 1989; Manchester-Neesvig and Andren 1989; Hoff et al. 1992). The process behind these results is the enhanced volatilization from the ground and water bodies as temperature increases. The pollutants are also partitioned from particles to the volatilized stage when temperature increases and the washout by rain is not so efficient. These processes are governed by the temperature-dependent vapor pressure, i.e. the volatilization ability, affecting both particle association and Henry's law constant.

No corresponding generally positive temperature effect could be found for HCHs. Instead, only four stations had a positive slope whereas nine stations had a regression curve with a negative slope. That the levels where higher during winter may be indicative of different source regions for HCHs during winter and summer.

There was a generally negative relationship between latitude and concentration of POPs in the atmosphere (Fig. 6.4). This may be due to the fact that

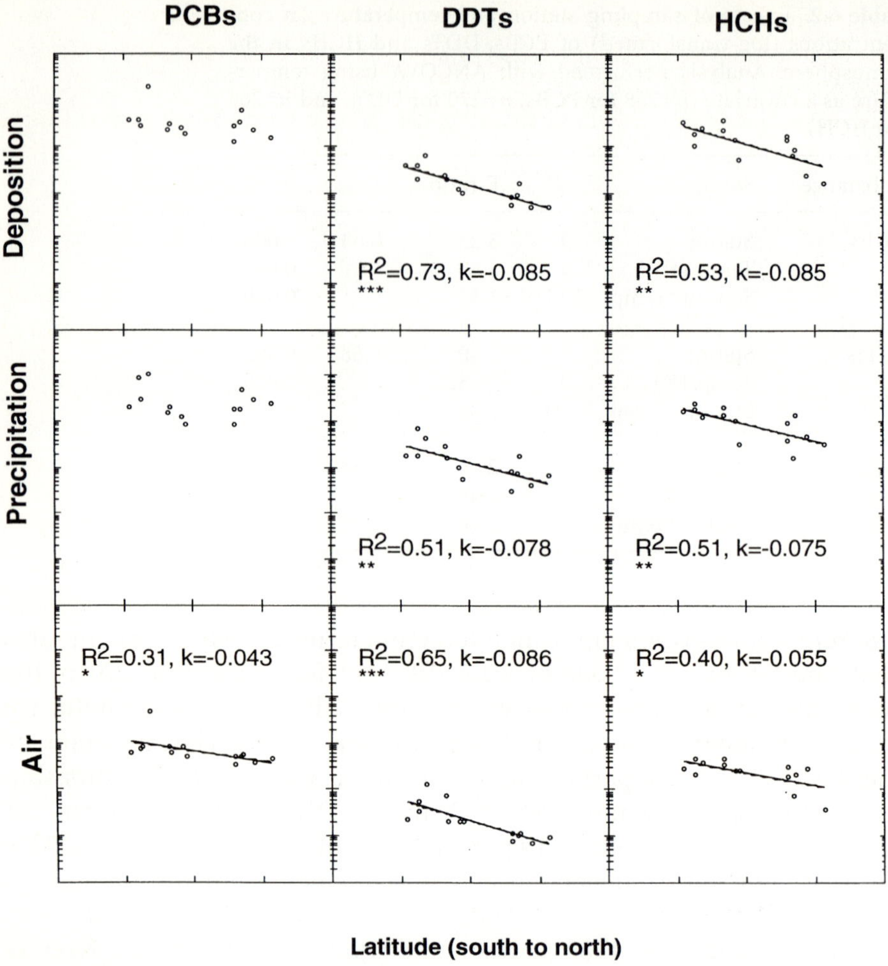

Fig. 6.4. Median concentration of PCBs, DDTs and HCHs in air (ng/m^3), rain (ng/l) and deposition (ng m^{-2} day) as a function of latitude (50–70°). Listed are linear regression coefficients, R^2, and slopes of regression curves

the stations are distributed at a distance of 1500 km from south to north, with higher mean temperature in the south. The higher mean temperature of the south will have the effect of "keeping the pollutants airborne as temperature affects the distribution of POPs in the atmosphere so that the proportion of chemicals in the atmosphere that condense onto soil, water, aerosols, snow and ice decrease with increasing temperatures" (Wania and Mackay 1993, 1995). Thus, even if the levels of POPs in the atmosphere are lower in the north, the number of local point sources and the total load of the ecosystem may not be less, only distributed towards condensed phases.

The south to north gradient of semivolatile POPs could also be the result of a "regional distillation" process, where POPs used in the warmer climate regions of the Baltic (or further south, in the equatorial region) are volatilized and driven by temperature gradients to colder areas, as described, with a global perspective, by Wania and Mackay (1993). Local point sources in the north are, in this model, not necessary to explain the levels found. Do the temperature differences between the southern and northern parts alone explain the observed results or is the assumption of "regional distillation" of POPs within the Baltic Sea valid? The sampling stations in the northern Baltic Sea may all have been influenced by local sources, which makes it possible that the lower mean temperature in the north explains the results. However, "regional distillation" effects and thus export of PCBs from the south may also be of importance. As the southern Baltic Sea is heavily industrialized and densely populated, the PCBs in the northern Baltic Sea may originate from this area.

HCHs and DDTs the assumption of "global" or "regional distillation" and a "common source" is probably more valid as the major consumption areas are situated in the temperate regions (Voldner and Li 1995). It is noteworthy that the strongest relationship between latitude and concentration was found for DDTs (R^2=0.65, k=–0.086; Fig. 6.4). This substance group has the lowest vapor pressure of the ones studied. It should, therefore, show the greatest response to a decreasing temperature gradient from south to north, because it is more easily scavenged by precipitation and adsorbed to particles that are deposited.

6.3.2 Concentrations of Persistent Pollutants in Precipitation

The median concentration in rain for all stations was 2.28 ng/l for PCBs, 0.12 ng/l for DDTs and 1.02 ng/l for HCHs (Fig. 6.5, Table 6.1). The ratio between maximum and minimum median concentration between stations in rain samples was slightly higher compared to air samples, but the same tendencies were found. Thus, the ratio was lowest for PCBs (10 times) and higher for the two pesticides (DDTs 41 and HCHs 28 times, respectively). Similar to air samples, generally, the variation in concentration within the stations was higher for the pesticides than for PCBs.

The local contamination of PCBs in Salaspils near Riga was also reflected in rain, where the highest level of the study was recorded (140 ng/l, median 10.7 ng/l). High levels of PCBs in precipitation were also recorded on the island of Öland (median 8.3 ng/l). Highest median concentrations of DDTs and HCHs in rain were found in Swibno (median 1.24 and 8.65 ng/l, respectively) and Öland (median 0.71 and 2.5 ng/l; Table 6.1).

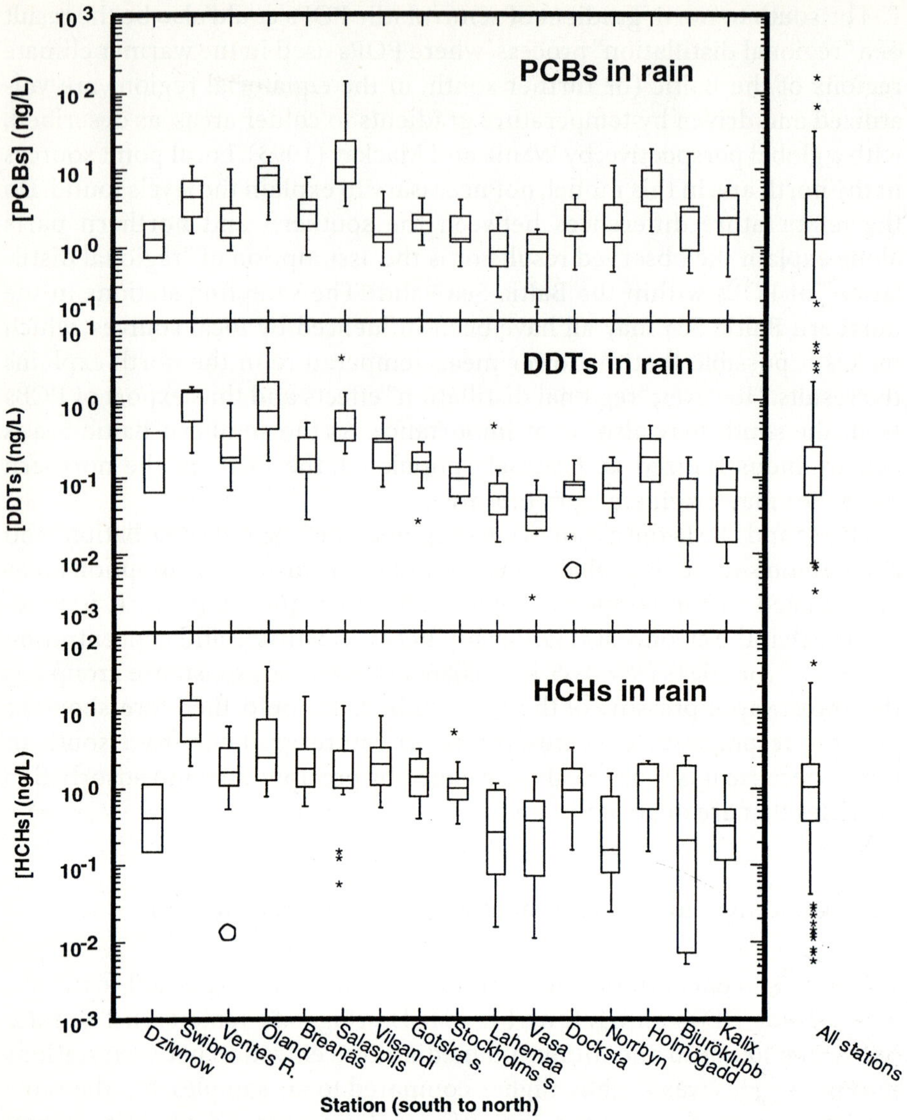

Fig. 6.5. Box plot showing concentration of PCBs, DDTs and HCHs in rain (ng/l) from sampling stations. Median, 25th and 75th percentiles and values that are outliers are plotted

These findings were consistent with the results for pesticides in air for Swibno, which also showed high levels of both compound groups. Öland had a high concentration of DDTs in the air samples but a low concentration of HCHs.

Few data on PCBs, DDTs and HCHs in rain (ng/l) from the Baltic region have been reported, although several studies on deposition exist (ng m^{-2} day; see next section). However, in the previously described study by Backe et al. (1994), concentrations were 3 ng/l for PCBs and 0.5 ng/l for DDTs. In a study from Lake Michigan in 1989–1990, Franz and Eisenreich (1993) found PCB concentrations of 2.7±2.1 ng/l. From the western Mediterranean, levels of PCBs in rain were 25 ng/l, for DDTs 5.5 ng/l and for HCHs about 37 ng/l, which is higher than from our study, especially for the DDTs and HCHs (Villeneuve and Cattini 1986). Generally, our data for the pesticides are slightly lower than those reported in the literature. This might be due to the fact that the reported data are older and thus representing a time period when the usage was higher, but they may also represent areas with higher usage than the countries surrounding the Baltic Sea.

As mentioned above, concentrations of POPs in air determine the concentration in rain by two processes: the dissolution of vapor, which is controlled by a compound's Henry's laws constant and the partitioning to particles which may be scavenged by raindrops. At low temperatures, both these processes are favored. Similar to concentrations in air, levels of PCBs in rain were dependent on sampling station but were negatively related to temperature, with a common slope for all stations (k=–0.020; Table 6.3). The same was valid for DDTs and HCHs (ten and eight stations, respectively, having negative slopes), although the data showed higher variation and the interaction term was significant, i.e. the generally negative relationship was different for the different stations. This implies that the local influence (e.g. precipitation amount, vegetation type and cover, closeness to sea) on the relationship between concentration in rain and temperature is important for DDTs and HCHs but not important for PCBs. Brun et al. (1991) found higher levels of PCBs in precipitation during winter and spring in Nova Scotia, but several studies have not recorded such a relationship (Swackhammer et al. 1988; Franz and Eisenreich 1993).

In principle, the differences between contaminant concentrations in rain may be attributed to the differences in air concentrations. Similar to the concentration in air, a negative relationship between concentration in rain and latitude was found for DDTs and HCHs (Fig. 6.4), implying that there are, relative to the south, less local sources. For PCBs there was no negative relationship with latitude. Thus, the colder climate in the north affects concentrations of pollutants in rain in a way that concentrations

Table 6.3. Effects of sampling station and temperature on concentrations (log-transformed) of PCBs, DDTs and HCHs in precipitation. Analysis performed with ANCOVA using temperature as a covariate (n=186 for PCBs, n=173 for DDTs and n=173 for HCH)

Substance	Source	Coefficient	*df*	F-ratio	R^2	*p*
PCBs	Station		13	17.4	0.60	0.000
	Temp. (K)	−0.020	1	32.1		0.000
DDTs	Station		13	2.04	0.66	0.022
	Temp. (K)		1	4.28		0.040
	Station * temp.		13	2.06		0.020
HCHs	Station		13	2.21	0.49	0.012
	Temp. (K)		1	3.97		0.048
	Station * temp.		13	2.27		0.009

increase relatively more than they do in the southern part of the Baltic. This, in turn, implies that there may be no difference between south and north in total contamination of PCBs.

6.3.3 Atmospheric Deposition

The yearly deposition of POPs to the Baltic Sea was calculated in the following way: for each station the total amount of pollutants deposited (ng) was divided by the length of the total sampling period and the sampler area (1 m^2). The daily mean deposition was then multiplied by 365 days to obtain the yearly deposition for each station. The median deposition was then calculated from these mean values, and multiplied by the sea surface area, 373,000 km^2.

Thus, the yearly deposition of POPs to the Baltic Sea was 387.1 kg for PCBs, 19.7 kg for DDTs and 226.8 kg for HCHs, reflecting the combined input through wet deposition and dry particle deposition. Overall, the deposition of PCBs showed small variation between stations (Fig. 6.6).The median values varied between 1.2 and 5.6 ng m^{-2} day (excluding Salaspils with the highest value of 17.9 ng m^{-2} day). Variability between stations was larger for the pesticides, about 10 times for DDTs and 25 times for HCHs, with median values of 0.15 and 1.30 ng m^{-2} day, respectively (Fig. 6.6, Table 6.1). For both of these substances, but especially for HCHs, the variation in deposition within a station was also greater than for PCBs. For the pesticides, the deposition from the sampling sites showed a negative rela-

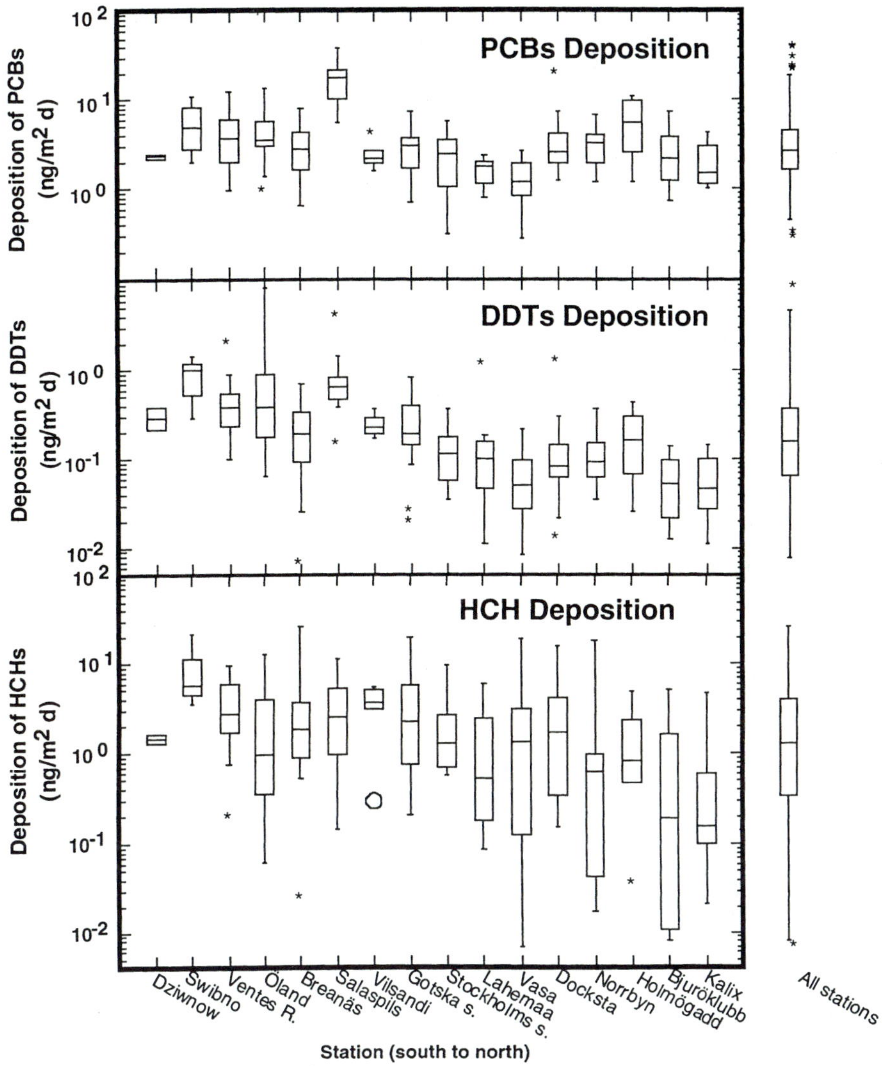

Fig. 6.6. Box plot showing deposition of PCBs, DDTs and HCHs (ng m^{-2} day) from sampling stations. Median, 25th and 75th percentiles and values that are outliers are plotted

tionship with latitude (Fig. 6.4). No such relationship was found for PCBs. Again, this implies that there may be no difference between south and north in total contamination of PCBs.

Brorström-Lundén (1995) found higher deposition of HCHs and PCBs over the Skagerak during 1990–1994, with a deposition of PCBs 10–15 ng m^{-2} day and of HCHs 5–15 ng m^{-2} day (α-, β- and γ-HCH; Table 6.4).

Table 6.4. Effects of sampling station and amount of precipitation on depositions (log-transformed) of PCBs, DDTs and HCHs. Analysis performed with ANCOVA using amount of precipitation as a covariate (n=186 for PCBs, n=173 for DDTs and n=173 for HCH)

Substance	Source	Slope	*df*	F-ratio	R^2	*p*
PCBs	Station		13	15.4	0.54	0.000
	Precip. amount	0.0041	1	13.3		0.000
DDTs	Station		13	13.4	0.54	0.000
	Precip. amount	0.0085	1	28.2		0.000
HCHs	Station		13	2.53	0.49	0.004
	Precip. amount		1	40.52		0.000
	Station * precip. amount		13	2.41		0.006

Higher deposition of HCHs was also found by Hilbert and Poulsen (1992) in a study in Denmark in 1990. However, during this time, lindane was used in Denmark (8.4 tons was sold in 1990). The deposition of PCBs in southern Sweden during 1993 was comparable with this study, but the deposition of DDTs was higher, possibly due to the southern location of the study area (Backe et al. 1994). The deposition rates of PCBs and DDTs recorded in the present study were lower than those measured at coastal stations of the Baltic Sea in 1984–1985 (Table 6.5; Larsson and Okla 1989). Since concentrations of PCBs in air were similar in both studies, methodological differences may explain the discrepancy. Larsson and Okla (1989) used silicon-impregnated nets to derive deposition estimates.

Table 6.5. Deposition of PCBs, DDTs and HCHs in five different studies. Calculated depositions are presented as µg m^{-2} year

Region	PCBs	DDTs	HCHs	Sampling year	Reference
Baltic Sea	1.0	0.05	0.5	1991–1993	This study
Baltic Sea	9.6	1.7		1985	Larsson and Okla (1989)
S. Sweden	4	1		1992–1993	Backe et al. (1994)
Denmark			15–17	1990	Hilbert and Poulsen (1992)
Skaggerak	3–5		2–5	1989–1994	Brorström-Lundén (1995)

Deposition of PCBs, DDTs and HCHs was dependent on sampling station and generally positively related to amount of precipitation (for PCBs, DDTs and HCHs; 11, 12 and 13 out of 14 stations, respectively, having positive slopes of the regression curve; Table 6.4).The relationship was expected as organochlorines are being washed out by rain, dissolved or associated to particles. Only HCHs had a significant interaction effect, implying that this substance group is more sensitive to local differences between the stations (e.g. type of precipitation, rain or snow, and the amounts of precipitation at rain events) than PCBs and DDTs. No direct effect of temperature on deposition of PCBs, DDTs or HCHs was found.

6.3.4 Concentration and Transport of Pollutants in Rivers

The concentration of PCBs and DDTs in the water from two rivers, Wistula and Odra, was relatively constant over time. Variation between maximum and minimum concentrations was about five to eight times. There was no apparent seasonal trend in the data (Figs. 6.7 and 6.8). Like the concentration in air and rain, the variation in HCHs was greater, about 50 times. In both years of study, levels of HCHs in the Wistula were high during late autumn, about 0.6 ng/l. A possible explanation for this may be a higher usage of these substances during summer and a "lag time" before they reach the rivers. Also, lower volatilization and/or reduced degradation

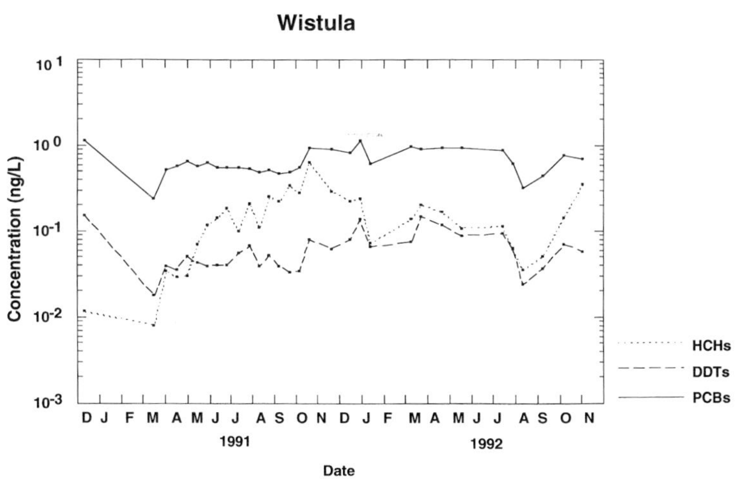

Fig. 6.7. Concentration of PCBs, DDTs and HCHs in water (ng/l) from River Vistula

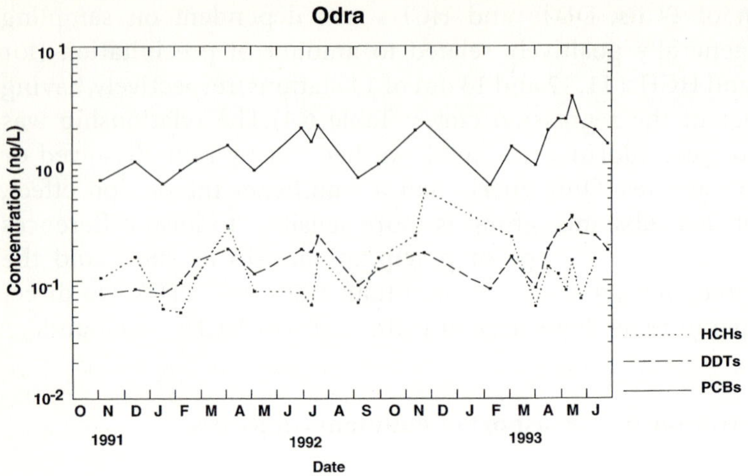

Fig. 6.8. Concentration of PCBs, DDTs and HCHs in water (ng/l) from River Odra

rates due to lower temperatures during autumn could account for an increase in concentration.

A comparison with the PCB-contaminated River Em (southern Sweden; Larsson et al. 1990; Bremle et al. 1995) showed that the presently investigated rivers had lower levels of PCBs. Depending on water discharge, the River Em contained 0.5–22 ng/l at the mouth. Concentration of PCBs, DDTs and HCHs in the Rivers Wistula and Odra showed no relationship with water discharge in contrast to the River Em, for which an inverse relationship between water discharge and concentrations of PCBs was recorded. The relationship was explained by the fact that PCBs were desorbed from contaminated sediments of a small lake, and these sediments constituted the major source to the river. The dominating source of POPs in Wistula and Odra could be precipitation, as the levels in the riverwater of all three compound groups approach and are somewhat lower than those in precipitation from the Polish stations Dziwnow and Swibno (Table 6.1, although note the few data points).

The River Mörrum had concentrations of PCBs and DDTs comparable to the Polish rivers, but showed the highest concentrations of HCHs in the study (although note the lower number of samples). The concentrations of PCBs and DDTs in the River Ume were the lowest in the investigation (Table 6.6). Levels of HCHs were comparable to the Polish rivers. In an international comparison, the contamination of all four rivers is low. In the Hudson River (USA), the PCB concentration in the water was between 50 and 100 ng/l (Brown et al. 1985). In the Shiawassee River, a smaller river in

Table 6.6. Median concentrations of PCBs, DDTs and HCHs in riverwater (ng/l)

River	PCBs	DDTs	HCHs
Wistula (*n*=31)	0.61	0.06	0.14
Odra (*n*=24)	1.44	0.17	0.10
Mörrum (*n*=11)	0.72	0.04	0.30
Ume (*n*=9)	0.21	0.02	0.11

the United States, the concentration of PCBs in the water was 650–1100 ng/l (Rice and White 1987). However, these rivers are considered to have extreme PCB levels due to heavy contamination from industrial discharge. The River Seine in France contained between 50 and 150 ng/l of PCBs during high water discharge, to 500 ng/l at the beginning of low water (Chevreuil et al. 1987), which according to the authors corresponds to the "moderate to high polluted" class of the WHO.

Previous measurements in the Wistula (Niemirycz et al. 1988) indicated PCB concentrations ranging from 1 to 15 ng/l and with an average of 5 ng/l. The results were considerably higher than the concentrations recorded in this study. During 1990–1992, the Institute of Meteorology and Water Management in Poland recorded average concentrations of 5.3 ng/l (range 0.1–40 ng/l) in the Wistula and 4.8 ng/l (range 0.1–15 ng/l) in the Odra. Differences in sampling and analytical techniques probably explain the different results.

Based on the pollutant concentration and the amount of water discharged, the transport of PCBs, DDTs and HCHs into the Baltic Sea via the rivers was calculated. Although concentrations of PCBs and DDTs were higher in the Odra, the largest transport of PCBs was recorded for the Wistula (Table 6.7), because it has a considerably higher water discharge (median discharge 780 m^3/s; Anonymous 1991, 1992, 1993) than the Odra (median discharge 270 m^3/s). This higher water flow did not compensate for the threefold higher concentration of DDTs in the Odra which thus had the largest transport of DDTs.

The seasonal transport of PCBs, DDTs and HCHs (only Odra) followed the water discharge, reaching a maximum in the spring and minimum during autumn (Figs. 6.9 and 6.10). In the Rivers Mörrum and Ume, the transport of PCBs and DDTs was low compared with the Polish rivers. These differences between Swedish and Polish rivers are partly related to differences in concentration and water flow, but other factors may also influence the results. For example, the drainage areas in Sweden are smaller and less populated than the drainage areas in Poland.

Table 6.7. Median transport of PCBs, DDTs and HCHs in river water (kg/month)

River	PCBs	DDTs	HCHs
Wistula	1.26	0.11	0.31
Odra	1.15	0.11	0.08
Mörrum	0.04	0.002	0.02
Ume	0.20	0.02	0.09

When calculated on a yearly basis, Wistula and Odra contributed about the same load of PCBs and DDTs to the Baltic Sea (Table 6.7), while for HCHs the Wistula River was more important. The yearly transport of PCBs and DDTs by the Swedish rivers was low compared to the Polish rivers, except for the transport of HCHs. For this compound group, transport in the northern River Ume was comparable to that of the River Odra. Both rivers have similar water discharges.

Using the total water discharge to the Baltic Sea via rivers (475 km^3/year) and using the median concentrations of 0.7 ng/l PCBs, 0.06 ng/l DDTs and 0.1 ng/l HCHs, the river transport results in a yearly amount of 332 kg PCBs, 2.8 kg DDTs and 47.5 kg HCHs to the Baltic Sea. Consequently, the rivers and the atmosphere contribute about equally in PCB load, while for the pesticides atmospheric deposition is about five to seven times more important.

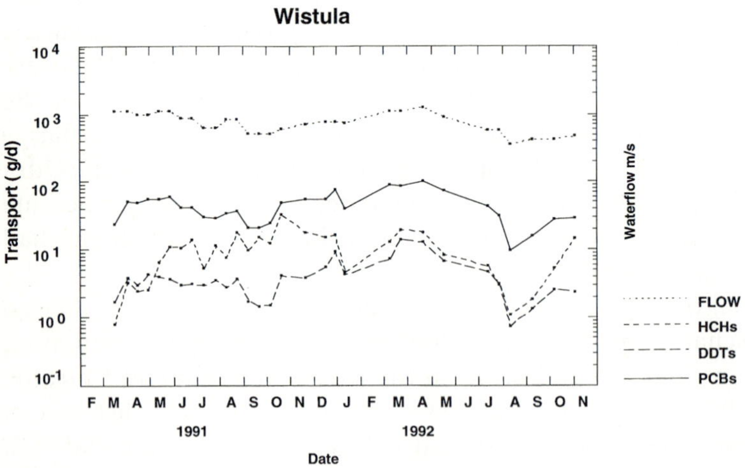

Fig. 6.9. Transport of PCBs, DDTs and HCHs (g/day) from River Vistula

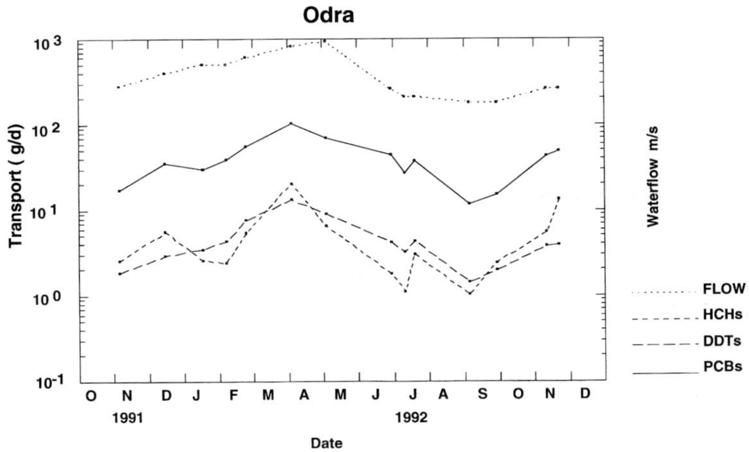

Fig. 6.10. Transport of PCBs, DDTs and HCHs (g/day) from River Odra

References

Andersson E, Gustafsson N, Meuller L, Omstedt G (1986) Development of meso-scale analysis schemes for now-casting and very short-range forecasting. PROMIS Rep 1. SMHI, Norrköping

Anonymous (1979) Polychlorinated biphenyls. National Academy of Sciences, Washington, DC

Anonymous (1991) Environmental conditions of the Polish zone of the southern Baltic in 1990. Institute of Meteorology and Water Management, Gdynia

Anonymous (1992) Environmental conditions of the Polish zone of the southern Baltic in 1991. Institute of Meteorology and Water Management, Gdynia

Anonymous (1993) Environmental conditions of the Polish zone of the southern Baltic in 1992. Institute of Meteorology and Water Management, Gdynia

Backe C, Larsson P, Okla L (1994) Undersökning av stabila organiska miljögifter i Skåne, atmosfär och nederbörd. Rapport till Skånes luftvårdsförbund. Lunds University

Ballschmiter K (1990) Distribution of organic compounds in the environment. In: Proc from Workshop on Long-Range Transport of Organic Pollutants. SNV Rep 3807. Swedish Environmental Protection Agency, Solna, Sweden

Barrie LA, Gregor D, Lake R, Muir D, Shearer R, Tracey B, Bidleman T (1992) Arctic contaminants: sources, occurrence and pathways. Sci Total Environ 122:1–74

Bidleman T (1988) Atmospheric processes; wet and dry deposition of organic compounds are controlled by their vapor-particle partitioning. Environ Sci Technol 22:361–367

Bignert A, Litzén K, Odsjö T, Olsson M, Persson W, Reutergårdh L (1995) Time-related factors influence the concentration of sDDT, PCBs and shell parameters in eggs of Baltic guillemot *(Uria aalge)*, 1861–1989. Environ Pollut 89:27–36

Borgmann U, Whittle DM (1991) Contaminant concentration trends in Lake Ontario lake trout (*Salvelinus namaycush*): 1977–1988. J Great Lakes Res 17:368–381

Braun T, Navratil J, Farag A (1985) Polyurethane foam sorbents in separation science. CRC Press, Boca Raton, 219 pp

Bremle G, Okla L, Larsson P (1995) Uptake of PCBs in fish in a contaminated river system: bioconcentration factors measured in the field. Environ Sci Technol 29: 2010–2015

Brorström-Lundén E (1995) Measurements of semivolatile organic compounds in air and deposition. PhD Thesis, Department of Analytical and Marine Chemistry, Chalmers University of Technology, Göteborg

Brown MP, Werner MB, Sloan RJ, Simpson KW (1985) Polychlorinated biphenyls in the Hudson River. Environ Sci Technol 19:656–661

Brun GL, Howell GD, O'Neill HJ (1991) Spatial and temporal patterns of organic contaminants in wet precipitation in Atlantic Canada. Environ Sci Technol 25: 1249–1261

Chevreuil M, Chesterikoff A, Letolle R (1987) PCB pollution behavior in the River Seine. Water Res 21:427–434

Dickhut M, Gustafson K (1995) Atmospheric inputs of selected polycyclic aromatic hydrocarbons and polychlorinated biphenyls to southern Chesapeake Bay. Mar Pollut Bull 30:385–396

Doskey P, Andrén A (1981) Modeling the flux of atmospheric polychlorinated biphenyls across the air/water interface. Environ Sci Technol 15:705–711

Eisenreich SJ, Looney BB, Thornton JD (1981) Airborne organic contaminants in the Great Lake ecosystem. Environ Sci Technol 15:30–38

Fisher J, Petty R, Lick W (1983) Release of polychlorinated biphenyls from contaminated sediments: flux and apparent diffuse of four individual PCBs. Environ Pollut 5:121–132

Franz TP, Eisenreich SJ (1993) Wet deposition of polychlorinated biphenyls to Green Bay, Lake Michigan. Chemosphere 26(10):1767–1788

Gandin LS (1963) Objective analysis of meteorological fields. Hydromet Press, Leningrad

Halsall CJ, Lee RG, Coleman PJ, Burnett V, Harding-Jones P, Jones K (1995) PCBs in UK urban air. Environ Sci Technol 29:2368–2376

Helander B, Olsson M, Reutergårdh L (1982) Residue levels of organochlorine and mercury compounds in unhatched eggs and the relationships to breeding success in white-tailed sea eagles *Haliateeus albicilla* in Sweden. Holarctic Ecol 5:349–366

Hermans JH, Smedes F, Hofstraat JH, Cofino WP (1992) A method for estimation of chlorinated biphenyls in surface water: influence of sampling method on analytical results. Environ Sci Technol 26:2028–2035

Hermanson FH, Hites RA (1989) Long term measurements of atmospheric polychlorinated biphenyls in the vicinity of superfund dumps. Environ Sci Technol 23: 1253–1258

Hilbert G, Poulsen M (1992) Bulk deposition of hexachlorocyclohexane. NERI Tech Rep 42, National Environmental Research Institute, Roskilde, Denmark, 29 pp

Hoff RM, Muir DCG, Grift NP (1992) Annual cycle of polychlorinated biphenyls and organohalogen pesticides in air in southern Ontario. 1. Air concentration data. Environ Sci Technol 26:266–275

Junge CE (1977) Basic considerations about trace constituents in the atmosphere as related to the fate of global pollutants. In: Sufflet IH (ed) Fate of pollutants in the air and water environments. Wiley, New York, pp 7–25

Jensen S, Jhonels AG, Olsson M, Otterlind G (1969) DDT and PCB in marine animals from Swedish waters. Nature 224:247–250

Larsson P, Okla L (1989) Atmospheric transport of chlorinated hydrocarbons to Sweden in 1985 compared to 1973. Atmos Environ 23:1699–1711

Larsson P, Okla L, Ryding S-O, Westöö B (1990) Contaminated sediment as a source of PCBs in a river system. Can J Fish Aquat Sci 47:746–754

Larsson P, Backe C, Bremle G, Eklöv A, Okla L (1996) Persistent pollutants in a salmon population (*Salmo salar*) of the southern Baltic Sea. Can J Fish Aquat Sci 53:62–69

Manchester-Neesvig JB, Andren AW (1989) Seasonal variations in the atmospheric concentration of polychlorinated biphenyl congeners. Environ Sci Technol 23: 1138–1148

Mullin MD, Pochini CM, McCrindle S, Romkes M, Safe SH, Safe LM (1984) High-resolution PCB analysis: synthesis and chromatographic properties of all 209 PCB congeners. Environ Sci Technol 18:468–476

Newton I (1988) Determination of critical pollutant levels in wild populations, with examples from organochlorine insecticides in birds of prey. Environ Pollut 55:29–40

Niemirycz E, Korzec E, Makowski Z (1988) Outflow of some organic substances transported by the Wistula River into the Baltic Sea. Institute of Meteorology and Water Pollution Control, Gdansk, Poland

Olsson M, Reutergårdh L (1986) DDT and PCB pollution trends in the Swedish aquatic environment. Ambio 15:2

Rice CP, White DS (1987) PCB availability assessment of river dredging using caged clams and fish. Environ Toxicol Chem 6:259–274

Rylander L, Hagmar L (1995) Mortality and cancer incidence among women with a high consumption of fatty fish contaminated with persistent organochlorine compounds. Scand J Work Environ Health 21:419–426

Rylander L, Strömberg U, Hagmar L (1995) Decreased birth weight among infants born to women with a high dietary intake of fish contaminated with persistent organochlorine compounds. Scand J Work Environ Health 21:368–375

Schulz DE, Petrick G, Duinker JC (1988) Complete characterization of polychlorinated biphenyl congeners in commercial Aroclor and Clophen mixtures by multidimensional gas chromatography-electron capture detection. Environ Sci Technol 23:852–859

Simmleit N, Herrmann R (1986) The behavior of hydrophobic, organic micropollutants in different karst water systems. Water Air Soil Pollut 34:79–95

Sokal RR, Rohlf FJ (1995) Biometry. The principles and practice of statistics in biological research, 3rd edn. Freeman, New York

Stow CA, Carpenter SR, Amrhein JF (1993) PCB concentration trends in Lake Michigan coho (*Oncorhynchus kisutch*) and Chinook salmon (*O. tshawytscha*). Can J Fish Aquat Sci 51:1384–1390

Swackhammer DL, Armstrong DE (1987) Distribution and characterization of PCBs in Lake Michigan water. J Great Lakes Res 13:24–36

Swackhammer DL, McVeety BD, Hites RA (1988) Deposition and evaporation of polychlorinated congeners to and from Siskiwit Lake, Isle Royale, Lake Superior. Environ Sci Technol 22:664–672

Villeneuve J-P, Catttini C (1986) Input of chlorinated hydrocarbons through dry and wet deposition to the Western Mediterranean. Chemosphere 2:115–120

Voldner EC, Li Y-F (1995) Global usage of selected persistent organochlorines. Sci Total Environ 160/161:201–210

Wania F, Mackay D (1993) Global fractionation and cold condensation of low volatility organochlorine compounds in Polar regions. Ambio 22:10–18

Wania F, Mackay D (1995) A global distribution model for persistent organic chemicals. Sci Total Environ 160/161:211–232

Zakharov VM, Yablokov (1990) Skull asymmetry in the Baltic gray seal: effects of environmental pollution. Ambio 19:266–269

7 Pelagic Plankton Growth and Resource Limitations in the Baltic Sea

Å. Hagström, F. Azam, J. Kuparinen, and U.-L. Zweifel

7.1 Introduction

The Baltic Sea is a large transition area between fresh and marine conditions with a drainage area that includes vast forested lands as well as some of the most densely populated regions in Europe. Substantial quantities of allochthonous material enter the Baltic via rivers and streams and the net outflow of water through the Danish straits is very large, being comparable to the Mississippi River (Bergström and Carlsson 1994). In view of such large terrestrial influences extensive monitoring programs exist whose goal is to gain predictive knowledge of ecosystem behavior and ecosystem response to perturbation.

A challenge to understanding the Baltic biogeochemistry is that this sea experiences great temporal and spatial variations in physical forcing which, in turn, contribute to spatial and temporal variability in the biogeochemical processes. Density gradients prevent annual mixing, except in the northernmost part, and consequently each sea basin harbors two or more water-bodies that have distinctive properties. Critical processes in one layer will influence the organisms of the next water mass at rates largely determined by climatic conditions. The most important driving force for this variation is the infrequent inflow of salt water via the Danish straits to the deep waters of the Baltic proper. These events have an effective life span of several years and can, in magnitude, overwhelm the anthropogenic impact on pelagic processes in the Baltic Sea proper (Matthaeus and Frank 1988). For instance, in 1993, a major salt-water inflow replaced the anoxic deep layer in the central basin with oxic water and greatly changed the stocks of ammonium and phosphate in the water mass underlying the photic zone. The ammonium from the basin storage roughly equaled the amounts of ammonium entering the eastern Gulf of Finland from land and atmospheric deposition during 2 years (Grönlund et al. 1996). Intuitively, it can be appreciated that such major redistribution events will affect Baltic

Ecological Studies, Vol. 148
F. Wulff et al. (eds.)
A Systems Analysis of the Baltic Sea

productivity and biogeochemistry over large areas (Nômmann et al. 1991). Thus, predictions of the future for the Baltic, and the management strategies, need to distinguish between the effects of climate change and anthropogenic loading and to give full attention to both.

In this chapter we discuss the biogeochemical cycling of carbon and other bioelements in the Baltic Sea ecosystems. Our emphasis here is on the significance of microbial processes. It is now generally recognized that the majority of biogeochemical cycling of materials in the ocean is mediated by microorganisms, but in this context, few attempts to present a synthesis on the microbial processes in the Baltic have been made. We will attempt such a synthesis and then use it as the basis to formulate hypotheses on Baltic ecological issues such as cyanobacterial blooms, limiting nutrients and organic carbon degradation.

7.2 Ecosystem Structure

7.2.1 Microbial Food Web

Figure 7.1 illustrates how processes that control plankton growth are linked within a microbial food web. In particular, the transformations between particulate organic matter (POM) and dissolved organic matter (DOM) are key factors in the pelagic energy flow. Often 10–50 % of the total primary production is channeled into DOM which only bacteria can use (Larsson and Hagström 1982; Cole et al. 1988). A significant flow of organic carbon enters the pathway: DOM→Bacteria→Protozoa→Metazoa (Azam et al. 1983) and, thus, primary producers simultaneously support the growth of osmotrophic bacteria as well as particle-eating herbivores via the microbial loop.

Bacteria are osmotrophs par excellence, a fact that explains their central role in the ocean's food web and biogeochemistry. With a radius of around 0.2 µm, they have the highest surface area to volume ratio and, due to their numerical abundance, they represent the majority of the biotic surface area in the sea (Williams 1981). Endowed with highly efficient and versatile permeases for nutrient uptake, bacteria are avid consumers of both organic and inorganic nutrients. They virtually monopolize DOM uptake as well as compete strongly with algae for inorganic nutrients and this has important implications for productivity and carbon cycling. In the ecosystem context, the superior uptake capacity of bacteria means that they control DOM concentration as well as serve to concentrate N and P into bacterial biomass, subject to remineralization by protozoan grazing (see

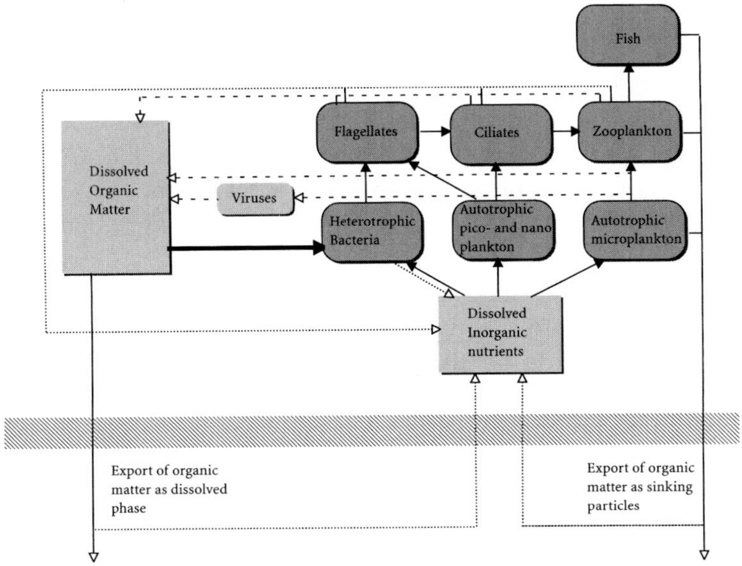

Fig. 7.1. Microbial food web. *Solid lines* Uptake or consumption; *dashed lines* release and transport

below). Both DOM and inorganic nutrients are released into the surrounding water when protozoa feed on bacteria (Johannes 1965; Caron et al. 1985).

A final adaptive strategy of marine bacteria that we wish to consider, and which has biogeochemical relevance, concerns the significance of hydrolytic ectoenzymes. These enzymes enable bacteria to hydrolyze POM and polymers (Hollibaugh and Azam 1983; Ammerman and Azam 1985) and thus cause POM→DOM phase transitions. This has consequences for biogeochemical cycling and food web structure.

While bacterial action on POM can produce DOM to fuel the microbial loop, further supply of DOM results from mortality within all groups of organisms. A recent important discovery is that viral lysis may be a significant cause of mortality of bacteria and phytoplankton (Preisig and Hibberd 1984; Proctor and Fuhrman 1990; Steward et al. 1992). In addition to affecting population and species dynamics of bacteria and algae, viral lysis also converts part of the prey POM into DOM. An interesting effect of virus attack on bacteria is that it increases the efficiency of DOM remineralization by mediating a futile cycle: DOM→Bacteria→DOM (Fuhrman 1992; Blackburn et al. 1996).

Nanoflagellates are often the most important consumers of bacteria and cyanobacteria and serve as a link in the transport of the bacterial biomass to the higher trophic level organisms (Fenchel 1982a–d; Kuosa and Kivi 1989). Second in the predator chain are the ciliates which prey on flagellates as well as differently sized algae. The ciliates, in turn, are fed upon by small Metazoa. Dinoflagellates do not conform to this size dependence but have exceptional properties allowing a prey to be much larger than themselves. All steps of predation result in release of DOM due to egestion from food vacuoles and this DOM can potentially enter the microbial loop through utilization by bacteria.

The above discussion illustrates that a large fraction of the carbon and energy fixed by phytoplankton, as well as organic matter derived from land, is subject to processing by the microbial food web. Further, the patterns of carbon and energy flow within the microbial food web, and their spatial and temporal variations, can strongly influence the overall biological and chemical dynamics of the ecosystem. These biological processes are then modified by strong spatial and temporal variability of physical forcing in the Baltic Sea, as mentioned earlier. It is from this perspective that we present the pelagic processes in the Baltic.

7.2.2 Temporal Changes in the Baltic Ecosystem

The Baltic surface waters experience large seasonal variations, from −0.3 to 20 °C. While the extreme values do not differ much from north to south, the length of each season varies greatly. The Bothnian Bay is covered by ice on average for 190 days per year (Leppäranta et al. 1988) and the productive season lasts only from June to September. Midway down the Baltic gradient, in the northern Baltic proper, 100 ice days per year may still occur, compared with an average of 10 days at the southern end. These temporal characteristics result in annual production profiles typical for each basin (Fig. 7.2).

A compilation of primary production data shows an obvious gradient, with primary production increasing from north to south (Table 7.1). Despite such large latitudinal variations in primary production, the food web structure for each season shows many similarities between basins. For example, during certain parts of summer all basins have low algal biomass, moderate productivity and a dominance of small algae. The similarity of the structure between seasons allows us to present the common "mechanics" of the microbial food web in the same format where the shifting dominance of organism groups can be demonstrated for each of the four seasons (Fig. 7.3).

Table 7.1. Annual primary (PP) and bacterioplankton (BP) production in different basins of the Baltic Sea

Basin	PP ($g\ C\ m^{-2}\ year^{-1}$)	Reference	BP ($g\ C\ m^{-2}\ year^{-1}$)	Reference
Bothnian Bay	16– 29	Lassig et al. (1978); Wikner (1996)	16–19	Å. Hagström (unpubl. data)
Bothnian Sea, coast	28– 90	Lassig et al. (1978); Zweifel et al. (1995); Wikner (1996)	18–21	Wikner and Hagström (1991); Zweifel et al. (1995)
Bothnian Sea, open	92	Zweifel et al. (1995)	22	Zweifel et al. (1995)
Gulf of Finland, coast	65–126	Kuparinen (1984); Kuparinen et al. (1984); Kuosa and Kivi (1989); Grönlund and Leppänen (1990); Lignell (1990)	9–19	Kuparinen (1984); Kuparinen et al. (1984); Kuosa and Kivi (1989); Lignell (1990)
N. Baltic proper, coast	93–160	Larsson and Hagström (1979, 1982)	29	Larsson and Hagström (1979)
N. Baltic proper, open	160–200	U. Larsson (unpubl. data)		
Gotland Basin	84–266	Wulff et al. (1987)		
Arkona Sea	92–366	Wulff et al. (1987)		
Kiel Bight			31	Ærtebjerg et al. (1996)
Bornholm Basin	132–256	Wulff et al. (1987)		

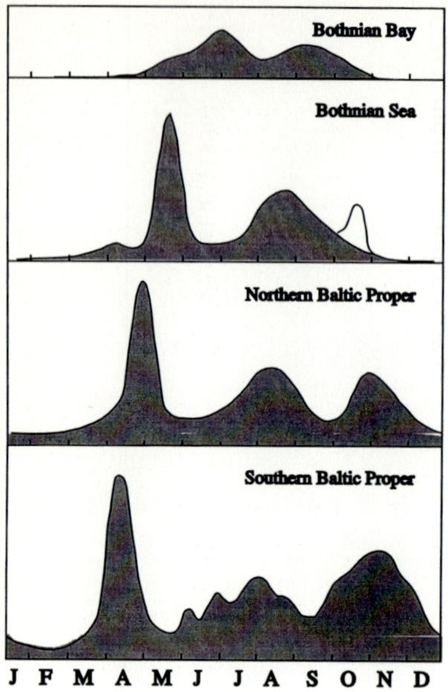

Fig. 7.2. Generalized annual cycles of phytoplankton development in different parts of the Baltic Sea (after Hällefors and Niemi 1986). Vertical scale is arbitrary. Specific studies of the included areas are presented, e.g. in Kononen and Niemi (1984); Smetacek et al. (1984); Kononen (1988); Kuosa and Kivi (1989); Andersson et al. (1996)

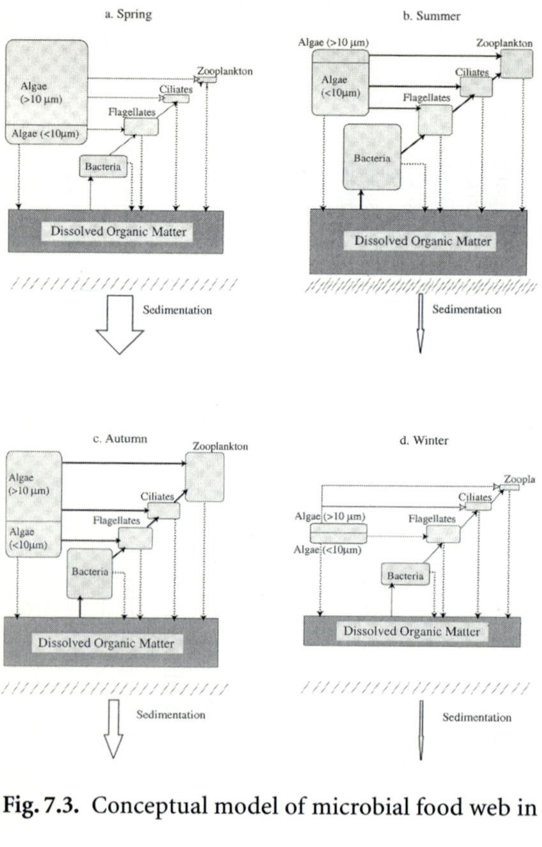

Fig. 7.3. Conceptual model of microbial food web in relation to temporal variation

7.2.2.1 Spring: High Nutrients, High Energy, Mixed Water Column

In early spring, nutrient concentrations in the water are elevated due to regeneration and mixing during winter. Nutrient repletion and sufficient light cause exponential growth of phytoplankton when low wind stress allows algae to accumulate in the upper layer. The phytoplankton that are favored by these conditions need not utilize the nutrient pools to the limit but can optimize their growth to outgrow the competition. The selective advantage for larger organisms at high nutrient levels and low temperature results in a dominance of large organisms and a short food chain (Suttle et al. 1990; Thingstad et al. 1993; Andersson et al. 1996; Grönlund et al. 1996). During spring, zooplankton are largely absent since the resting eggs of copepods hatch much later. This results in high sedimentation of the algal biomass, which is subsequently consumed by the benthic community, with an effective transfer to higher trophic levels. In the spring scenario (Fig. 7.3a) the microbial loop plays only a minor role in the microbial food web. Nevertheless, a large variety of organisms continue to be present in low numbers poised to participate in a future climax succession.

7.2.2.2 Summer: Low Nutrients, High Energy, Stratified Water Column

Characteristic of the summer situation is the low transfer of biomass to higher trophic levels and the low loss of biomass through sedimentation. During summer, the water column is stratified and upward transport of nutrients is restricted. After the spring bloom has exhausted the nutrients from the surface layer, the summer community is driven by regeneration of nutrients. Thus, a strong selection pressure towards small organisms with low sinking rates is established (Fig. 7.3b). Production of heterotrophic bacteria increases during summer and, subsequently, protozoan predation on the bacteria reaches the highest level of the year. Flagellates and larger phytoplankton are subject to predation from microzooplankton, but the net amount of biomass that is transferred via this route is small due to respiration in each of several trophic layers (Hagström et al. 1988). The carbon flow is largely a loop of respiratory losses and resupply of fixed carbon, and while gross carbon fixation is coupled to the inflow of energy, the availability of mineral nutrients determines the net production of new biomass (Eppley and Peterson 1979). The same dependency on available mineral nutrients or substrate for regeneration of nutrients is valid for heterotrophic bacteria. During summer, lack of mineral nutrients and high predation on the bacteria are causes of an accumulation of dissolved organic carbon (DOC) (Zweifel et al. 1995; Thingstad et al. 1997).

7.2.2.3 Autumn: High Nutrients, Medium Light, Mixed Water Column

During autumn, the physical conditions resemble the spring situation (Fig. 7.3c). Frequent storm activity erodes the density gradient and the increased nutrient levels give rise to a community with large phytoplankton such as cold-water diatoms and dinoflagellates throughout the Baltic Sea (Niemi 1973; Hällfors and Niemi 1981; Kononen and Niemi 1984; Smetacek et al. 1984; Andersson et al. 1996). However, the zooplankton populations that have evolved during summer are able to control the algal biomass to a larger extent than during spring. The bacterial production remains high and a significant flow of carbon is transferred to higher trophic levels via the microbial loop.

7.2.2.4 Winter: High Nutrients, Low Light, Mixed Water Column

During winter, low light and high wind energy slow down primary production to a marginal level except in the southernmost part of the Baltic. Heterotrophic organisms are therefore limited to utilizing existing organic matter (Fig. 7.3d). The standing stocks of microorganisms decrease drastically during this period and this degradation maintains a slow bacterial production throughout winter. A continuous advection from the deeper layers returns nutrients to the photic layer, setting the scene for the next spring bloom. As the water becomes ice-covered, light levels become even lower and primary production in the water comes to a complete halt. However, already in early spring, primary production takes place in the pores of the ice. In both the coastal and the offshore ice a distinct ice-biota is developed (Norrman and Andersson 1994). As the ice melts and becomes more porous the algal community washes out of the ice and mixes with the initial spring bloom. The contribution of algal production from the ice to the annual production is quite small (approx. 1 %), but its seeding effect may be important for the development of the spring bloom (Norrman and Andersson 1994).

7.3 Pelagic Production in the Baltic

7.3.1 Seasonal Cycles Along the Salinity Gradient

The shallow Bothnian Bay is characterized by low salinity (2–4 ‰), giving rise to weak stratification and poor water-column stability. After the winter period, the mixed layer contains low amounts of PO_4 (<0.1 μM PO_4-P). The low level has been attributed to the continuous inflow of iron hydroxides via the rivers that precipitate the PO_4, resulting in a high inorganic N:P ratio (>70 by atoms) (Wulff and Rahm 1988). Also, since the water column is unstable during most of the year, resuspension of the sediments effectively traps available PO_4. Thus, the annual phytoplankton production is limited by both the short growth season and a chronic shortage of PO_4. No obvious spring bloom occurs in this area, but as a temperature-induced density gradient becomes established around mid-summer a peak in algal growth follows in June–July (Fig. 7.2). During August, autotrophic production slows while heterotrophic bacterial production remains high (Kuparinen et al. 1994). Multicellular cyanobacteria blooms do not occur (Niemi 1979). Due to increased mixing, a slight increase in primary production can be seen in September–October before light becomes limiting (Andersson et al. 1993, 1996).

In the Bothnian Sea higher salinity contributes to a more stable deep-water layer. The winter level of PO_4 is approx. 0.3 μM PO_4-P and the N:P ratio is close to 10 (Wulff and Rahm 1988). This allows a larger algal biomass after the ice cover disappears. Due to the high insulation in late spring, the water temperature rises quickly and a density gradient becomes established in late May. At this time the winter supply of nutrients has been exhausted by algal growth. Sedimentation of the spring bloom results in a clear-water period (Fig. 7.4). The same clear-water period can be seen after the spring bloom in most of the Baltic Sea. After the spring bloom primary producers undergo a succession leading to the dominance of unicellular cyanobacteria (Fig. 7.5; Andersson et al. 1996). A few late warm summer weeks are sufficient to establish a regenerative microbial food web resembling a Southern oligotrophic sea area (Fig. 7.3b). Increasing storm activity in August brings nutrients up to the surface and, depending on temperature and wind conditions, larger algae gain momentum, though rarely forming blooms.

In the northern Baltic proper/Gulf of Finland, the salinity ranges from 6 to 8‰ in the surface layer. The carrying capacity of the water during spring, in terms of mineral nutrients in the winter storage of phosphate, is twice that of the Bothnian Sea (Sandén and Rahm 1993) and the inorganic N:P ratio is approx. 3 (Wulff and Rahm 1988). Similar to the Bothnian Sea, the spring bloom ends when all the mineral nutrients from above the thermocline have been assimilated and the presence of a density gradient prevents further influx. High sedimentation at the end of the spring bloom results in an obvious clear-water period (Larsson and Hagström 1982; Kuparinen et al. 1984). The clear-water period is followed by succession of small algae until late summer when dense cyanobacterial blooms of large

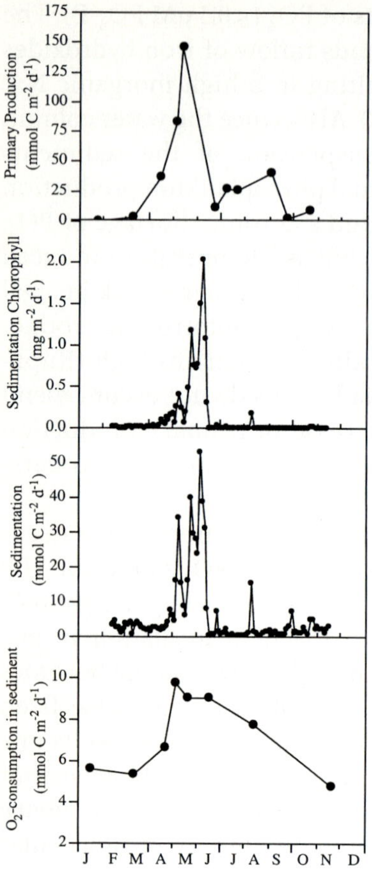

Fig. 7.4. Case study (Gulf of Bothnia 1991) of primary production, sedimentation and oxygen consumption in the southern Bothnian Sea (station SR5). (Redrawn from Kuparinen et al. 1994)

trichomatous species may develop. The intensity and areal distribution of these blooms vary greatly from year to year. These "mosaic" summer blooms resemble coastal, nutrient-spiked eutrophic localities, and contribute to the productivity of the open Baltic Sea. Contrary to what might be expected, the primary production of the offshore stations is higher than at the coastal stations in both the Bothnian Sea and the Baltic proper (Table 7.1). One explanation for this situation is the relatively shallow pycnocline allowing frequent erosion of the density gradient, which causes a patchy supply of nutrients. In the Baltic proper the occurrence of an autumn bloom is a regular feature.

In the southern part of the Baltic Sea proper, the most significant difference in mineral nutrient composition is a lower dissolved silicate (DSi) to dissolved inorganic nitrogen (DIN) ratio. Over the last 10-year period, the DSi:DIN ratio showed decreasing trends in all Baltic Sea basins.

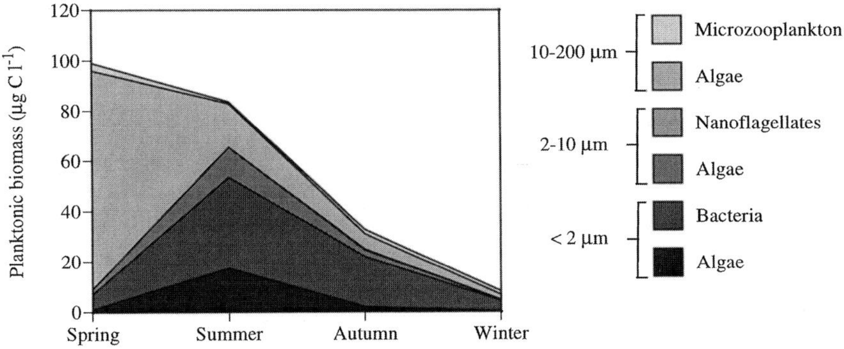

Fig. 7.5. Annual size distribution of autotrophic and heterotrophic plankton groups in the northern Bothnian Sea (station US5b), 1991–1992. (After Andersson and Rudehäll 1993)

Diatoms require a DSi:DIN ratio of about 1 and a DSi concentration above 2 µM for unlimited growth. A shift from diatoms to species that do not require DSi for growth is therefore not an unlikely scenario in the southern part of the Baltic (Rahm et al. 1996). Similar trends and development in species composition have been observed in other sea areas apart from the Baltic (Conley et al. 1993). The annual production cycle resembles that of the northern Baltic with the exception that short-lived mixing events occurring during autumn and winter are more frequent. As in the northern Baltic proper, dense chain-forming cyanobacterial blooms occur at the peak of summer. These cyanobacterial blooms are visible phenomena that can be observed on both a small scale as stringy particles in the water and in mesoscale features on satellite imagery. The importance of this locally produced carbon which remains in the pelagic systems due to low sedimentation will be discussed in a later section (see also Laakkonen et al. 1981; Kuparinen et al. 1984; Smetacek et al. 1984; Heiskanen and Kononen 1994).

7.3.2 Algal and Bacterioplankton Coupling

In open ocean ecosystems, where the bulk of organic carbon originates from phytoplankton, a close coupling between phyto- and bacterioplankton has been demonstrated (Cole et al. 1988). One would therefore expect that the north to south gradient of primary production would be reflected in bacterial production. The bacterial production in the Baltic Sea, however, shows strikingly similar levels in the different basins (Table 7.1).

Several mechanisms may interact to create this condition. In the Bothnian Bay and the coastal Bothnian Sea organic matter of riverine origin may serve to compensate low primary production in the northern Baltic. In support of this argument the carbon budgets found in these areas show a higher bacterial carbon demand than can be supported by primary production (Kuparinen et al. 1994; Zweifel et al. 1995). The data compiled here also demonstrate this feature. In the Bothnian Bay the bacterial carbon demand (calculated with conservative estimates of growth efficiency) is four-fold the available carbon produced by autotrophs on an annual basis (Table 7.2). In fact, the Bothnian Bay resembles a subarctic lake more than a marine environment and should be considered as net heterotrophic. In the Bothnian Sea the ratio becomes closer to one and in the Baltic proper the primary production may well support the bacterial carbon demand.

Another factor that may serve to even the bacterial production in the Baltic is the uniform summer condition. During spring when high nutrient levels occur, the water temperature is too low to allow significant bacterial production (Hagström and Larsson 1984). Thus, during the spring bloom the winter storage of mineral nutrients becomes exhausted, with very little accompanying bacterial production. Since most of the spring bloom is transferred to the benthos through sedimentation the pelagic system becomes "reset" to a low nutrient status before the onset of summer production. This means that primary and bacterial production during summer start with about equal conditions in all basins.

Table 7.2. Bacterial carbon demand in relation to primary production in the Baltic basins

	Bothnian Bay	Bothnian Sea	Gulf of Finland	Baltic proper
Area (km^2)	35,100	73,400	26,600	178,300
Primary production (million tons C $year^{-1}$)	0.6–1	2.1–6.8	1.7–3.4	12–29
Bacterial production (million tons C $year^{-1}$)	0.6–0.7	1.3–1.6	0.2–0.5	1.6–5.2
Carbon demand ratio (bacterial carbon demand: primary production[a])	3–4	1–3	0.5–0.6	0.5–0.7

[a] Calculated with a bacterial growth efficiency of 25 %

7.3.3 Patchiness; Temporal Versus Spatial Variation

In the previous section we have demonstrated the great temporal variation in the different basins, based on the seasonal development shown in Fig. 7.2. These annual cycles were all obtained from selected single stations along the Baltic Sea studied through at best biweekly sampling. It follows that when conclusions are drawn for the whole of each of the sea basins, questions may be raised regarding the representation of these measurements. These questions have become increasingly relevant with continuous monitoring of surface-water plankton. Especially unattended monitoring on merchant ships, combined with satellite imaginary, has revealed significant phytoplankton patchiness in both spatial and temporal scales of kilometers and days (http://www.fimr.fi/algaline/; Kahru et al. 1990; Kahru and Nômmann 1990; Leppänen and Rantajärvi 1995; Leppänen et al. 1995a,b; E. Rantajärvi and J.-M. Läppinen 1994, unpubl. data) Thus, we need to evaluate data obtained from various scales in time and space to see whether the general conclusions are affected by different sampling scales.

The shipboard analysis of chlorophyll in the whole Baltic proper transect allows us to show variation in the relevant spatial scales (Fig. 7.6). The difference in the timing of the spring bloom along the 1000-km-long transect between Helsinki and Travemünde causes a 40-fold variation in chlorophyll levels during the spring period. The spring bloom propagates in a wave-like manner basically from south to north, although local variation may change the direction of propagation over smaller areas (Kahru and Nômmann 1990). During summer and autumn, however, the spatial variation is only two- to three-fold and thus lower than the temporal variations within each basin. This means that in any limited geographical area the annual temporal variation is much greater than the variation that can be found by moving the sampling point a few kilometers in one or the other direction.

Still, the spatial variation is highly significant and it has been the object of especially Estonian and Finnish researchers to determine the causes of this variation. During the warm-water period in late summer, two fundamentally different hydrographical conditions lead to spatial variation. The first kind is found as rich mesoscale plankton patches in association with quasi-persistent salinity fronts and upwelling areas (Kahru et al. 1986; Kononen et al. 1996). The distribution of these events depends largely on the bottom topography. Nômmann et al. (1991) demonstrated the importance of coastal upwellings for the productivity of the Baltic near-shore ecosystems in the summer stage. These authors found a fast biological response to upwelling resulting in enhanced primary production and a

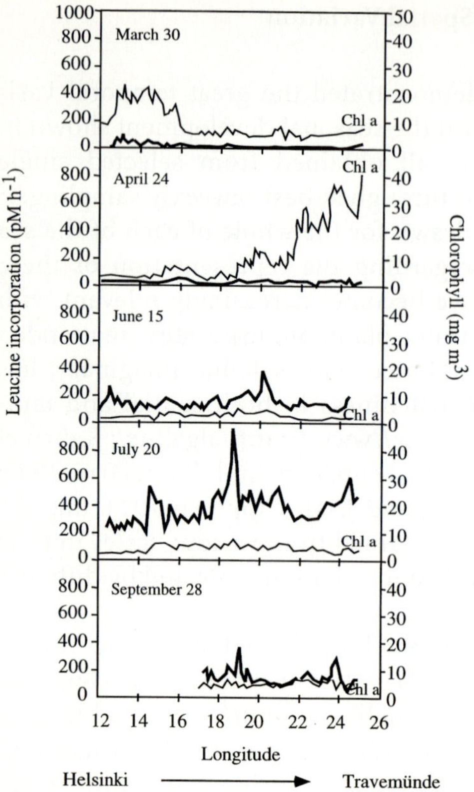

Fig. 7.6. Variability in chlorophyll *a* concentrations and bacterial production (leucine uptake) in surface layer of the Baltic Sea proper between Helsinki and Travemünde during selected transects in spring 1993. *Thin line* chl *a*; *thick line* bacterial production. Chlorophyll data redrawn from Leppänen et al. (1995a)

four- to seven-fold increase in the standing crop of some phytoplankton populations within 4 days.

A much more subtle effect of mixing is caused by breaking internal waves or shear-induced turbulence. Kahru et al. (1981) showed that these events can be recorded as the number of density inversions (>0.005 sigma-t units) in the water column. In the open Baltic this number ranges between 0 and 10 inversions in the depth interval 20–60 m. However, as these density inversions are ephemeral they merely indicate pulses of nutrient-rich water. Much work therefore remains before accurate statistical models can relate driving forces to the net flow of nutrients through the density gradient. Only through such models can we estimate the effects of changing climatic pattern. However, at present, our understanding of these processes helps to explain the mosaic pattern of phytoplankton biomass.

Leucine incorporation has been used as a measure of bacterial production on the Helsinki-Travemünde transect (Fig. 7.6). During spring, the variability is low, although mesoscale differences in bacterial abundance coupled with the developmental stage of phytoplankton blooms have been shown by Heinänen et al. (1995). During the clear-water period, spatial variability of bacterioplankton is remarkably low in the open Baltic (Fig. 7.6; see also Autio 1990; Gocke et al. 1990; Gocke and Rheinheimer 1991; Heinänen and Kuparinen 1991). Still, during the summer period bacterial production may show five-fold variation along the 1000-km transect. The levels are, however, 100-fold higher than during spring and winter and the spatial variation is thus lower than the temporal variation.

7.4 Organic Matter Distribution and Degradation

A fundamental difference between the Baltic Sea and the open oceans is the high concentration of dissolved organic carbon (DOC) in the Baltic. Compared with open sea areas, the DOC concentrations in the Baltic are 3–4 times higher. Since this feature has overall bearing on pelagic processes in the Baltic this section intends to review basic organic matter dynamics.

7.4.1 Utilizable Versus Non-degradable Organic Matter

The dissolved organic phase in seawater can be described as a dilute nutrient broth. It consists of all the organic molecules common to living matter, although in low amounts. Monomers and polymers of amino acids, carbohydrates, lipids and nucleotides have been chemically defined and quantified. The origin of these molecules is mainly "healthy" processes (exudation from algal cells and sloppy feeding during predation) but also mortality (lysis induced by virus and senescence).

A considerable amount of the organic matter in the sea, however, cannot be chemically characterized. This fraction includes "humics" and "fulvic acids" that are not pure compounds but heterogeneous mixtures of many compounds with generally similar chemical properties, typically defined as elution products from ion-exchange columns at a defined pH. Humics and fulvic acids are thought to be produced by abiotic reactions that link low-molecular-weight compounds by condensations, polymerization, oxidation and reduction (Libes 1992). Humics can be of both marine and terrestrial and origin. In areas with large influence of river inflow such as the Baltic the contribution from terrestrial humics is dominating.

From the consumers' point of view (the bacteria), DOM can be divided into two pools: utilizable and non-degradable (Fig. 7.7). The utilizable pool consists of either compounds that are readily accessible for direct assimilation by the bacteria or compounds that can be hydrolyzed by enzymes.

The individual components of non-degradable DOM, when chemically hydrolyzed, are also simple molecules such as sugars, amino acids and various ring structures and yet degradation is impossible. This contradiction can now be better understood since it has been shown that peptides added to seawater become slow to degrade within hours to days, possibly because of condensation with carbohydrates (Keil and Kirchman 1993). Bacterial enzymes have evolved towards specific types of molecular bonds, while random condensation reactions may create "new" bonds, not alike any biological entities. The number of possible combinations available for chemical condensation reactions is immense, far beyond the biological diversity determined by a set of precise biochemical reactions mediated by enzymes (Azam et al. 1993). The non-degradable pool of organic matter thus includes molecules in a size range from hundreds to hundreds of thousands of daltons.

Terrestrial carbon which largely consists of humic substances has traditionally been regarded as non-degradable regarding bacterial utilization. Recent investigations have shown, however, that humics may stimulate

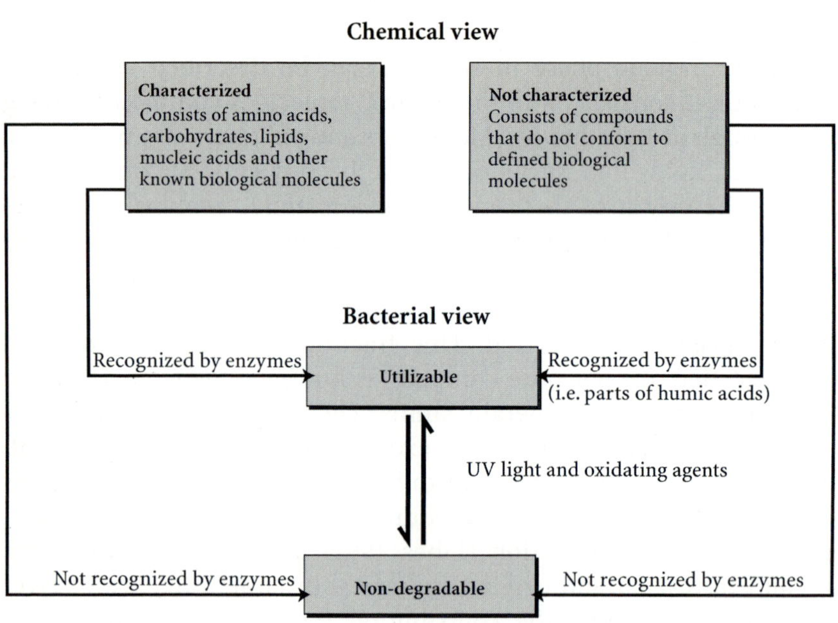

Fig. 7.7. Dissolved organic matter partitioned in functional biological and chemical compartments

bacterial production (Moran and Hodson 1990; Carlsson and Granéli 1993). This is conceivable since "humics" is a broad definition for compounds with similar chemical characteristics and may comprise both utilizable and non-degradable fractions of carbon.

The non-degradable DOM is not lost forever from the system. In the surface layer of the sea, photochemical reactions mediated by solar UV-B radiation (280–320 nm) may play a key role in converting non-degradable DOC into degradable products (Kieber et al. 1989). When irradiated with UV light, photons may be absorbed by DOM and some of these interactions lead to direct photochemical transformations of the DOC, producing low-molecular-weight (LMW) compounds. This provides a removal pathway for otherwise recalcitrant material into utilizable substrate. On the other hand, free radicals may also accelerate polymerization reactions. The relative importance of the two processes and the controlling factors in situ are presently poorly understood. These processes may be of major importance to the Baltic microbial community due to the large amount of DOM present (Allard et al. 1994; Wedborg et al. 1994).

7.4.2 The Dissolved Phase in the Baltic

The concentration of bulk DOC in the Baltic ranges from 325 μM C in the Bothnian Bay to 200 μM C in the Kattegat (Wedborg et al. 1994). Surface water values between 300 and 325 μM C are observed all the way from the Bothnian Bay down to the Kattegat where a rapid decrease in DOC is observed concurrent with a rapid increase in salinity due to mixing with Skagerrak water (Fig. 7.8). Thus, the DOC concentrations in the Bothnian Bay, Bothnian Sea and Baltic Sea proper are significantly influenced by freshwater input. In the Bothnian Bay and the Bothnian Sea, humics make up on average 64 and 43 % of the total organic carbon in the respective basin (Mäkinen et al. 1994).

Although the Baltic has a high concentration of DOC, the utilizable pool (determined from bioassays) constitutes only 4–10 % of the total DOC pool and amounts to 10–30 μM C (Zweifel et al. 1993). Both the Bothnian Bay and coastal areas of the Bothnian Sea have been shown to be net heterotrophic since bacterial production is partially fueled by carbon sources other than those produced by the autotrophs (Kuparinen et al. 1994; Zweifel et al. 1995). Temporal variations in DOC have been observed in the Bothnian Bay and the Bothnian Sea (Fig. 7.9). During summer, the DOC concentration increased by 20 % compared to the annual average. This is likely caused by mineral nutrient limitation which slows down consumption of utilizable riverine material, leading to accumulation of

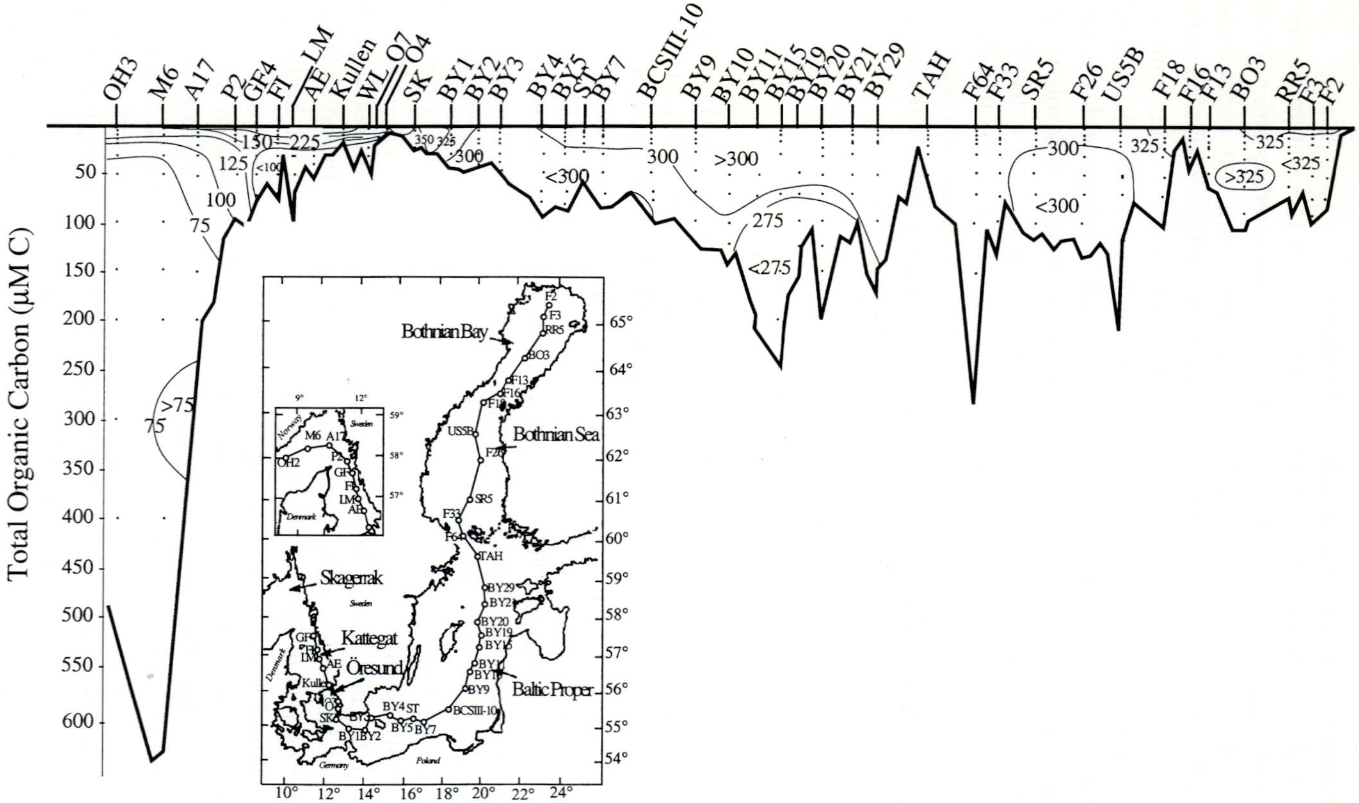

Fig. 7.8. Concentration of total organic carbon along a Baltic transect extending into the Skagerrak area, May–June 1992. *Insert* shows cruise track. (Redrawn from Wedborg et al. 1994)

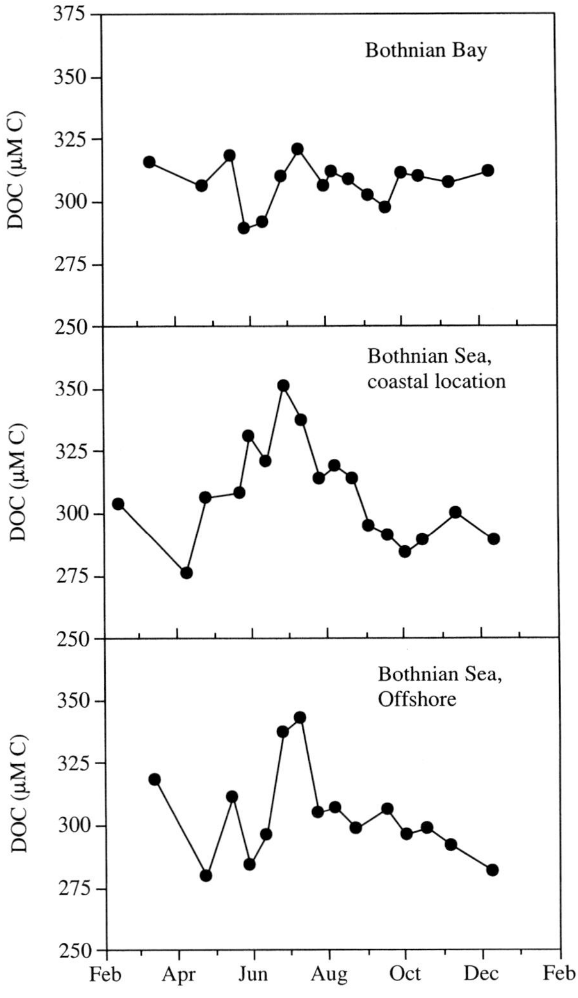

Fig. 7.9. Annual variation of dissolved organic carbon in the Bothnian Bay and northern Bothnian Sea, 1991. *Symbols* represent mean DOC concentration in water column (After Zweifel et al. 1995)

DOC (Zweifel et al. 1995). During winter, a basic DOC level of about 300 µM C is prevalent in the Bothnian Bay and Bothnian Sea. This value may reflect a non-degradable pool that is reached every year during low autotrophic production.

7.4.3 Particulate Organic Matter in the Sea

Compared to the huge pool of DOC, the particulate organic carbon (POC) pool in the Baltic constitutes only a minor fraction of total organic matter (on average 3–7 % in the open sea) (Andersson and Rudehäll 1993). In addition, less then 50 % of this pool is composed of living organisms. The remaining detritus represent an important transition in the biogeochemical pathway.

A number of "new" types of particles have been discovered during the 1990s at abundances that are orders of magnitude greater than the "classical" particles and this changes our perception of the structure of the organic matter field at the micrometer scale. Koike et al. (1990) discovered that surface seawater contains 0.3–1.3 μm organic particles in the order of 10^7 ml^{-1} not previously observed by light microscopy. Alldredge et al. (1993) discovered that surface seawater contains particles (10^2–10^4 ml^{-1}) which stain with the acid mucopolysaccharide stain alcian blue (transparent exopolymer particles, TEP). TEPs are highly variable in size (three to hundreds of micrometers) and their forms include sheets, strings and bundles of filaments. TEPs are often colonized by bacteria; 28–68 % of the bacteria were found attached to TEPs (Alldredge et al. 1993). Wells and Goldberg (1991) discovered highly abundant (10^7–10^{10} ml^{-1}) colloidal-sized (<120 nm) particles in seawater. Lately, a class of protein-rich particles staining with Comassie blue (CSP) has been described (Long and Azam 1996).

The pools of all these "new" particles are highly dynamic and the origins and degradation of these particles are believed to be dominantly biological. In the Baltic, where additional large amounts of particles enter via river inflow, much remains to be learned about the chemical composition, origins and fates of the particulate matter. Andersson and Rudehäll (1993) demonstrated that POC was dominated by unidentified detritus material at both a coastal and an offshore station in the Bothnian Sea. The degree to which this material is mineralized in the pelagic or benthic community will have important bearing on the oxygen levels in the bottoms of the Baltic. Another important consequence of sedimenting particles is the action as a biochemical filter removing organic pollutants (Andersson et al. 1993). The lipid-rich organisms serve as an efficient extraction phase accumulating xenobiotica of various origins. This effect is of special importance in relation to the vertical transport of the spring bloom.

7.5 Inorganic Nutrient Limitation

7.5.1 State of the Problem

Control of pelagic production can in most cases be viewed as a mix of two extreme conditions: bottom-up and top-down control. During bottom-up control, the availability of nutrient (food) determines the standing stock of the organism in focus, while during top-down control the grazing pressure from higher trophic levels controls the standing stocks. In the southern Kattegat, Nielsen and Kiørboe (1994) demonstrated that the ciliate community rarely or never became limited by food availability and similar results have been recorded in the Baltic in mesocosm experiments (Kuuppo-Leinikki et al. 1994). The herring→mesozooplankton→phytoplankton food web is another example of top-down control in the Baltic Sea (Rudstam et al. 1992; Arrhenius and Hansson 1993). During the spring bloom, growth of grazers is controlled due to low temperature and thus there is no predation of the algae. Instead, a bottom-up control can be observed where the winter storage of nutrients allows exponential growth of phytoplankton until nutrient exhaustion (Larsson and Hagström 1982; Kuparinen et al. 1984; Lignell et al. 1993). The least available mineral nutrient will determine the spring bloom production. To evaluate the effects of anthropogenic inputs and management measures limiting nutrients are determined.

The terms systemic and physiological limitations are often found in reference to these investigations (Thingstad and Rassoulzadegan 1995). Physiological limitation is a state where the existing organisms are unable to grow at maximum growth rate. Shortage of phosphorus, for instance, may be observed as an increased expression of the P-sequestering enzyme alkaline phosphatase. Continuous physiological limitation is restricted to areas with a very low level of the limiting substance where the prevailing species coexist under starvation conditions. Under these conditions, succession has promoted the most efficient species or group of species which do their best to grow. An increase in the supply of the limiting substance will allow the organisms to increase the per cell growth rate and an increased biomass production will result even if the number of organisms is controlled by predation. A further increase in nutrient supply will inevitably lead to a shift in species composition and a new set of equilibria between organisms will be established.

Systemic limitation on the other hand reflects the balance of nutrients in a given environment compared to the substrate requirements of the microorganisms. In this situation the organisms may grow at maximum growth rate but will respond with an increased biomass to an addition of the least available substrate. Figure 7.10 is a compilation of data from the entire Baltic depicting status of the limiting nutrient in the different sea areas. Experiments monitoring the increase in biomass (chl *a*) as the result of nutrient additions were used as a measure of systemic limitation (Granéli et al. 1990; Andersson et al. 1996; T. Tamminen et al. 1993, unpubl. data).

In most cases either N or P addition results in a clear quantitative response, but often a somewhat confusing effect of the non-limiting nutrient is also recorded. If we regard the competition for nutrients between bacteria and algae as shown in the outline of the microbial food web

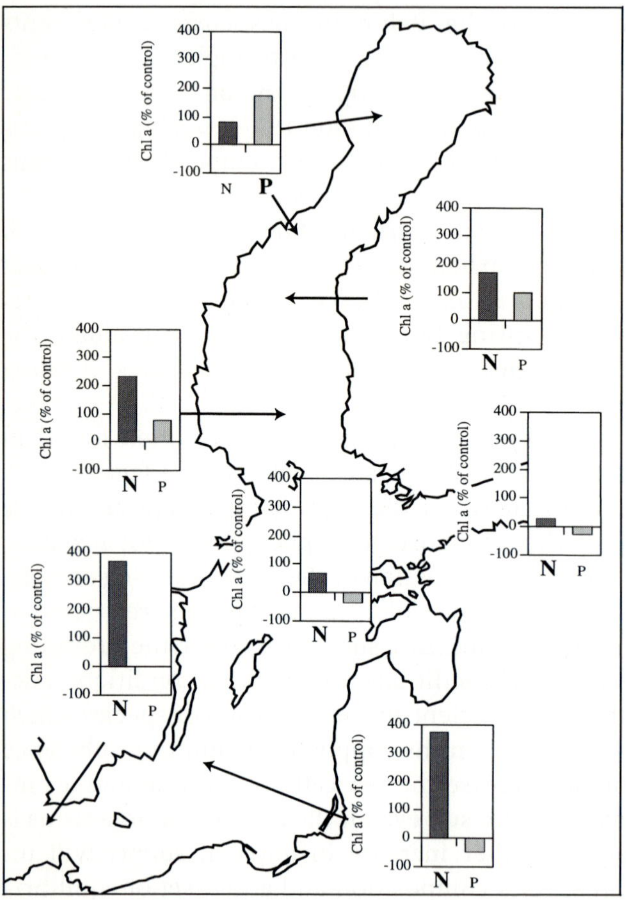

Fig. 7.10. Analysis of limiting nutrients in the Baltic Sea. *Bars* represent percent increase in chl *a* concentration compared with a control with no nutrient addition. Early summer and summer data compiled from Granéli et al. (1990); Kivi et al. (1993); Andersson et al. (1996)

(Fig. 7.1), it becomes clear that depending on the prevailing conditions several outcomes of the addition experiments are possible. In the central Baltic proper, for instance, it is quite possible that by increasing the PO_4 level the heterotrophic bacteria are more successful in the competition for the available nitrogen and thus the growth of algae measured as increased chl *a* will be lower than in the control with no additions. Another scenario is seen in the Bothnian Sea where an addition of nitrogen gives the highest response but phosphate additions also produce a lower but significant growth compared to the control. This can be ascribed to at least two equally plausible explanations: (1) heterotrophic bacteria respond to the phosphate addition by increased exoenzyme activity which results in release of mineral nitrogen from dissolved organic nitrogen which can be used by the algae; (2) fast-growing algae can out-compete heterotrophic bacteria for nitrogen when given high concentrations of PO_4. Thus, the outcome of an increased nutrient supply is the result of a complex interaction within the microbial food web.

7.5.2 Nutrient Limitation Along the Salinity Gradient

Let us examine the situation as we pass from north to south. A gradient is clearly visible in the DIN:DIP ratios, which can reach 250 in the Bothnian Bay compared to 10 in the southern Bothnian Sea and 3 in the Baltic proper. In the Bothnian Bay low phosphate concentrations and high nitrate levels exist throughout the year since phosphorus is preferentially removed from the water column by precipitation. In most cases, enrichment studies performed in the Baltic have assayed the gross increase in primary production and/or chlorophyll, thus determining the systemic limitations. In the Bothnian Bay and coastal sites in the Bothnian Sea the autotrophic community is phosphate limited (Fig. 7.10; Andersson et al. 1996). Further into the Baltic proper tests for nutrient limitations constantly show nitrogen to be the least available nutrient (Granéli et al. 1988, 1990; Andersson et al. 1996). Extensive series of nutrient-addition experiments in the coastal Gulf of Bothnia, Archipelago Sea and the Gulf of Finland demonstrate the same trend of increasing nitrogen deficiency towards the south (T. Tamminen and K. Kivi 1996, unpubl. data). A similar trend can be seen from the pollution-affected Gulfs (Gulf of Finland, Gulf of Riga) out to the open sea (T. Tamminen and K. Kivi 1996, unpubl. data).

Experiments with bacterioplankton predator-free cultures have shown that during the season of regenerated production, bacterial growth is limited by phosphorus in coastal areas of the Bothnian Sea (Zweifel et al. 1993) and by nitrogen in the Gulf of Finland and in the Kiel Bight

(Kuparinen and Heininen 1993). Thus, although the data are limited, the bacterial community in the Baltic is likely to experience the same north to south gradient of P to N limitation as the phytoplankton community.

7.6 Why Cyanobacteria Blooms in the Baltic?

Planktonic filamentous nitrogen-fixing cyanobacteria are common in eutrophic lakes, rare in open oceans and nearly absent in estuaries and coastal seas (Pearl 1988). Yet, significant estuarine blooms of cyanobacteria are regularly recorded in the Baltic Sea and in a few other sites such as the Peel-Harvey estuary in Australia (Niemi 1979; Huber 1986; Howarth et al. 1988; Kononen 1992). The mass appearance of nitrogen-fixing cyanobacteria in eutrophic lakes is related to the ratio of nitrogen to phosphorus loading. An N:P ratio near or below the Redfield ratio of 16:1 gives the selective advantage for the phytoplankton species that may fix nitrogen. Although this requirement can be found in many estuaries, planktonic nitrogen fixation in estuaries is generally low even when the N:P input ratio is low. One explanation that has been offered is the additional requirement for trace elements like molybdenum (Capone and Carpenter 1982) and iron (Ferreira and Staus 1993). These elements are lacking in most estuarine and coastal seas, but the high concentration of dissolved organic matter in the Baltic Sea and Peel-Harvey may increase the availability of iron and molybdenum through chelation (Pearl et al. 1987).

At the onset of blooms of filamentous cyanobacteria the phytoplankton community in the Baltic Sea is nitrogen limited (Tamminen et al. 1985) and dominated by pico- and nano-sized organisms. The competitive advantage for pico-sized organisms is broken when physical forcing generates nutrient pulses into the heated shallow upper layer (Rothhaupt and Güde 1992; Thingstad et al. 1993; Kononen et al. 1996). Of the two filamentous cyanobacteria *Nodularia spumigena* and *Aphanizomenon flos-aquae*, the former has the highest affinity for phosphate, but with its greater ability to store phosphorus *Aphanizomenon flos-aquae* initially benefits from the nutrient pulses and is usually the first nitrogen fixer to occur in large numbers (Sommer 1985; Wallström et al. 1992).

Later, however, *Nodularia* takes over and the massive cyanobacterial blooms that are formed in the frontal system at the mouth of the Gulf of Finland and elsewhere consist of *Nodularia spumigena* (Kononen et al. 1996). The scales of appearance of *Nodularia spumigena* in kilometers and hundreds of kilometers shortly after the initialization of the bloom cannot be explained by localized nutrient pulses and advection velocities. Instead, the selection for *Nodularia* is generated by the ability to maintain an active microbial food web within the aggregates. Since phosphorus input to the planktonic community via turbulent mixing is a temporary event, the persistence and development of the bloom must rely on regeneration of phosphorus. Phosphorus is stored in dissolved and particulate organic phosphorus and its regeneration is controlled by the turnover in the microbial food web.

At the initiation of a bloom both *Nodularia spumigena* and *Aphanizomenon flos-aquae* appear as uncoiled, long chains of cells. In later stages of the bloom development, *Nodularia* forms large aggregates due to coiling (Fig. 7.11). These coils provide the means for the *Nodularia* to support an active microbial food web in order to benefit from phosphorus regeneration events. *Aphanizomenon flos-aquae* which produces aggregates of tight bundles of cell treads cannot harbor the same diverse microscopic "epifauna" (Fig. 7.12). Therefore, in absence of further hydrodynamic events, *Aphanizomenon flos-aquae* has to give room to *Nodularia spumigena* due to the lack of effective phosphorus regeneration.

The argument that aggregates harbor an active microbial food web was initially presented in relation to degradation of marine snow. Marine particulate phase harbors rich bacterial populations in the order of 10^8–10^9 ml^{-1}, three orders of magnitude greater than in the surrounding waters (Alldredge and Gotschalk 1990). Because of high activities of hydrolytic enzymes and loose hydrolysis-uptake coupling, particle-bound bacteria

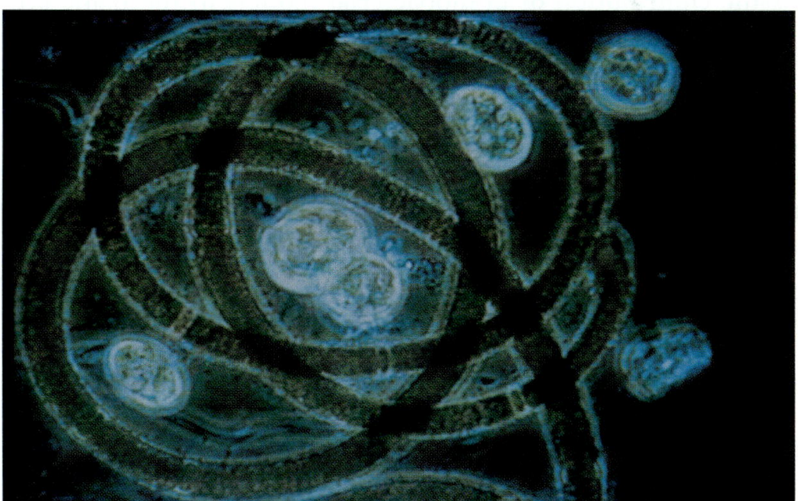

Fig. 7.11. Aggregates of coiled *Nodularia spumigena* with attached microorganisms. (Photo: K. Kononen)

Fig. 7.12. Tight bundles of *Aphanizomenon flos-aquae*. (Photo: K. Kononen)

can render the particle a source of DOM and mineral nutrients (Smith et al. 1992). Aggregates can contain very high concentrations of nutrients (Shanks and Trent 1979), rendering them and the surrounding microenvironment a "hot spot" of remineralized nutrients.

What is the contribution of cyanobacterial blooms to the Baltic ecosystem? In the nitrogen flux cyanobacteria are responsible for most of the planktonic nitrogen fixation, but the rates are high only when these organisms make up a majority of the biomass (Howarth et al. 1988). In eutrophic lakes planktonic nitrogen fixation may account for 6–82 % of the nitrogen input into the system. In the Baltic Sea, however, planktonic nitrogen fixation provides 15 % of the nitrogen input to the Baltic proper (Larsson et al. 1985). This amount is less than the nitrogen deposition from the air, but in the annual budget it is significant and must be considered when trends in the nitrogen pools are considered.

Apart from the toxicity aspect of the cyanobacterial blooms, little is known about the contribution of blooms to the carbon flux. Although visually massive, the cyanobacteria may contribute only little to the pelagic carbon flux as the two most common cyanobacteria species, *Nodularia spumigena* and *Aphanizomenon flos-aquae*, are poorly eaten (only a few percent of the biomass) by mesozooplankton (Haney 1987; Sellner et al. 1994). Little or negligible sedimentation of the summer plankton including cyanobacteria has been recorded in various basins (Smetacek 1980;

Kuparinen et al. 1984; Blomqvist and Larsson 1994; Heiskanen and Kononen 1994); thus the impact on benthic biota seems negligible compared to spring and autumn blooms (Smetacek 1980; Kuparinen et al. 1994). During the *Aphanizomenon flos-aquae* and *Nodularia spumigena* bloom period <1 % of the suspended concentrations settled daily (Heiskanen and Kononen 1994), which made the authors conclude that the cyanobacterial bloom was disintegrated and decomposed in the water column into the dissolved and particulate organic pools and thus incorporated into the food web via a microbial loop. However, in *Nodularia spumigena*-dominant blooms, a highly active microbial food web may develop within *Nodularia* aggregates and thus contribute to pelagic plankton community (Tamminen et al. 1985).

7.7 Management of Nutrient Release

Over the last 25 years, the countries around the Baltic have battled eutrophication by wastewater treatment, restricted use of fertilizers and catalytic cleaning of combustion. Still the total deposition to the Baltic Sea has increased (Sandén and Rahm 1993). The question facing governments and environmental protection agencies is what to do next. In the past, lake management has successfully handled eutrophication in lakes based on the concept of reducing the limiting nutrient and thus reducing excess growth. Is this concept valid also for the treatment of the Baltic? The answer is – probably not. On a small scale, the result of such action has been demonstrated already when the water treatment in the Stockholm area was tightened and phosphorus effectively removed (Brattberg 1986). The result was an export of nitrogen out in the archipelago towards the open sea. The excess nitrogen then met phosphorus further out in the archipelago, causing massive growth of annual green alga covering the *Fucus*. This front could be observed in enrichment tests (Brattberg 1986). These results can be transferred directly to basin-scale considerations. In the Bothnian Bay, where phosphate is limiting, it may be argued from a management point of view that, if the phosphate release in kept low, the system will remain in balance. In the prevailing condition, however, excess nitrogen is transported to the Bothnian Sea. If instead a monitored increase in phosphate would be supplied to the Bothnian Bay, this might result in larger standing stocks of organisms or a selection towards new organisms and reduced levels of inorganic nitrogen in the water. If such an increased biomass also resulted in an increased burial of sedimenting material, a reduction in the transport of nitrogen from the Bothnian Bay to the Bothnian Sea would occur. Whether such an experiment would result in changes towards "desired" species is, however, not possible to predict.

For the Baltic as a whole the most challenging task is to accurately determine a nutrient budget. Two major subjects make estimates of biological production uncertain. First, the net regeneration of organic matter rich in nitrogen and phosphorus being transported from land to the sea is difficult to determine. Complex models are needed to resolve the competition between autotrophs and heterotrophs for inorganic nutrients and thus whether or not increased secondary production via the microbial loop may accelerate eutrophication. Second, during summer, advection and upwelling events are important factors controlling the flow of nutrients through the system. Presently, no models exist that allow an accurate estimate of this inflow. Thus, trends based on productivity assays are virtually impossible to interpret without the possibility to tell anthropogenic input apart from the effects of short-term changes in climate.

References

Ærtebjerg G, Albjerg J, Andersson L, Behrends G, Dahl K, Giesenhagen H, Hansen H-P, Josefson AB, Powilleit M, Rydberg L, Wikner J (1996) State of the marine environment of the Kattegat and Belt Sea. In: Ærtebjerg G (ed) Third periodic assessment of the state of the marine environment of the Baltic Sea area 1989–1993; background document. Baltic Marine Environment Protection Commission–Helsinki Commission, Helsinki

Allard B, Borén H, Pettersson C, Zhang G (1994) Degradation of humic substances by UV irridation. Environ Int 20:97–101

Alldredge AL, Gotschalk CC (1990) The relative contribution of marine snow of different origins to biological processes in coastal waters. Cont Shelf Res 10:41–58

Alldredge AL, Passow U, Logan BE (1993) The abundance and significance of a class of large, transparent organic particles in the ocean. Deep Sea Res 40:1131–1140

Ammerman JW, Azam F (1985) Bacterial 5′-nucleotidase in aquatic ecosystems: a novel mechanism of phosphorus regeneration. Science 227:1338–1340

Andersson A, Rudehäll Å (1993) Proportion of plankton biomass in particulate organic carbon in the northern Baltic Sea. Mar Ecol Prog Ser 95:133–139

Andersson A, Selstam E, Hagström Å (1993) Vertical transport of lipid in seawater. Mar Ecol Prog Ser 98:149–155

Andersson A, Hajdu S, Haecky P, Kuparinen J, Wikner J (1996) Succession and growth limitation of phytoplankton in the Gulf of Bothnian (Baltic Sea). Mar Biol 126:791–801

Arrhenius F, Hansson S (1993) Food consumption of larval, young and adult herring and sprat in the Baltic Sea. Mar Ecol Prog Ser 96:125–137

Autio RM (1990) Pelagic bacterioplankton in the Baltic proper. Arch Hydrobiol Beih 34:217–226

Azam F, Fenchel T, Field JG, Gray JS, Meyer-Reil LA, Thingstad F (1983) The ecological role of water-column microbes in the sea. Mar Ecol Prog Ser 10:257–263

Azam F, Smith DC, Steward GF, Hagström Å (1993) Bacteria-organic matter coupling and its significance for oceanic carbon cycling. Microbial Ecol 28:167–179

Bergström S, Carlsson B (1994) River runoff to the Baltic Sea: 1950–1990. Ambio 23:280–287

Blackburn N, Zweifel UL, Hagström Å (1996) Cycling of dissolved organic matter: II. A model analysis. Aquat Microb Ecol 11:79–90

Blomqvist S, Larsson U (1994) Detrital bedrock elements as tracers of settling resuspended particulate matter in a coastal area of the Baltic Sea. Limnol Oceanogr 39:880–896

Brattberg G (1986) Decreased phosphorus loading changes phytoplankton composition and biomass in the Stockholm archipelago. Vatten 42:141–153

Capone DG, Carpenter EJ (1982) Nitrogen fixation in the marine environment. Limnol Oceanogr 39:1140–1142

Carlsson P, Granéli E (1993) Availability of humic bound nitrogen for coastal phytoplankton. Estuarine Coast Shelf Sci 36:433–447

Caron DA, Goldman JC, Andersen OK, Dennett MR (1985) Nutrient cycling in a microflagellate food chain: II. Population dynamics and carbon cycling. Mar Ecol Prog Ser 24:243–254

Cole JJ, Findlay S, Pace ML (1988) Bacterial production in fresh and saltwater ecosystems: a cross-systems overview. Mar Ecol Prog Ser 43:1–10

Conley DJ, Schelske CL, Stoermer EF (1993) Modification of the biogeochemical cycle of silica with eutrophication. Mar Ecol Prog Ser 101:179–192

Eppley RW, Peterson BJ (1979) Particulate organic matter flux and planktonic new production in the deep ocean. Nature 282:677–680

Fenchel T (1982a) Ecology of heterotrophic microflagellates. I. Some important forms and their functional morphology. Mar Ecol Prog Ser 8:211–223

Fenchel T (1982b) Ecology of heterotrophic microflagellates. II. Bioenergetics and growth. Mar Ecol Prog Ser 8:225–231

Fenchel T (1982 c) Ecology of heterotrophic microflagellates. III. Adaption to heterogeneous environments. Mar Ecol Prog Ser 9:25–33

Fenchel T (1982d) Ecology of heterotrophic microflagellates. IV. Quantitative occurrence and importance as bacterial consumers. Mar Ecol Prog Ser 9:35–42

Ferreira F, Straus NA (1993) Iron deprivation in cyanobacteria. In: Parker BC (ed) Microalgal biotechnology and commercial applications. Colloq 47th Annu Meet Phycological Society of America, Inc, 3–4 August, 1993. J Appl Phycol 6:199–210

Fuhrman JA (1992) Bacterioplankton roles in cycling of organic matter: the microbial food web. In: Falkowski PG, Woodhead AD (eds) Primary productivity and biogeochemical cycles in the sea. Plenum Press, New York, pp 361–383

Gocke K, Rheinheimer G (1991) Influence of eutrophication on bacteria in two fjords of the Western Baltic. Int Rev Gesamten Hydrobiol 76:371–385

Gocke K, Heinänen A, Kirstein K-O, Maciejowska M, Panov G, Tsiban A (1990) Second periodic assessment of the state of the marine environment of the Baltic Sea 1984–1988: microorganisms. Baltic Sea Environ Proc 35B. Helsinki Commission, pp 303–329

Granéli E, Schulz S, Schiewer U, Gedziorwska D, Kaiser W, Plinski M (1988) Is the same nutrient limiting potential phytoplankton biomass formation in different coastal areas of the southern Baltic? Kieler Meeresforsch Sonderh 6:191–202

Granéli E, Wallström K, Larsson U, Granéli W, Elmgren R (1990) Nutrient limitation of primary production in the Baltic Sea area. Ambio 19:142–151

Grönlund L, Leppaenen J-M (1990) Long-term changes in the nutrient reserves and pelagic production in the western Gulf of Finland. Finn Mar Res 257:15–27

Grönlund L, Kononen K, Lahdes E, Maekelae K (1996) Community development and modes of phosphorus utilization in a late summer ecosystem in the central Gulf of Finland. Hydrobiologia 331:97–108

Hagström Å, Larsson U (1984) Diel and seasonal variation in growth rates of pelagic bacteria. In: Hobbie JE, Williams PJL (eds) Heterotrophic activity in the sea. Plenum, New York, pp 249–262

Hagström Å, Azam F, Andersson A, Wikner J, Rassoulzadegan F (1988) Microbial loop in an oligotrophic pelagic marine ecosystem: possible roles of cyanobacteria and nanoflagellates in the organic fluxes. Mar Ecol Prog Ser 49:171–178

Hällfors G, Niemi Å (1981) Vegetation and primary production. In: Voipio A (ed) The Baltic Sea. Elsevier, Amsterdam, pp 220–238

Hällfors G, Niemi A (1986) Views on the use of phytoplankton as a parameter in monitoring the state of the Baltic Sea. Baltic Sea Environ Proc 19:246–258

Haney JF (1987) Field studies on zooplankton–cyanobacteria interactions. NZ J Mar Freshwater Res 21:467–475

Heinänen A, Kuparinen J (1991) Horizontal variation of bacterioplankton in the Baltic Sea. Appl Environ Microbiol 57:3150–3155

Heiskanen A-S, Kononen K (1994) Sedimentation of vernal and late summer phytoplankton communities in the coastal Baltic Sea. Arch Hydrobiol 131:175–198

Heinänen A, Kononen K, Kuosa H, Kuparinen J, Mäkelä K (1995) Bacterioplankton growth associated with physical fronts during a cyanobacterial bloom. Mar Ecol Prog Ser 116:233–245

Hollibaugh JT, Azam F (1983) Microbial degradation of dissolved proteins in seawater. Limnol Oceanogr 28:1104–1116

Howarth RW, Marino R, Lane J, Cole JJ (1988) Nitrogen fixation in freshwater, estuarine and marine ecosystems. 1. Rates and importance. Limnol Oceanogr 33:669–687

Huber AL (1986) Nitrogen fixation by *Nodularia spumigena* Mertens (Cyanobacteriaceae). 1. Field studies and the contribution of blooms to the nitrogen budget of the Peel-Harvey Estuary, Western Australia. Hydrobiology 131:193–203

Johannes RE (1965) Influence of marine protozoa on nutrient regeneration. Limnol Oceanogr 10:434–442

Kahru M, Nômmann S (1990) Phytoplankton spring bloom in the Baltic Sea in 1985–1986: multitude of spatio-temporal scales. J Cont Shelf Res 10:329–354

Kahru M, Aitsam A, Elken J (1981) Coarse-scale spatial structure of phytoplankton standing crop in relation to hydrography in the open Baltic Sea. Mar Ecol Prog Ser 5:311–318

Kahru M, Nômmann S, Simm M, Vilbaste K (1986) Plankton distributions and processes in the Baltic boundary zones. In: Nihoul JCJ (ed) Marine interfaces ecohydrodynamics. Elsevier, Amsterdam, pp 273–293

Kahru M, Leppänen J-M, Nômmann S, Passow U, Postel L, Schulz S (1990) Spatio-temporal mosaic of the phytoplankton spring bloom in the open Baltic Sea in 1986. Mar Ecol Prog Ser 66:301–309

Keil RG, Kirchman DL (1993) Dissolved combined amino acids: chemical form and utilization by marine bacteria. Limnol Oceanogr 38:1256–1270

Kieber DJ, McDaniel J, Mopper K (1989) Photochemical source of biological substrates in seawater: implications of carbon cycling. Nature 341:637–639

Kivi K, Kaitala S, Kuosa H, Kuparinen J, Leskinen E, Lignell R, Marcussen B, Taminnen T (1993) Nutrient limitation and grazing control of the Baltic plankton community during annual succession. Limnol Oceanogr 38:893–905

Koike I, Shigemitsu H, Kazuki T, Kogure K (1990) Role of submicrometer particles in the ocean. Nature 345:242–244

Kononen K (1988) Phytoplankton summer assemblages in relation to environment factors at the entrance to the Gulf of Finland during 1972–1985. Kieler Meeresforsch Sonderh 6:281–294

Kononen K (1992) Dynamics of the toxic cyanobacterial blooms in the Baltic Sea. Finn Mar Res 261:3–36

Kononen K, Niemi Å (1984) Long-term variation of the phytoplankton composition at the entrance to the Gulf of Finland. Ophelia 3:101–110

Kononen K, Kuparinen J, Maekelae K, Laanemets J, Pavelson J, Nômmann S (1996) Initiation of cyanobacterial blooms in a frontal region at the entrance to the Gulf of Finland, Baltic Sea. Limnol Oceanogr 41:98–112

Kuosa H, Kivi K (1989) Bacteria and heterotrophic flagellates in the pelagic carbon cycle in the northern Baltic Sea. Mar Ecol Prog Ser 53:93–100

Kuparinen J (1984) Annual and seasonal fluctuation of primary productivity and overall respiration in a pelagic plankton community off Tvärminne, SW coast of Finland. Ophelia 3:111–122

Kuparinen J, Heinänen A (1993) Inorganic nutrient and carbon controlled bacterioplankton growth in the Baltic Sea. Estuarine Coast Shelf Sci 37:271–285

Kuparinen J, Leppänen J-M, Sarvala J, Sundberg A, Virtanen A (1984) Production and utilization of organic matter in a Baltic ecosystem off Tvärminne, southwest coast of Finland. Rapp P-v Réun Cons Int Explor Mer 183:180–192

Kuparinen J, Leonardsson K, Mattila J, Wikner J (1994) Food web structure, carbon flow and trends in the Gulf of Bothnia, Baltic Sea. Vatten 50:201–219

Kuuppo-Leinikki P, Autio R, Hällfors S, Kuosa H, Kuparinen J, Pajuniemi R (1994) Trophic interactions and carbon flow between picoplankton and protozoa in pelagic enclosures manipulated with nutrients and a top predator. Mar Ecol Prog Ser 107:89–102

Laakkonen A, Mälkki P, Niemi Å (1981) Studies on the sinking, degradation and sedimentation of organic matter off Hanko peninsula, entrance to the Gulf of Finland. Meri 9:1–42

Larsson U, Hagström Å (1979) Phytoplankton exudate release as an energy source for the growth of pelagic bacteria. Mar Biol 52:199–206

Larsson U, Hagström Å (1982) Fractionated phytoplankton primary production, exudate release and bacterial production in a Baltic eutrophication gradient. Mar Biol 67:57–70

Larsson U, Elmgren R, Wulff F (1985) Eutrophication and the Baltic Sea: causes and consequences. Ambio 14:9–14

Lassig J, Leppaenen J-M, Niemi A, Tamelander G (1978) Phytoplankton primary production in the Gulf of Bothnia in 1972–1975 as compared with other parts of the Baltic Sea. Finn Mar Res 244:101–115

Leppänen JM, Rantajärvi E (1995) Unattended recording of phytoplankton and supplemental parameters on board merchant ships – an alternative to the conventional algal monitoring programmes in the Baltic Sea. In: Lassus P, Arzul G, Erard E, Gentien T, Marcaillore C (eds) Harmful marine algal blooms. Lavoisier Science, Paris, pp 719–724

Leppänen JM, Rantajärvi E, Hällefors S, Kruskopf M, Laine V (1995a) Unattended monitoring of potentially toxic phytoplankton species in the Baltic Sea in 1993. J Plankton Res 17:891–902

Leppänen JM, Rantajärvi E, Maunumaa M, Larinmaa M, Pajala J (1995b) Unattended algal monitoring system – a high resolution method for detection of phytoplankton blooms in the Baltic Sea. Ocean 94 Proc, IEEE, New York, pp 461–463

Leppäranta M, Palosuo E, Grönvall H, Kalliosaari S, Seinä A, Peltola J (1988) Phases of the ice season in the Baltic Sea (north of latitude 57°N). Finn Mar Res 254(Suppl 2):1–83

Libes SM (1992) The production and destruction of organic compounds in the sea. In: An introduction to marine biogeochemistry. Wiley, New York, pp 394–422

Lignell R (1990) Excretion of organic carbon by phytoplankton: its relation to algal biomass, primary productivity and bacterial secondary productivity in the Baltic Sea. Mar Ecol Prog Ser 68:85–99

Lignell R, Heiskanen A-S, Kuosa H, Gundersen K, Kuuppo-Leinikki P, Pajuniemi R, Uitto A (1993) Fate of a phytoplankton spring bloom: sedimentation and carbon flow in the planktonic food web in the northern Baltic. Mar Ecol Prog Ser 94:239–252

Long RA, Azam F (1996) Abundant protein-containing particles in the sea. Aquat Microb Ecol 10:213–221

Mäkinen I, Poutanen E-L, Manninen P (1994) Distribution of organic matter in water and sediment in the northern Baltic Sea. In: Senesi N, Miano TM (eds) Humic substances in the global environment and implications on human health. Elsevier Science, Amsterdam, pp 857–862

Matthaeus W, Frank H (1988) The seasonal nature of major Baltic inflows. Kieler Meeresforsch Sonderh 6:64–72

Moran MA, Hodson RE (1990) Bacterial production on humic and nonhumic components of dissolved organic carbon. Limnol Oceanogr 35:1744–1756

Nielsen TG, Kiørboe T (1994) Regulation of zooplankton biomass and production in a temperate, coastal ecosystem. 2. Ciliates. Limnol Oceanogr 39:508–519

Niemi Å (1973) Ecology of phytoplankton in the Tvärminnne area, SW coast of Finland. Acta Bot Fenn 100:1–68

Niemi Å (1979) Blue-green algal blooms and N:P ratio in the Baltic Sea. Acta Bot Fenn 110:57–61

Nômmann S, Sildam J, Nôges T, Kahru M (1991) Plankton distribution during a coastal upwelling event off Hiiumaa, Baltic Sea: impact of short-term flow field variability. Cont Shelf Res 11:95–108

Norrman B, Andersson A (1994) Development of ice biota in a temperate sea area (Gulf of Bothnia). Polar Biol 14:531–537

Pearl HW (1988) Nuisance phytoplankton blooms in coastal, estuarine, and inland waters. Limnol Oceanogr 33:823–847

Pearl HW, Crocker KM, Prufert LE (1987) Limitation of N_2 fixation in coastal marine waters: relative importance of molybdenum, iron, phosphorus, and organic matter availability. Limnol Oceanogr 32:525–536

Preisig HR, Hibberd DJ (1984) Virus-like particles and endophytic bacteria in *Paraphysomonas* and *Chromophysomonas* (Chrysophyceae). Nord J Bot 4:279–285

Proctor LM, Fuhrman JA (1990) Viral mortality of marine bacteria and cyanobacteria. Nature 343:60–62

Rahm L, Conley D, Sandén P, Wulff F, Stålnacke P (1996) Time series analysis of nutrient inputs to the Baltic Sea and changing DSi:DIN ratios. Mar Ecol Prog Ser 130: 221–228

Rothhaupt KO, Güde H (1992) The influence of spatial and temporal concentration gradients on phosphate partitioning between different size fractions of plankton: further evidence and possible causes. Limnol Oceanogr 37:739–749

Rudstam LG, Hansson S, Johansson S, Larsson U (1992) Dynamics of planktivory in a coastal area of the northern Baltic Sea. Mar Ecol Prog Ser 80:159–173

Sandén P, Rahm L (1993) Nutrient trends in the Baltic Sea. Environmetrics 4:75–103

Sellner KG, Olson MM, Kononen K (1994) Copepod grazing in a summer cyanobacteria bloom in the Gulf of Finland. Hydrobiology 292/293:249–254

Shanks AL, Trent JD (1979) Marine snow: microscale nutrient patches. Limnol Oceanogr 24:850–854

Smetacek V (1980) Annual cycle of sedimentation in relation to plankton ecology in western Kiel Bight. Ophelia (Suppl 1):65–76

Smetacek V, von Bodungen B, Knoppers B, Peinert R, Pollehne F, Stegman P, Zeitzschel B (1984) Seasonal stages in characterizing the annual cycle of an inshore pelagic system. Rapp P-v Réun Cons Int Explor Mer 183:125–135

Smith DC, Simon M, Alldredge AL, Azam F (1992) Intense hydrolytic enzyme activity on marine aggregates and implications for rapid particle dissolution. Nature 359:139–142

Sommer U (1985) Comparison between steady state and nonsteady state competition: experiments with natural phytoplankton. Limnol Oceanogr 30:335–346

Steward GF, Wikner J, Cochlan WP, Smith DC, Azam F (1992) Estimation of virus production in the sea: II. Field results. Mar Microb Food Webs 6:79–90

Suttle CA, Fuhrman JA, Capone DG (1990) Rapid ammonium cycling and concentration-dependent partitioning of ammonium and phosphate: implications for carbon transfer in planktonic communities. Limnol Oceanogr 35:424–433

Tamminen T, Kaitala S, Kivi K, Kuparinen J (1985) Response of a planktonic brackish water community to single and combined additions of ammonium and phosphate in a factorial mesocosm experiment. In: Gray JS, Christiansen ME (eds) Marine biology of polar regions and effects of stress on marine organisms. Wiley, New York, pp 363–378

Thingstad TF, Rassoulzadegan F (1995) Nutrient limitations, microbial food webs, and 'biological C-pumps': suggested interactions in a P-limited Mediterranean. Mar Ecol Prog Ser 117:299–306

Thingstad TF, Skjoldal EF, Bohne RA (1993) Phosphorus cycling and algal–bacterial competition in Sandsfjord, western Norway. Mar Ecol Prog Ser 99:239–259

Thingstad TF, Hagström Å, Rassoulzadegan F (1997) Accumulation of degradable DOC in surface waters: caused by a "malfunctioning" microbial loop? Limnol Oceanogr 42:398–404

Wallström K, Johansson S, Larsson U (1992) Effects of nutrient enrichment on planktonic blue–green algae in the Baltic Sea. Acta Phytogeogr Suec 78:25–31

Wedborg M, Skoog A, Fogelqvist E (1994) Organic carbon and humic substances in the Baltic Sea, Kattegat, and the Skagerrak. In: Sense N, Miano TM (eds) Humic substances in the global environment and implications on human health. Elsevier, Amsterdam, pp 917–924

Wells ML, Goldberg ED (1991) Occurrence of small colloids in sea water. Nature 353: 342–344

Wikner J (1996) State of the marine environment of the Baltic Sea regions. Gulf of Bothnia. HELCOM, 1996. Third periodic Assessment of the State of the Marine Environment of the Baltic Sea, 1989–1993; background document. Baltic Sea, Environ Proc no 64B

Wikner J, Hagström Å (1991) Annual study of bacterioplankton community dynamics. Limnol Oceanogr 36:1313–1324

Williams P (1981) Microbial contributions to overall marine plankton metabolism: direct measurements of respiration. Oceanol Acta 4:359–370

Wulff F, Rahm L (1988) Long-term, seasonal and spatial variations of nitrogen, phosphorus and silicate in the Baltic: an overview. Mar Environ Res 26:19–37

Wulff F, Ærtebjerg G, Nicolaus G, Niemi Å, Ciszewski P, Shulz S, Kaiser W (1987) Pelagic biology. In: Lassig J (ed) First periodic assessment of the state of the marine environment of the Baltic Sea area 1980–1985; background document. Baltic Sea Environ Proc 17B. Baltic Marine Environment Protection Commission – Helsinki Commission, Helsinki, pp 216–255

Zweifel UL, Norrman B, Hagström Å (1993) Consumption of dissolved organic carbon by marine bacteria and demand for inorganic nutrients. Mar Ecol Prog Ser 101:23–32

Zweifel UL, Wikner J, Hagström Å, Lundberg E, Norrman B (1995) Dynamics of dissolved organic carbon in a coastal ecosystem. Limnol Oceanogr 40:299–305

8 The Challenge of Sedimentation in the Baltic Sea

S. Blomqvist and A.-S. Heiskanen

8.1 Introduction

This chapter focuses on the sinking of particulate matter through the water column of the Baltic Sea. This downward flux represents a loss from the pelagic system, as well as a link between the pelagic and benthic realms, transferring matter and energy to the deep water and to the sediments. From an applied perspective, knowledge of the factors governing formation and subsequent sedimentation of the particulate matter is vital for a proper understanding of changes in the pelagic and benthic realms of the Baltic Sea. We provide here a brief overview of those processes of sedimentation that have received particular attention in Baltic studies, and address some critical current issues related to this topic, and hope to encourage further efforts in this field of Baltic Sea research.

8.2 Retrospect in Brief

Since the early pioneer studies of settling fine-particulate matter in the Baltic proper by Zeitzschel (1965), numerous measurements have been reported. In practice, studies on this topic have been based on interception by means of sediment traps. The objectives of these investigations have varied, as reflected in a wide variety of analyses of the intercepted matter, for instance, of the content of nutrients, metals, organic compounds and various contaminants.

8.2.1 Annual Dynamics

In the 1970s, sedimentation of particulate matter was recognized as a key process in ecosystem research and modeling of the Baltic Sea (e.g., Jansson

Ecological Studies, Vol. 148
F. Wulff et al. (eds.)
A Systems Analysis of the Baltic Sea

1972), and sediment traps were deployed to investigate the temporal variability and the magnitude of the material flux between the pelagic and benthic ecosystems. Such studies revealed a strong seasonality in the organic matter settling from the euphotic zone. It became evident that the sedimentation of the spring phytoplankton bloom does not occur as a steady loss, but rather as a fairly short event, during which a considerable portion of the phytoplankton biomass sinks out rapidly, en masse (von Bodungen et al. 1981). Also, it was realized that the distinct event of the spring bloom settling acted as a trigger to growth season activity of the benthos in the tropholytic zone (Graf et al. 1982, 1984), and provided an essential supply of nutrition and energy to these communities (see reviews by Elmgren 1978; Smetacek 1984; Graf 1987, 1992). During the rest of the seasonal cycle, the sinking flux of newly formed organic matter from the photic production was found to be much smaller.

Still, most studies reporting annual sediment trap fluxes in the Baltic are from coastal areas (e.g., Lax et al. 1993; Blomqvist and Larsson 1994; Heiskanen and Leppänen 1995; Ólafsson and Elmgren 1997; Heiskanen et al. 1998; Heiskanen and Tallberg 1999), which are easier to sample frequently than the open sea, where studies covering the whole seasonal cycle are few (Kuparinen et al. 1996; Kankaanpää et al. 1997). Today, the paucity of data from the open Baltic restricts our understanding of the overall settling dynamics of particulate matter in the Baltic Sea.

Much attention has been given to factors governing sedimentation and pelagic nutrient cycling (e.g., Heiskanen 1998). However, a thorough review of all findings reported is beyond the scope of this chapter. In the following, we therefore discuss only a selection of studies dealing with central issues.

8.3 Current Topics

8.3.1 Fate of Phytoplankton Blooms

The spring bloom in the Baltic Sea is dominated by diatoms and dinoflagellates (e.g., Hobro 1979; Kell et al. 1982; Müller-Haeckel 1983; Kononen and Niemi 1984; Stegmann and Peinert 1984; Niemi and Åström 1987; Andersson et al. 1996; Wasmund et al. 1998), causing variations in the export flux rates. In diatoms, the sinking loss rates of a suspended population can be as high as 30–60% day^{-1} (Heiskanen and Kononen 1994; Heiskanen 1998). Therefore, diatoms are generally the major vehicle for the downward transport of organic matter, resulting in high *e*-ratios (proportion of the total primary production exported from the pelagic system), and a seasonally highly variable supply of organic matter to the benthos (Figs. 8.1 and 8.2). It is thereby evident that the *e*-ratio is strongly influenced by the abundance of diatoms.

Dinoflagellates are frequently reported as a dominant phytoplankton group during the spring bloom in the Baltic (Fig. 8.2; e.g., Heiskanen and Kononen 1994; Andersson et al. 1996; Wasmund et al. 1998), except in the open sea of the northern Bothnian Bay where diatoms predominate (Alasaarela 1979). The sinking loss rate of dinoflagellates is commonly less than 10% day^{-1}, and few vegetative cells are recorded in sediment trap samples (Heiskanen 1998). Still, the cyst formation of dinoflagellates can represent a considerable loss from the pelagic population (Kremp and Heiskanen 1999), reflected in high numbers of cysts in the settled material

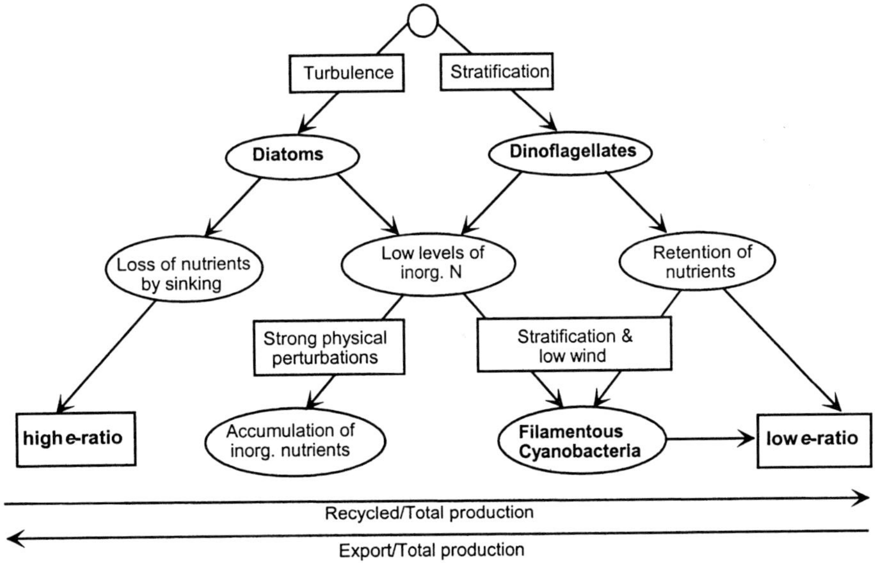

Fig. 8.1. Bifurcation scenario for alternative pathways of seasonal phytoplankton development in the Baltic Sea (i.e., spring bloom dominated by co-occurring diatom species or dinoflagellates and late summer bloom of filamentous cyanobacteria) and their impact on seasonal export flux and retention of nutrients in the pelagic system (inspired by Legendre and Le Fèvre 1989). Scenario identifies physical forcing as the chief factor for the selection of dominant groups (cf. Margalef 1978), while other factors, such as nutrient availability and grazing control, are not considered here. A dominance of dinoflagellates is likely to promote retention of nutrients in the pelagic system and result in lower export to total production ratio (i.e., *e*-ratio) than a vernal bloom dominated by diatoms. Also, cyanobacterial blooms tend to retain nutrients within the pelagic system, and reduce export of organic matter and nutrients to the benthos. (Modified from Heiskanen 1998)

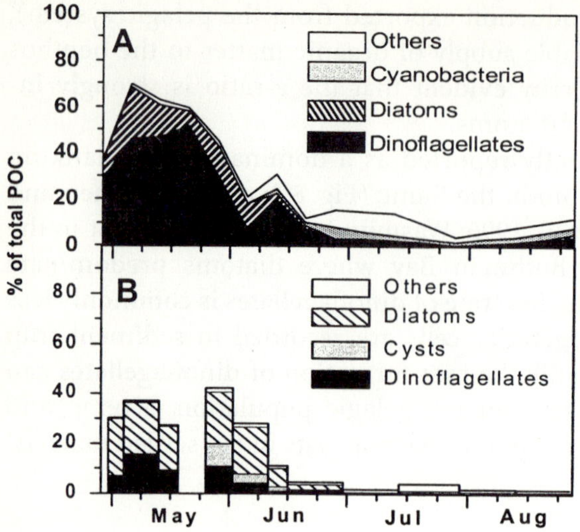

Fig. 8.2. A Contribution of different phytoplankton groups to total particulate organic carbon (*POC*) suspended in surface layer (1–15 m) of water column. **B** Gross sedimentation of POC below this layer during spring and summer of 1988, at entrance of the Gulf of Finland, Baltic Sea. (Heiskanen 1998)

(Heiskanen 1993). However, most of the dinoflagellate biomass forms slowly settling phytodetrital material after the bloom has terminated, resulting in low *e*-ratios (Figs. 8.1 and 8.2). Therefore, the organic matter and nutrients in dinoflagellates are retained in the planktonic food web to a greater extent than for diatoms.

Blooms of filamentous, diazotrophic cyanobacteria (in certain coastal areas also aggregate-forming unicellular cyanobacteria) are a nuisance problem in the Baltic Sea (e.g., Persson et al. 1984; Sivonen et al. 1989; Kononen 1992; Edler et al. 1996). Although these blooms may be visually spectacular (Kahru et al. 1994), due to massive surface accumulations, the fraction of cyanobacteria in the overall export production during summer has often been found to be relatively small (Heiskanen and Kononen 1994; Olli and Heiskanen 1999). In fact, when *Nodularia spumigena* is the dominant species in the water column, it was not found to settle in measurable quantities (Heiskanen and Kononen 1994). Likewise, the biomass of *Aphanizomenon* sp. has been estimated to constitute only a small part of the total settling flux of particulate organic carbon (Heiskanen and Kononen 1994), although the opposite is sometimes also reported (Tallberg and Heiskanen 1998; Olli and Heiskanen 1999). Thus, commonly the cyanobacterial biomass appears to disintegrate in the water column, resulting in low *e*-ratios (Figs. 8.1 and 8.2). It consequently contributes little to the organic matter loading below the euphotic zone.

8.3.2 Fecal Material

Feces produced by pelagic heterotrophs seem to be a minor component (<10 %, often much less) of the total downward flux of particulate matter in the Baltic Sea (Smetacek 1980; Viitasalo et al. 1999). However, the ecological significance of settling fecal pellets remains equivocal. For instance, summer records from the Gulf of Riga suggest a considerable variation in the ratio of fecal pellet carbon to amorphous detrital carbon in the vertical flux rate (Lundsgaard et al. 1999). In the Kiel Bight, most (>90 %) of the fecal pellets produced by mesozooplankton were mineralized in the water column in summer, whereas in autumn the recycling capacity of the pelagic system deteriorated, and 76 % of the pellets produced settled out and were recorded in sediment traps (Voss 1991). Likewise, a study in the southern Kattegat reported that sedimentation of fecal pellets nearly equalled the estimated copepod defecation, suggesting low retention of fecal matter in the water column (Lundsgaard and Olesen 1997).

8.3.3 Resuspension

Numerous studies in the Baltic Sea suggest that resuspension of bottom sediments is a key process in the annual mass dynamics of particulate matter, in both the pelagic and benthic environments. In fact, resuspended sediment may at times be a significant source of energy to the pelagic community (Witek et al. 1997), as well as to the benthos (Graf et al. 1982, 1984). In the open Baltic, resuspension has not been much studied, but appears to be an important topic for the future.

The largest data set on sediment resuspension in the Baltic Sea is from a coastal study in the archipelago of Trosa (five stations, 5 years), northwestern Baltic proper (Blomqvist and Larsson 1994). A considerable variation within and between years, seasons and nearby stations was revealed for gross sedimentation rate, and for the ratio between primary settling matter and resuspended sediment. A predominance of primary settling matter was recorded mainly in spring. Resuspended matter was found to be a large component in the annual settling flux budget for all years and stations (Fig. 8.3). In fact, the resuspended portion commonly exceeded 50 %, and periods without some resuspension seem rarely or never to occur in the investigated area. The main part of the resuspended matter had a bedrock (petrogenic) origin.

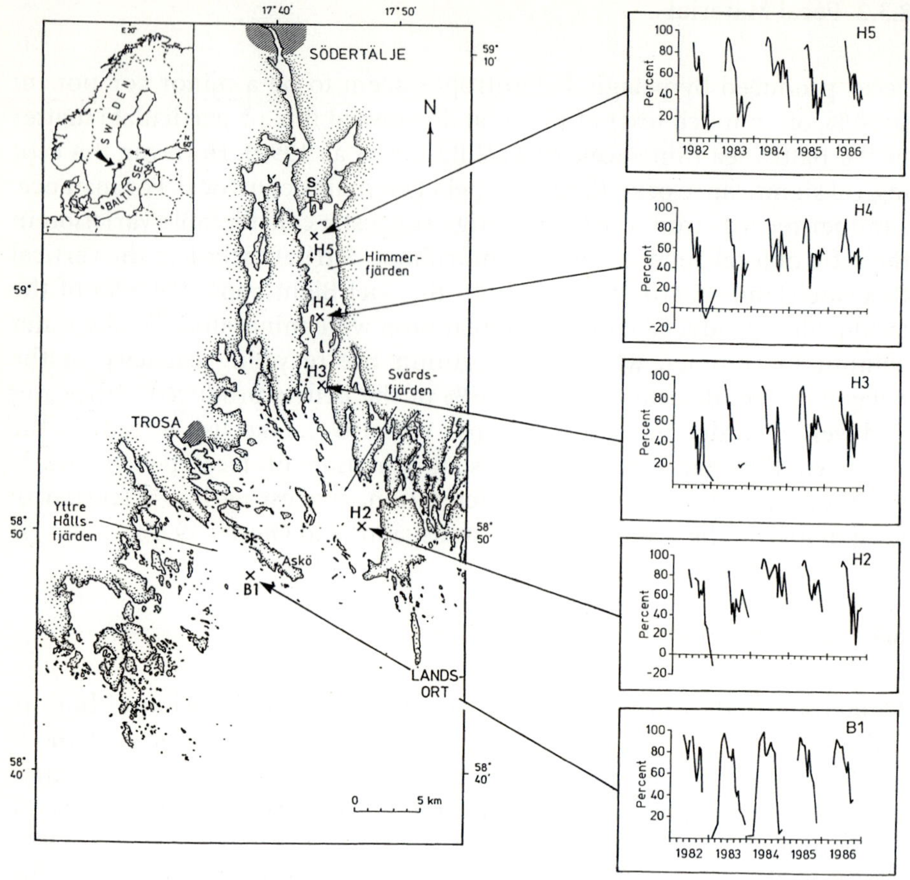

Fig. 8.3. Primary sedimentation of organic carbon expressed as percent of gross organic carbon sedimentation, as estimated from content of aluminium, at five stations (*B1–H5*) in the archipelago of Trosa, north-western Baltic proper, 1982–1986. (Blomqvist and Larsson 1994)

8.4 Critical Issues

8.4.1 Anthropogenic Elevation of Sedimentation in the Open Baltic Sea?

A common notion is that eutrophication and increased phytoplankton production ultimately lead to a high sedimentation rate in many aquatic environments. Several algorithms compiled from simultaneous measurements of pelagic primary production and sedimentation have revealed an exponentially increasing relationship between export flux of carbon from the euphotic zone and total primary production (Wassmann 1990a). Accordingly, with increasing primary productivity, the load of organic carbon to the aphotic zone should be proportionally larger in more eutrophied areas (Wassmann 1990b). It is, however, worth noting that the flux data used by Wassmann (1990b) are gross sedimentation rates (Wassmann 1990a), i.e., include not only primary settling matter, but also carbon which has been resuspended from the bottoms. Moreover, the locally varying grazing impact by microzooplankton and mesozooplankton has a considerable influence on the vertical export flux, implying that such general algorithms may not be applicable in all coastal environments (Wassmann 1998).

8.4.1.1 Direct Measurements

Despite the number of studies focusing on relationships between primary production and sedimentation of carbon in the Baltic (e.g., Lignell et al. 1993; Jonsson and Carman 1994; Heiskanen and Leppänen 1995; Kuparinen et al. 1996; Uitto et al. 1997; Tallberg and Heiskanen 1998), few investigations provide direct evidence that the trends of increase in the concentration of dissolved nitrogen and phosphorus in the open Baltic Sea since the 1950s (Elmgren 1989; Chap. 12) have resulted in a related elevation of primary production. Perhaps the most convincing is the demonstration of an approximate doubling of the export primary production in the 25-year period of 1957–1982 in the Arkona basin (southern Baltic proper), calculated from increasing oxygen supersaturation in the photic zone during productive conditions (Stigebrandt 1991; Rahm et al. 1995). However, congruent increases in sedimentation of organic matter from the open sea pelagic system cannot be taken for granted, since the pelagic food web structure may also have changed (cf. Peinert et al. 1989; Kiørboe 1993, 1998; Witek et al. 1997; Schiewer 1998; Wassmann 1998). Also, despite the clear increase in external nutrient input to the Baltic Sea during this century (Larsson et al. 1985; Granat 1990; Chap. 5), in fact, mainly before 1980 – and most likely even before 1970 (Stålnacke et al. 1999; Chap. 4) – it is still unclear to what degree the oxygen deficiency in the deep sedimentary basins (Jonsson et al. 1990; HELCOM 1996), as well as in more shallow coastal areas and embayments (Morris et al. 1988; Rosenberg and Diaz 1993; Blomqvist and Larsson 1996; Persson and Jonsson 2000), is a result of increased export production of organic matter from the pelagic system, or of hydrodynamic changes, i.e., long temporal residence of high-density deep water due to lack of major water exchanges in the Baltic during recent decades, and locally hampered physical water exchange processes, respectively.

8.4.1.2 Interpretation of Sedimentary Records

Sedimentary records provide an opportunity to track the environmental history of the Baltic Sea (e.g., Sohlenius et al. 1996; Neumann et al. 1997; Westman and Sohlenius 1999; Bianchi et al. 2000). However, evaluation of sediment data is tricky. There is an obvious risk of misinterpretation of the records when comparing data from different depth levels in the sediment bed, especially if different devices have been used for the sediment sampling (Blomqvist 1991). In addition to methodological biasing, factors such as lateral transport and diagenesis complicate the interpretation of sediment sequences (e.g., Emeis et al. 1998).

From studies on sediments of the northern Baltic proper, a more than 1.7-fold increase in the deposition of organic matter has been reported from the late 1920s to the late 1980s (Jonsson and Carman 1994). Basically, this inference was based on a comparison of measurements of weight loss on ignition (LOI=2×C; see Persson and Jonsson 2000) in samples of surficial sediments collected in 1987–1988, compared with literature records of sediments collected in 1925–1930 (Gripenberg 1934) and in 1970 (Hasselrot 1971, 1972), respectively. In fact, in the total amount of carbon (5 g C m^{-2} $year^{-1}$) sequestered in offshore accumulation areas in the 1920s, 80% (4 g C m^{-2} $year^{-1}$; from Jonsson et al. 1990) was presumed to have originated from continuing erosion and resuspension of land uplift (crustal rebound), following the retreat of the last (Weichselian) glaciation, whereas in the 1980s the portion originating from erosion constituted 31% of the total amount of carbon found sequestered (13 g C m^{-2} $year^{-1}$).

Based on sediment data in Jonsson et al. (1990), Jonsson and Carman (1994) related the recorded 1.7-fold increase in sediment organic matter (LOI) to a five- to tenfold enhancement in the annual sequestering of carbon originating from primary production and external sources (i.e., 9 g C m^{-2} $year^{-1}$). However, a five- to tenfold enhancement of the burial rate suggests a considerable increase in the total primary production, or large changes in the ecological cycling of carbon, e.g., export production, import of allochthonous matter, decomposition, burial efficiency/sediment accumulation rate, etc. In fact, the percentage (5.6%) of annual burial rate versus total primary production reported by Jonsson and Carman (1994) for the open sea is even higher than the 4.4% reported for a coastal environment in the Kiel Bight, in the southern Baltic proper (Balzer 1984). The latter environment is shallower, has a higher sediment accumulation rate, and large lateral supply of carbon (Balzer et al. 1986); all factors conducive to high sequestering of organic matter. On the other hand, the deep Baltic sediments of Jonsson and Carman (1994) are more or less permanently euxinic and azoic, a condition found only seasonally in the

Kiel Bight (Gerlach 1990), and often considered to improve preservation of organic matter in marine sediments (Canfield 1993).

In order to judge the reliability of their sedimentary records, Jonsson and Carman (1994) related the enhanced sedimentary sequestering of carbon during the twentieth century to certain quantitative alterations in the ecological carbon flux of the pelagic zone. Hence, a thorough evaluation of their conclusions should look closely at assumptions and auxiliary data, as well as at their sampling procedures. The applied empirical power model (Wassmann 1990b) may have resulted in overestimation of the export production. Still, the estimate of Jonsson and Carman (1994) on the annual export production at the beginning of the twentieth century was comparatively low (21 g C m^{-2} $year^{-1}$), since they selected the lowest estimate of total primary production (73 g C m^{-2} $year^{-1}$) from a range stretching up to 98 g C m^{-2} $year^{-1}$, given by Elmgren (1989), when running the power model of Wassmann (1990b). Sediment sampling methods may also have biased the picture of the alterations in the carbon flux during the twentieth century given in the study of Jonsson and Carman (1994). The old sediment cores were collected by means of narrow coring tubes (Gripenberg 1934; Hasselrot 1971, 1972) and, therefore, were probably biased by a hydraulic shock wave in front of the orifice, as well as by entry deficiency from inner friction of the coring tube, both causing discrimination of sediments upon sampling (Blomqvist 1991). Thus, sediment layers obtained often represented carbon-poor levels deeper down in situ than apparently recorded in the coring tube. Moreover, there were methodological differences between the studies in the analysis of organic carbon, which lower the comparability between the data sets evaluated by Jonsson and Carman (1994).

Given these uncertainties, the degree to which the downward settling flux of particulate matter has been anthropogenically enhanced in the open Baltic Sea is still unclear. In order to understand system changes in the Baltic Sea, and evaluate the effect of anthropogenic loading on export production and sequestering in the sediments of organic matter, collection of reliable field data relating nutrient dynamics, new production and primary sedimentation of autochthonously formed organic matter is urgently needed.

8.4.2 Sequestering of Carbon in Baltic Sea Sediments: A Missing Carbon Sink?

Marine continental shelf areas, including nutrient-enriched epicontinental seas (Kempe and Pleger 1991) such as the Baltic (Shaffer 1987), have been suggested as significant repositories of organic carbon (Berner 1982; Romankevich 1984; Wollast 1991). In fact, better knowledge of ecological interactions in coastal seas (Smith and Hollibaugh 1993) and adjacent land-water interface zones (Downing et al. 1993) appears crucial for proper understanding of the global carbon cycle at large, and for resolving the carbon dioxide issue in particular (Intergovernmental Panel on Climate Change 1996).

Carbon dioxide is the most significant greenhouse gas emitted from Sweden (Rodhe et al. 1991). In the national carbon dioxide budgets of Sweden, uptake by forests and sequestration in marine bottom sediments have been proposed as the main sinks (Table 8.1). However, some of the estimated sinks of carbon have wide limits of uncertainty. The greatest divergence is reported for the sedimentation of carbon in the marine environment. Within the Swedish territorial water, the annual sedimentation has been estimated to be 7 Tg C year^{-1} (Jernelöv 1992) and

Table 8.1. Anthropogenic carbon dioxide budgets of Sweden in 1990. Carbon unit: Tg C year^{-1} (million metric tons of carbon per year)

Source	Eriksson (1991)	Jernelöv (1992)	Rodhe et al. (1995)
I *Sum of carbon emissions to the atmosphere*	18.4	18.4	18.7
II *Carbon uptake from the atmosphere*			
Export of forest products	–	5	–
Stocking of forest biomass	9	9	9
Peat accumulation	1.2	1.2	–
Sedimentation in lakes	1.5	1.5	1.5
Sedimentation of organic matter from drainage areas	1.3	1.3	0.7
Sedimentation of primary production in marine areas (Swedish territorial water)	–	7	1.2
Accumulation in waste deposits and buildings	–	0.7	0.4
Sum	13	25.7	12.8
I–II *Net flux to the atmosphere*	5.4	–7.3	5.9

1.2 Tg C year^{-1} (Rodhe et al. 1995), respectively, while one budget (Eriksson 1991) did not include marine sequestration of carbon at all. An ecological analysis of the trophic dynamics of the Baltic (inside the Danish straits) suggested a low annual rate of 0.6 Tg C year^{-1} in sedimentary sequestering (Elmgren 1984). According to the calculations of Eriksson (1991) and Rodhe et al. (1995), a net emission of 5.4 and 5.9 Tg C year^{-1}, respectively, is lost to the atmosphere from Sweden, whereas the estimates of Jernelöv (1992) suggest a net accumulation of 7.3 Tg C year^{-1} from national Swedish sources (Table 8.1).

These divergences and uncertainties are remarkable, particularly since Sweden has adopted the United Nations Framework Convention on Climate Change (UNFCCC 1992), implying no significant net increase in the national carbon dioxide emission. In fact, the very different balances of the national carbon budget (Table 8.1) cause confusion, with bearing on international affairs. Presently, the best estimates on mean burial rates of organic carbon refer to studies of the Baltic proper; primarily based on modelling (Shaffer 1987) and sediment sampling (Jonsson et al. 1990; Jonsson and Carman 1994), respectively. From the estimated burial rate of 10 g C m^{-2} year^{-1} reported by Shaffer (1987), multiplied by the total area of the Baltic proper (Ehlin and Mattison 1976) we have calculated a sequestering rate of 2.1 Tg C year^{-1}. Based on studies of laminated sediments, a higher carbon sink of 2.7 Tg C year^{-1} has been proposed (Jonsson et al. 1990). Both these estimates fall within the range of 1.6–3.0 Tg C year^{-1} calculated by Jonsson and Carman (1994) for 70,000 km^2 of accumulation bottoms in the Baltic proper.

However, all these estimates on organic carbon burial rate are beset by large uncertainties. For instance, the significance of different physical bottom conditions, including varying resuspension and redeposition/redistribution pattern, resulting in various potential for burial, is not considered. Also, regarding what is more generally reported on the residual fraction of carbon buried and preserved in marine sediments versus deposition flux rate of carbon to the sediment surface (Canfield 1993), the yield of carbon sequestering in the open Baltic proper discussed above appears (perhaps with the exception of Elmgren 1984) fairly high. All in all, further research is urgently called for, in order to provide accurate estimates of the annual settling of particulate organic carbon in different regions of the Baltic Sea, including differences in transport processes and diagenesis, which are factors affecting the subsequent sequestering in the sediments.

8.5 Looking Ahead

For reliable construction of material balances and for dynamic modelling of the Baltic Sea ecosystem, a deeper and more quantitative understanding of the temporal and spatial variability of sedimentation is needed. In particular, studies are needed on controlling mechanisms, such as: (1) regulation of biological formation, retention and vertical flux of various particles, e.g., phytoplankton, fecal pellets, zooplankton corpses/exuviae, and components of the microbial food web; (2) different forms of particle aggregation by physical and chemical processes, i.e., coagulation by cohesion, for instance, salt flocculation by rivers entering coastal areas and coalescence by biological glues; (3) diagenesis during settling transit; (4) conditions for resuspension; as well as (5) long-term, base-line measurements of the seasonal and interannual variability of the downward flux in open sea areas.

Acknowledgements. We are grateful to Ragnar Elmgren and Christoph Humborg for constructive comments on this chapter. Financial support has been provided by the Swedish Environmental Protection Agency (Naturvårdsverket), the Swedish Natural Science Research Council (NFR), the research programme "Sustainable Coastal Zone Management" of the Swedish Foundation for Strategic Environmental Research (MISTRA), the Finnish Environment Institute, and the Nordic Council of Ministers (Hav- och luftgruppen).

References

Alasaarela E (1979) Ecology of phytoplankton in the north of the Bothnian Bay. Acta Bot Fenn 110:63–70

Andersson A, Hajdu S, Haecky P, Kuparinen J, Wikner J (1996) Succession and growth limitation of phytoplankton in the Gulf of Bothnia (Baltic Sea). Mar Biol 126:791–801

Balzer W (1984) Organic matter degradation and biogenic element cycling in a nearshore sediment (Kiel Bight). Limnol Oceanogr 29:1231–1246

Balzer W, Pollehne F, Erlenkeuser H (1986) Cycling of organic carbon in a coastal marine system. In: Sly PG (ed) Sediment and water interactions. Springer, Berlin Heidelberg New York, pp 325–330

Berner RA (1982) Burial of organic carbon and pyrite sulfur in the modern ocean: its geochemical and environmental significance. Am J Sci 282:451–473

Bianchi TS, Engelhaupt E, Westman P, Andrén T, Rolff C, Elmgren R (2000) Cyanobacterial blooms in the Baltic Sea: natural or human-induced? Limnol Oceanogr 45:716–726

Blomqvist S (1991) Quantitative sampling of soft-bottom sediments: problems and solutions. Mar Ecol Prog Ser 72:295–304

Blomqvist S, Larsson U (1994) Detrital bedrock elements as tracers of settling resuspended particulate matter in a coastal area of the Baltic Sea. Limnol Oceanogr 39:880–896

Blomqvist S, Larsson U (1996) Sediment och sedimentprocesser. In: Elmgren R, Larsson U (eds) Himmerfjärden. Förändringar i ett näringsbelastat kustekosystem i Östersjön. Rapport 4565. Naturvårdsverket Förlag, Stockholm, pp 97–109

Canfield DE (1993) Organic matter oxidation in marine sediments. In: Wollast R, Mackenzie FT, Chou L (eds) Interactions of C, N, P and S biogeochemical cycles and global change. Springer, Berlin Heidelberg New York, pp 333–363

Downing JP, Meybeck M, Orr JC, Twilley RR, Scharpenseel H-W (1993) Land and water interface zones. Water Air Soil Pollut 70:123–137

Edler L, Kononen K, Kuosa H (1996) Harmful algae. In: HELCOM (ed) Third periodic assessment of the state of the marine environment of the Baltic Sea 1989–1993; background document. Baltic Sea Environ Proc 64B. Helsinki Commission, Helsinki, pp 192–194

Ehlin U, Mattisson I (1976) Volymer och areor i Östersjöområdet (in Swedish with English summary). Vannet i Norden 1:16–20

Elmgren R (1978) Structure and dynamics of Baltic benthos communities, with particular reference to the relationship between macro- and meiofauna. Kieler Meeresforsch Sonderh 4:1–22

Elmgren R (1984) Trophic dynamics in the enclosed brackish Baltic Sea. Rapp P-v Réun Cons Int Explor Mer 183:152–169

Elmgren R (1989) Man's impact on the ecosystem of the Baltic Sea: energy flows today and at the turn of the century. Ambio 18:326–332

Emeis K-C, Neumann T, Endler R, Struck U, Kunzendorf H, Christiansen C (1998) Geochemical records of sediments in the Eastern Gotland basin – products of sediment dynamics in a not-so-stagnant anoxic basin? Appl Geochem 13:349–358

Eriksson H (1991) Sources and sinks of carbon dioxide in Sweden. Ambio 20:146–150

Gerlach SA (1990) Nitrogen, phosphorus, plankton and oxygen deficiency in the German Bight and in Kiel Bay. Kieler Meeresforsch Sonderh 7:1–341

Graf G (1987) Benthic response to the annual sedimentation patterns. In: Rumohr J, Walger E, Zeitzschel B (eds) Seawater–sediment interactions in coastal waters. Springer, Berlin Heidelberg New York, pp 84–92

Graf G (1992) Benthic–pelagic coupling: a benthic view. Oceanogr Mar Biol Annu Rev 30:149–190

Graf G, Bengtsson W, Diesner U, Schulz R, Theede H (1982) Benthic response to sedimentation of a spring phytoplankton bloom: process and budget. Mar Biol 67:201–208

Graf G, Bengtsson W, Faubel A, Meyer-Reil L-A, Schulz R, Theede H, Thiel H (1984) The importance of the spring phytoplankton bloom for the benthic system of Kiel Bight. Rapp P-v Réun Cons Int Explor Mer 183:136–143

Granat L (1990) Estimating atmospheric nitrogen loading to the Baltic Sea. Ambio Spec Rep 7:20–21

Gripenberg S (1934) A study of the sediments of the north Baltic and adjoining areas. Merentutkimuslait Julk/Havsforskningsinst Skr 96:1–231

Hasselrot TB (1971) Kvicksilverundersökningar 1966–1970. SNV PM 147. National Swedish Environment Protection Board, Research Lab, Drottningholm, 781 pp

Hasselrot TB (1972) Mercury investigations 1966–1970. SNV PM 239. National Swedish Environment Protection Board, Research Lab, Drottningholm, 781 pp

Heiskanen A-S (1993) Mass encystment and sinking of dinoflagellates during a spring bloom. Mar Biol 116:161–167

Heiskanen A-S (1998) Factors governing sedimentation and pelagic nutrient cycles in the northern Baltic Sea. Monogr Boreal Environ Res 8:1–80

Heiskanen A-S, Kononen K (1994) Sedimentation of vernal and late summer phytoplankton communities in the coastal Baltic Sea. Arch Hydrobiol 131:175–198

Heiskanen A-S, Leppänen J-M (1995) Estimation of export production in the coastal Baltic Sea: effect of resuspension and microbial decomposition on sedimentation measurements. Hydrobiologia 316:211–224

Heiskanen A-S, Tallberg P (1999) Sedimentation and particulate nutrient dynamics along a coastal gradient from a fjord-like bay to the open sea. Hydrobiologia 393:127–140

Heiskanen A-S, Haapala J, Gundersen K (1998) Sedimentation and pelagic retention of particulate C, N and P in the coastal northern Baltic Sea. Estuarine Coast Shelf Sci 46:703–712

HELCOM (1996) Third periodic assessment of the state of the marine environment of the Baltic Sea 1989–1993; background document. Baltic Sea Environ Proc 64B. Helsinki Commission, Helsinki, 252 pp

Hobro R (1979) Annual phytoplankton successions in a coastal area in the northern Baltic. In: Naylor E, Hartnoll RG (eds) Cyclic phenomena in marine plants and animals. Pergamon Press, Oxford, pp 3–10

Intergovernmental Panel on Climate Change (1996) Technical summary. In: Houghton JT, Meira Filho BA, Callander BA, Harris N, Kattenberg A, Maskell K (eds) Climate change 1995. The science of climate change. Second assessment report of the Intergovernmental Panel on Climate Change. Cambridge University Press, Cambridge, pp 9–49

Jansson B-O (1972) Ecosystem approach to the Baltic problem. Bull Ecol Res Comm/NFR no 16, Stockholm, 82 pp

Jernelöv A (1992) Miljöskulden. En rapport om hur miljöskulden utvecklas om vi ingenting gör. SOU 1992:58. Miljövårdsberedningen, Stockholm, 40 pp

Jonsson P, Carman R (1994) Changes in deposition of organic matter and nutrients in the Baltic Sea during the twentieth century. Mar Pollut Bull 28:417–426

Jonsson P, Carman R, Wulff F (1990) Laminated sediments in the Baltic – a tool for evaluating nutrient mass balances. Ambio 19:152–158

Kahru M, Horstmann U, Rud O (1994) Satellite detection of increased cyanobacteria blooms in the Baltic Sea: natural fluctuation or ecosystem change? Ambio 23:469–472

Kankaanpää H, Korhonen M, Heiskanen A-S, Suortti A-M (1997) Seasonal sedimentation of organic matter and contaminants in the Gulf of Finland. Boreal Environ Res 2:257–274

Kell V, Bloß W, Börner R, Hagemann M, Vökler T (1982) Die Sukzession des Phytoplankton in der Mecklenburger Bucht (südliche Ostsee) in den Jahren von 1977–1980. Wiss Z Wilhelm-Pieck-Univ Rostock 31, Wissenschaftl Reihe 6:27–30

Kempe S, Pleger K (1991) Sinks and sources of CO_2 in coastal seas: the North Sea. Tellus 43B:224–235

Kiørboe T (1993) Turbulence, phytoplankton cell size, and the structure of pelagic food webs. Adv Mar Biol 29:1–72

Kiørboe T (1998) Population regulation and role of mesozooplankton in shaping marine pelagic food webs. Hydrobiologia 363:13–27

Kononen K (1992) Dynamics of the toxic cyanobacterial blooms in the Baltic Sea. Finn Mar Res 261:3–36

Kononen K, Niemi Å (1984) Long-term variation of the phytoplankton composition at the entrance to the Gulf of Finland. Ophelia Suppl 3:101–110

Kremp A, Heiskanen A-S (1999) Sexuality and cyst formation of the spring-bloom dinoflagellate *Scrippsiella hangoei* in the coastal northern Baltic Sea. Mar Biol 134:771–777

Kuparinen J, Leonardsson K, Mattila J, Wikner J (1996) Food web structure and function in the Gulf of Bothnia, the Baltic Sea. Ambio Spec Rep 8:13–21

Larsson U, Elmgren R, Wulff F (1985) Eutrophication and the Baltic Sea: causes and consequences. Ambio 14:9–14

Lax H-G, Kangas P, Storgård-Envall C (1993) Spatio-temporal variations of sedimentation and soft bottom macrofauna in the coastal waters of the Gulf of Bothnia. Aq Fenn 23:177–186

Legendre L, Le Fèvre J (1989) Hydrodynamic singularities as controls of recycled versus export production in oceans. In: Berger WH, Smetacek VS, Wefer G (eds) Productivity of the ocean: present and past. Wiley, Chichester, pp 49–63

Lignell R, Heiskanen A-S, Kuosa H, Gundersen K, Kuuppo-Leinikki P, Pajuniemi R, Uitto A (1993) Fate of a phytoplankton spring bloom: sedimentation and carbon flow in the planktonic food web in the northern Baltic. Mar Ecol Prog Ser 94:239–252

Lundsgaard C, Olesen M (1997) The origin of sedimenting detrital matter in a coastal system. Limnol Oceanogr 42:1001–1005

Lundsgaard C, Olesen M, Reigstad M, Olli K (1999) Sources of settling material: aggregation and zooplankton mediated fluxes in the Gulf of Riga. J Mar Syst 23:197–210

Margalef R (1978) Life-forms of phytoplankton as survival alternatives in an unstable environment. Oceanol Acta 1:493–509

Morris RJ, Niemi Å, Niemistö L, Poutanen E-L (1988) Sedimentary record of seasonal production and geochemical fluxes in a nearshore coastal embayment in the northern Baltic Sea. Finn Mar Res 256:77–94

Müller-Haeckel A (1983) Winter–spring development of phytoplankton near the coast of the Bothnian Sea. Aquilo Ser Zool 22:139–146

Neumann T, Christiansen C, Clasen S, Emeis K-C, Kunzendorf H (1997) Geochemical records of salt-water inflows into the deep basins of the Baltic Sea. Cont Shelf Res 17:95–115

Niemi Å, Åström A-M (1987) Ecology of phytoplankton in the Tvärminne area, SW coast of Finland. IV. Environmental conditions, chlorophyll *a* and phytoplankton in winter and spring 1984 at Tvärminne Storfjärd. Ann Bot Fenn 24:333–352

Ólafsson E, Elmgren R (1997) Seasonal dynamics of sublittoral meiobenthos in relation to phytoplankton sedimentation in the Baltic Sea. Estuarine Coast Shelf Sci 45:149–164

Olli K, Heiskanen A-S (1999) Seasonal stages of phytoplankton community structure and sinking loss in the Gulf of Riga. J Mar Syst 23:165–184

Peinert R, von Bodungen B, Smetacek V (1989) Food web structure and loss rate. In: Berger WH, Smetacek VS, Wefer G (eds) Productivity of the ocean: present and past. Wiley, Chichester, pp 35–48

Persson P-E, Sivonen K, Keto J, Kononen K, Niemi M, Viljamaa H (1984) Potentially toxic blue-green algae (Cyanobacteria) in Finnish natural waters. Aq Fenn 14:147–154

Persson J, Jonsson P (2000) Historical development of laminated sediments – an approach to detect soft sediment ecosystem changes in the Baltic Sea. Mar Pollut 40:122–134

Rahm L, Sturesson L, Danielsson A, Sandén P (1995) Oxygen saturation trends in the Baltic Sea. Environ Monit Assess 35:13–25

Rodhe H, Eriksson H, Robertson K, Svensson BH (1991) Sources and sinks of greenhouse gases in Sweden: a case study. Ambio 20:143–145

Rodhe H, Hedlund T, Eriksson H, Jonsson P, Klemedtsson L, Nilsson M (1995) Sveriges växthusgasbudget. Klimatdelegationen, Stockholm, 30 pp

Romankevich E.A (1984) Geochemistry of organic matter in the ocean. Springer, Berlin Heidelberg New York, 334 pp

Rosenberg R, Diaz RJ (1993) Sulfur bacteria (*Beggiatoa* spp.) mats indicate hypoxic conditions in the Inner Stockholm Archipelago. Ambio 22:32–36
Schiewer U (1998) 30 years' eutrophication in shallow brackish waters – lessons to be learned. Hydrobiologia 363:73–79
Shaffer G (1987) Redfield ratios, primary production, and organic carbon burial in the Baltic Sea. Deep Sea Res 34:769–784
Sivonen K, Kononen K, Carmichael WW, Dahlem AM, Rinehart KL, Kiviranta J, Niemelä SI (1989) Occurrence of hepatotoxic cyanobacteria *Nodularia spumigena* in the Baltic Sea and structure of the toxin. Appl Environ Microbiol 55:1990–1995
Smetacek VS (1980) Zooplankton standing stock, copepod faecal pellets and particulate detritus in Kiel Bight. Estuarine Coast Mar Sci 11:477–490
Smetacek V (1984) The supply of food to the benthos. In: Fasham MJR (ed) Flows of energy and materials in marine ecosystems. Plenum Press, New York, pp 517–547
Smith SV, Hollibaugh JT (1993) Coastal metabolism and oceanic organic carbon balance. Rev Geophys 31:75–89
Sohlenius G, Sternbeck J, Andrén E, Westman P (1996) Holocene history of the Baltic Sea as recorded in a sediment core from the Gotland Deep. Mar Geol 134:183–201
Stålnacke P, Grimvall A, Sundblad K, Tonderski A (1999) Estimation of riverine loads of nitrogen and phosphorus to the Baltic Sea, 1970–1993. Environ Monit Assess 58:173–200
Stegmann P, Peinert R (1984) Interrelationships between herbivorous zooplankton and phytoplankton and their effect on production and sedimentation of organic matter in Kiel Bight. Limnologica 15:487–495
Stigebrandt A (1991) Computations of oxygen fluxes through the sea surface and the net production of organic matter with application to the Baltic and adjacent seas. Limnol Oceanogr 36:444–454
Tallberg P, Heiskanen A-S (1998) Species-specific phytoplankton sedimentation in relation to primary production along an inshore–offshore gradient in the Baltic Sea. J Plankt Res 20:2053–2070
Uitto A, Heiskanen A-S, Lignell R, Autio R, Pajuniemi R (1997) Summer dynamics of the coastal planktonic food web in the northern Baltic Sea. Mar Ecol Prog Ser 151: 27–41
UNFCCC (1992) United Nations Framework Convention on Climate Change. United Nations Environment Programme, Châtelaine, Geneva, 29 pp
Viitasalo M, Rosenberg M, Heiskanen A-S, Koski M (1999) Sedimentation of copepod fecal material in the coastal northern Baltic Sea: where did all the pellets go? Limnol Oceanogr 44:1388–1399
von Bodungen B, von Bröckel K, Smetacek V, Zeitzschel B (1981) Growth and sedimentation of the phytoplankton spring bloom in the Bornholm Sea (Baltic Sea). Kieler Meeresforsch Sonderh 5:49–60
Voss M (1991) Content of copepod faecal pellets in relation to food supply in Kiel Bight and its effect on sedimentaion rate. Mar Ecol Prog Ser 75:217–225
Wasmund N, Nausch G, Matthäus W (1998) Phytonplankton spring blooms in the southern Baltic Sea – spatio-temporal development and long-term trends. J Plank Res 20:1099–1117
Wassmann P (1990a) Relationship between primary and export production in the boreal coastal zone of the North Atlantic. Limnol Oceanogr 35:464–471
Wassmann P (1990b) Calculating the load of organic carbon to the aphotic zone in eutrophicated coastal waters. Mar Pollut Bull 21:183–187
Wassmann P (1998) Retention versus export food chains: processes controlling sinking loss from marine pelagic systems. Hydrobiologia 363:29–57

Westman P, Sohlenius G (1999) Diatom stratigraphy in five offshore sediment cores from the northwestern Baltic proper implying large scale circulation changes during the last 8500 years. J Paleolimnol 22:53–69

Witek Z, Ochocki S, Maciejowska M, Pastuszak M, Nakonieczny J, Podgórska B, Kownacka JM, Mackiewicz T, Wrzesińska-Kwiecień M (1997) Phytoplankton primary production and its utlilization by the pelagic community in the coastal zone of the Gulf of Gdańsk (southern Baltic). Mar Ecol Prog Ser 148:169–186

Wollast R (1991) The coastal organic carbon cycle: fluxes, sources, and sinks. In: Mantoura RFC, Martin J-M, Wollast R (eds) Ocean margin processes in global change. Wiley, Chichester, pp 365–381

Zeitzschel B (1965) Zur Sedimentation von Seston, eine produktionsbiologische Untersuchung von Sinkstoffen und Sedimenten der Westlichen und Mittleren Ostsee. Kieler Meeresforsch 21:55–80

9 Role of Sediments in the Nutrient Dynamics of the Baltic

M. Jansson

9.1 Introduction

The Baltic Sea is an intercontinental sea with a large supply of nutrients from the surrounding terrestrial areas forming the catchment of the Baltic Sea. Rivers contribute with approximately 70 % of the annual phosphorus and 50 % of the annual nitrogen input. In total, 70,000 tons year^{-1} of phosphorus and 1,100,000 tons year^{-1} of nitrogen are supplied to the Baltic Sea from external sources (Wulff et al. 1990). Slightly more than one half of these nutrients is supplied to the Baltic proper. The present nitrogen loading on the Baltic proper is about four times the estimated natural loading and the phosphorus loading has probably increased by about eight times (Larsson et al. 1985). The increased loading on the Gulf of Bothnia is less dramatic. The larger increase in phosphorus supply compared to nitrogen supply in the Baltic proper is regarded as the reason why nitrogen is the key limiting element for primary production (Larsson et al. 1985). Bothnian Bay is still phosphorus-limited while the Bothnian Sea is a transition zone where both phosphorus and nitrogen are limiting (Larsson et al. 1985).

The net flow of water in the Baltic Sea is from north to south and the Danish Sounds constitute an "outlet" to the Kattegat and the North Sea even if fluxes of water occur in both directions. However, the exchange of water with the Kattegat and the North Sea through the Danish Sounds is intermittent and small compared to the total volume of the Baltic (Stigebrandt 1983). Therefore, the Baltic Sea is similar to large lakes in the respect that most nutrients are supplied from the catchment and a major part of the nutrient import is retained within the basin. A mass balance for the period 1971–1981 (Wulff et al. 1990) demonstrated that the net export through the Danish Sounds was only about 10 % of the import for both phosphorus and nitrogen. Twenty five percent of the phosphorus retention and 13 % of the nitrogen retention were accumulated in the water, i.e. reflected in increased concentrations of nutrients, while the rest was

Ecological Studies, Vol. 148
F. Wulff et al. (eds.)
A Systems Analysis of the Baltic Sea

buried in the sediments (phosphorus and nitrogen) or denitrified (nitrogen). Since denitrification proceeds mainly in the surface sediments (Rönner 1985; Stigebrandt and Wulff 1987), the nutrient balance clearly demonstrates the importance of the sediments in the biogeochemical cycling of elements.

Even though the sediments of the Baltic Sea are a net sink for both phosphorus and nitrogen on an annual basis, fluxes from the sediment to the water mass can occur from limited areas for periods of variable length (Stigebrandt and Wulff 1987; Persson 1990). The magnitude and possible consequences of such release on the productivity of the Baltic Sea are largely unknown. Sediments also constitute a transition environment for nitrogen where transformation of dissolved nitrogen to gaseous nitrogen takes places by denitrification. The long-term development of the Baltic Sea in a situation with increasing nutrient supply and related consequences for production and degradation of organic material is, therefore, to a large extent dependent on the nutrient turnover in the sediment. The sediment nutrient dynamics will be expressed in different ways because of different morphometry, salinity, external loading of nutrients, presence or absence of archipelagos, etc. in different parts of the Baltic Sea.

The purpose of this chapter is to present an outline of how sedimentological processes affect the turnover of phosphorus and nitrogen, to discuss how these processes are expressed in different parts of the Baltic Sea and how they affect the turnover of nutrients. Relevant data assessing sediment nutrient dynamics are scarce, especially from the Gulf of Bothnia. Most of the chapter, therefore, concerns the Baltic proper and the focus is on the role of the sediment in the large scale, long-term turnover of phosphorus and nitrogen in the Baltic Sea ecosystem.

9.2 Physical Conditions Affecting Sediment Nutrient Dynamics

The nutrient dynamics in the sediment of the Baltic depend on the composition of the sediment particles and the chemical environment in the surface sediment and bottom waters. In turn, the distribution and translocation of particulate material as well as the chemical conditions in the sediment are determined by physical conditions and processes of which water depth, salinity and temperature-dependent stratification and wind and wave action are the most critical (Brydsten 1993; Sand-Jensen et al. 1994). The net impact of these processes is reflected in the distribution of erosion, transport and accumulation bottoms. The Baltic Sea consists of the three main basins, the Bothnian Bay, the Bothnian Sea and the Baltic proper. Areas with accumulation bottoms are found mainly in the central parts of these basins in depths that exceed 60–80 m (Freden 1994). Accumulation bottoms are covered with postglacial deposits. Bottoms at depths between 40 and 100 m are to a large extent transport bottoms with discontinuous deposition/erosion, a surface sediment dominated by glacial clays and temporal covers of postglacial material. The peripheral areas at depths below 40 m are mostly erosion bottoms dominated by coarse material. The coastal zones, especially those with archipelagos, have more complicated bottom dynamics and contain both erosion, transport and accumulation bottoms in comparatively shallow areas.

A perennial and stable halocline exists in the Baltic proper, generally at depths between 60 and 100 m. In the Gulf of Bothnia there is a vertical saline gradient but no pronounced halocline (Kullenberg 1981). Erosion and transport bottoms are located mainly above 60–100 m, i.e. above the halocline in the Baltic proper. The deposition and resuspension above the halocline are largely determined by the seasonal temperature-dependent stratification of the upper water mass (in principle the water above the halocline circulates like a temperate dimictic lake), the wind intensity and the external and internal (primary production) supply of particulate material. Oxygen consumption and oxygenation in the sediment, which are critical for the sediment phosphorus and nitrogen dynamics, are to a large extent governed by these physical prerequisites. Above the halocline, anoxic conditions occur mainly in accumulation bottoms in the coastal areas and in temporal assemblages of degradable organic material on transport bottoms (cf. Nedwell et al. 1983). Though it has not been studied in the Baltic Sea, it is likely that organic material is gathered as ephemeral blankets with a life length from weeks to several years, similar to what has been described for transport bottoms in the Kattegat (Floderus and Håkansson 1989). The life length of such blankets is determined by stratification and weather conditions. Anoxic conditions develop rapidly in these highly organic assemblages (Nedwell et al. 1983). Oxygenation of the sediments above the halocline should occur, above all, when the thermal stratification is broken in the autumn and spring. Below the halocline the stratification is more stable. There is a comparatively strong saline stratification also below the halocline because of the intermittent inflow of salt water through the Danish Sounds (Stigebrandt and Wulff 1987). In the Baltic proper, the sediments in, and immediately below, the halocline are oxidized most of the time (Carman and Wulff 1989), while anoxic conditions prevail in deeper sediments and also in deep water for extended periods (Stigebrandt and Wulff 1987; Jonsson and Jonsson 1988). Oxygenation of deep-water sediments only occurs when the inflow of salt water is very great, which happens with intervals of about 5–20 years. In the Gulf of Bothnia the vertical distribution of erosion, transport and accumulation bottoms is similar to the situation in the Baltic proper. The oxygen consumption is much less and anoxia seldom or never occurs in the bottom water.

9.3 Phosphorus Dynamics in Sediments

Phosphorus exchange across the sediment–water interface and the role of retention and release of phosphorus have been recognized as extremely important for the nutrient metabolism of lake ecosystems ever since the papers of Einsele and Mortimer were published around 1940 showing the substantial P release from anoxic sediments (Einsele 1938; Mortimer 1941, 1942). Net phosphorus release, commonly termed internal loading, thus accelerates eutrophication and delays recovery of polluted lakes after the external nutrient loading has been reduced (Boström et al. 1982). The extent and consequences of decreased phosphorus retention capacity of marine sediments are less known, probably because marine eutrophication is a younger environmental problem than lake eutrophication. However, the literature on phosphorus in marine sediments especially from estuarine and coastal environments is steadily increasing. A limited number of these studies have been carried out in the Baltic Sea.

The phosphorus binding capacity of a sediment is determined by the amounts and distribution of phosphorus binding species in the particles that form the sediment, the degree to which phosphorus binding sites in different adsorbents are occupied and the dissolution and formation of particulate phosphorus due to changes of the chemical environment. Sediments with high phosphorus binding capacity generally have a high content of iron-oxy-hydroxides or calcium minerals (Boström et al. 1982). Iron (III) hydroxides sorb phosphorus efficiently and keep their binding capacity as long as the redox potential exceeds ca. 200 mV (Mortimer 1941, 1942). Phosphorus can therefore settle out of the water column as iron precipitates (Boström et al. 1982; Lucotte and d'Anglejan 1988; Forsgren et al. 1994). In oxygenated sediments flocs of iron hydroxides form an efficient barrier, which prevents the upward flux of phosphorus from the sediment to the overlying water (Boström et al. 1982). Phosphorus can also sediment out of the water column sorbed to calcite (Carman 1990) or as a structure element in allogenic apatite particles. Adsorption of phosphate aggregates to inorganic particles (clays, silt) can cause a rapid sedimentation in estuaries (Forsgren and Jansson 1992; Forsgren et al. 1994). Since phosphorus is an essential nutrient for pelagic organisms, a large share of the total phosphorus in the trophogenic layer is taken up by plankton and incorporated in organic structures. Organic phosphorus is to a large extent mineralized and reassimilated in the water column, but a net loss by sedimentation of organic phosphorus always occur. The more productive the water is, the more phosphorus will reach the sediment in organic form. Transfer of phosphorus between different distribution forms in the sediment occurs (Boström et al. 1982; Carman 1990). A common sequence is labile organic phosphorus→phosphorus adsorbed to iron→apatite.

In order to characterize sediment phosphorus a large variety of extraction schemes have been developed. All of these use sequential extraction where different chemical treatments remove one fraction after another. Needless to say, none of these extraction procedures identifies discrete phosphorus forms, they only define the different fractions by their sensitivity to different extractants. Nevertheless, characterization of phosphorus in sediments is a useful tool for the interpretation of phosphorus dynamics, both in lake and marine sediments (Boström et al. 1988; Carman and Jonsson 1991; Ruttenberg 1992; Jensen and Thamdrup 1993). Most extraction schemes derive from the procedures developed for characterization of phosphorus in soils (e.g. Chang and Jackson 1957), which have been subsequently adopted and modified for use in lake and marine sediments (Hieltjes and Lijklema 1980; Lucotte and d'Anglejan 1985; Psenner et al. 1985; Balzer 1986; Carman and Jonsson 1991; Ruttenberg 1992; Jensen and Thamdrup 1993).

Extraction procedures generally discriminate between loosely sorbed phosphorus, phosphorus associated with iron and aluminum, phosphorus sorbed to, or bound in, calcium minerals, and organic phosphorus. Loosely sorbed phosphorus is extracted, e.g. by NaCl or NaOAc, and usually constitutes only at the most a few percent of the total phosphorus content of the sediment. Phosphorus associated with iron can be discriminated as a reducible fraction by using H_2S (Carman and Jonsson 1991) or bicarbonate-dithionite (Psenner et al. 1986; Jensen and Thamdrup 1993; Ruttenberg 1992) as reductant. Phosphorus associated with clays or aluminum hydroxides can be further separated by extraction with NaOH. Calcium-bound phosphate is extracted by HCl, giving both apatite and phosphorus sorbed to calcite. Using NaOAc adjusted to pH 5.0 can separate calcite phosphorus. Though somewhat different in the procedure, which makes direct comparison between different investigations difficult, most extractions used on marine sediments so far aim at discrimination between loosely sorbed phosphorus, reducible phosphorus supposed to be associated with iron, aluminum-bound phosphorus and calcium-bound phosphorus. The sum of these fractions never equals the total phosphorus content of the sediment. The difference between total phosphorus and total extractable phosphorus is often called "residual phosphorus" and is taken as a measure of organic phosphorus content. It should be noted that this fraction is quite resistant and does not reflect the labile phosphorus incorporated in living or newly dead organic structures.

The most labile fraction, i.e. the fraction that to the largest extent participates in adsorption or release processes, with relevance to marine sediments is iron-bound phosphorus extracted by a reductant or by NaOH. For example, in a detailed study on the role of iron-bound phosphorus in sediments of the Aarhus Bight and the Kattegat, Jensen and Thamdrup (1993) demonstrated that $^{32}PO_4$ added to oxic sediments was almost totally recovered in the redox-sensitive fraction, supposed to represent phosphorus sorbed to iron. The same fraction by definition releases phosphorus when the sediment turns anoxic. Jensen and Thamdrup (1993) concluded that amorphous FeOOH controlled the concentration of dissolved phosphorus in the sediment. The important role of iron in controlling adsorption and mobilization of soluble phosphate in surface marine sediments, especially coastal-near sediments, has also been emphasized in several other investigations (Holm 1978; Krom and Berner 1981; Balzer 1986; Yamada and Kayama 1987; Carman and Jonsson 1991; Sundby et al. 1992; Ruttenberg 1992). In fact, few if any literature data exist that convincingly demonstrate that any other fraction than reducible amorphous iron to any significant extent takes part in both adsorption and desorption processes of phosphorus in marine sediments. Marine sediments thus resemble fresh-water sediments where the redox-controlled turnover of phosphorus is a major controlling factor for the internal phosphorus loading in lakes (Boström et al. 1988). The classical iron-controlled redox model of lake sediment (Einsele 1938; Mortimer 1941, 1942) has been challenged by theories suggesting that sorption and release of dissolved phosphorus in the sediment surface at shifts between oxic and anoxic conditions could be controlled by incorporation in, and release from, living bacterial cells (Fleischer 1983; Gächter and Meyer 1993). However, so far no evidence has been presented that release of bacterial cell phosphorus at the shift from oxic to anoxic conditions quantitatively influences the sediment phosphorus dynamics (Waara et al. 1993).

Two important differences in the phosphorus turnover in marine and fresh-water sediments should be pointed out. In fresh-water sediments, especially in shallow lakes, large variations in pH due to photosynthetic activity influence the phosphorus binding capacity of iron hydroxides. This type of influence is of minor importance in well-buffered saline water. In marine sediments the anaerobic respiration by sulfate reduction is much more intensive than in fresh-water sediments (Jörgensen et al. 1990). Reduction of sulfate produce sulfides which to a large extent are precipitated as iron sulfides, e.g. pyrite (Boesen and Postma 1988). Since sulfides are more or less insoluble, the precipitation of iron sulfides withdraws iron from further cycling in redox-controlled processes, which in turn means that sulfate reduction contributes to decrease the phosphate sorption capacity of marine sediments.

In addition to the redox-controlled release of dissolved phosphate, mineralization of organic phosphorus liberates dissolved phosphorus at the sediment surface. Especially in productive coastal areas, considerable amounts of labile organic phosphorus can settle into the sediment together with dead plankton. Such material liberates phosphorus, which to a large extent is rapidly sorbed to iron hydroxides (cf. Jensen and Thamdrup 1993) but can also contribute to phosphorus release from sediments in experimental systems (Slomp et al. 1993). However, in oxygenated sediments, all liberated phosphate is probably sorbed to iron hydroxides (Jensen and Thamdrup 1993) and direct release of organic phosphorus most likely occurs only from reduced sediments.

Diagenetic processes tend to move phosphorus from organic compounds and iron or aluminum compounds to apatite (Williams et al. 1971; Carman 1990). Aged sediments therefore contain mostly apatite phosphorus and residual organic phosphorus. However, diagenetic translocation to apatite is a slow process compared to the comparatively rapid dynamics governed by redox processes. Apatite formation in the sediment is, therefore, a process that causes a net immobilization of phosphorus but which does not to any large extent affect the conditions in the surface layer and the exchange of phosphorus between the sediment and the overlying water.

In short, marine sediments contain phosphorus bound to iron and aluminum hydroxides, apatite, calcite and organic structures. Iron-bound phosphorus, apatite phosphorus and non-labile organic phosphorus are the dominant distribution forms. Iron phosphorus is mostly endogenic or authigenic, apatite phosphorus is allogenic or of diagenetic origin. Organic phosphorus is allogenic or endogenic. In most sediments, iron controls the short-term dynamics in the surface sediments and regulates the phosphorus sorption capacity of the sediment. The role of iron as a buffer for dissolved phosphorus is largely dependent on the redox state of the sediment (Fig. 9.1).

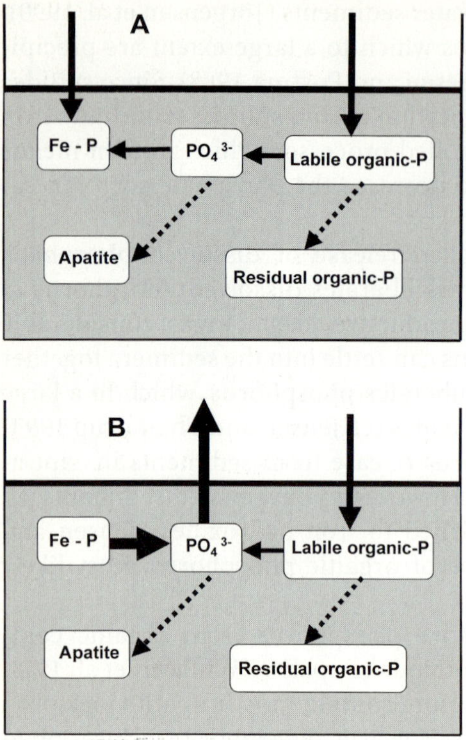

Fig. 9.1. Schematic presentation of transfer of phosphorus between different distribution forms at sediment-water interface of oxygenated (**A**) and anoxic (**B**) sediments

9.4 Character and Turnover of Phosphorus in Sediments of the Baltic Sea

Very little is known about the quality of the particulate phosphorus load on the Baltic Sea. Characterization of river-borne particles has been made during the spring flood on samples from the River Öre in the central parts of the Gulf of Bothnia (Forsgren and Jansson 1993). About 65 % of the input to the Öre Estuary consisted of HCl-extractable phosphorus (apatite phosphorus), 25 % residual (organic) phosphorus and 15 % iron-bound phosphorus. It is not possible to say to what extent this composition is representative of the external input to the different parts of the Baltic Sea. The River Öre is a large, non-regulated forest river with a high specific transport of suspended solids. Most of the riverine input to the Gulf of Bothnia is by large rivers exploited for hydroelectric power purposes. The large reservoirs in these rivers are efficient traps for particulate phosphorus and reduce the phosphorus supply to the Gulf of Bothnia (Brydsten and Jansson 1989). Both apatite phosphorus and iron phosphorus are retained in reservoirs while phosphorus associated with colloidal organic material remains unaffected (Brydsten and Jansson 1989). The supply of phosphorus to the Gulf of Bothnia should have been altered somewhat both in quality and quantity due to construction of reservoirs, but the catchment export is not dramatically changed. The largest increase in phosphorus loading to the Gulf of Bothnia is the inflow of phosphorus-rich surface water from the Baltic proper (Wulff et al. 1990).

The Baltic proper receives about 65 % of its present phosphorus supply from sewage water and industrial effluents, while 25 % comes from arable land. Approximately 85 % of the total supply flows into the south-eastern parts of the Baltic proper with large rivers in Poland, Lithuania, Latvia and Estonia (Miljödepartementet 1990). Quite different in origin from the particulate phosphorus draining into the Gulf of Bothnia, it is plausible that the fractional composition of riverine particles in the Baltic proper is different from that in the northern parts of the Baltic Sea. Characterizations of river-suspended matter thus show a considerable variation in the distribution of individual fractions depending on the character of the river catchment (Broberg and Persson 1988). A relevant characterization of particulate phosphorus draining to the different basins of the Baltic Sea should be a valuable achievement since it could be compared with the fractional distribution of the phosphorus in the sediment in order to reveal the magnitude of transformation processes within the basins.

Characterization of sediment phosphorus in the Baltic Sea has been made in a few investigations, unfortunately mostly in small restricted areas like the Kiel bight in the south (Balzer 1986) and the Öre Estuary in the north (Forsgren and Jansson 1993). The most extended investigation was carried out by Carman and Jonsson (1991) who characterized 25 samples from different depths in the area between Stockholm and Gotland by sequential extraction. The HCl-extractable phosphorus dominated and made up 40–70 % of the total phosphorus content. The amount of reducible phosphorus was between a few percent in anoxic sediments and up to 30–40 % in oxic sediments. Organic phosphorus was generally between 10 and 20 %. Though the number of characterizations is too small to permit general conclusions, it still contains enough substance for at least tentative reasoning. Thus, all characterizations indicate that the phosphorus in the sediments is bound mainly to calcium and iron carriers. The content of organic phosphorus generally does not exceed 20 %. In the latter respect, the sediments of the Baltic resemble other marine sediments and also those in more eutrophic areas (e.g. Jensen and Thamdrup 1993). On the other hand, the Baltic Sea sediments are different from lake sediments, where residual organic phosphorus both in oligotrophic and eutrophic lakes often makes up 50 % or more of the total phosphorus content

(Boström et al. 1988). It is possible that the saline environment to some extent prevents the formation of stable organic phosphorus compounds. The studies in the low saline (2–6‰) Öre Estuary (Forsgren and Jansson 1992; Forsgren et al. 1994) showed that when river water was mixed with saline water in the estuary, phosphorus aggregated with iron in a process in which intensity was proportional to the salinity but which did not include organic material. Iron–phosphorus aggregates rapidly formed sediments, while organic material remained in the water column. A comparison with lake sediments in the River Öre system also demonstrated that the content of organic phosphorus was considerably higher in lake sediment traps and lake sediments than in traps and sediments in the estuary, though both systems received the same types of particulate phosphorus (Forsgren and Jansson 1993).

The content of HCl-extractable phosphorus in sediments of the Baltic proper is much higher than in most lake sediments where values exceeding 50% are found only in calcareous lakes (Boström et al. 1988). The high content of HCl-phosphorus in coarse sediments on erosion and transport bottoms could be explained by the presence of allogenic, or diagenetically formed, apatite minerals in these sediments, which to some extent contain old deposits, also at the sediment surface. The high content of HCl-phosphorus in accumulation bottom sediments could result from a high yearly deposition of particles of old postglacial origin which are continuously being subjected to erosion by resuspension due to the isostatic land uplift (Jonsson and Carman 1992). Another possibility is endogenic-authigenic formation of apatite. Both hydroxy- and fluoroapatite are saturated on both sides of the halocline in the Baltic (Carman 1990). Considering the difference compared to fresh-water sediments, the latter explanation is perhaps more relevant. If so, apatite formation binds considerable amounts of phosphorus in Baltic Sea sediments by which process phosphate is largely withdrawn from exchange processes with overlying water. This type of withdrawal can be especially pronounced in environments with low redox potential where high concentrations of dissolved phosphate can prevent the dissolution of calcite (Walter and Burton 1986), which is a pre-state in the diagenesis of calcium minerals to apatite.

Apatite formation is probably not so rapid, or so complete, that the short-term phosphorus exchange with iron in the surface sediment is affected to any large extent. From the characterizations made by Carman and Jonsson (1991), it is also obvious that considerable amounts of phosphorus can be exchanged between particulate and dissolved forms depending on the redox conditions. The content of reducible phosphorus in oxidized sediments from different depths in the Baltic proper varies between 10 and 30 μmol g^{-1}. Considering that the mean total phosphorus content of the surface sediment (0–2 cm) is about 30–36 µmol g^{-1} (Holm 1978; Carman and Wulff 1989), it is possible that as much as 30 % of the phosphorus in oxidized surface sediments can be released during a negative redox turnover.

The capacity of the sediment to retain phosphorus is therefore to a very large extent dependent on the redox conditions. The only part of the Baltic Sea where the sediment phosphorus retention capacity is similar to the situation in the pre-industrial era is the Gulf of Bothnia, where the sediments are still covered by well-oxygenated water throughout the year. The situation in the Baltic proper is widely different. The bottoms below the halocline (one third of the total bottom area) are today more or less permanently reduced (Stigebrandt and Wulff 1987; Jonsson and Jonsson 1988) and bottoms in and above the halocline containing organic material are most likely subjected to several shifts between oxic and anoxic conditions annually. The sorption capacity of sediments has thus decreased below the halocline and is variable above the halocline.

The present situation with discontinuously reduced sediments above the halocline and more or less permanently reduced bottoms below the cline is likely to persist. Therefore, in the large-scale, long-term perspective, the phosphorus retention capacity of the deep bottoms will be greatly reduced compared to the natural state. Labile phosphorus settling through the halocline can to a large extent be stored as dissolved phosphorus in the deep water, a development that was quite evident from field data during the 1970s (Stigebrandt and Wulff 1987). After large inflows of oxygen-rich water from the Kattegat, dissolved phosphate disappears from the deep water (Stigebrandt and Wulff 1987), which probably depends on resorption to the sediment. The major part of the phosphorus cycling in deep water of the Baltic proper is thus internal, since the halocline is never broken. There is no evidence from field data that longer periods with anoxia below the halocline contribute to increased phosphorus concentrations in the water above the halocline, at least not in the central parts of the Baltic proper (Stigebrandt and Wulff 1987). Exchange between deep water and coastal water can occur by episodic upwelling. Such episodes have been calculated to occur as frequently as 20–25 % of the time during a 10-year period along limited coastal reaches (Gidhagen 1987). Upwelling is also reported to have caused plankton blooms in coastal regions (Jansson et al. 1984). However, very little is known about the large-scale impact of upwelling on phosphorus utilization in the Baltic Sea. It is very likely that a major share of the phosphorus in upwelling water is reprecipitated in the oxic sediments immediately above the halocline (Holm 1978). Phosphorus settled through the halocline and cycled between the sediment and water below the halocline should therefore not to any large extent affect the primary productivity of the Baltic proper. Even if the sorption capacity of the surface sediment is greatly reduced, the deep waters are still an effective phosphorus sink due to the halocline.

The possible exchange between particulate and dissolved phosphorus in surface sediments above the halocline is more critical in terms of consequences for primary productivity in the trophogenic layer. In principle, the water column above 60 m behaves like a dimictic lake and is thermally stratified during summer. This stratification is more pronounced in the Baltic proper than in the Gulf of Bothnia due to higher temperatures in the surface water. A significant phosphorus release can occur from sediments superimposed by temporarily stagnant water. The phosphorus release rate from sediments subjected to anoxia above the halocline of the Baltic proper has experimentally been determined to be between 2 and 40 μmol P m^{-2} h^{-1} during 1 month of induced anoxia. Compared with the estimated phosphorus demand in phytoplankton, this release was large enough to more than cover the present net uptake by phytoplankton (Koop et al. 1990). These calculations were based on very limited material and must be interpreted with great caution, but they stress that the phosphorus release potential in shallow sediments can be important in affecting the ongoing eutrophication of the Baltic Sea. The question is probably not if the sediments of the shallow areas contain enough potentially mobilizable phosphorus to seriously affect the conditions of the overlying water, but rather to what extent the water and sediment dynamics permit release of phosphorus to the trophogenic layer.

Regeneration of phosphorus in shallow sediments followed by phosphate release to the overlying water can be especially pronounced from organic assemblages containing the rest of the diatom bloom in the spring (Persson 1990). It is possible that phosphate is enriched in the bottom water, especially if a thermocline has developed. Phosphate-rich bottom water trapped between the thermocline and the halocline during the summer can be mixed with surface water after the breakup of the thermocline in the autumn.

However, since phosphorus released from sediments above the halocline is likely to be reprecipitated rather rapidly (Holm 1978; Koop et al. 1990), the bottoms above the halocline can still be considered as bottoms from which the net release of phosphorus to the overlying water is low. Furthermore, bottoms above the halocline, especially erosion and transport bottoms, have a high areal phosphorus sorption capacity compared to bottoms below the halocline in the Baltic proper (Carman and Wulff 1989). Phosphorus release contributing to significantly increased concentrations in the water is, therefore, probably a manifest problem only from bottoms below the halocline.

A most important feature in the sediment phosphorus dynamics of the Baltic Sea is the continuous translocation of particles from shallow sediments to deep-sea sediments by resuspension. Resuspension and its consequences have been studied in several different investigations during recent years. In the Öre Estuary with depths between 20 and 40 m, most of the particles in the surface sediments are transported to the deep-sea sediments within 10–15 years (Malmgren and Brydsten, unpubl. data). The net retention of total phosphorus in the Gulf of Riga was close to zero while the net retention of dissolved phosphate was very high, which indicates a large export of particulate phosphorus from this shallow area (Yurkovskis et al. 1993). Sedimentation studies in different archipelagos in the Baltic proper indicated that as much as 50–60 % of material caught in sediment traps was resuspended particles (Wallin and Håkansson 1992). Resuspension is a quantitatively important process for the removal of sediment also in very closed estuaries (Blomquist and Larsson 1994; Nilsson et al. 1997). In a model simulation of resuspension frequencies at different depths in the Gulf of Bothnia (Brydsten 1993) it was demonstrated that resuspension occurred several times a year on bottoms above 30 m. Between 30 and 40 m, resuspension occurred every second year and even at depths down to 60 m, resuspension occurred with less than 10-year intervals. The fate of resuspended particles is largely unknown, but in many cases they are probably redeposited close to the spot from where they were resuspended (cf. Brydsten and Jansson 1989). However, the net effect of resuspension is a more, or less, continuous relocalization to deeper bottoms. In the long-term perspective, focusing of particulate-bound phosphate occurs in the whole Baltic Sea, and also around archipelagos. Resuspension therefore increases the yearly loading of phosphorus on deep-sea bottoms.

The influence of resuspension on the large-scale dynamics is realized by the fact that approximately 60 % of the bottoms is located above the halocline (60 m) and affected by resuspension. The share of the total primary sedimentation on these bottoms is probably much greater than 60 %. The largest areal primary sedimentation should take place in the coastal areas due to aggregation and sorption processes (Forsgren et al. 1994) and due to the incorporation of phosphorus in plankton which forms sediments after death. Since the resuspension activity is proportional to water depth, resuspension causes a concentration of particulate material to deeper bottoms, i.e. resuspension does not simply relocate particles from above the halocline to sediments below the halocline, it also creates a gradient on the bottoms above the halocline. Since the resuspension critical depth roughly corresponds to the depth of the halocline, it is likely that a concentration of material will occur in sediments at depths from 60 to 100 m. This concentration should be amplified by the fact that the halocline in itself constitutes a barrier which slows down the sinking velocity of sedimenting particles. Fine particulate matter settling to the halocline can therefore be trapped on the halocline and transported by horizontal water movements to sediments at the same depth as the cline (cf. Nilsson et al. 1997). The high phosphorus concentration in sediments approximately at the halocline depth (Holm 1978) in the Baltic proper can be a result of the processes discussed above. However, since the halocline in itself has an eroding effect due to seiches which occur in the Baltic proper (Kullenberg 1981), the enrichment on bottoms in connection with the halocline is counteracted by erosion.

For reasons discussed above, it is likely that most of the accumulation of phosphorus in deep-water sediments below the halocline is by phosphorus relocated from shallower sediments. Considering the increase in external phosphorus supply during this century, it is an open question whether a steady state in the supply of phosphorus loading on the deep bottoms related to the external supply has been established or whether the transport of particulate phosphorus to the deeper sediments will continue to increase. The delay between the onset of increased sedimentation of phosphorus and the time for a balance between sedimentation and accumulation in deep-sea sediments will be a direct function of the mean retention time of particulate phosphorus above the halocline, which at present cannot be estimated. A simple way to obtain a fairly accurate estimate could be to look at the distribution of radio-cesium after the Chernobyl accident in the Baltic Sea sediment.

However, a rough estimate of the present-day importance of resuspension on the phosphorus sediment dynamics can be made for the Baltic proper. Wulff et al. (1990) calculated the average annual phosphorus loss to the sediments in the Baltic proper for the period 1971–1981 to be 36,700 tons. The annual burial of phosphorus during roughly the same period in laminated sediments which approximately corresponds to sediments below the halocline was calculated to be 33,000 tons (Jonsson and Carman 1992). If it is assumed that primary sedimentation occurred in proportion to the areal distribution of sediment above and below the halocline, the sediments below the halocline should receive about $0.3 \times 36{,}700$ tons=12,233 tons by direct sedimentation. As discussed above, this figure is probably too high since the areal primary sedimentation should be higher above than below the halocline. However, even with the conservative estimate of about 12,000 tons, there is a great discrepancy between this mass-balance-calculated sedimentation and the 33,000 tons of sedimentation calculated from analyses of the sediment phosphorus content. Jonsson et al. (1990) suggested that erosion of preglacial material exposed on transport and erosion bottoms above the halocline due to the isostatic land uplift was part of the explanation for the high burial of phosphorus in deep-sea sediments. However, the burial of phosphorus in deep-sea sediments was twice as high in 1980 as around 1900 (Jonsson and Carman 1992), which means that phosphorus of recent origin makes up a dominant share of the total content in deep-sea sediments. Independent of whether the source of phosphorus to the sediments below the halocline in the Baltic proper comprises "new" or "old" particles, the discrepancy between the two different ways of estimating phosphorus supply to deep-sea sediments emphasizes the role of resuspension as a force of redistribution. A net balance including sedimentation and resuspension can be established if it is assumed that resuspension occurs above but not below the halocline. The difference between 33,000 tons calculated from the sediment analysis and the 12,000 tons given by the mass balance must then be accounted for by redistribution due to resuspension. Thus, 21,000 tons/year is redistributed from erosion and transport bottoms to deep-sea sediments. This translocation means that the net deposition on bottoms above the halocline is only 25,000 tons (deposition on bottoms above the halocline) minus 21,000 tons (erosion from bottoms above the halocline) = 4000 tons. According to this calculation, the net accumulation of phosphorus on bottoms above the halocline corresponds to only 15–20 % of the primary sedimentation.

Part of the phosphorus transported through the halocline will be dissolved due to anoxic conditions in sediments and deep water. The concentration of dissolved phosphorus in waters below the halocline generally increases by about 0.5 mmol m^{-3} $year^{-1}$ during the first years of periods when the absence of salt-water intrusions from the Kattegat permits anoxia to be developed (data from Stigebrandt and Wulff 1987). The volume below the halocline in the Baltic proper is approximately 1500 km^3 which means that about 20,000 tons/year is released during such periods. This release, which originates both from previous deposits and freshly supplied material thus seems to correspond to about 60–70 % of the annual supply of phosphorus to deep waters. If the calculation is correct, it shows that the sediments of the deep bottoms are a net sink for phosphorus also during periods of anoxia. It should be stressed that this comparison assumes that the previous statement that phosphate released from sediments below the halocline is reprecipitated when oxygenated water flows in from the Kattegat is reasonably correct.

9.5 Nitrogen Transformations in Surface Sediments

Nitrogen in aquatic ecosystems occurs as reduced nitrogen, e.g. amino groups in organic structures, and as dissolved inorganic nitrogen of different oxidation number. Nitrate is typical of oxidized environments while ammonium occurs in reduced water and sediment pore water. Organic nitrogen is transformed to inorganic nitrogen primarily through ammonification, which is release of ammonium from organic nitrogen by autolysis, bacterial mineralization or excretion. Release of dissolved organic nitrogen and ammonification are the first steps in recycling of nitrogen which allows reutilization of nitrogen assimilated by biota. When organic particles, e.g. dead plankton, settle to the sediment, the nitrogen content of the particles has most likely been depleted by ammonification in the water column. During the further degradation in the sediment, the fate of released ammonium depends on the redox state of the sediment surface-bottom water environment. In the presence of oxygen, ammonium is rapidly oxidized by nitrification to nitrate via nitrite. The formed nitrate can be transported upwards through the water column and again be assimilated by plankton, or it can be reduced by dissimilative nitrate reduction which usually occurs by denitrification. Denitrification is the bacterial utilization of nitrate as electron acceptor in the oxidation of reduced organic material and generally leads to the formation of nitrogen gas, which is lost to the atmosphere. Denitrification takes place in reduced environments which means that the process generally occurs in transition zones between oxidized and reduced environments where supply of nitrate as well as absence of oxygen are at hand. In marine environments, the surface sediment is the prime site for denitrification, but anoxic zones in the water column also promote denitrification (Seitzinger 1988). The major source of nitrate for denitrifying bacteria in marine environments is nitrification (Rönner 1985; Nielsen et al. 1994). Since denitrification produces nitrogen gas, the process allows the ecosystem to dismiss excess nitrogen. In nitrogen balances of water bodies, denitrification plus permanent burial in the sediment therefore make up the total nitrogen retention.

The key processes in the sediment turnover of nitrogen are thus liberation of ammonium from organic nitrogen, release of ammonium to anoxic bottom water or oxidation of ammonium to nitrate, reassimilation of nitrate or denitrification (Fig. 9.2). The magnitude and direction of the nitrogen flux in these processes are governed by the supply of organic material, which provides chemically bound energy, generating oxygen consumption, and determines the redox potential. The organic material

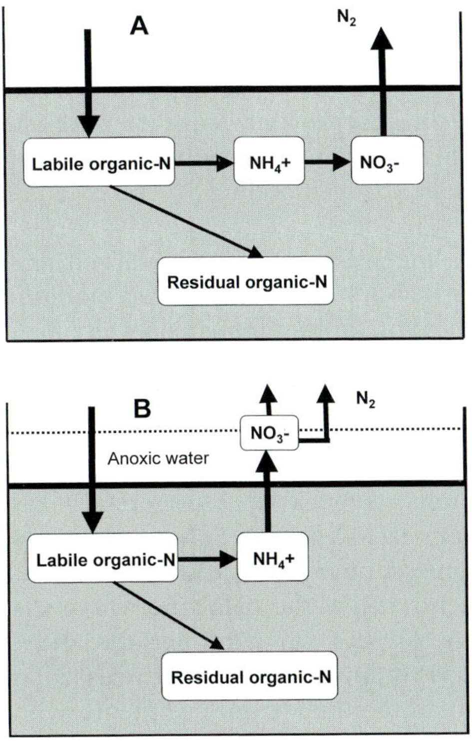

Fig. 9.2. Schematic presentation of transfer of nitrogen between different distribution forms at sediment-water interface of oxygenated (**A**) and anoxic (**B**) sediments

also supplies the nitrogen that is to be transformed in sedimentological processes. The physical conditions providing or preventing supply of oxygenated water to the sediment surface are of paramount importance for the nitrogen cycling within the ecosystem.

9.6 Nitrogen Turnover in Sediments of the Baltic Sea

If data on phosphorus dynamics in the Baltic Sea sediments are few, investigations on the nitrogen metabolism are even less. For example, in the Gulf of Bothnia there is probably only one single report from a few experiments with coastal near sediments (Nedwell et al. 1983) plus measurements on denitrification (Stockenberg and Johnstone 1997). In the Baltic proper a handful of estimates of denitrification losses (Rönner 1985; Stigebrandt and Wulff 1987; Brettar and Rheinheimer 1992; Jonsson and Carman 1992), plus a few studies on the nitrogen flux from different types of sediments (e.g. Koop et al. 1990) are reported. In addition, the budget calculations of Wulff et al. (1990) provide useful information on the net effects of different processes in the cycling of nitrogen. The following discussion on the nitrogen dynamics of the sediments of the Baltic Sea will therefore be speculative and based on general knowledge of nitrogen turnover in different types of sediments in combination with the scattered information on the intensity of different nitrogen cycling processes in the Baltic.

Organic nitrogen, which settles to the sediments, is mineralized and ammonium is released. In sediments above the halocline ammonium released in the surface sediment will be more or less immediately nitrified. Formed nitrate can be denitrified at the sediment surface or reach the trophogenic layer where it can be reassimilated. Similar to lakes, inorganic nitrogen regenerated in the surface sediment, may be an important complement to mineralization in the water column. However, the superior importance of sediment mineralization reported from eutrophic lakes (Ahlgren et al. 1994) is probably not reached in the Baltic Sea simply because of a much greater water volume:sediment surface ratio.

In reduced sediments, e.g. sediments below the halocline where the sediment is superimposed by more or less stagnant bottom water, ammonium can be adsorbed to particles in the sediment or be released to the overlying water. The release of ammonium from sediments with experimentally induced anoxia in the Baltic proper has been measured to be in the order of 15–35 $\mu M\ m^{-2}\ h^{-1}$ (Koop et al. 1990). Upwards transport occurs primarily by diffusion, which is a slow process, and considerable amounts of ammonia can, therefore, be built up in deep water. The ammonium concentrations in the Gotland Deep can increase from close to zero up to 10 mmol m^{-3} in deep water below 200 m and to between 1 and 5 mmol m^{-3} in layers between 100 and 150 m after 2–5 years without supply of oxygenated water from the Danish Sounds (Stigebrandt and Wulff 1987). However, part of this increase should be from ammonification in the water column.

When the ammonium concentrations increase in stagnant deep water it means that ammonium production exceeds the ammonium consumption by nitrification. Nitrification is an oxygen-consuming process, but it can proceed also at low concentrations of oxygen. Nitrification in the Baltic proper seems to be prevented only at totally anoxic conditions indicated by the presence of H_2S (Brettar and Rheinheimer 1992). A continuous transformation of the released ammonium to nitrate should therefore occur in the deep water below the halocline. In strongly reduced environments, nitrification, and consequently also denitrification, are inhibited (Nielsen et a1. 1994). The prime site for nitrification in deep water is therefore the transition zone between oxic and anoxic water and this zone is located at the sediment surface after intrusion of oxygenated water, after which it can be progressively moved upwards towards the lower part of the halocline. Field data and model calculations imply that it takes about 5 years after an intrusion of oxygenated water until the nitrification zone has reached the lower part of the halocline (Stigebrandt and Wulff 1987). The fact that increases in ammonium concentration appear to stop, or at least proceed at a much slower rate, after several years of stagnation indicates that a balance between ammonium release and nitrification is formed in the waters below the halocline.

A large part of the nitrified nitrogen produced below the halocline is bound to be reduced and exported to the atmosphere by denitrification since the cline in itself is an effective barrier against upwards transport to the trophogenic layer. Denitrification below the halocline takes places in both water and the sediment. Significant denitrification in the water column is likely to occur only below the halocline in the Baltic proper. Above the halocline, or in the water column of the Gulf of Bothnia, the oxygen concentration in the water should be too high to permit denitrification losses. Brettar and Rheinheimer (1992) measured in situ denitrification during summer in vertical profiles below the halocline in the Gotland Deep and at one station in the northern Baltic proper. In the Gotland Deep, denitrification occurred only in a layer at 130–140 m depth located at the oxic-anoxic interface, while denitrification at the station in the northern Baltic proper took place only in the water near the sediment. Denitrification in the water column at both stations was restricted by low carbon supply which directed the denitrification to carbon-rich water with low oxygen concentrations. The authors suggested that denitrification in the water column was carbon limited most of the year and highly variable due to differences in carbon supply. It was also suggested that the energetically more favorable reduction of N_2O to N_2 was important in addition to denitrification in carbon-limited water. To this can be added that the very large variation in oxygen concentration and nitrate supply in the deep waters, which occurs due to the intermittent intrusions of oxygenated water, is probably a more important governing factor for the magnitude of denitrification in the water column than the carbon supply. The investigation of Brettar and Rheinheimer was carried out after an extremely long period with stagnant water below the halocline and the results must be interpreted with regard to this situation.

Very few direct measurements of denitrification rates in the sediments in the Baltic Sea exist. Estimates of denitrification losses are therefore indirect and based on models or balance calculations. Denitrification estimates are made for the Baltic proper only. Rönner (1985) calculated denitrification losses from the total volume below 60 m depth, which roughly corresponds to layers in and below the halocline, to be 410,000 tons/year. About 80 % of this amount was proposed to be lost due to denitrification in the surface sediment. Rönner's calculation was based on a model of Shaffer and Rönner (1984) which calculated that denitrification below the halocline from production and consumption of nitrate was a function of salinity and oxygen consumption. The conclusion that 80 % of the denitrification took place in the sediment was based on the result that the model predicted that denitrification occurred at depths where the oxygen concentration in the water was too high to permit denitrification. Another important conclusion was that about 90 % of the denitrification was supplied with nitrate from nitrification. Rönner (1985) also compared denitrification losses with sedimentation of nitrogen through the halocline, which was estimated to be 571,000 tons/year. The difference between sedimentation and denitrification corresponds to permanent burial in the sediments, which according to this calculation was 160,000 tons/year. This figure falls well in the range of sediment nitrogen burial in Baltic proper offshore sediments (91,000–245,000 tons N/year) given by Jonsson and Carman (1992). If Rönner's calculation of denitrification losses is correct, about 70–75 % of the nitrogen reaching the halocline would be mineralized, nitrified and denitrified. However, as discussed below, Rönner probably overestimated the denitrification below the halocline, which means that a considerably smaller share of the settling nitrogen will be metabolized in microbial processes.

In a model of the nutrient dynamics in the Baltic proper (Stigebrandt and Wulff 1987), the denitrification losses below the halocline were given approximately the same figure as in Rönner's model. In addition, the model estimated that another 500,000–600,000 tons N/year was lost by denitrification in the sediments above the halocline. The total losses were therefore about 1,000,000 tons/year, which is nearly the same as the total yearly nitrogen supply to the Baltic proper. However, this estimate of the total denitrification was probably too high. In a later model (Wulff et al. 1990), the total annual loss (sediment burial plus denitrification) was calculated to be 729,000 tons/year. Furthermore, the burial of nitrogen in offshore sediments calculated from data of sediment cores was estimated to be 91,000–245,000 tons/year (Jonsson and Carman 1992). Since this range has to be considered as the most accurate available measure of burial in sediments in and below the halocline, it is obvious that the denitrification estimate of Stigebrandt and Wulff does not fit in the frames set by input, output through the Danish Sounds and permanent burial in the sediments.

Jonsson and Carman (1992) used the total nitrogen retention of 729,000 (Stigebrandt and Wulff 1987) for the whole Baltic proper and their own figure on the permanent burial in the offshore sediments to arrive at an estimate of 387,000–587,000 tons/year for total denitrification losses plus burial in the sediments of coastal areas (i.e. areas above the halocline). Though the total denitrification is not given by this figure, it nevertheless sets an upper limit of close to 600,000 tons/year. Thus, so far, one critical piece of information is missing, preventing the construction of an accurate nitrogen balance equation which can be used to estimate denitrification losses: the net burial of nitrogen in sediments above the halocline.

However, since most of the bottom area above the halocline in the Baltic proper consists of erosion and transport bottoms (Freden 1994), the mean areal net burial of nitrogen must be much lower than in the accumulation bottoms below the halocline. The bottoms above the halocline occupy approximately 60 % of the total bottom area. If it is assumed that the sedimentation on these bottoms is to a large extent balanced by resuspension losses, the net burial should be less or much less than the 91,000 tons/year which is the lower limit for burial in the deeper accumulation bottoms. If it is accepted that the total denitrification losses are 387,000–587,000 tons/year minus burial in shallow sediments, which most likely is less than 91,000 tons/year, it appears as if the total denitrification losses are in the range of 300,000–500,000 tons/year. This estimate corresponds to about 30–50 % of the total load on the Baltic proper given by Wulff et al. (1990) and about equals the permanent burial in the sediments. Considering that the denitrification losses in estuaries and highly eutrophic lakes seldom exceed 50 % and usually are considerably less than that (Ahlgren et al. 1994; Nielsen et al. 1994), denitrification losses for the Baltic proper of at the most 50 % of the loading seem reasonable.

If the range for denitrification losses given above is reasonably correct it means that previous estimates by Rönner (1985) and Stigebrandt and Wulff (1987) are wrong. Rönner's estimate falls within the range but only accounts for denitrification below 60 m and the estimate of Stigebrandt and Wulff is obviously too high.

If Rönner's estimate was correct it presumes that denitrification above the halocline is more or less non-existent, which is not likely. Measurements of denitrification rates in coastal near sediments thus often yield high rates (Andersen et al. 1984; Jensen et al. 1988; Nielsen et al. 1994), and in a detailed study of the nitrogen metabolism in stratified and non-stratified lakes it was clearly demonstrated that denitrification rates were much higher in shallow sediments than in deep sediments below the thermocline (Ahlgren et al. 1994). Large amounts of organic nitrogen must be deposited and further metabolized in sediments above the halocline and ammonification, nitrification and denitrification must occur above the halocline also in sediments superimposed by oxygenated water (cf.

Nielsen et al. 1994). Most of the nitrogen introduced to the Baltic proper must go through the sequence of assimilation–ammonification–nitrification–assimilation a large number of times before being linked off to burial in the sediment or denitrification. After sedimentation, the "average" particles are subjected to several steps of resuspension before being trapped below the halocline. Thus, most of the nitrogen will probably be exposed to environments suitable for denitrification several times before being permanently lost from biological processing. Similar to in-lake sediments, nitrification is most likely the limiting step for denitrification in the sediments above the halocline in any given moment. However, in a long-term perspective it is probably the retention time of nitrate above the halocline that determines the total denitrification losses (cf. Ahlgren et al. 1994; Jansson et al. 1994). Because of the very effective utilization–reutilization of nitrogen in a nitrogen-limited system like the Baltic proper, the retention time should be long enough to permit high denitrification losses. When denitrification in the future is defined by accurate measurements, it is likely that denitrification losses will be higher above the halocline than below the halocline. This distribution of denitrification losses is also indicated in the nitrogen model (Stigebrandt and Wulff 1987), which gave somewhat higher losses above the halocline than below. At present that model offers the best estimate of the relative denitrification losses from shallow and deep areas of the Baltic proper and a reasonable guess based on present knowledge is therefore that denitrification losses in the Baltic proper are about 300,000–500,000 tons/year and that more than half of this denitrification occurs above the halocline. With these figures the average denitrification rate in the Baltic proper is 12–20 kg ha^{-1} $year^{-1}$, which is similar to the rather stable rate of 20 kg ha^{-1} $year^{-1}$ measured in Danish estuaries (Nielsen et al. 1994).

In the oligotrophic Gulf of Bothnia, where the permanent halocline is absent and where deep-water anoxia is seldom developed, the sediment nitrogen metabolism should be less complicated than in the Baltic proper. Most of the sediment areas are likely to behave similar to the nearshore sediment at 20 m, as investigated by Nedwell et al. (1983). Ammonium is formed within the sediment and oxidized to nitrate in the sediment surface. Most likely, denitrification proceeds in the sediment surface. However, a large share of the formed nitrate can escape to the overlying water column. In the summer, local assemblages of organic material from sedimenting plankton generate ammonium release from the sediment. The ammonium is, however, rapidly oxidized after escaping to the water. Denitrification is probably increased at the sediment–water interface. These should be the dominating "typical situations", but it is not possible to say how they are distributed in time or space. The first situation is probably much more predominant than the second, which should be restricted to shallow areas during the period from May to October.

The magnitude of denitrification losses can only be guessed. Recent measurements (Stockenberg and Johnstone 1997) gave average rates between 10 and 20 kg ha^{-1} $year^{-1}$. These rates multiplied by the total area (116,300 km^{-2}) give denitrification losses of about 120,000–240,000 tons/year. These figures correspond to about 80–160 % of the total annual loading and 85–165 % of the average annual total retention using the nitrogen balance of Wulff et al. (1990). Even at the lower end of these ranges such high relative denitrification losses are less plausible when compared with other marine systems with more favorable conditions for denitrification (e.g. the Baltic proper). It is likely that the average denitrification rate is clearly less than 10 kg ha^{-1} $year^{-1}$. For example, if the denitrification losses correspond to 30 % of the annual input, the average denitrification rate should be 4 kg ha^{-1} $year^{-1}$.

9.7 Concluding Remarks

The sediments of the Baltic Sea offer efficient sink mechanisms for both phosphorus and nitrogen, which to a large extent can buffer the consequences of the increased input of nutrients. The permanent burial of phosphorus in the sediments thus corresponds to 70 % of the external input. Even in the now moderately eutrophic Baltic proper, the net sedimentation is about 60 % of the calculated input. The corresponding share for nitrogen is probably about 25–45 % for the Baltic proper. If denitrification losses are added, the total nitrogen retention is 75 % in the Baltic proper and almost 80 % for the entire Baltic Sea.

The ongoing eutrophication of the Baltic proper has reduced the phosphorus retention efficiency of the sediments by the extension of areas with reduced sediments. This is particularly evident below the halocline. However, the phosphorus cycling between water and sediment below the halocline appears to be internal in the sense that most phosphorus released to and accumulated in the water is trapped below the halocline. Reprecipitation of phosphorus after intrusion of oxygenated water from the Kattegat regularly restores the water quality below the halocline. The sediments above the halocline are also a net sink for phosphorus, but release to the overlying water should occur in certain areas and at certain times of the year in connection with degradation of sedimented organic material.

An important process in the Baltic Sea which counteracts extended development of phosphorus-saturated and -reduced bottoms above the halocline is resuspension, which translocates considerable amounts of sedimented material from shallow bottoms to deep-sea areas. A simple balance calculation indicates that 80–85% of the phosphorus settling above the halocline is redistributed to deeper bottoms. The average transit time is unclear but should be in the order of several decades. The resuspension thus keeps the phosphorus retention capacity of the shallow bottoms high and also increases the phosphorus loading on deep sediments. If this translocation takes at least several decades for an average phosphorus-containing particle, then it is an open question whether the present phosphorus loading on deep-sea sediments is in balance with the present external input of phosphorus. It is likely that this is not the case (see below) and that the continuous scavenging of particles from sediment above the halocline will increase the phosphorus supply to deep-sea sediments for some time, even if the present external loading is reduced. A very clear indication that the Baltic proper is not in a steady state situation where input, output and retention are in balance is given by the fact that a considerable part of the phosphorus input is accumulated in the water column (Wulff et al. 1990). It means that the sedimentation of phosphorus in the Baltic proper has not increased as much as the input to the system. Two possible explanations exist. Phosphorus sedimentation can be partly counteracted by a return flow from the sediments. Though this explanation cannot be rejected, it appears less plausible against the background of the situation discussed in this chapter. An alternative explanation is that phosphorus at present is accumulated in the Baltic proper water because the biological utilization–reutilization of phosphorus is very efficient above the permanent halocline. Particulate phosphorus, which settles to the halocline, can be mineralized and returned to the trophogenic layer without being trapped by sediments. This type of internal circulation was evident at a much smaller scale in experimental lake fertilization experiments with stratified and non-stratified lakes (Jansson 1984). The burial of phosphorus and nitrogen in the sediments was significantly lower in a lake with a thermocline than in a lake without a thermocline. Added nutrients were to a large extent accumulated in the water above the thermocline in the stratified lake and in the sediment of the non-stratified lake.

The most important role of the sediments in the nitrogen balance of the Baltic Sea is to supply the conditions where denitrification can proceed. The denitrification losses estimated for the Baltic proper in this chapter thus correspond to 30–50% of the external input of nitrogen. Without this denitrification, the Baltic proper should have been a phosphorus-limited system. However, the denitrification appears to be a rather constant share of available nitrate, at least in nitrogen-limited systems, which have been stressed by experimental work (Seitzinger and Nixon 1985) and data from estuaries (Seitzinger 1988; Nielsen et al. 1994). If the N:P ratio of the external supply is increased, which is a possible future scenario for the Baltic, the amounts of nitrogen available for assimilation can therefore increase relative to phosphorus in spite of the denitrification. Phosphorus can then substitute nitrogen as the key limiting nutrient in the Baltic proper. The N:P ratio in the external input of today is about 15 (weight), which can be compared with the Redfield ratio of 7, expressing the approximate ratio in biota. The net result of phosphorus sedimentation and denitrification means that the average N:P ratio in the Baltic proper is quite close to 7 (Wulff et al. 1990). It is therefore tempting to suggest that sedimentological processes at present create a rather ideal balance between phosphorus and nitrogen content within the Baltic proper. However, it could just as well be the other way around, so that biological utilization in the water determines the amount of nutrients available for sedimentation and denitrification.

References

Ahlgren I, Sörensson F, Waara T, Vrede K (1994) Nitrogen budgets in relation to microbial transformations in lakes. Ambio 23:367–377

Andersen TK, Jensen MH, Sörensen J (1984) Diurnal variation of nitrogen cycling in coastal marine sediments. Mar Biol 83:171–176

Balzer W (1986) Forms of phosphorus and its accumulation in coastal sediments in Kielerbucht. Ophelia 26:19–35

Blomquist S, Larsson U (1994) Detrital bedrock element as tracers of settling suspended particulate matter in a coastal area of the Baltic Sea. Limnol Oceanogr 39:80–896

Boesen C, Postma D (1988) Pyrite formation in the anoxic environments of the Baltic. Am J Sci 288:575–603

Boström B, Jansson M, Forsberg C (1982) Phosphorus release from lake sediments. Arch Hydrobiol Beih Ergebn Limnol 18:5–59

Boström B, Andersen JM, Fleischer S, Jansson M (1988) Exchange of phosphorus across the sediment–water interface. In: Persson G, Jansson M (eds) Phosphorus in freshwater ecosystems. Developments in hydrobiology 48. Kluwer, Dordrecht, pp 229–244

Brettar I, Rheinheimer G (1992) Influence of carbon availability on denitrification in the central Baltic Sea. Limnol Oceanogr 37:1146–1163

Broberg O, Persson G (1988) Particulate and dissolved phosphorus forms in freshwater: composition and analysis. In: Persson G, Jansson M (eds) Phosphorus in freshwater ecosystems. Developments in hydrobiology 48. Kluwer, Dordrecht, pp 61–90

Brydsten L (1993) Characterization of transport bottoms in the Gulf of Bothnia – a model approach. Aqua Fenn 23(2):153–164

Brydsten L, Jansson M (1989) Studies of estuarine sediment dynamics using 137Cs from the Tjernobyl accident as a tracer. Estuarine Coast Shelf Sci 28:249–259

Brydsten L, Jansson M, Andersson T, Nilsson Å (1990) Element transport in regulated and non-regulated rivers in northern Sweden. Regul Rivers Res Manage 5: 167–176

Carman R (1990) Biogeochemical constraints on phosphorus exchange and related diagenetic processes in Baltic Sea sediments. PhD Thesis, Medd Stockholm Univ Inst Geologi Geokemi 282, 38 pp

Carman R, Jonsson P (1991) Distribution patterns of different forms of phosphorus in some surficial sediments of the Baltic Sea. Chem Geol 90:91–106

Carman R, Wulff F (1989) Adsorption capacity of phosphorus in Baltic Sea sediments. Estuarine Coast Shelf Sci 29:447–456

Chang C, Jackson ML (1957) Soil phosphorus fractions in some representative soils. J Soil Sci 9:109–119

Einsele W (1938) Über chemische und kolloidchemische Vorgänge in Eisen-Phosphat-systemen unter limnologischen und limnogeologischen gesichtspunkten. Arch Hydrobiol 33:361–387

Fleischer S (1983) Microbial phosphorus release during enhanced glycolysis. Naturwissenschaften 70:415–416

Floderus S, Håkansson L (1989) Resuspension, ephemeral mud blankets and nitrogen cycling in Laholmsbukten, south east Kattegat. Hydrobiologia 176/177:61–75

Freden C (ed) (1994) Berg och Jord. Sveriges Nationalatlas. SNA Förlag, Stockholm

Forsgren G, Jansson M (1992) The turnover of river-transported iron, phosphorus and organic carbon in the Öre Estuary, northern Sweden. Hydrobiologia 235/236: 585–596

Forsgren G, Jansson M (1993) Sedimentation of phosphorus in limnetic and estuarine environments in the River Öre system, northern Sweden. Hydrobiologia 253:23–248

Forsgren G, Jansson M, Nilsson P (1994) Aggregation and sedimentation of iron, phosphorus and organic carbon in experimental mixtures of fresh water and estuarine water. Estuarine Coast Shelf Sci 43:259–268

Gächter R, Meyer JS (1993) The role of microorganisms in mobilization and fixation of phosphorus in sediments. Hydrobiologia 253:103–121

Gidhagen L (1987) Coastal upwelling in the Baltic Sea – a presentation of satellite and in situ measurements of sea surface temperatures indicating coastal upwelling. Estuarine Coast Shelf Sci 24:449–462

Hieltjes AHM, Lijklema L (1980) Fractionation of inorganic phosphates in calcareous sediment. J Environ Qual 9:405–407

Holm NG (1978) Phosphorus exchange through the sediment–water interface. Mechanism studies of the dynamic processes in the Baltic Sea. Contributions in microbial geochemistry. PhD Thesis, Department of Geology, University of Stockholm, no 3, 149 pp

Jansson BO, Wilmot W, Wulff F (1984) Coupling the subsystems – the Baltic Sea as a case study. In: Fasham MJR (ed) Flows of energy and material in marine ecosystems. Plenum, New York, pp 549–595

Jansson M (1984) Experimental lake fertilization: turnover of nitrogen and phosphorus in stratified and non-stratified subarctic lakes in northern Sweden. Verh Int Verein Limnol 22:708–711

Jansson M, Leonardson L, Fejes J (1994) Denitrification and nitrogen retention in a farmland stream in southern Sweden. Ambio 23:326–331

Jensen MH, Andersen TK, Sörenson J (1988) Denitrification in coastal bay sediment. Regional and seasonal variation in Aarhus Bight, Denmark. Mar Ecol Prog Ser 48:155–162

Jensen HS, Thamdrup B (1993) Iron bound phosphorus in marine sediments as measured by bicarbonate-dithionite extraction. Hydrobiologia 253:47–59

Jonsson P, Carman R (1992) Changes in sediment deposition of organic matter and nutrients in the Baltic Sea during the twentieth century. In: Jonsson P (ed) Large scale

changes of contaminants in the Baltic Sea during the twentieth century. Acta Univ Ups 407
Jonsson P, Jonsson B (1988) Dramatic changes in Baltic sediments during the last three decades. Ambio 17:158–160
Jonsson P, Carman R, Wulff F (1990) Laminated sediments in the Baltic – a tool for evaluating nutrient mass balances. Ambio 19:152–158
Jörgensen BB, Bang M, Blackburn TH (1990) Anaerobic mineralization in marine sediments from the Baltic Sea–North Sea transition. Mar Ecol Prog Ser 59:39–54
Koop K, Boynton WR, Wulff F, Carman R (1990) Sediment-water oxygen and nutrient exchanges along a depth gradient in the Baltic Sea. Mar Ecol Prog Ser 63:65–77
Krom MD, Berner RA (1981) The diagenesis of phosphorus in a nearshore marine sediment. Geochim Cosmochim Acta 45:207–216
Kullenberg G (1981) Physical oceanography. In: Vuopio A (ed) The Baltic Sea. Elsevier Oceanogr Ser 30:135–182
Larsson U, Elmgren R, Wulff F (1985) Eutrophication and the Baltic Sea: causes and consequences. Ambio 14:9–14
Lucotte M, d'Anglejan B (1985) A comparison of several methods for the determination of iron hydroxides and associated orthophosphate in estuarine particulate matter. Chem Geol 48:257–264
Lucotte M, d'Anglejan B (1988) Processes controlling phosphate adsorption by iron oxides in estuaries. Chem Geol 67:75–83
Miljödepartementet (1990) Vem förorenar Sverige? Allmäna Förlaget, Stockholm
Mortimer CH (1941) The exchange of substances between mud and water in lakes. I. J Ecol 29:280–329
Mortimer CH (1942) The exchange of substances between mud and water in lakes. II. J Ecol 30:147–201
Nedwell DB, Hall SE, Andersson A, Hagström ÅF, Lindström EB (1983) Seasonal changes in the distribution and exchange of inorganic nitrogen between sediment and water in the northern Baltic (Gulf of Bothnia). Estuarine Coast Shelf Sci 17: 169–179
Nielsen LP, Christensen PB, Rysgaard S (1994) Denitrifikation i fjorde og kystnaere farvande. Havforskning fra Miljöstyrelsen 50. Miljöministeriet Miljöstyrelsen, Köbenhamn
Nilsson P, Brydsten L, Jansson M (1997) Turnover of particulate matter in a Baltic estuary. Estuarine Coast Shelf Sci 44:107–116
Persson G (1990) Växtnäringsämnen och eutrofiering i havet (in Swedish). Swedish Environmental Protection Board,Rep 3964, Stockholm, 47 pp
Psenner R, Pucsko R, Sager M (1985) Fraktionierung organischer und anorganischer Phosphorverbindungen von Sedimenten. Versuch einer Definition ökologisch wichtiger Fraktionen. Arch Hydrobiol Suppl 70:11–115
Rönner U (1985) Nitrogen transformations in the Baltic proper: denitrification counteracts eutrophication. Ambio 14:135–148
Ruttenberg KC (1992) Development of a sequential extraction method for different forms of phosphorus in marine sediments. Limnol Oceanogr 37:1460–1482
Sand-Jensen K, Borumm J, Geertz-Hansen H, Nörrevang-Jensen J, Josefson AB, Möhlenberg F, Rieman B (1994) Resuspension og stofomsaetning i Roskilde Fjord. Havforskning Fra Miljöstyrelsen 51. Danmarks Miljöstyrelse, Copenhagen
Shaffer G, Rönner U (1984) Denitrification in the Baltic proper deep water. Deep Sea Res 31:197–220
Seitzinger SP (1988) Denitrification in freshwater and coastal marine ecosystems. Limnol Oceanogr 33:702–724

Seitzinger SP, Nixon SW (1985) Eutrophication and the rate of denitrification and N_2O production in coastal marine sediments. Limnol Oceanogr 30:1332–1339

Slomp CP, van Raaphorst W, Malschaert JFP, Kok A, Sandee JJ (1993) The effect of decomposition of organic matter on phosphorus dynamics in experimental marine sediment systems. Hydrobiologia 253:83–98

Stigebrandt A (1983) A model for the exchange of water and salt between the Baltic and the Skaggerak. J Phys Oceanogr 13:411–427

Stigebrandt A, Wulff F (1987) A model for the dynamics of nutrients and oxygen in the Baltic proper. J Mar Res 45:729–759

Stockenberg A, Johnstone RW (1997) Denitrification in the Gulf of Bothnia. Estuarine Coast Shelf Sci 45:835–843

Sundby B, Gobeilm C, Silverberg N, Mucci A (1992) The phosphorus cycle in coastal marine sediments. Limnol Oceanogr 37:1129–1145

Waara T, Jansson M, Pettersson K (1993) The relative importance of biological and chemical processes in the release of phosphorus from a highly organic sediment. Hydrobiologia 253:131–140

Wallin M, Håkanson L (1992) Morphometry and sedimentation as regulating factors for nutrient recycling and trophic state in coastal waters. Hydrobiologia 235/236:33–45

Walter LM, Burton EA (1986) The effect of orthophosphate on carbonate dissolution rates in seawater. Chem Geol 56:313–323

Williams JDH, Syers JK, Armstrong DE, Harris F (1971) Fractionation of inorganic phosphate in calcareous lake sediments. Soil Sci Soc Am Proc 35:250–255

Wulff F, Stigebrandt A, Rahm L (1990) Nutrient dynamics of the Baltic Sea. Ambio 19:16–133

Yamada H, Kayama M (1987) Distribution and dissolution of several forms of phosphorus in coastal marine sediments. Oceanol Acta 10:311–321

Yurkovskis A, Wulff F, Rahm L, Andruzaitis A, Rodrigues-Medina M (1993) A nutrient budget of the Gulf of Riga; Baltic Sea. Estuarine Coast Shelf Sci 37:133–127

10 Dynamics and Distribution of Hydrophobic Organic Compounds in the Baltic Sea

J. Axelman, C. Näf, C. Bandh, R. Ishaq,
H. Pettersen, Y. Zebühr, and D. Broman

10.1 Introduction

The main objectives of this study were to quantify the distribution of polycyclic aromatic hydrocarbons (PAHs) and polychlorinated biphenyls (PCBs) between air, water and sediments and to estimate the most important fluxes between these compartments by synthesizing all data in a semi-empirical mass balance of PAHs and PCBs for the Baltic Sea.

The measurements have taken into account not only the pollutant distribution between air, water and sediments but also the phase distribution of these compounds within these compartments. The concentrations of PAHs and PCBs reported here distinguish between dissolved (for seawater and river water)/vapor phase (air) and particle bound (all samples) hydrophobic organic compounds (HOCs). Different mechanisms act upon HOCs in the different phases which makes this distinction necessary to assess the intercompartmental exchange processes. Fluxes between the compartments have been studied both directly by measuring sedimentation and indirectly as for calculation of deposition from air and diffusive exchange processes between atmosphere/water and sediment/water.

PAHs and PCBs represent two different types of contaminants. Both groups of compounds have well-characterized physicochemical properties (Tables 10.1 and 10.2) and have been considered to pose an ecotoxicological hazard. Some of the PCBs are highly persistent to degradation and can thus serve as model compounds for other persistent compounds, such as DDTs and PCDD/Fs. The persistent nature of PCBs causes them to accumulate in biota. PAHs are also relatively persistent in the abiotic phases of the environment, but although they are emitted in large amounts they are not found in correspondingly high levels in higher organisms, due to biotransformation (Broman et al. 1990a). However, the

Ecological Studies, Vol. 148
F. Wulff et al. (eds.)
A Systems Analysis of the Baltic Sea

Table 10.1. Physicochemical and environmental properties of PCBs

PCB	Notes	Unit	#52	#77	#101	#118	#153	#138	#180
Log K_{ow}			5.84[a]	6.63[a]	6.38[a]	6.74[a]	6.92[a]	6.83[a]	7.36[a]
p°		Pa	0.016[b]	0.0021[c]	0.0032[b]	0.0032[d]	0.00068[b]	0.00051[b]	0.00013[b]
K_H		Pa m^3 mol^{-1}	20.3[e]	20.3[f]	9.1[e]	9.1[d]	2.3[e]	2.1[e]	1.0[e]
K'_H	25 °C		0.0082	0.0082	0.0037	0.0037	0.00094	0.00086	0.00041
ΔH_{vap}		kJ mol^{-1}	52[g]	52[g]	52[g]	52[g]	52[g]	52[g]	52[g]
K'_H	8 °C		0.0023	0.0023	0.0010	0.0010	0.00026[h]	0.00024	0.00011
W_{tot}[i]	Min		8000	8000	8000	8000	8000	8000	8000
	Max		40,000	40,000	40,000	40,000	40,000	40,000	40,000
V_{dp}[j]	Min	cm s^{-1}	0.1	0.1	0.1	0.1	0.1	0.1	0.1
	Max	cm s^{-1}	0.5	0.5	0.5	0.5	0.5	0.5	0.5
k_{OL}		m h^{-1}					0.009		
$^{In\text{-}situ}$log K_{oc}			6.84	7.51	6.96	7.41	7.65	7.52	7.85
$^{Calc.}$log K_{oc}[k]			5.45	6.24	5.99	6.35	6.53	6.44	6.97

[a] Hawker and Connell (1988)
[b] Bidleman (1984)
[c] Foreman and Bidleman (1985)
[d] Value for PCB #101 used
[e] Brunner et al. (1990)
[f] Value for PCB #52 used
[g] ten Hulscher et al. (1992) (value for PCB #52)
[h] Mean value for PCB 101 and PCB 153 used in gas-exchange calculations
[i] Bidleman (1988)
[j] Eisenreich et al. (1981)
[k] According to Karickhoff (1981)

Table 10.2. Physico-chemical and environmental properties of PAHs. *Flu* Fluoranthene; *Pyr* pyrene; *2mePyr* 2-methylpyrene; *1mePyr* 1-methylpyrene; *B(ghi)P* benzo(ghi)fluoranthene; *B(a)A* benzo(a)anthracene; *Chr* chrysene; *B(k)F* benzo(k)fluoranthene; *B(e)P* benzo(e)pyrene; *B(a)P* benzo(a)pyrene; *Per* perylene; *Ind* indeno(1,2,3-cd)pyrene; *B(ghi)P* benzo(ghi)-perylene; *Cor* coronene

PAH	Notes	Unit	Flu	Pyr	2 mePyr	1 mePyr	B(ghi)F	B(a)A
Log K_{ow} [a]			5.22	5.18	5.18	5.18	5.65	5.91
p^{o} [b]		Pa	0.00068	0.000413				1.47×10^{-05}
K_H [c]		Pa m^3 mol^{-1}	2.38					
K'_H	25 °C		0.000961					
ΔH_{vap} [c]		kJ mol^{-1}	57					
K'_H	8 °C		0.000184					
W_{tot} [d]	Min		3500	3000	3000[e]	3000[e]	60,000[f]	60,000
	Max		8000	7000	7000[e]	7000[e]	60,000[f]	60,000
V_{dp} [i]	Min	cm s^{-1}	0.1	0.1	0.1	0.1	0.1	0.1
	Max	cm s^{-1}	0.5	0.5	0.5	0.5	0.5	0.5
k_{OL}		m h^{-1}						
$^{in\text{-}situ}\log K_{oc}$			6.51	6.45	7.25	7.04	6.57	7.09
$^{calc.}\log K_{oc}$ [j]			4.83	4.79	4.79	4.79	5.26	5.52

Table 10.2. *(continued)*

PAH	Notes	Unit	Chr	B(k)F	B(e)P	B(a)P	Per	Ind	B(ghi)P	Cor
Log K_{ow}[a]			5.86	6	6.2	6.2	6.25	6.5	6.5	6.75
p^{o}[b]		Pa			3.2E $\times 10^{-07}$	3.2×10^{-07}		1.35×10^{-08}	1.35×10^{-08}	
K_H[c]		Pa m^3 mol^{-1}		0.138		0.092		0.061	0.054	
K'_H	25 °C			5.57×10^{-05}		3.71×10^{-05}		2.46×10^{-05}	2.18×10^{-05}	
ΔH_{vap}[c]		kJ mol^{-1}		49		39		30	26	
K'_H	8 °C			1.34×10^{-05}		1.2×10^{-05}		1.03×10^{-05}	1.02×10^{-05}	
W_{tot}[d]	Min		60,000[f]	200,000	200,000	200,000	200,000[g]	200,000[g]	250,000	250,000[h]
	Max		60,000[f]	200,000	200,000	200,000	200,000[g]	200,000[g]	250,000	250,000[h]
V_{dp}[i]	Min	cm s^{-1}	0.1	0.1	0.1	0.1	0.1	0.1	0.1	0.1
	Max	cm s^{-1}	0.5	0.5	0.5	0.5	0.5	0.5	0.5	0.5
k_{OL}		m h^{-1}				0.0003				
$^{in\text{-}situ}\log K_{oc}$			6.78	7.42	7.65	8.10	8.17	8.37	7.91	8.33
$^{calc.}\log K_{oc}$[j]			5.47	5.61	5.81	5.81	5.86	6.11	6.11	6.36

[a] Mackay et al. (1991)
[b] Pearlman et al. (1984)
[c] ten Hulscher et al. (1992)
[d] Bidleman (1988)
[e] Value for pyrene used
[f] Value for benzo(a)anthracene used
[g] Value for benzo(a)pyrene used
[h] Value for benzo(ghi)perylene used
[i] Eisenreich et al. (1981)
[j] according to Karickhoff (1981)

biotransformation process itself may be harmful to the organism due to the fact that metabolites may be toxic.

It is worth pointing out a difference between PCBs and PAHs. The sources of PCBs are much less complex than for PAHs. Most PCB congeners originate from the technical products Aroclor, Clophen, Kanechlor, Phenoclor or Sovol, while PAHs are generally formed in various combustion processes generating different PAH compositions. In addition, PAHs are often released associated to soot particles from which PAHs have been shown to desorb so slowly that equilibrium partitioning in water is not always applicable (McGroddy et al. 1996).

All concentrations, fluxes and total contents of PAHs and PCBs given in the text are sums of the following compounds:

PCBs Congeners with IUPAC numbers 52, 101, 118, 138, 153, 180 which are the congeners found in highest concentrations in most environmental matrices. In addition, the non-ortho substituted congener 77 was studied which is one of the PCBs with the highest toxicity. In the technical PCB mixture Aroclor 1254, these congeners constitute approximately 27 % of the total weight (Schulz et al. 1989).

PAHs Fluoranthene, pyrene, 2-methylpyrene, 1-methylpyrene, benzo-[*ghi*]fluoranthene, benzo[*a*]anthracene, chrysene, benzo[*k*]fluoranthene, benzo[*e*]pyrene, benzo[*a*]pyrene, perylene, indeno[1,2,3-*cd*]pyrene, benzo[*ghi*]perylene and coronene.

10.2 Materials and Methods

10.2.1 Sampling

Samples of the different environmental matrices were taken at locations shown in Fig. 10.1. Air samples were taken with an air sampler similar to that of Broman et al. (1991a). Five different remote locations were chosen and samples were taken during 11–34 days at two occasions: one winter and one summer sample at each location. Thirteen water filtrations were carried out in the three major sub-basins (Bothnian Bay, Bothnian Sea and Baltic proper) during both summer and winter with a technology similar to that described by Broman et al. (1991b). River samples were taken from one major river in four of the five major drainage areas (Bergström and Carlsson 1994): Umeälven in northern Sweden, Neva in the Gulf of Finland, Daugava in the Gulf of Riga and Vistula in the southern Baltic proper. One sample was taken at high flow during the spring and one at low flow during

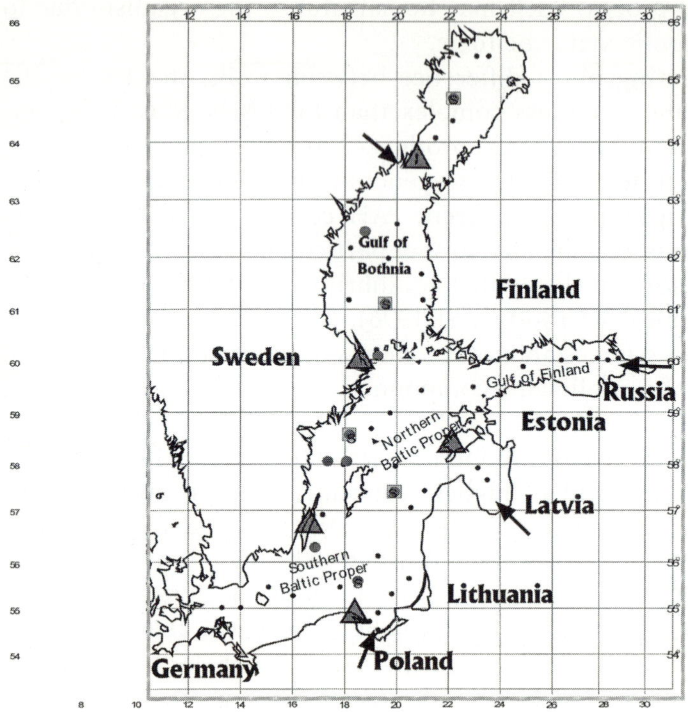

Fig. 10.1. Map of sampling area with location of the different samples taken indicated: sediments (*dots*), water (*circles*), sediment traps (*squares*), air (*triangles*) and rivers (*arrows*)

late summer at all locations except Vistula. In the southern rivers the highest flow occurs during autumn since there is no significant snow melting during the spring in these drainage areas. From the River Vistula only one sample during high flow in the autumn was taken. Samples were taken by pumping river water through precleaned 1.2-μm Sartopure filter capsules after which the water passed through PUF adsorbent. Filters and PUF adsorbent were analyzed separately. Sediment samples were taken with Kajak-type gravity corers from 44 locations. From the sediment core only the upper 1 cm of the sediment was further processed. Sediment traps were placed in all the major sub-basins below the euphotic zone (26–155 m depth) for 0.5–1.5 years. The hydrodynamically stable (Broman et al. 1990b) sediment traps had an opening diameter of 98–160 mm and all had a height diameter ratio of >3. All samples were stored in the freezer during the time between sampling and chemical analysis.

10.2.2 Chemical Analysis

Prior to extraction, all samples were spiked with internal standards, ^{13}C-labeled PCBs of all analyzed congeners for the PCBs and D_{12}-perylene for PAHs. All samples except river filter capsules were extracted with toluene in Soxhlet extractors for 24 h. Filter capsules of river samples were extracted with 2 × 200 ml acetone for 2×2 h in an ultrasonic bath, then 2 × 200 ml hexane for 2 × 2 h in ultrasonic bath, after which the solvents were dried with precombusted $NaSO_4$. The volume-reduced extracts were eluted on 100 × 10-mm columns containing deactivated SiO_2 (10 % water w/w) with *n*-hexane as mobile phase (Zebühr et al. 1989; Broman et al. 1991a). Thereafter, the extracts were run on an HPLC system using two different columns similar to that described by Bandh et al. (1994) for separation into three fractions: mono- and di-ortho PCBs, non-ortho PCBs and PAHs. The mono-/di-ortho PCB and the non-ortho PCB fractions were analyzed on gas chromatograph/mass spectrometer, Hewlett Packard 5890/5971 A and Fisons GC8000/MD800 respectively. The PAH fraction was analyzed on a GC Varian 3300/FID. A standard mixture containing all PAH compounds analyzed together with the internal standards was extracted and cleaned-up according to the same procedure as for the sediment samples for determination of response factors and retention times for the analyzed compounds relative to the internal standards.

10.2.3 Statistics

Concentrations of pollutants in the environment have been suggested to be log normal distributed (Ott 1990). Shapiro-Wilk's tests on individual data sets (for air, water and sediments) for normality confirmed this to be valid also for the present data while all data were log transformed before statistical treatment including linear regressions. Analysis of variance (ANOVA) was used to test differences in concentrations of pollutants between winter and summer samples. Linear regression p values were calculated using Statistica 4.5 software and ANOVA was calculated using Microsoft Excel 7.0.

10.3 Results and Discussion

10.3.1 Concentrations and Phase Distribution of HOCs in Air, Water and Sediments

10.3.1.1 Seawater

The average concentrations of PCBs were 8.4 pg l^{-1} (range 2.4–24) for the dissolved fraction and 15 pg l^{-1} (range 3.1–70) for the particle-bound fraction. These data are in accordance with what has been found by others in the Baltic Sea surface waters (Schulz-Bull et al. 1995). For PAHs the corresponding concentrations were 2.1 ng l^{-1} (range 0.1–8.9) and 0.9 ng l^{-1} (range 0.3–2.7) respectively. For PCBs the particle-bound form dominates, whereas the PAHs occur mainly in the dissolved phase. The difference in distribution between the two groups of compounds cannot be explained by the difference in hydrophobicity for the two groups. The range of log K_{ow} is 5.48–7.36 (Hawker et al. 1988) for the PCBs and 5.16 (De Bruijn et al. 1989) to 7.64 (Rekker et al. 1979) for the PAHs, i.e. they are relatively similar. The explanation can instead be found in the relative abundance of the different compounds. Among the PAHs the two most abundant compounds are fluoranthene and pyrene, the two least hydrophobic compounds with log K_{ow} of 5.15 (De Bruijn et al. 1989) and 5.18 (Karickhoff et al. 1979) respectively, whereas the two most abundant PCB congeners in most samples were the more hydrophobic #138 and #153 with log K_{ow} of 6.83 and 6.92 respectively (Hawker and Connell 1988). The concentrations of the dissolved phase did not correlate with latitude either for PCBs ($p<0.22$) or for PAHs ($p<0.35$). The overall variation in the concentration in both the particle-bound and the dissolved phase was relatively low, 95 % confidence limits were ±50 % of the average value for PCBs and ±40 % for PAHs, which means that the total inventory in the water column in the mass balance in the last section of this chapter can be assessed relatively accurately.

The concentrations in suspended matter can also be expressed as carbon normalized concentrations. For PCBs the average concentration was 44 ng C g^{-1} (range 17–80) and for PAHs 10 µg C g^{-1} (range 4–22).

10.3.1.2 Air

The observed average fraction bound to aerosols (ϕ) for PCBs was 0.14 and 0.67 for PAHs, with a broader range within the PAH samples due to the broader range in vapor pressure (p°). ϕ, based on an average of all samples, varied from 0.45 for fluoranthene to 0.99 for coronene, while the corresponding range for PCBs was found to be from 0.11 for PCB #52 to 0.28 for PCB #180. This can be compared with calculated aerosol-vapour distributions using the equation of Junge (1977):

$$\phi = \frac{c \cdot S_T}{p^\circ + c \cdot S_T} \tag{10.1}$$

where p° is the saturation vapor pressure (Pa) of the subcooled liquid compound. p° varies with temperature; at low temperatures p° is lower. S_T is the average total surface area of the aerosols ($cm^2 \cdot cm^{-3}$) in air and c a factor depending on the adsorbing capacity of the aerosol as well as the properties of the compound. Junge (1977) proposes a typical value for c of 17 Pa cm. S_T can vary by an order of magnitude from day to day and between different locations. Background S_T is $1.5 \times 10^{-6}\,cm^2\ cm^{-3}$ (Whitby 1978). The calculated ϕ for fluoranthene was 0.0038 and for coronene >0.9999.

For PCBs, calculations yield ϕ of 0.0024 for PCB #52 and 0.16 for PCB #180. It is difficult to draw any conclusions from comparison with calculated values due to the variations in S_T and possibly c as well as the uncertainties in measurements of p°, but the general trend seems to be that the lighter PAHs as well as both lighter and heavier PCB congeners are more bound to particles than would be expected from their vapor pressures. Previous studies of air phase distribution of PAHs in the Baltic Sea area (Broman et al. 1991) and in the Great Lakes area (McVeety and Hites 1988) also showed a higher fraction bound to aerosols than calculated values. This implies that the empirically derived values of 17 Pa cm^{-1} for the factor c as well as $1.5 \times 10^{-6}\,cm^{-2}\ cm^{-3}$ for S_T may not be valid in field situations at these remote locations. Separating the samples into summer and winter samples reveals that temperature can have an effect on the distribution as would be expected according to Eq. (10.1). For all compounds the average fraction bound to aerosols is lower in the summer than in the winter (Fig. 10.2).

The total air concentrations of PCBs ranged from 5 to 40 pg m^{-3} with an average value of 21 pg m^{-3}. There were no differences in concentrations between seasons ($p<0.074$), nor did the concentrations correlate with latitude ($p<0.57$). The lack of seasonal variation of total air concentration is probably due to the limited number of samples. A marked peak in air concentrations of PCBs in the warmest month of the year was observed in a study by Hoff et al. (1992) who used samples with 14-day resolution. The higher air concentrations during the warm period were explained as an effect of increased volatilization from soil, vegetation and water as a result

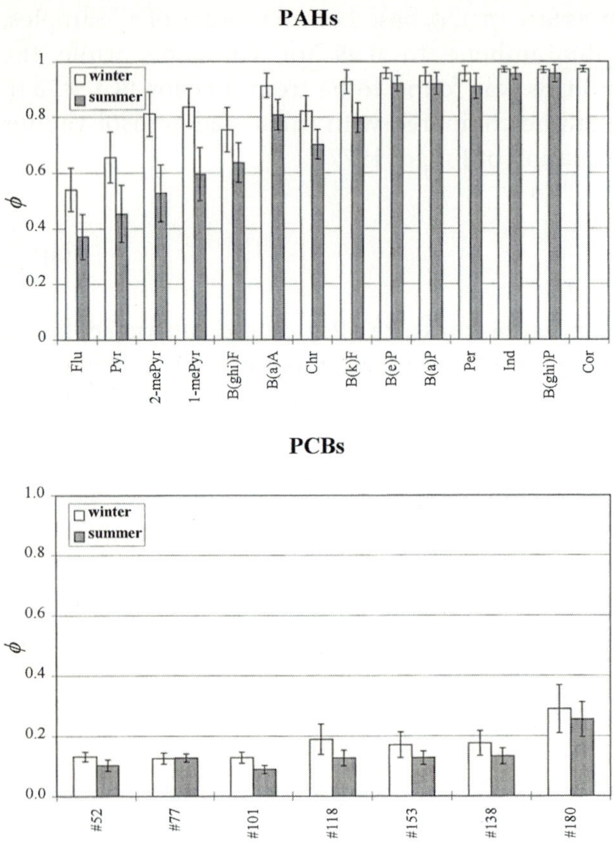

Fig. 10.2. Average fraction bound to aerosols, ϕ, in air samples during winter and summer for PAHs and PCBs

of the increased vapor pressure at the high temperature. During the rest of the year, the variations were considerably smaller.

For PAHs the concentrations ranged from 1.3 to 9.8 ng m^{-3}, with an average value of 4.9 ng m^{-3}. PAHs exhibited a latitudinal trend. A linear regression between latitude and log total concentrations for all ten air samples yielded an r^2=0.61 (p<0.01). A negative relationship was found between latitude and air concentration of PAHs which is consistent with population density and thus emission of PAHs which are most likely higher in the southern parts of the Baltic Sea. Interestingly, the air concentrations are also higher during the winter at all locations, which could be explained by increased domestic heating activity during this period. However, there were no significant statistical differences between the summer and winter.

Finally, the low overall variation in concentration between samples may be explained by the fact that the samples are time integrated, on average 24 days. Integrated representative air concentrations are valuable when put together in a semi-empirical mass balance in the last section of this chapter.

10.3.1.3 Sediments

Within the sediment, HOCs partition between the sediment particles, the interstitial water and DOC in the interstitial water. In the sediment the ratio between particulate organic carbon and water/DOC is so high that virtually all of the HOCs occur in particulate phase. Nevertheless, the amount of HOC in the pore water is still important since it is this fraction that accounts for the diffusive exchange of HOCs with the overlying water (see section below). Due to the importance of the sediments as being the dominant pool of HOCs in the aquatic environment a more extensive sampling program was performed to investigate the spatial distribution of PAHs and PCBs in the Baltic Sea sediments. The sediment samples analyzed for PAHs and PCBs revealed that the distribution of these compounds is uniform at locations far from point sources. The average concentration of PCBs was 204 ng C g^{-1} (range 78–794) and for PAHs 33 µg C g^{-1} (range 11–91). For both groups of compounds the overall variation is approximately one order of magnitude when normalized to carbon content but more than two orders of magnitude when normalized to dry weight, which supports the theory that the organic carbon is a controlling factor for the sediments' capacity to sorb HOCs. For PAHs there was a weak negative correlation between latitude and log carbon normalized concen-

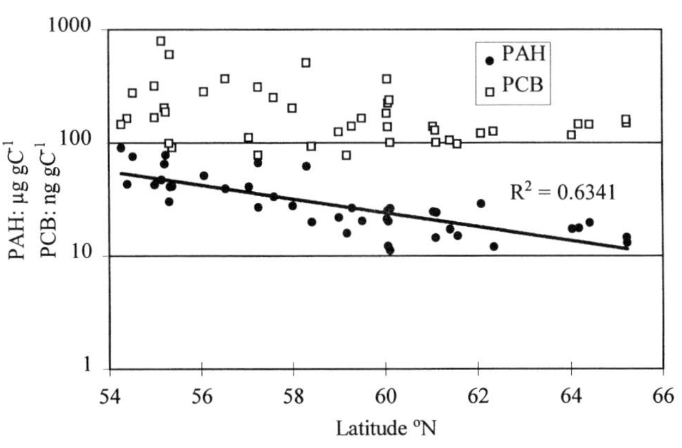

Fig. 10.3. Relationship between logarithm of PAH (*solid circles*) and PCB (*open quadrats*) concentrations in sediments plotted against latitude. *Solid line* represents linear regression line of significant correlation between PAH concentration and latitude

tration (Fig. 10.3), which is consistent with findings in the atmospheric concentrations. The samples taken south of Öland show a significantly higher average value than those north of Öland ($p<0.01$; two extremely high values south of Öland are excluded). For the PCBs there was no correlation between latitude and concentration.

10.3.1.4 Distribution Coefficients Between the Different Organic Carbon Matrices and Water

Average values for the entire Baltic Sea were used to calculate log K_{oc}, which is the in situ partition coefficient for the distribution of PAHs and PCBs between the dissolved phase in the water column and the three different organic carbon matrices: POC (suspended particles), sediment trap material and the sediments. K_{oc} values for all individual PAHs and PCBs are plotted against their log K_{ow} value in Fig. 10.4. Log K_{oc} values for POC were calculated from winter occasions only since it has been shown that growth dilution may reduce concentrations in suspended particulate matter (SPM) during the productive season, e.g. Swackhamer and Skoglund (1993) and Axelman et al. (1997). In Fig. 10.4 the relationship between log K_{oc} and log K_{ow} of Karickhoff (1981) is also shown, which is based on laboratory studies of the sorption of HOCs to suspended sediments. All K_{oc}s measured in situ within the project are 1–2 orders of magnitude higher than the calculated K_{oc}s, for both PCBs and PAHs (Fig. 10.4). It has been suggested that partition coefficients for PAHs between soot and water are orders of magnitude higher than partition coefficients between natural organic matter and water and that soot serves as a significant phase for the distribution of PAHs in natural aquatic systems (Gustafsson and Gschwend 1997). The structure of soot at the molecular level resembles that of PAH, so a strong binding can be expected. The three-dimensional structure of soot varies depending on the origin; however, at the molecular level, soot consists of large flakes of multi-ringed aromatics. A strong interaction between the electron pi-clouds of the PAHs and the soot can therefore be expected (Gustafsson and Gschwend 1997). It is also possible that soot may be an important matrix for PCBs, though the same strong binding to soot as for PAHs can probably not be expected since the large chlorine atoms will sterically hinder the interactions between the pi-clouds of the aromatic rings in the PCBs and the aromatic rings of the soot. High-equilibrium organic carbon–water partition in suspended particles for PCBs has, however, been observed in a field study in the Baltic by Axelman et al. (1997). They proposed another explanation and argued that the equation of Karickhoff (1981) is not

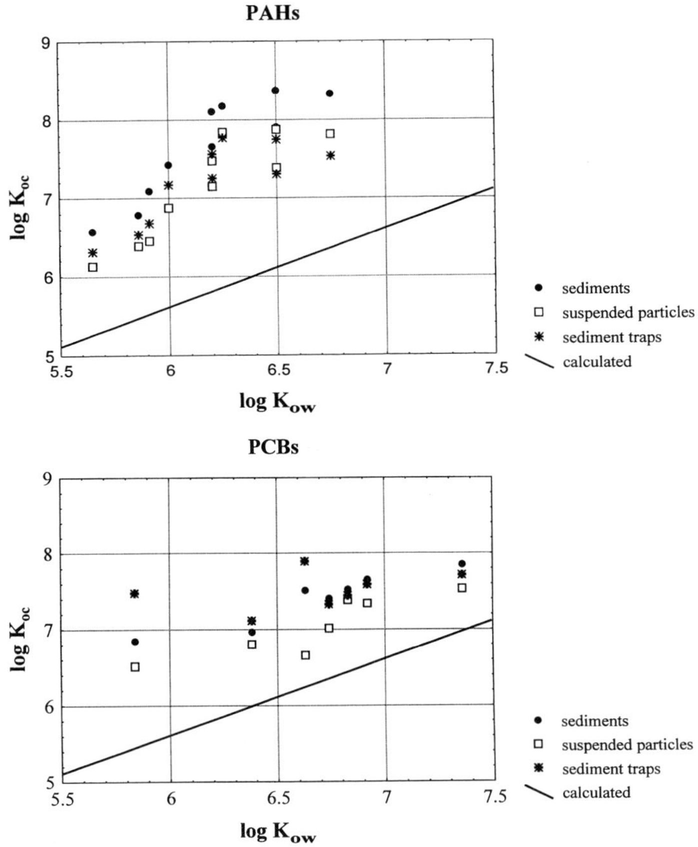

Fig. 10.4. Carbon-normalized partition coefficients of individual PAHs (*above*) and PCBs (*below*) for distribution between the dissolved phase in the water column and sorbing matrices. *Solid line* represents calculated K_{oc} values according to empirical relationship between log K_{ow} and log K_{oc} derived by Karickhoff (1981)

applicable on compounds with log K_{ow} >6. They argued that solvent–solute interactions of highly hydrophobic compounds could be significantly different in a octanol–water system than in a planktonic organism–water system in the field. Another possible explanation they gave was that the relationship of Karickhoff (1981) was based on laboratory set-ups with concentrations of HOCs much higher than under natural conditions which could lead to saturation of the most favorable binding sites in the natural organic matrix. So far, only little work has been done regarding field partition coefficients and more work is needed especially in the light of the wide use of the equation of Karickhoff (1981) in environmental modeling.

10.3.2 Inputs of HOCs to the Baltic Sea and Exchange Processes Between the Different Compartments

10.3.2.1 Exchange Between the Atmosphere and Water Column

The aim of measuring PAHs and PCBs in air was not only to obtain average concentrations of PAHs and PCBs in the air over the Baltic but also to estimate the atmospheric input of these compounds to the Baltic waters, i.e. the dry and wet deposition. Dry deposition of HOCs is dependent on the average settling velocity of the aerosols and the concentration of HOCs bound to the aerosols. Briefly, there are several size classes of aerosols with different origins and properties. Large particles settle faster than small particles and the size spectrum of aerosols will clearly affect the efficiency with which settling aerosols remove HOCs from air, i.e. dry deposition. Though differences in aerosol size distribution between urban air and air at background locations have been observed, it is still difficult to directly relate a given aerosol size composition to a deposition velocity. However, Eisenreich et al. (1981) reported a deposition velocity range of 0.1–0.5 cm s^{-1} for the Great Lakes area which we have chosen to use for both PAHs and PCBs. The flux of HOC per unit area, *F*, is calculated as:

$$F_{dry}=C_p \cdot V_{dp} \tag{10.2}$$

where C_p is the concentration of aerosol-bound HOC (pg m^{-3}) and V_{dp} is the settling velocity (m s^{-1}).

Apart from the continuous removal from air by particle dry deposition, HOCs are scavenged episodically by precipitation. During the time a raindrop travels towards the ground, HOCs in the vapor phase partition into the rainwater. Due to the large surface to volume ratio of raindrops, equilibrium is believed to be reached within the time it takes for a raindrop to fall 10 m (Slinn et al. 1978). The vapor washout ratio, W_v, i.e. the ratio between dissolved phase concentration in the raindrop and the vapor concentration in air, may be estimated using the Henry's law constant, K_H,

$$W_v = \frac{RT}{K_H} \tag{10.3}$$

where *R* is the gas constant (J K^{-1} mol^{-1}) and *T* is the temperature (K). The raindrops also scavenge aerosols from the air, so the total washout ratio is dependent on both particle washout, W_p, and vapor washout, W_v, as follows:

$$W_{tot}=W_v(1-\sigma)+W_p\sigma \tag{10.4}$$

The particle washout component, W_p, of the total washout ratio, W_{tot}, is highly dependent on particle size spectrum and meteorological factors. Typical W_ps lie within the range 10^5–10^6 (Bidleman 1988). From Eqs. (10.1) and (10.4), it follows that the relative importance of W_p increases for less volatile compounds. For PCBs, vapor and particle washout is of equal importance while for PAHs washout of particles dominates. The total wet deposition flux would consequently be calculated as:

$$F_{wet}=\varrho \cdot W_{tot} \cdot (C_{vapor}+C_{aerosol}) \tag{10.5}$$

where ϱ is the precipitation rate (m year^{-1}). Deviations from these relationships occur because natural conditions are far more complex than assumed in the models. For example, heavier PAHs do not seem to equilibrate between the vapor phase and the raindrop during events with high precipitation intensity, as observed by Dickhut and Gustafson (1995). The authors also noticed a rainfall intensity dependence of W_p, implying that particles are more efficiently scavenged during intense rainfall.

To estimate the deposition from air to the Baltic Sea, aerosol-bound, $C_{aerosol}$, and total air ($C_{aerosol}$ + C_{vapor}) concentrations of PCBs and PAHs on a volume basis were calculated. For PAHs we used washout ratios given by Bidleman (1988), in which a reasonably good consistency between calculated and field measurements was found and typical deviation of calculated values from field measurements was less than 30 %. For the individual PAHs, for which no washout values could be found in the literature, values for compounds with the same number of aromatic rings were used. For PCBs the values calculated by Bidleman (1988) deviated so much from field data that we preferred to use the field values of W_{tot}, ranging from 8000–40,000 for Aroclor 1254. Some of the discrepancy may have been caused by the fact that the values reported were for Aroclor mixtures and not for individual congeners. The dry deposition flux was calculated according to Eq. (10.2), where the dry deposition velocity range of 0.1–0.5 cm s^{-1} of Eisenreich et al. (1981) was used; the calculated ranges for PCBs and PAHs thus reflect both uncertainties in V_{dp} and variations in $C_{aerosol}$. For PCBs the average dry deposition over the Baltic was calculated to be 213 ng m^{-2} year^{-1} (range 44–669) and for PAHs 462 µg m^{-2} year^{-1} (range 41–1545). The wet deposition flux was calculated by the using the values listed for PAHs in Bidleman (1988). For those PAHs in the present study not listed in Bidleman (1988), a value for a PAH with the same number of rings was selected. Precipitation was assumed to be 0.60 m year^{-1} (Anonymous 1986). The average wet deposition of PCBs was thus calculated to be 250 ng m^{-2} year^{-1}

(range 44–669) and 231 μg m^{-2} year^{-1} (range 20–556) for PAHs. For PCBs the variation in estimates of average air concentration over the Baltic was as high as the variation in V_d and W_{tot} (Table 10.1). The maximum/minimum ratio of PCB concentration in air was 3 for the aerosol-bound fraction and 8 for total air. The maximum/ minimum ratios for PAHs were 12 and 8 respectively. These variations in pollutant concentrations are probably the main cause of uncertainties in the estimates of dry and wet deposition for PAHs.

In addition to deposition, surface water absorption/volatilization is another important flux of HOCs to and from the water-body. The volatilization flux is calculated according to the following equation:

$$F_{vol}=k_{OL}C_w \tag{10.6}$$

where C_w is the concentration of HOC in the dissolved phase in water and k_{OL} is the mass transfer coefficient. The absorption flux is similarly calculated as:

$$F_{ads} = k_{OL}\frac{C_{vapor}RT}{K_H} \tag{10.7}$$

According to Eqs. (10.6) and (10.7), the direction of the net flux is dependent on the accuracy of K_H. K_H can either be measured directly by measuring concentrations of a compound in the air and water phases in an air-water system in equilibrium, or be calculated from measured vapor pressure and water solubility, C^{sat}_w, as follows:

$$K_H = \frac{p^\circ}{C_w^{sat}} \tag{10.8}$$

Schwartzenbach et al. (1993) pointed out that the two different methods may yield different K_H values, up to 30–100% for the two PAHs phenanthrene and anthracene. The uncertainty can be expected to be yet even higher for PAHs with higher molecular weight and lower C^{sat}_w and $p°$. Phenanthrene and anthracene are the two most volatile PAHs in this study which implies that the K_H data for the PAHs in this study are even more uncertain and should be used with care and that the calculated direction is very uncertain, especially if the difference $\left(\frac{C_{vapor}RT}{K_H} - C_w\right)$ is small. However, since the concentration of high-molecular-weight PAHs in gas phase is small in comparison to the aerosol-bound phase, gas exchange is a relatively insignificant process compared to deposition and sedimentation (see last section of this chapter). K_H has been shown to be inversely related to temperature. ten Hulscher et al. (1992) found that a decrease in temperature of 10 °C will cause reduction in K_H of approximately 50 %. In Tables 10.1 and 10.2 K_Hs measured at 25 °C are given together with K's corrected for the assumed average environmental temperature of 8 °C for all the PCBs and a selection of the PAHs.

The gas exchange mass transfer coefficient is typically in the order of 10^{-4} to 10^{-2} cm^{-1} s^{-1}, and is dependent on the properties of the compound and on the wind speed (Schwartzenbach et al. 1993). Mass transfer coefficients were calculated for two compounds with intermediate properties – PCB #153 and benzo(a)pyrene (Table 10.1) – according to the equation of Schwartzenbach et al. (1993):

$$\frac{1}{k_{OL}} = \frac{1}{\left(\frac{D_a}{D_a^{H_2O}}\right)^{0.67} \frac{K_H}{RT}(2u_{10}+0.3)} + \frac{1}{\left(\frac{D_w}{D_w^{O_2}}\right)^{0.57} (4\times10^{-5}u_{10}^2+4\times10^{-4})} \qquad (10.9)$$

where D_a is the diffusion coefficient of the compound in air, $D_a^{H_2O}$ the diffusion coefficient of water in air (0.26 cm^{-2} s^{-1}), u_{10} the wind speed in m s^{-1}, D_w the diffusion coefficient of the compound in water and $D_a^{H_2O}$ the diffusion coefficient of oxygen in water (2.1×10^{-5} cm^{-2} s^{-1}). Schwartzenbach et al. (1993) give estimates of D_w for PCB #153 and benzo(a)pyrene of 6.3×10^{-6} and 7.5×10^{-6} cm^{-2} s^{-1} respectively, and for D_a 0.058 and 0.069 cm^{-2} s^{-1}. With an assumed average wind speed of 6 m s^{-1} and an average temperature of 8 °C the calculated k_{OL} for PCB #153 was 0.005 m h^{-1} which is 50 % less than the value of 0.01 m h^{-1} used by Wania et al. (Chap. 15). For benzo(a)pyrene, k_{OL} was 0.002 m h^{-1}. The volatilization fluxes of PCBs and PAHs were calculated according to Eq. (10.6) using the mass transfer rates and K_H for PCB #153 and benzo(a)pyrene. For PCBs the volatilization flux was 331 ng m^{-2} year^{-1}.

In both air and water, fluoranthene and pyrene occur mostly in the free form, i.e. in vapor phase and truly dissolved, and are the two PAHs present in the highest concentrations in these two phases. Fluoranthene and pyrene are also much more volatile and have accordingly higher K'_H than the rest of the PAHs, e.g. K'_H for fluoranthene is 25 times higher than for benzo(a)pyrene. Fluoranthene and pyrene were therefore excluded from the calculations of the diffusive exchange between air and water together with the related 1-methylpyrene and 2-methylpyrene. The volatilization flux for PAHs was calculated to be 2 µg m^{-2} year^{-1}. The adsorption fluxes for PCBs and PAHs were calculated according to Eq. (10.7) as 2.7 µg m^{-2} year^{-1} and 0.67 ng m^{-2} year^{-1}.

10.3.2.2 River Input

The annual average river input of water to the Baltic Sea is approximately 480 km^3 (Bergström and Carlsson 1994), which is equal to ~1/50 of its total water volume. This is comparable in magnitude to the input by precipitation, which is approximately 220 km^3. The discharge of individual rivers varies greatly between years as well as seasons (Bergström and Carlsson 1994). In the northern rivers there is a pronounced increase during the snow-melting period in the spring, although some of these rivers are subjected to substantial regulation where up to 70 % of the annual flow is stored for hydroelectric power usage.

The average concentration of the dissolved fraction in the river water was 88 pg l^{-1} (range 4–231) for PCBs and 1.2 ng l^{-1} (0.04–2.5) for PAHs. The particle-bound fraction had an average concentration of 426 pg l^{-1} (range 18–807) for PCBs and 11 ng l^{-1} (range 0.2–33) for PAHs. The average particle-bound fraction of PCBs in individual samples was 77 % for PCBs and 83 % for PAHs. For both PCBs and PAHs the concentrations were considerably higher in river water than in seawater. The dissolved concentrations of PCB were ten times higher in river water than in the open Baltic, while for PAHs the difference was six times. The particulate fraction cannot be compared directly on a volumetric basis since the suspended particle concentrations are generally much higher in rivers than in the open sea. However, the carbon-normalized concentrations of the pollutants in the particulate phase may be compared. The carbon-normalized concentration in the suspended particulate phase was 540 ng C g^{-1} (range 17–1000) for PCBs which is also ten times higher than in the open Baltic sea. The carbon-normalized concentration of PAHs in the rivers was 16 μg C g^{-1} (range 0.2–47) which does not differ much from the average value of 10 μg C g^{-1} for the open sea. This means that, in contrast to the PCBs, the carbon-normalized partition coefficients of PAHs between water and suspended particles are higher in the open sea. The reason for this can only be speculated. However, one explanation could be that soot plays a more important role in the partitioning for PAHs than for PCBs (McGroddy et al. 1996). In the open Baltic Sea, direct air deposition is the dominant input route for PAHs. Most of the precipitation water is infiltrated through the ground before reaching the rivers. There is thus a possibility that a large fraction of the soot particles deposited with rain are trapped in the soil. The partitioning of PAHs in groundwater and river water may therefore instead be controlled by natural organic matter, which has less sorbing capacity than soot (Gustafsson and Gschwend 1997).

10.3.2.3 Exchange Between Sediment and Water

The total content of HOCs in the sediments is large compared to that of the other compartments (see last section of this chapter). The exchange processes between sediments and water have therefore been considered as very important and the sediments have traditionally been viewed as a buffering reservoir of HOCs, causing the aquatic system to respond more slowly to changed inputs, e.g. Mackay (1989). However, the exchange processes between sediments and water are difficult to study. Investigations face several methodological problems in terms of quantifying the fluxes between the two compartments, even when other substances than trace organic compounds are in focus. The exchange of HOCs between sediments and water has two major pathways: particle-mediated transport (i.e. sedimentation and resuspension) and various types of diffusive exchange.

10.3.2.3.1 Sedimentation/Resuspension

Particle settling is an important process for removal of HOCs from the water column. The average value for the sedimentation of PCBs was 11 μg m^{-2} $year^{-1}$ (range 3.8–21) and for PAHs it was 0.47 mg m^{-2} $year^{-1}$ (range 0.09–0.73). Once the particulate material has reached the sediment surface three processes may occur. It may be (1) *mineralized*, i.e. degraded primarily by microorganisms and invertebrates, or (2) *resuspended*. Both processes mean a transport of HOCs back to the water column. In the first case, this reflux is accounted for by a diffusive flux since a positive fugacity gradient builds up in the remaining parts of the particles subjected to mineralization. In the second case, HOCs associated to the resuspended particles will be brought back to the water column. Also, some part of the material will be (3) *buried* by overlying particles, a process that together with degradation will serve as a final removal of HOCs. The net removal of particle-associated HOCs from the water column cannot be measured directly because an unknown amount of the material collected in sediment traps will be previously resuspended material. An indirect method to measure resuspension by the use of the content of detrital bedrock elements such as Al, Fe and Ti in the sedimenting material has been used to quantify influence of resuspension in a coastal area of the Baltic (Blomqvist and Larsson 1994); however, to use the method offshore in the Baltic requires a more extensive sampling program than has been done in the present study. In the present study we have compared two methods of estimating the burial of HOCs. First, a ^{210}Pb-dated core from an accumulation bottom in the northern Baltic proper gave a rough estimate of the burial capacity of the sediments for the last 10 years, where the burial of the same PCBs as in this study was reported to be in the order of 2 μg m^{-2} $year^{-1}$ and the burial of PAHs in the order of 250 μg m^{-2} $year^{-1}$ (Axelman et al. 1995). Assuming that half of the bottom sediments of the Baltic Sea are accumulation-focusing areas, this gives, for the entire Baltic Sea, an average burial rate of 1 and 125 μg m^{-2} $year^{-1}$ for PCBs and PAHs respectively. Secondly, the average carbon content of 5 % in the 42 sediment samples analyzed yields an average carbon inventory of 77 g C m^{-2} in the top 1 cm of the sediment. In Axelman et al. (1995), the sediment accumulation rate in accumulation areas was estimated to be between 0.8 and 1.7 mm $year^{-1}$ for the last three decades, and in the Gotland Basin east of Gotland, an accumulation rate of 0.8–2.9 mm $year^{-1}$ was found (Neumann et al. 1997). The average sediment accumulation rate was assumed to be 1 mm $year^{-1}$ so the average carbon burial rate was calculated to be 3.9 g C m^{-2} $year^{-1}$. This estimate is higher than the estimate of 1.5 g C m^{-2} by Elmgren (1984) but lower than 13 g C m^{-2} reported by Jonsson (1992); 3.9 g C m^{-2} $year^{-1}$ in turn yields a burial of PCBs and PAHs of 0.8 and 125 μg m^{-2} $year^{-1}$, figures relatively well in accordance with the figures derived from Axelman et al. (1995). The assessment of the sediment carbon burial is uncertain, e.g. it is difficult to estimate how large the area of true accumulation bottoms is. The burial fluxes of PCBs and PAHs deduced from Axelman et al. (1995) were used for the discussion regarding the difference between PCBs and PAHs in internal recycling and for the mass balance in the last section of this chapter.

For PCBs the sedimentation is 11 times the burial, which means that >90 % of the settling PCBs is either resuspended or, as a result of mineralization of organic carbon, released back to the water. There was a negative relationship between the degree of chlorination of the PCBs and the degree of recycling. The tetrachlorinated congener #52 had an almost 50 times higher sedimentation rate than burial rate while for the heptachlorinated congener #180 the corresponding ratio was an order of magnitude less. A similar relationship was found by Baker et al. (1991) in a study in Lake Superior where the sedimentation/burial ratio ranged from 10 to 1000 for different PCBs. Part of the difference can be explained by the fact that the traps of that study were deployed at a depth of only 35 m, which leads to an overestimate of the flux near the bottom since a large part of the organic carbon degradation takes place during settling. Also, sediment traps can over-trap settling matter under conditions with extreme water currents, and especially if the design or deployment of the traps are non-optimal. However, the hydrodynamic properties of the sediment traps used in this study have been carefully evaluated (Broman et al. 1990a).

For PAHs the sedimentation was approximately the same as the burial. There was no relationship between degree of recycling and hydrophobicity or molecular size for PAHs. The ratio between sedimentation and burial was between 2 and 5 for individual PAHs. The lower recycling rate of PAHs compared to PCBs, which was also observed in the above-mentioned Great Lakes study of Baker et al. (1991), is an additional indication of the PAHs being distributed in the environment in association with soot particles, which are resistant to degradation and which decrease the subsequent release of PAHs back to the water. It also implies that most of the reflux of PCBs from the sediments is an effect of particle mineralization rather than particle resuspension.

10.3.2.3.2 Diffusive Exchange

In addition to sedimentation/resuspension processes, diffusive exchange of HOCs between the sediments and water occurs. The type of sediment and its fauna affect the rates at which the exchange occurs. Bioturbation and pore-water flushing are examples of processes that would increase the exchange rates significantly compared to unmixed sediments with no pore-water flushing. One of the reasons for quantifying the exchange rate of HOCs is to assess the risk that the sediments start acting as sources of HOCs when release of HOCs to the environment has diminished. Therefore, the major interest should be to quantify the possible diffusive release of HOCs from accumulation bottoms where the sediment particles are finally deposited. During the last century large parts of the accumulation bottoms in the Baltic have been laminated (Jonsson 1992) with little or no bioturbation due to periodically anoxic bottom conditions. To our knowledge there are no studies of bioturbation mixing coefficients specific for Baltic conditions; however, the ^{210}Pb-dating method in a study by Axelman et al. (1995) yielded a mixing coefficient of 3.8×10^{-7} for the surface sediment which is at the higher end of the range of typical bioturbation coefficients of 10^{-7} to 10^{-11} cm^2 s^{-1} for the Great Lakes given by Eisenreich (1987). Below, a model approach is presented to assess the exchange rates in sediments under conditions with or without bioturbation.

10.3.2.3.4 Model Approach to Diffusive Sediment/Water Exchange of HOCs

Sediment-water exchange of HOCs has been studied in several laboratory experimental setups (Formica et al. 1988; Baron et al. 1990; Booij et al. 1992), all focusing on the intrasediment processes, i.e. the resistance to the flux caused by the buildup of concentration gradients within the sediment. It was concluded that the exchange process can be modeled by using a few properties of the sediment and the compound of interest. A model based on the equation of Fick's second law of diffusion is used to calculate the release of an HOC as a function of time:

$$\frac{\delta C}{\delta t} = D_{eff} \frac{\delta^2 C}{\delta x^2} \tag{10.10}$$

where D_{eff} is the effective diffusion coefficient, which is dependent on the diffusion coefficient of the compounds in pure water, $D_{w;}$ the equilibrium particle partition coefficient of the compound, K_{oc} (l kg^{-1}); the fraction of organic carbon in the sediment, f_{oc}; the density of the sediment solids, ϱ_{solids} (kg l^{-1}); the porosity of the sediment, θ; and finally, the tortuosity of the sediment, τ, which is a factor correcting for the elongated diffusion path the molecule has to travel due to the structure of the sediment. The tortuosity is related to sediment porosity as $\tau=\theta^2$ for sediments where $\theta>0.7$ (Ullman and Aller 1982). D_{eff} is described as:

$$D_{eff} = \frac{D_w \tau}{\theta + K_{oc}(1-\theta) f_{oc} \varrho_{solids}} \tag{10.11}$$

D_w (cm^2 s^{-1}), in turn, can be estimated by the following equation from Schwartzenbach et al. (1993):

$$D_w = \frac{13.26 \times 10^{-5}}{(V)^{0.589} \mu^{1.14}} \tag{10.12}$$

where V is the molar volume of the compound in cm^3 mol^{-1} and μ the viscosity of water in centipoise (10^{-2} g s^{-1}). Using the sediment-water log K_{oc} in Table 10.1 for PCBs, Eqs. (10.9) and (10.10) yield D_{eff}s in the range 5×10^{-13} to 6×10^{-12} cm^2 s^{-1}. A characteristic transport time for diffusion over a distance of 0.1 cm can be calculated according to Schwartzenbach et al. (1993):

$$t = \frac{x^2}{2D_{eff}} \tag{10.13}$$

where x is the distance (cm) and t the time (s). The given range of D_{eff}s for PCBs corresponds to a characteristic transport time of more than 100 years. For PAHs the range of D_{eff}s for sediment–water log K_{oc}s is 7.8×10^{-14} to 1.1×10^{-11} cm^2 s^{-1} which corresponds to 10–1000 years. Since the sediment accumulation rate is approximately 1 year for 0.1 cm, it appears that from a completely unmixed sediment the diffusive flux from older underlying sediment layers will be negligible.

Chen (1993) developed a model taking into account pore-water flushing, bio-irrigation (animals pumping water in the sediments), and bioturbation as processes governing the resistance within the sediment. The influence of the immobile layer of water above the sediment, Z (μm), was also included in the estimation of the total resistance to the flux. The thickness of the immobile layer is dependent on sediment topology and current speed of the overlying water. Chen (1993) gives Z of 100–300 μm for currents of 6–17 cm s^{-1}. Data on bottom currents in the Baltic are scarce; however, Leppänen and Alenius (1988) reported near-bottom currents to be typically in the range 1–10 cm s^{-1}. Therefore, it can be assumed that common current speed over accumulation bottoms in the Baltic is a few centimeters per second for most areas. A very rough estimate of a typical range of Z would then be 200–1000 μm.

Chen (1993) concluded that the effect of bio-irrigation was negligible under most environmental conditions and that the following parameters were governing: bioturbation coefficient, D_b, thickness of immobile layer, Z, fraction of organic carbon in the sediment, f_{oc}, and K_{oc}. If $D_b > 10^{-7}$ cm^2 s^{-1} the resistance in the immobile layer is significantly higher than in the sediment and hence there will be no buildup of concentration gradients in the sediment. During such conditions, the diffusive flux can be estimated by:

$$F = \frac{\frac{C_{sed}}{K_{oc}} - C_w}{\frac{Z}{\overline{D}_w} - \frac{L}{\varepsilon\psi(\overline{D}_b + \overline{D}_m)}} \tag{10.14}$$

where a positive F equals a net flux from the sediments to the water. C_{sed} is the organic carbon-normalized concentration of the compound in the sediment and C_w the concentration in the overlying water. $\overline{D}_w$ is the diffusion coefficient in the boundary layer corrected for diffusion of the part of the compound bound to DOC. $\overline{D}_w$ equals $D_w + D_{coll}[\mathrm{DOC}]_w K_{DOC}$, where D_{coll} is the diffusion coefficient of organic colloids (~3×10^{-6} cm^2 s^{-1}; Chin et al. 1991) and $[\mathrm{DOC}]_w$ is the concentration of DOC in water. $\overline{D}_b$ is the diffusion coefficient of bioturbation corrected for the mixing of DOC-bound fraction of the compound. $\overline{D}_b$ equals $D_b(1+K_{DOC}[\mathrm{DOC}]_{sed})$ where $[\mathrm{DOC}]_{sed}$ is the concentration of DOC in the sediment (assumed to be the same as $[\mathrm{DOC}]_w$, i.e. 4 mg l^{-1}). $\overline{D}_m$ is the diffusion coefficient for diffusion in the pore water of the truly dissolved and the DOC-bound fraction.

$\overline{D_m}$ equals D_w+$D'_{coll}K_{DOC}[\mathrm{DOC}]_{sed}$ where $'_w$ and $'_{coll}$ are the diffusion coefficients corrected for tortuosity; D'_w equals $D_w\tau$. and D'_{coll}=$D_{coll}\tau$ respectively. The factor ψ is according to Chen (1993) dependent on $\overline{D_b}$, $\overline{D_m}$ and a few sediment characteristics, i.e. density of the solids which was assumed to be 2 g cm^{-3}, the particle radius which was assumed to be 0.01 cm, the sediment's porosity which was assumed to be 0.85, fraction of the organic carbon of the sediment which was measured to be on average 5 %, and the intraparticle porosity for which 0.13 was chosen, a typical value for silt (Wu and Gschwend 1986).

Equation (10.14) can be used to roughly assess the diffusive flux between sediments and water in the Baltic Sea of PCBs and PAHs. D_bs are assumed to be in the range 10^{-6} to 10^{-8}. The average sediment concentrations together with the sediment–water log K_{oc}values from Tables 10.1 and 10.2 yield a flux from the sediments to water from bioturbated bottoms of approximately 1–5 μg m^{-2} year^{-1} for PCBs and 100–700 μg m^{-2} year^{-1} for PAHs, and an equal flux back since the system is assumed to be in equilibrium with the sediments. K_{oc} values were calculated from the measured sediment and water concentration. This corresponds to transfer velocities for the dissolved phase (Chap. 15) of 0.01–0.5 m h^{-1} for PCBs and 0.004–23 m h^{-1} for PAHs. The reason for the high transfer velocities compared to those of Wania et al. (Chap. 15) is the high K_{oc}s used in these model calculations; only a very small fraction will be in the dissolved phase or bound to colloids and subjected to transport. If instead K_{oc}s are calculated according to Karickhoff (1981) as in Chapter 15, the gross flux from the sediment to the water will still be of the same magnitude, 4–11 μg m^{-2} year^{-1} for PCBs and 300–2500 μg m^{-2} year^{-1} for PAHs. However, the water to sediment flux would be insignificant (less than 10 %), so net flux from these bottom sediments would be so high that they would rapidly equilibrate with the overlying water. The average total content in the upper 1 cm in 1 m^2 of the sediment is 13 μg of PCB and 2.7 mg of PAH, which means that the top 1 cm reaches a state close to equilibrium within a few years. If that were the case, sediment concentrations 1.5–2 orders of magnitude lower in sediments with a well-mixed biologically active layer could be expected than in non-mixed sediments. However, no such effect could be found. In summary, the sediments appear to be in equilibrium with the overlying water. Still, the diffusive flux is important when describing the behavior of HOCs in aquatic systems. An effective exchange between water and sediments will increase the system's response time to changed inputs.

10.3.3 A Semi-Empirical Mass Balance of PAHs and PCBs in the Baltic Sea

All data presented above were combined in a semi-empirical mass balance (Fig. 10.5) to obtain an overall picture of the relative importance of the different compartments as pools for PAHs/PCBs and the relative importance of the different fluxes of PAHs/PCBs between these compartments.

Sedimentation, burial, dry deposition, wet deposition and the diffusive exchange between sediment/water and air/water was calculated by multiplying the rates given in the previous section with the total area of the Baltic Sea, 373,000 km^2, and converted to kg $year^{-1}$ for PCBs and t $year^{-1}$ for PAHs. Resuspension/mineralization was calculated as the difference between sedimentation and burial.

Diffusive exchange between water and the sediments was calculated using the average value of the ranges of diffusive fluxes given in the previous section: 6 μg m^{-2} $year^{-1}$ for PCBs and 4 mg m^{-2} $year^{-1}$ for PAHs. The accumulation bottoms were assumed to be half of the area and approximately 25% of these bottoms were assumed to be effectively bioturbated, i.e. the area for which these fluxes are applicable would then be 47,000 km^2. Both the fluxes and the area are based on several uncertain assumptions so the given total diffusive exchanges in Fig. 10.5 should only be considered as estimates of the order of magnitude of these fluxes.

Content of PCBs and PAHs of the different compartments was calculated by using the given average concentrations multiplied with the volumes of the compartments. For air the selected boundary layer was

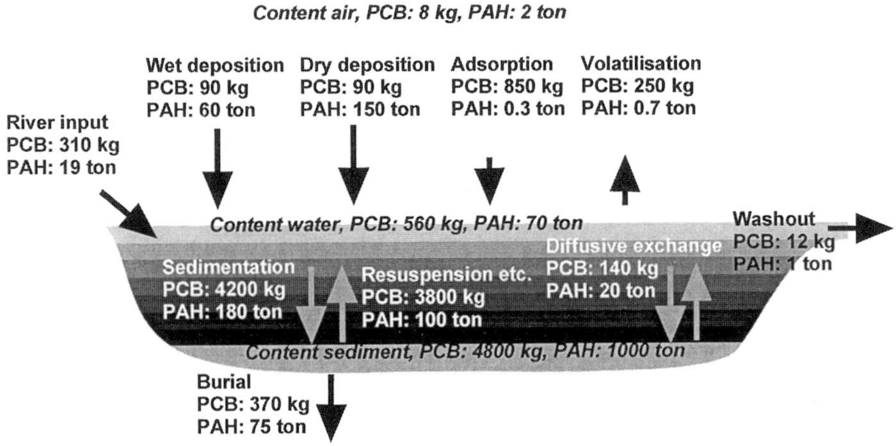

Fig. 10.5. Semi empirical mass balance of PAHs and PCBs for the Baltic Sea

selected (373,000 km^3); land-air interactions affect the atmosphere up to approximately 1000 m. The average depth of the Baltic Sea is 60 m which yields a total volume of 22,000 km^3. For the sediments the top 1 cm was selected (3.7 km^3) which typically corresponds to the last 10 years in accumulation areas. For inventory calculations, the top 1 cm was selected since it was this part of the sediment that was analyzed.

The discharge of particles from rivers is highly dependent on water velocity which is why particle concentrations are highly variable according to the season. The best way of assessing the discharge of particle-bound HOCs was therefore considered to be the use of literature data covering the discharge of particles (Anonymous 1986) in combination with the organic carbon-normalized concentrations of particle-bound PCBs and PAHs in rivers. Total discharge of dissolved PCBs and PAHs was calculated using literature data of water discharge (Bergström and Carlsson 1994). Weighted (for each drainage area) average concentrations of PCBs and PAHs were used to calculate the total annual inflow of these compounds to rivers, which was estimated to be approximately 311 kg for PCBs and approximately 19 t for PAHs.

For both PAHs and PCBs the inputs to the system (wet and dry deposition, rivers and adsorption) are greater than the losses (burial, volatilization and washout). The net increase compared to the total content of the investigated system (water column + upper 1 cm of the sediments) corresponds to an annual concentration increase rate for the entire system of 10–15% for both groups of compounds. For PCBs this is not in accordance with time trend data of concentrations in herring that show a clear decrease (Bignert et al. 1993). However, taking into account the uncertainties with which the different fluxes were calculated, it is obvious that mass balance also could produce a decrease. Even with knowledge of time trends of the system's concentration of HOCs, the ranges are too large to use the balance to identify any unknown processes such as internal sinks or emissions from point sources.

However, there are a few conclusions to be drawn. The sediment is the environmental matrix containing most of the contaminants. The sampling strategy of the project was to collect only the upper 1 cm of the sediment since it was believed that only the upper part of the sediment is relevant in terms of exchange with the overlying water. In a perspective of total pool of contaminants a much larger part of the sediment must be taken into account. In studies by Axelman et al. (1995) and Kjeller and Rappe (1995) significant levels of PCBs and PAHs were found down to 7 cm depth. Therefore the total content of PCBs and PAHs in the whole sediment could be as much as one order of magnitude higher than reported in (Fig. 10.5), i.e. up to 50 and 10,000 t respectively. Compared to the estimates of total world production of PCBs even 50 t is fairly small, less than 0.1 % of the total amount released to the environment of the world which is estimated to be approximately 100,000 t (National Academy of Science 1997). This is somewhat surprising since it could be expected that the level of contamination in the Baltic Sea was high due to the dense population around its borders. The surface area of the Baltic Sea is approximately 0.1 % of the Earth's total surface area, which actually implies that the contamination level of PCBs in the Baltic Sea resembles the average contamination level of the entire world.

Air is, not unexpectedly, an insignificant pool of contaminants. The annual downward flux from air is orders of magnitude greater than the total content, implying a very short residence time of these contaminants in this compartment – approximately 3 days for both PAHs and PCBs. This is much less than the reported residence times of 50–60 days for the tetra- to heptachlorinated PCBs by Panshin and Hites (1994). Panshin and Hites (1994) calculated the residence time using Junge's relationship between atmospheric concentration variability and residence time (Junge 1974). There are a few explanations for this large discrepancy. In this study we selected the Ekman layer (1 km) as the air compartment for HOCs. If the entire troposphere (10 km) is taken into account, the total content of PCBs will be higher and thus the estimated residence times will be higher. In addition, the largest atmospheric removal process in our mass balance is the gas phase adsorption, which is probably also the most uncertain estimate. An overestimated gas adsorption also means underestimated residence times.

The water column has intermediate properties in terms of fluxes and content. The content of the water column is estimated to be about half of the sedimentary flux for PAHs which indicates that practically all of the contaminants in the water column are replaced during the period of half a year. For PCBs the residence time is even shorter; the cumulative annual sedimentation flux is almost one order of magnitude higher than the content of the water column. A similar calculation of the residence time of PCBs in the water column of Lake Superior, which has an average depth of more than twice that of the Baltic Sea, gave a value of 1.5 years (Baker et al. 1991). Finally, for PCBs the net direction of the gas exchange with the atmosphere appears to go from air to water, which is the opposite to what has been found in the Great Lakes area, e.g., by Baker and Eisenreich (1990) and Hornbuckle et al. (1995).

There was an interesting difference between PAHs and PCBs in terms of routes of input to the system. Rivers account for a significant input of PCBs (26 %), while they account for only 10 % of the input of the PAHs. This can probably be explained as an effect of PAHs being released directly to air bound to aerosols to a higher extent than PCBs which also leak via groundwater from dumps, contaminated soil, etc. The air-water gas exchange appears to be negligible for PAHs, while it is an important exchange process for PCBs. In summary, PAHs enter the system bound to particles via wet and dry deposition, practically nothing is transported back to air, and thereafter they settle out within half a year. PCBs enter the system mainly via gas adsorption but also to some extent via rivers and wet and dry deposition, they settle out rapidly and are intensively recycled between the surface sediments and the water column before finally being buried. Again, these results imply that the transport of PAHs in the environment is mainly mediated by soot particles while it is the free form (truly dissolved and vapor phase) that is important for most transport processes regarding PCBs.

References

Anonymous (1986) Water balance of the Baltic Sea. Helsinki Commission, Helsinki

Axelman J, Bandh C, Broman D, Carman R, Jonsson P, Larsson H, Linder H, Näf C, Pettersen H (1995) Time-trend analysis of PAH and PCB sediment fluxes in the northern Baltic proper using different dating methods. Mar Freshwater Res 46:137–144

Axelman J, Broman D, Näf C (1997) Field measurements of the partitioning of PCBs in planktonic organisms – effects of growth, particle size and solvent-solute interactions. Environ Sci Technol 31:665–669

Axelman J, Näf C, Broman D (2000) Vertical flux and particulate/water dynamics of polychlorinated biphenyls (PCBs) in the open Baltic Sea. Ambio 29:210–216

Baker JE, Eisenreich SJ (1990) Concentrations and fluxes of polycyclic aromatic hydrocarbons and polychlorinated biphenyls across the air-water interface of Lake Superior. Environ Sci Technol 24:342–352

Baker JE, Eisenreich SJ, Eadie BJ (1991) Sediment trap fluxes and benthic recycling of organic carbon, polycyclic aromatic hydrocarbons, and polychlorobiphenyl congeners in Lake Superior. Environ Sci Technol 25:500–509

Bandh C, Ishaq R, Broman D, Näf C, Rönnqvist-Nii Y, Zebühr Y (1994) Separation for subsequent analysis of PCBs, PCDD/Fs, and PAHs according to aromaticity and planarity using a two-dimensional HPLC-system. Environ Sci Technol 30:214–219

Baron JA, Thibodeaux LJ, Reible DD, Templet P-H, Henry CB Jr (1990) Laboratory simulation of diffusion in contaminated sediments. Estuaries 13:81–88

Bergström S, Carlsson B (1994) River runoff to the Baltic Sea: 1950–1990. Ambio 23:280–287

Bidleman TF (1984) Estimation of vapor pressures nonpolar organic compounds by capillary gas chromatography. Anal Chem 56:2490–2496

Bidleman TF (1988) Atmospheric processes. Environ Sci Technol 22:361–367

Bignert A, Göthberg A, Jensen S, Litzén K, Odsjö T, Olsson M, Reutergårdh L (1993) The need for adequate biological sampling in ecotoxicological investigations: a retrospective study of twenty years' pollution monitoring. Sci Total Environ 128:121–139

Blomqvist S, Larsson U (1994) Detrital bedrock elements as tracers of settling resuspended particulate matter in a coastal area of the Baltic Sea. Limnol Oceanogr 39:880–896

Booij K, Achterberg EP, Sundby B (1992) Release rates of chlorinated hydrocarbons from contaminated sediments. Neth J Sea Res 29:297–310

Broman D, Näf C, Lundbergh I, Zebühr Y (1990a) An in-situ on the distribution, biotransformation and flux of polycyclic aromatic hydrocarbons (PAHs) in an aquatic food chain (seston-*Mytilus edulis* L.-*Somateria mollissima* L.) from the Baltic: an ecotoxicological perspective. Environ Tox Chem 9:429–442

Broman D, Kugelberg J, Näf C (1990b) Two hydrodynamically stable self-suspended buoyant sediment traps. Estuar Coast Shelf Sci 30:429–436

Broman D, Näf C, Zebühr Y (1991a) Long term high- and low-volume air sampling of polychlorinated dibenzo-p-dioxins and dibenzofurans and polycyclic aromatic hydrocarbons along a transect from urban to remote areas on the Swedish Baltic coast. Environ Sci Technol 25:1841–1850

Broman D, Näf C, Rolff C, Zebühr Y (1991b) Occurrence and dynamics of polychlorinated dibenzo-p-dioxins and dibenzofurans and polycyclic aromatic hydrocarbons in the mixed surface layer of remote coastal and offshore waters of the Baltic. Environ Sci Technol 25:1850–1864

Brunner S, Hornung E, Santi H, Wolff E, Piringer OG, Altschuh J, Brüggemann R (1990) Henry's law constants for polychlorinated biphenyls: experimental determination and structure–property relationships. Environ Sci Technol 24:1751–1754

Chen H-W (1993) Fluxes of organic pollutants from the sediments in the Boston Harbour. Massachusetts Institute of Technology, Boston

Chin Y-P, McNichol AP, Gschwend P (1991) Quantification and characterization of porewater organic colloids. In: Baker RA (ed) Organic substances in sediments and water. Lewis, Chelsea, MI

De Bruijn J, Busser F, Seinen W, Hermens J (1989) Determination of octanol/water partition coefficients for hydrophobic organic chemicals with the "slow-stirring" method. Environ Tox Chem 8:499–512

Dickhut RM, Gustafson KE (1995) Atmospheric washout of polycyclic aromatic hydrocarbons in the southern Chesapeake Bay region. Environ Sci Technol 29:1518–1525

Eisenreich SJ (1987) The chemical limnology of nonpolar organic contaminants: polychlorinated biphenyls in Lake Superior. In: Hites RA (ed) Source and fates of aquatic pollutants. American Chemical Society, Washington, DC

Eisenreich SJ, Looney BB, Thornton JD (1981) Airborne organic contaminants in the Great Lakes ecosystem. Environ Sci Technol 15:30–38

Elmgren R (1984) Trophic dynamics in the enclosed, brackish Baltic Sea. Rapp P-v Réun Cons Int Explor Mer 183:152–163

Foreman WT, Bidleman TF (1985) Vapor pressure estimates of individual polychlorinated biphenyls and commercial fluids using gas chromatographic data. J Chromatogr 330:203–216

Formica SJ, Baron JA, Thibodeaux LJ, Valsaraj KT (1988) PCB transport into lake sediments. Conceptual model and laboratory simulation. Environ Sci Technol 22:1435–1440

Gustafsson Ö, Gschwend PM (1997) Soot as strong partition medium for polycyclic aromatic hydrocarbons in aquatic systems. In: Eganhouse R (ed) Molecular markers in environmental geochemistry. American Chemical Society, Washington, DC

Hawker DW, Connell DW (1988) Octanol–water partition coefficients of polychlorinated biphenyl congeners. Environ Sci Technol 22:382–387

Hoff RM, Muir DCG, Grift NP (1992) Annual cycle of polychlorinated biphenyls and organohalogen pesticides in air in southern Ontario. 1. Air concentration data. Environ Sci Technol 26:266–275

Hornbuckle KC, Sweet CW, Pearson RF, Swackhamer DH, Eisenreich SJ (1995) Assessing annual water–air fluxes of polychlorinated biphenyls in Lake Michigan. Environ Sci Technol 29:869–877

Jonsson P (1992) Large-scale changes of contaminants in Baltic Sea sediments during the twentieth century. Doctoral Thesis, Uppsala University

Junge CE (1974) Residence time and variability of tropospheric trace gases. Tellus 26:477–488

Junge CE (1977) Basic considerations about trace constituents in the atmosphere as related to the fate of global pollutants. In: Suffet IH (ed) Fate of pollutants in the air and water environments. Wiley-Interscience, New York, pp 7–25

Karickhoff SW (1981) Semi-empirical estimation of sorption of hydrophobic pollutants on natural sediments and soils. Chemosphere 10:833–846

Kjeller L-O, Rappe C (1995) Time trends in levels, patterns, and profiles for polychlorinated dibenzo-p-dioxins, dibenzofurans, and biphenyls in a sediment core from the Baltic proper. Environ Sci Technol 29:346–355

Karickhoff SW, Brown DS, Scott A (1979) Sorption of hydrophobic pollutants on natural water sediments. Water Res 13:241–248

Leppänen J-M, Alenius P (1988) Cycling of organic matter during the vernal growth period in the open northern Baltic proper. I. Hydrography, currents and related factors. Finn Mar Res 255:3–18

Mackay D (1989) Modeling the long-term behavior of a contaminant in a large lake. J Great Lakes Res 15:183–297

Mackay D, Shiu WY, Ma KC (1991) Illustrated handbook of physical-chemical properties and environmental fate for organic chemicals, vols I–II. Lewis, Chelsea, MI

McGroddy SE, Farrington JW, Gschwend PM (1996) Comparison of the in-situ and desorption sediment–water partitioning of polycyclic aromatic hydrocarbons and polychlorinated biphenyls. Environ Sci Technol 30:172–177

McVeety BD, Hites RA (1988) Atmospheric deposition of polycyclic aromatic hydrocarbons to water surfaces: a mass balance approach. Atmos Environ 22:511–536

National Academy of Science (1979) Polychlorinated biphenyls. National Academy of Science, Washington, DC

Neumann T, Christiansen C, Clasen S, Emeis K-C, Kunzendorf H (1997) Geochemical records of salt-water inflows into deep basins of the Baltic Sea. Cont Shelf Res 17:95–115

Ott WR (1990) A physical explanation of the log normality of pollutant concentrations. J Air Waste Manage Assoc 40:1378–1383

Panshin SY, Hites RA (1994) Atmospheric concentrations of polychlorinated biphenyls at Bermuda. Environ Sci Technol 28:2001–2007

Pearlman RS, Yalkowsky SH, Banerjee S (1984) Water solubilities of polynuclear aromatic and heteroaromatic compounds. J Phys Chem Ref Data 13:555–562

Rekker RF, De Kort NN (1979) The hydrophobic fragment constant; an extension to a 1000 point data set. Eur J Med Chem Chim Ther 14:479–488

Schulz DE, Petrick G, Duinker JC (1989) Complete characterization of polychlorinated biphenyl congeners in commercial Aroclor and Clophen mixtures by multidimensional gas chromatography-electron capture detection. Environ Sci Technol 23:852–859

Schulz-Bull DE, Petrick G, Kannan N, Duinker JC (1995) Distribution of individual chlorobiphenyls (PCB) in solution and suspension in the Baltic Sea. Mar Chem 48:245–270

Schwartzenbach RP, Gschwend PM, Imboden DM (1993) Environmental organic chemistry. Wiley Interscience, New York
Slinn WNG, Hasse L, Hicks BB, Hogan AW, Lal D, Liss PS, Munnich KO, Sehmel GA, Vittor O (1978) Some aspects of the transfer of atmospheric trace constituents past the air-sea interface. Atmos Environ 12:2055–2087
Swackhamer DL, Skoglund RS (1993) Bioaccumulation of PCBs by algae: kinetics versus equilibrium. Environ Tox Chem 12:831–838
ten Hulscher TEM, van der Velde LE, Bruggeman WA (1992) Temperature dependence of Henry's law constants for selected chlorobenzenes, polychlorinated biphenyls and polycyclic aromatic hydrocarbons. Environ Tox Chem 11:1595–1603
Ullman WJ, Aller RC (1982) Diffusion coefficients in nearshore marine sediments. Limnol Oceanogr 27:552–556
Whitby KT (1978) The physical characteristics of sulfur aerosols. Atmos Environ 12:135–159
Wu S, Gschwend PM (1986) Sorption kinetics of hydrophobic organic compounds to natural sediments and soils. Environ Sci Technol 20:717–725
Zebühr Y, Näf C, Broman D, Lexén K, Colmsjö A, Östman C (1989) Sampling techniques and clean up procedures for some complex environmental samples with respect to PCDDs and PCDFs and other organic contaminants. Chemosphere 19:34–44

11 Sediments and Macrofauna in the Baltic Sea – Characteristics, Nutrient Contents and Distribution

R. Carman and H. Cederwall

11.1 Introduction

The nutrient load to the Baltic Sea has increased markedly in the last century (Larsson et al. 1985; Elmgren 1989; Wulff et al. 1990; Sandén et al. 1991; Jonsson and Carman 1994). This increase has resulted in greatly extended bottom areas with hypoxia or anoxia, and large-scale elimination of the macrofauna in the Baltic proper (Cederwall and Elmgren 1990). One consequence is that bioturbation has disappeared, and an expansion of laminated sediments has followed (Jonsson and Jonsson 1988; Jonsson et al. 1990). Fluxes of dissolved nutrients across the sediment-water interface and burial of different chemical substances fluctuate on a seasonal or annual basis, in response to variable inputs of minerogenic material and organic detritus. By a combination of diagenetic processes the nutrients are either retained in the sediments or released to the overlying water. The interplay between burial and release is determined by complex interactions between different biogeochemical processes (see Chap. 9).

Most of the carbon in the sediments of the Baltic Sea is inorganic, and mainly of terrigenous origin (Emelyanov 1988; Carman and Rahm 1997). This is because inorganic carbon species are normally undersaturated in the waters of the Baltic Sea. However, during certain anoxic or euxinic conditions (oxygen-free bottom-water masses usually with high contents of hydrogen sulfide) authigenic carbonate formation has been observed (Suess 1979; Jakobsen and Postma 1989; Carman 1996; Sternbeck and Sohlenius 1997). In such environments (e.g. the eastern Gotland basin) relatively high concentrations of carbonates (10 % dry wt.) are observed.

Most of the nitrogen in the Baltic Sea sediments is organically bound (Müller 1977; Carman et al. 1996; Carman and Rahm 1997; see Table 11.3). However, during anoxic conditions large amounts of ammonia accumulate in the interstitial water. It means that the possibility for ion exchange reactions between ammonium and preferably potassium in certain clay

Ecological Studies, Vol. 148
F. Wulff et al. (eds.)
A Systems Analysis of the Baltic Sea

structures would increase the amount of inorganic nitrogen (Müller 1977; Rosenfeldt 1979; Mackin and Aller 1984). Furthermore, prefixation (Müller 1977; Rosenfeld 1979) and precipitation of authigenic minerals (Martens et al. 1978; Elderfield et al. 1981) are other possible reaction pathways to increase the amount of inorganic nitrogen in the sediment.

While most of the carbon and nitrogen in the sediments of the Baltic proper are organically bound, high concentrations of inorganic phosphorus are normal in all kinds of sediments (Holm 1978; Balzer 1984; Carman and Jonsson 1991; Carman et al. 1996; Carman and Rahm 1997). However, the concentration of inorganically bound phosphorus is highly variable, mainly depending on prevailing redox conditions (Holm 1978; Balzer 1984; Carman and Wulff 1989; Carman and Jonsson 1991; Carman et al. 1996; Carman and Rahm 1997; Chap. 9).

The composition of infauna, and also abundance of animals, are very important for the microbial and chemical processes in the sediment, and consequently also for the exchange between sediment and water (e.g. Anderson and Macfayden 1976; Blackburn 1986; Blackburn and Sörensen 1988; Graf 1991).The deposit-feeders take part in the remineralization of organic matter, partly by excreting some of the assimilated nutrients, partly by breaking down the organic matter to smaller particles. Both processes facilitate microbial degradation.

Bioturbation is a collective term for those processes (burrowing, water pumping and search for food in the sediment, as well as sorting, consuming and defecating sediment particles) by which the benthic fauna influences the biogeochemical conditions of the sediment. The benthic fauna also makes up a variable storage of nutrients (biomass), which may exchange depending on the environmental conditions. The depth of the bioturbated layer in the sediments is dependent on the composition of the benthic fauna. In *Macoma* communities, which dominate the soft bottoms of the Baltic Sea, the depth of the bioturbated layer normally is about 5 cm (Ankar 1977; Hill and Elmgren 1987).

In order to be able to establish trustworthy nutrient budgets and develop reliable large-scale models of the biogeochemical cycles of the Baltic Sea, it is essential, to estimate the amount and distribution of nutrients and fauna in the sediments in combination with studies of certain biogeochemical diagenetic processes within the sediment-water strata (Wulff et al. 1990). By estimating the areas covered by different types of sediment, with their particular physical, chemical and biological composition, the results from site process studies can then be extrapolated to entire basins of the Baltic Sea.

The aim of this project has been to produce reliable estimates on the amounts of carbon, nutrients and macrofauna in the Baltic Sea sediments, as well as their horizontal and vertical distribution (carbon and nutrients) and an estimate of their burial rate.

11.1.1 Historical Background

The first attempts to estimate the amount of macrobenthic biomass in the Baltic Sea were made by fishery biologists in the early twentieth century (Petersen 1913; Thulin 1922; Hessle 1924; Hagmeier 1930). A broadly arranged international survey of the macro- and meiofauna of the Baltic Sea soft bottoms (Joint Macro and Meiofauna Sampling Programme) has been carried out more recently (Elmgren 1978; Elmgren et al. 1984). In the present survey, also some sediment data were collected (water content and loss on ignition). From the mid-1970s other data on macrozoobenthos are also available (Andersin et al. 1978; Järvekülg 1979; Cederwall and Elmgren 1980; Persson 1982; Zmudzinski and Ostrowski 1982; Lagzdins et al. 1987).

The sediment survey by Gripenberg (1934) was the first major report on the sediment characteristics of the Baltic Sea and its areal and vertical variation in nutrient concentrations. Extensive sediment surveys were later carried out by Soviet scientists during the 1950s and 1960s (Gorshkova 1972). Additional sediment investigations in the 1960s were, for instance, done by Manheim (1961) and Hartmann (1964). The amount of sediment investigations increased in the mid-1970s (County Administrative Board of Norrbotten 1974; Suess and Erlenkeuser 1975; Blashchishin and Pustelnikov 1977; Bloomquist 1977; Lithner 1974; Olausson et al. 1977; Suess and Djafari 1977; Boström et al. 1978; Engwall 1978; Holm 1978; Niemistö et al. 1978; Suess 1979; Höglund 1984) followed by additional investigations in the 1980s and 1990s (Balzer 1982, 1984, 1986; Pustelnikov et al. 1982; Emelyanov and Lakashin 1986; Boesen and Postma 1988; Emelyanov 1988; Carman and Wulff 1989; Jakobsen and Postma 1989; Jonsson et al. 1990; Koop et al. 1990; Carman and Jonsson 1991; Jonsson and Carman 1994; Conley and Johnstone 1995; Carman et al. 1996; Conley et al. 1997; Carman and Aigars 1997; Carman and Rahm 1997). Comparisons could sometimes be made with present data and data from the above-mentioned studies.

11.2 Materials and Methods

11.2.1 General Considerations

This study presents the results of several regional sediment surveys of selected physical, chemical and biological parameters. The investigation includes spatial distributions and concentration measurements within all the main sub-basins of the Baltic Sea (Fig. 11.1).

The number of sediment samples within each sub-basin is based on the area, bottom types and depth distribution. However, due to logistic problems, combined with the sampling technique used (gravity coring), we have sometimes obtained fewer samples than required from the original monitoring criteria. We have for the same reason not entirely achieved an even geographical distribution of stations (Fig. 11.1). Some areas, where special scientific studies were made, are over-represented, e.g. the Gulf of Riga.

Altogether 147 stations have been used for the physical and chemical characterization. Two physical and 12 chemical parameters have been measured as described below. In total, 8652 quantifications have been carried out, of which 5768 originate from measurements done in the laboratory.

The choice of stations for the macrobenthic investigation was not made at random for the same reason as for sediment sampling. Hence the stations used may not be fully representative for the respective subareas (Fig. 11.1). This is a trade-off for getting a large number of data to base the estimations on. Furthermore, the data were collected by different laboratories, which have not used identical methods, something that may lead to under- or overestimates. The wet weight data are mainly collected during the period of late spring-summer. Since the biomass changes over the year (Cederwall 1978) the seasonal distribution of samples probably does not give representative annual means. The material for analysis of the carbon and nitrogen content of benthic fauna was only collected during summertime. If the contents of these elements vary considerably over the year, our data may not equal annual means. However, studies carried out in other parts of the Baltic Sea (Cederwall and Jermakovs 1999; K. Leonardsson, pers. comm.) have shown that carbon as well as nitrogen content does not vary much over the year.

To calculate the wet weights of macrofauna presented in this report, the pooled data material from the 1970s and 1980s, stored in our database, were used together with recent data collected within the National Swedish Monitoring Programme. The monitoring data cover the Swedish coastal

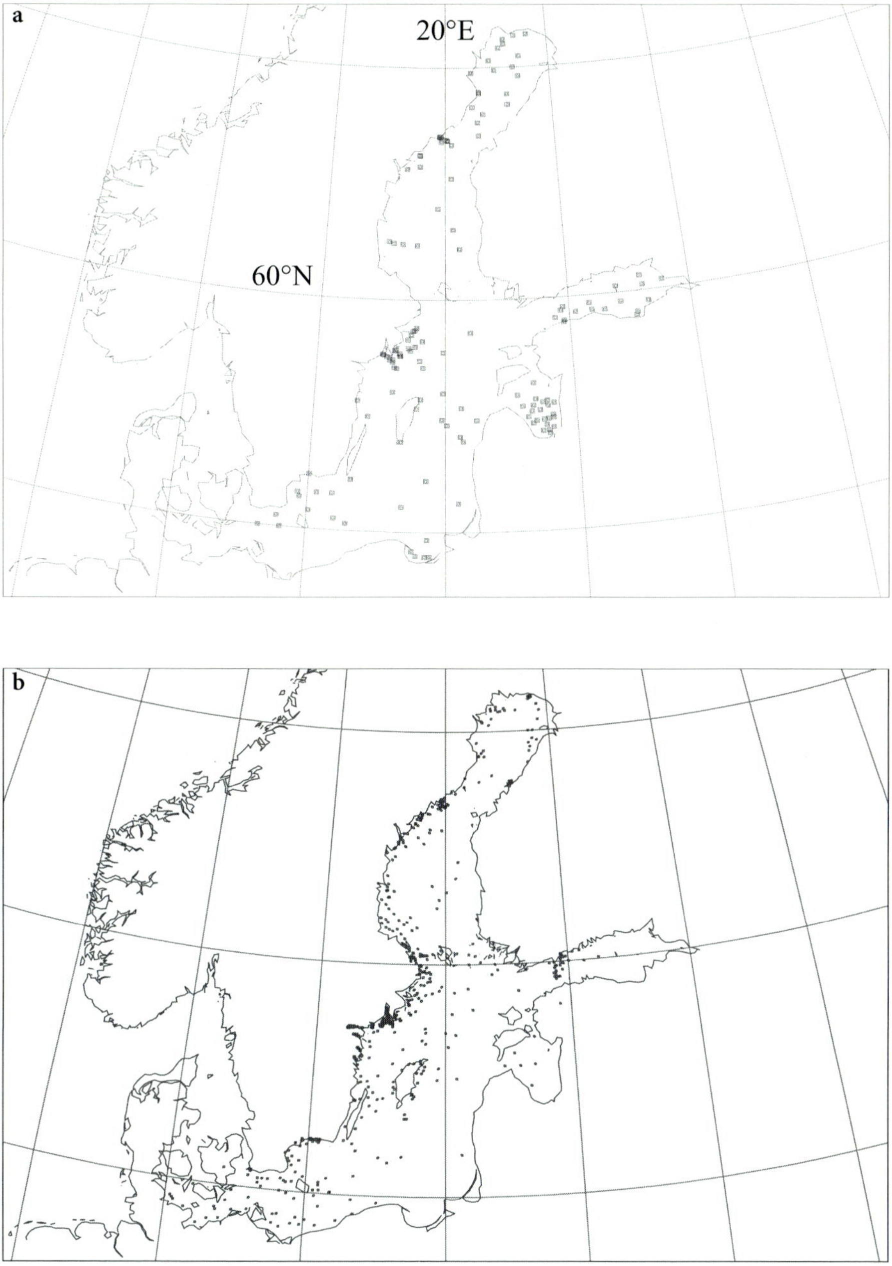

Fig. 11.1. Location of sites used in the sediment investigation (**a**) and macrofauna investigation (**b**)

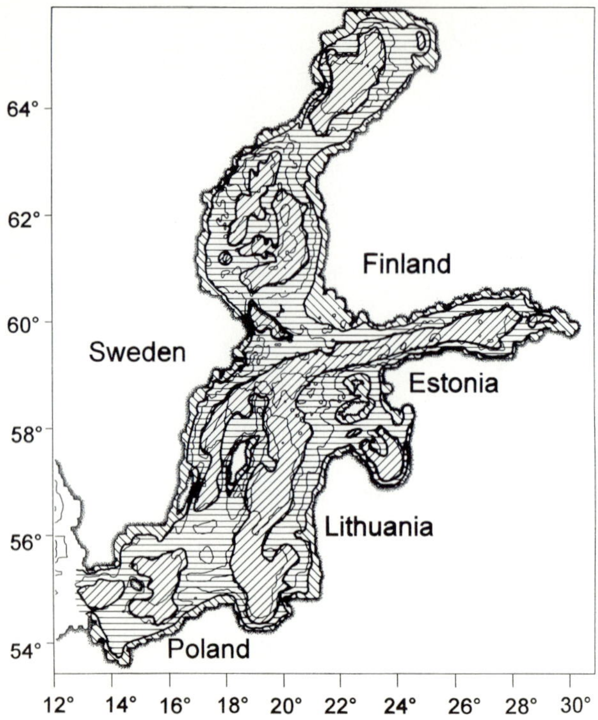

Fig. 11.2. Distribution of different bottom types in the Baltic Sea: *Horizontal lines* transportation bottoms; *left tilted lines* accumulation bottoms; *right tilted lines* erosion bottoms. The different water depth intervals (30, 70 and 120 m) used in calculation of storage amounts are also shown. Adapted from similar maps of Winterhalter et al. (1981), Emelyanov (1988), Juškevics et al. (1994), Raudsep et al. (1994) and sediment data from present study

zone of the Baltic Sea plus some offshore stations. Additional data published from local monitoring programs in Sweden and Finland have also been used. Most of the data in the database have been published. Since there were no data in the database from the deepest parts (>70 m) of the Gulf of Finland, data published by Seire (1992) were used to calculate mean biomass for this area. Altogether, data from 1294 samples from 752 stations were used.

Four different depth intervals (0–30, 30–70, 70–120 and >120 m; Fig. 11.2, Table 11.1) and three different bottom types (Table 11.2) were used to calculate average carbon and nutrient concentrations in the uppermost 1-cm layer of the sediments (Table 11.3) as well as the total amounts at deposition bottoms in the different sub-basins (Table 11.4). The extension of erosion bottoms may be underestimated since all bottoms characterised

Table 11.1. Areas of different depth zones of the Baltic Sea. Calculation of the different areas is based on data originally generated by Stigebrandt and Wulff (1987)

Bothnian Bay		
Total area	36,000 km^2	
Depth interval (m)	Area (km^2)	(%)
0– 30	18,000	50
30– 70	9 400	26
70–115	8 600	24
Bothnian Sea (incl. Åland Sea)		
Total area	70,000 km^2	
Depth interval (m)	Area (km^2)	(%)
0– 30	15,000	21
30– 70	25,000	36
70–120	22,500	32
120–250	7 500	11
Gulf of Finland		
Total area	30,000 km^2	
Depth interval (m)	Area (km^2)	(%)
0– 30	13,900	46
30– 70	14,500	48
70–100	1 600	6
Gulf of Riga		
Total area	19,200 km^2	
Depth interval (m)	Area (km^2)	(%)
0–30	13,000	68
30–62	6 200	32
Baltic proper		
Total area	206,000 km^2	
Depth interval (m)	Area (km^2)	(%)
0– 30	53,000	26
30– 70	72,000	35
70–120	59,000	28
120–459	22,000	11

as "sand bottoms" have been assigned to the transportation bottom category. Furthermore, it must also be mentioned that in the coastal areas, especially those within an archipelago, a mosaic of different bottom types occurs and not only erosion bottoms which is the case in Fig. 11.2. The four depth zones (Table 11.1) were used for calculation of the average macrobenthos biomass (Table 11.5) as well as macrofaunal wet weight. Even the

Table 11.2. Percentage composition of bottom types in the main sub-basins of the Baltic Sea (estimated from Fig. 11.2)

Subarea	Bottom type Accumulation	Transportation	Erosion
Bothnian Bay	33	43	24
Bothnian Sea (incl. Åland Sea)	30	46	24
Gulf of Finland	41	39	19
Gulf of Riga	28	39	33
Baltic proper	38	38	24

carbon, nitrogen and phosphorus storage (Table 11.6) were studied in the different sub-basins. The data given for the four depth zones represent roughly:

1. The bottoms down to the deepest part of the summer thermocline.
2. The bottoms between the thermocline and the primary halocline (permanent halocline; Kullenberg 1981).
3. The bottoms between the primary and the secondary halocline (the depth below which water is only renewed during major inflows of saline water; see Fonselius 1974).
4. The bottoms below the secondary halocline.

It should be noted that permanent haloclines only exist in the Baltic proper and the Gulf of Finland. The bottom type classification (accumulation, transportation and erosion bottoms; Fig. 11.2 and Table 11.2) is based on the criteria formulated by Håkanson and Jansson (1983), i.e.

- *Accumulation* areas prevail where fine materials, medium silt with grain sizes of less than 0.006 mm, can be deposited *continuously.*
- *Transportation* zones appear where there is a *discontinuous* deposition of fine particles/aggregates.
- *Erosion* areas prevail where there is *no deposition* of fine materials.

Table 11.3. Concentration of total, organic and inorganic carbon and nutrients (in uppermost centimetre) in the different basins of the Baltic Sea. Ratios are by atom weight

Baltic proper										
Total concentration										
Depth	Carbon *A-bottoms*[a] (mmol/g)	SD	Nitrogen (mmol/g)	SD	Phosphorus (µmol/g)	SD	No. of sites	Ratio C/N/P	Sediment weight (kg/m²)	SD
0– 30	5.7	2	0.62	0.2	48	14	4	119:12.8:1	1.46	0.44
30– 70	4.1	0.93	0.45	0.1	46	9.5	9	88:9.7:1	1.98	0.63
70–120	4.7	1	0.51	0.1	53	14	10	90:9.8:1	1.34	0.48
120–459	6.9	1.2	0.68	0.12	39	15	9	178:17.7:1	0.88	0.7
Sum	**5.3**	**1.6**	**0.56**	**0.15**	**46**	**14**	**32**	**115:12.1:1**	**1.41**	**0.58**
	T-bottoms									
0– 30	3.6	0.9	0.4	0.1	27	0.8	2	133:14.7:1	3.2	1
30– 70	3	1.3	0.32	0.13	38	13	15	79:8.5:1	3.4	1.5
70–120	1.9	1.1	0.21	0.13	30	10.5	5	63:6.9:1	4.2	1.8
120–459	–	–	–	–	–	–	–	–	–	–
Sum	**2.8**	**1.3**	**0.3**	**0.13**	**35**	**12.2**	**22**	**80:8.7:1**	**3.54**	**1.53**
	E-bottoms									
0– 30	0.51	0.19	0.04	0.25	12	3.7	2	41:3.2:1	11.4	3.9
30– 70	0.56	0.18	0.06	0.21	23	7.7	5	24:2.4:1	12	2.6
70–120	–	–	–	–	–	–	–	–	–	–
120–459	–	–	–	–	–	–	–	–		–
Sum	**0.55**	**0.14**	**0.05**	**0.02**	**22**	**8.7**	**7**	**25:2.4:1**	**11.8**	**2.7**

Table 11.3. (*continued*)

Baltic proper

Organic concentration

Depth	Carbon (mmol/g)	SD	Nitrogen (mmol/g)	SD	Phosphorus (µmol/g)	SD	No. of sites	Ratio C/N/P	Sediment weight (kg/m²)	SD
	A-bottoms[a]									
0– 30	5.7	2	0.59	0.02	17	4.9	4	33 5:34:1	1.46	0.44
30– 70	4	0.9	0.43	0.09	13	3.3	9	303:3 3:1	1.98	0.63
70–120	4.7	1	0.48	0.1	18	3.6	10	260:27:1	1.34	0.48
120–459	6.4	1.3	0.65	0.12	17	4.1	9	3 79:39:1	0.88	0.7
Sum	**5.1**	**1.5**	**0.53**	**0.14**	**16**	**4.6**	**32**	**319:33:1**	**1.41**	**0.58**
	T-bottoms									
0– 30	3.5	0.9	0.37	0.08	10	4.5	2	3 73:39:1	3.2	1
30– 70	2.9	1.2	0.3	0.12	9.3	4.1	15	312:32:1	3.4	1.5
70–120	1.9	1.1	0.19	0.12	8.2	3.3	5	227:23:1	4.2	1.8
120–459	–	–	–	–	–	–	–	–		
Sum	**2.7**	**1.3**	**0.28**	**0.13**	**9.1**	**3.8**	**21**	**298:31:1**	**3.54**	**1.53**
	E-bottoms									
0– 30	0.46	0.23	0.04	0.03	1.2	0.4	2	3 85:29:1	11.4	3.9
30– 70	0.54	0.18	0.05	0.02	1.9	0.8	5	284:27:1	12	2.6
70–120	–	–	–	–	–	–	–	–	–	–
120–459	–	–	–	–	–	–	–	–	–	–
Sum	**0.52**	**0.18**	**0.05**	**0.02**	**1.7**	**0.8**	**7**	**304:27:1**	**11.8**	**2.7**

Bothnian Sea

Total concentration

Depth	Carbon *A-bottoms*[a] (mmol/g)	SD	Nitrogen (mmol/g)	SD	Phosphorus (µmol/g)	SD	No. of sites	Ratio C/N/P	Sediment weight (kg/m^2)	SD
0– 30	3.2	0.36	0.28	0.04	41	2.6	3	78:6.9:1	1.86	0.38
30– 70	2.5	–	0.23	–	42	–	1	60:5.5:1	2.07	–
70–120	2.7	–	0.26	–	46	–	1	58:5.6:1	1.86	–
120–250	2.8	0.06	0.25	0.01	57	2.5	4	49:4.3:1	2	0.4
Sum	**2.9**	**0.32**	**0.26**	**0.03**	**49**	**8**	**9**	**59:5.3:1**	**1.96**	**0.34**
	T-bottoms									
0– 30	1.7	–	0.14	–	36	–	1	47:3.8:1	1.9	-
30– 70	2.1	0.13	0.19	0.13	40	6.9	2	53:4.7:1	3.6	2.6
70–120	2	0.38	0.19	0.05	37	6.6	5	54:5.2:1	3.3	0.6
120–250	0.7	–	0.06	–	46	–	1	15:1.4:1	8.3	–
Sum	**1.9**	**0.62**	**0.17**	**0.07**	**38**	**6.2**	**9**	**49:4.4:1**	**3.7**	**2.1**
	E-bottoms									
0– 30	–	–	–	–	–	–	–	–	–	-
30– 70	0.9	0.14	0.09	0.02	22	2.8	3	42:4:1	9	2
70–120	0.6	–	0.06	–	37	–	1	17:1.5:1	9.5	–
120–250	–	–	–	–	–	–	–	–	–	-
Sum	**0.9**	**0.19**	**0.08**	**0.02**	**26**	**7.9**	**4**	**33:3.1:1**	**9.2**	**1.7**

Table 11.3. *(continued)*

Bothnian Sea

Organic concentration

Depth	Carbon *A-bottoms*[a] (mmol/g)	SD	Nitrogen (mmol/g)	SD	Phosphorus (µmol/g)	SD	No. of sites	Ratio C/N/P	Sediment weight (kg/m^2)	SD
0– 30	3.2	0.35	0.26	0.04	13	3.2	3	242:20:1	1.86	0.38
30– 70	2.5	–	0.21	–	13	–	1	199:17:1	2.07	–
70–120	2.7	–	0.23	–	15	–	1	180:16:1	1.86	–
120–250	257	0.06	0.23	0.01	16	3.8	4	169:14:1	2	0.4
Sum	**258**	**0.31**	**0.24**	**0.03**	**15**	**3.3**	**9**	**194:16:1**	**1.96**	**0.34**
	T-bottoms									
0– 30	1.7	–	0.13	–	9	–	1	184:14:1	1.9	–
30– 70	2.1	1.3	0.18	0.12	9	652	2	227:20:1	3.6	256
70–120	2	0.38	0.18	0.04	8	3.4	5	220:20:1	3.3	0.6
120–250	0.7	–	0.06	–	3	–	1		8.3	–
Sum	**1.8**	**0.68**	**0.16**	**0.07**	**8**	**3.6**	**9**	**221:20:1**	**3.7**	**2.1**
	E-bottoms									
0– 30	–	–	–	–	–	–	–		–	–
30– 70	0.9	0,15	0.08	0.02	4	1.5	3	228:21:1	9	2
70–120	0.6	–	0.05	–	6.1	–	1	99:8.5:1	9.5	–
120–250	–	–	–	–	–	–	–	–	–	-
Sum	**0.8**	**0.2**	**0.08**	**0.02**	**4.6**	**1.6**	**4**	**181:16:1**	**9.2**	**1.7**

Bothnian Bay

Total concentration

Depth	Carbon *A-bottoms*[a] (mmol/g)	SD	Nitrogen (mmol/g)	SD	Phosphorus (µmol/g)	SD	No. of sites	Ratio C/N/P	Sediment weight (kg/m^2)	SD
0– 30	4.8	0.32	0.43	0.03	53	13	2	90:8.1:1	1.39	0.1
30– 70	3.6	–	0.33	–	42	–	1	86:7.8:1	1.17	–
70–115	3.4	0.27	0.28	0.02	45	1.6	5	76:6.1:1	1.57	0.12
Sum	**3.8**	**0.67**	**0.32**	**0.07**	**48**	**6.5**	**8**	**79:6.7:1**	**1.47**	**0.18**
	T-bottoms									
0– 30	2.3	0.33	0.18	0.03	43	4.2	3	54:4.3:1	2.6	0.3
30– 70	2.3	0.73	0.19	0.07	36	7.9	7	63:5.3:1	3	1.5
70–115	–	–	–	–	–	–	–	–	–	–
Sum	**2.3**	**0.61**	**0.19**	**0.06**	**38**	**7.4**	**10**	**60:4.9:1**	**2.9**	**1.3**
	E-bottoms									
0– 30	1.4	–	0.11	–	33	–	1	41:3.2:1	3.7	–
30– 70	0.8	–	0.06	–	82	–	1	10:0.7:1	10	–
70–115	–	–	–	–	–	–	–	–	–	–
Sum	**1.1**	**0.4**	**0.08**	**0.04**	**58**	**33**	**2**	**18:1.4:1**	**6.9**	**4.6**

Table 11.3. (*continued*)

Bothnian Bay										
Organic concentration										
Depth	Carbon *A-bottoms*[a] (mmol/g)	SD	Nitrogen (mmol/g)	SD	Phosphorus (µmol/g)	SD	No. of sites	Ratio C/N/P	Sediment weight (kg/m²)	SD
0– 30	4.8	0.32	0.41	0.04	12	3.3	2	383:33:1	1.39	0.1
30– 70	3.6	–	0.31	–	16	–	1	228:19:1	1.17	–
70–115	3.4	0.27	0.26	0.02	19	6.6	5	177:13:1	1.57	0.12
Sum	**3.8**	**0.66**	**0.3**	**0.07**	**17**	**6**	**8**	**220:18:1**	**1.47**	**0.18**
	T-bottoms									
0– 30	2.3	0.34	0.18	0.03	7	2.3	3	325:25:1	2.6	0.3
30– 70	2.3	0.72	0.18	0.06	8.2	2.5	7	274:22:1	3	1.5
70–115	–	–	–	–	–	–	–	–	–	–
Sum	**2.3**	**0.61**	**0.18**	**0.05**	**7.8**	**2.4**	**10**	**288:23:1**	**2.9**	**1.3**
	E-bottoms									
0– 30	1.3	–	0.1	–	5	–	1	266:20:1	3.7	–
30– 70	0.66	–	0.05	–	5.8	–	1	114:9.2:1	10	–
70–115	–	–	–	–	–	–	–	–	–	–
Sum	**1**	**0.47**	**0.08**	**0.04**	**5.4**	**0.6**	**2**	**185:15:1**	**6.9**	**4.6**

Gulf of Finland

Total concentration

Depth	Carbon *A-bottoms*[a] (mmol/g)	SD	Nitrogen (mmol/g)	SD	Phosphorus (μmol/g)	SD	No. of sites	Ratio C/N/P	Sediment weight (kg/m²)	SD
0– 30	4.6	–	0.48	–	67	–	1	69:7.1:1	1.5	
30– 70	5	1	0.54	0.13	78	34	6	64:6.9:1	1.8	0.3
70–110	4.4	1.1	0.44	0.12	87	23	3	50:5.1:1	1.53	0.6
Sum	**4.8**	**1**	**0.51**	**0.12**	**80**	**28**	**10**	**60:6.3:1**	**1.66**	**0.39**
	T-bottoms									
0– 30	2.3	0.36	0.21	0.03	49	14	3	46:4.3:1	7	1.8
30– 70	3.2	0.43	0.32	0.07	41	2.7	2	78:7.9:1	2.5	0.3
70–110	–	–	–	–	–	–	–		–	–
Sum	**2.6**	**0.61**	**0.26**	**0.07**	**46**	**10**	**5**	**57:5.6:1**	**5.2**	**2.8**
	E-bottoms									
0– 30	1.5	–	0.14	–	32	–	1	48:4.3:1	4.6	–
30– 70	0.64	0.4	0.05	0.03	23	0.3	2	28:2.4:1	11.5	4.5
70–110	–	–	–	–	–	–	–		–	–
Sum	**0.93**	**0.6**	**0.08**	**0.05**	**26**	**5**	**3**	**36:3.1:1**	**9.2**	**4.9**

Table 11.3. (*continued*)

Gulf of Finland

Organic concentration

Depth	Carbon *A-bottoms*[a] (mmol/g)	SD	Nitrogen (mmol/g)	SD	Phosphorus (μmol/g)	SD	No. of sites	Ratio C/N/P	Sediment weight (kg/m²)	SD
0– 30	4.6	–	0.48	–	15	-	1	300:31:1	1.5	–
30– 70	5	1	0.52	0.04	20	4.5	6	254:26:01	1.8	0.3
70–100	4.3	1.2	0.42	0.12	17	2.9	3	255:25:1	1.53	0.6
Sum	**4.7**	**1**	**0.48**	**0.12**	**18**	**3.7**	**10**	**257:26:1**	**1.66**	**0.39**
	T-bottoms									
0– 30	2.2	0.36	0.19	0.04	6.4	1.8	3	345:30:01	7	1.8
30– 70	3.2	0.43	0.3	0.07	11	2	2	294:28:01	2.5	0.3
70–100	–	–	–	–	–	–	–	–	–	–
Sum	**2.6**	**0.61**	**0.23**	**0.07**	**8.1**	**2.9**	**5**	**320:29:1**	**5.2**	**2.8**
	E-bottoms									
0– 30	1.5	–	0.12	–	4.9	–	1	296:25:01	4.6	–
30– 70	0.52	0.31	0.04	21	1.8	1.6	2	289:23:01	11.5	4.5
70–100	–	–	–	–	–	–	–	–	–	–
Sum	**0.83**	**0.54**	**0.07**	**50**	**2.8**	**2.1**	**3**	**296:25:1**	**9.2**	**4.9**

Gulf of Riga

Total concentration

Depth	Carbon *A-bottoms*[a] (mmol/g)	SD	Nitrogen (mmol/g)	SD	Phosphorus (µmol/g)	SD	No. of sites	Ratio C/N/P	Sediment weight (kg/m^2)	SD
0–30	–	–	–	–	–	–	–	–	–	
30–62	4.6	0.52	0.45	0.05	64	12.7	17	86:7.8:1	1.37	0.28
Sum	**4.6**	**0.52**	**0.45**	**0.05**	**64**	**12.7**	**17**	**79:6.7:1**	**1.37**	**0.28**
	T-bottoms									
0–30	2.9	1.2	0.2	0.1	35	11.3	3	54:4.3:1	4.2	2.1
30–62	2.7	0.73	0.23	0.03	48	14.5	2	63:5.3:1	3.44	0.86
Sum	**2.8**	**0.95**	**0.21**	**0.07**	**40**	**12.8**	**5**	**70:5.2:1**	**3.9**	**1.6**
	E-bottoms									
0–30	0.5	0.01	0.04	0.01	18	11	2	41:3.2:1	12.5	0.1
30–62	0.46	0.02	0.04	0.01	9	1.8	2	10:0.7:1	12.1	0
Sum	**0.48**	**0.03**	**0.04**	**0.01**	**14**	**4.3**	**4**	**18:1.4:1**	**12.3**	**0.2**

Table 11.3. (*continued*)

Gulf of Riga

Organic concentration

Depth	Carbon *A-bottoms*[a] (mmol/g)	SD	Nitrogen (mmol/g)	SD	Phosphorus (µmol/g)	SD	No. of sites	Ratio C/N/P	Sediment weight (kg/m^2)	SD
0–30	–	–	–	–	–	–	–	–	–	–
30–62	4.5	0.52	0.41	0.05	18	8.8	17	251:23:01	1.37	0.28
Sum	**4.5**	**0.52**	**0.41**	**0.05**	**18**	**8.8**	**17**	**251:23:01**	**1.37**	**0.28**
	T-bottoms									
0–30	2.3	1.2	0.18	91	8.7	3.5	3	264:21:01	4.2	2.1
30–62	2.4	0.4	0.21	30	11	5.7	2	224:20:1	3.44	0.86
Sum	**2.3**	**0.86**	**0.19**	**68**	**9.5**	**3.9**	**5**	**245:20:1**	**3.9**	**1.6**
	E-bottoms									
0–30	0.47	0.02	0.04	3	1.7	0.5	2	276:21:01	12.5	0.1
30–62	0.41	0.02	0.03	3.5	1.7	0.2	2	243:20:1	12.1	0
Sum	**0.44**	**0.04**	**0.04**	**2.8**	**1.7**	**0.3**	**4**	**258:21:1**	**12.3**	**0.2**

[a] For bottom type classification, see Håkanson and Jansson (1983)

Table 11.4. Total amounts of carbon, nitrogen and phosphorus (ton/km² and totals for whole deposition bottom area) in the different sub-basins of the Baltic Sea

	Carbon (ton/km²)	Total (/10⁻⁶)	Nitrogen (ton/km²)	Total (/10⁻⁶)	Phosphorus (ton/km²)	Total (/10⁻³)
Bothnian Bay						
0–1 cm	67	1.21	6.6	0.12	1.5	27
0–5 cm	288	5.2	27.5	0.5	11.4	205
Bothnian Sea						
0–1 cm	68	1.42	7.1	0.13	3	62
0–5 cm	309	5.6	31.7	0.57	15.6	281
Baltic proper						
0–1 cm	90	6.67	11	0.81	2	149
0–5 cm	404	30	48.2	3.6	7.6	564
Gulf of Finland						
0–1 cm	96	1.18	11.7	0.14	4.1	51
0–5 cm	415	5.1	51	0.63	13.9	171
Gulf of Riga						
0–1 cm	76	0.41	8.6	0.05	2.7	14.5
0–5 cm	445	2.4	52.8	0.28	13.8	74
Whole Baltic Sea						
0–5 cm		48.3		5.6		1295

Table 11.5. Depth distribution of macrobenthos biomass in different subareas of the Baltic Sea (measured as g wet weight, carbon and nitrogen) with standard errors of mean

Depth zone (m)	Wet weight (g/m²)	Carbon (g/m²)	Nitrogen (g/m²)	No. of samples
Bothnian Bay and the Quark				
0– 30	3.25±0.85	0.378±0.099	0.048±0.013	44
30– 70	0.61±0.51	0.182±0.152	0.020±0.017	9
70–100	2.69±0.78	0.772±0.224	0.085±0.025	6
Bothnian Sea, Åland Sea and Archipelago Sea				
0– 30	96.90±13.11	5.742±0.777	0.790±0.107	82
30– 70	57.84±8.11	6.568±0.921	0.816±0.114	71
70–120	16.95±2.18	3.305±0.425	0.429±0.055	37
120–250	14.03±2.96	2.329±0.491	0.323±0.068	26
Gulf of Finland				
0– 30	93.69±19.44	5.971±1.239	0.800±0.166	11
30– 70	82.07±42.59	5.966±3.096	0.785±0.407	7
70–100	4.00±1.66[a]	0.800±0.332	0.086±0.036	15

Table 11.5. (*continued*)

Depth zone (m)	Wet weight (g/m²)	Carbon (g/m²)	Nitrogen (g/m²)	No. of samples
Baltic proper, Mecklenburg Bight and Kiel Bight				
0– 30	117.20±7.92	5.498±0.372	0.833± 0.068	528
30– 70	57.83±3.32	4.244±0.244	0.561± 0.032	352
70–120	1.87±0.42	0.181±0.041	0.024± 0.005	156
120–350	0.10±0.10	0.002±0.002	0.0004±0.0004	8

[a] Estimated from Seire (1992)

Table 11.6. Storage of macrofaunal wet weight, carbon, nitrogen and phosphorus (calculated from C/P ratio published by Vink and Atkinson 1985) in different subareas and depth strata of the Baltic Sea

Depth zone (m)	Wet weight (tons)	Carbon (tons)	Nitrogen (tons)	Phosphorus (tons)
Bothnian Bay and the Quark				
0– 30	5.85×10^4	6.8×10^3	8.6×10^2	1.2×10^2
30– 70	5.7×10^3	1.7×10^3	1.9×10^2	0.3×10^2
70–115	2.3×10^4	6.6×10^3	7.3×10^2	1.2×10^2
Total	8.7×10^4	1.5×10^4	1.8×10^3	2.7×10^2
Bothnian Sea, Åland Sea and Archipelago Sea				
0– 30	1.45×10^6	8.6×10^4	1.2×10^4	1.5×10^3
30– 70	1.45×10^6	1.6×10^5	2.0×10^4	2.9×10^3
70–120	3.8×10^5	7.4×10^4	9.6×10^3	1.3×10^3
120–250	1.0×10^5	1.7×10^4	2.4×10^3	0.3×10^3
Total	3.4×10^6	3.4×10^5	4.4×10^4	6.0×10^3
Gulf of Finland				
0– 30	1.1×10^6	6.9×10^4	9.6×10^3	1.2×10^2
30– 70	1.2×10^6	8.9×10^4	1.2×10^4	1.6×10^3
70–100	1.2×10^4	2.4×10^3	2.6×10^2	4.0×10^1
Total	2.3×10^6	1.6×10^5	2.2×10^4	1.8×10^3
Baltic Proper				
0– 30	1.0×10^7	4.7×10^5	7.2×10^4	8.5×10^3
30– 70	3.5×10^6	2.5×10^5	1.7×10^4	4.5×10^3
70–120	1.1×10^5	1.1×10^4	1.4×10^3	2.0×10^2
120–320	2.0×10^3	4.0×10^1	8.0×10^0	7.2×10^{-1}
Total	1.35×10^7	7.3×10^5	9.0×10^4	1.3×10^4

11.2.2 Field Methods

A modified Kajak corer (Blomqvist and Abrahamsson 1985) or a multicorer (Barnett et al. 1984) was used to collect sediment cores for measurement of physical and chemical analysis. The cores were sectioned in 1-cm sections down to 5 cm depth except for a few locations where the cores were cut only in two sections, i.e. between 0–1 and 1–5 cm below seafloor. In-situ measurements of bottom water temperature and oxygen concentration were taken simultaneously. At the same time, in-situ measurements of redox conditions were taken at several stations. On many stations samples for both sediment chemistry and macrofauna were taken at the same time. The main sampling gear for collecting macrofauna has been the van Veen-grab (Dybern et al. 1976). In a few local programs the Ekman grab (Birge 1921) has been used. The sieves used had a mesh-size of 1 × 1 mm, except for a few local programs, where a mesh size of 0.5 × 0.5 mm was used.

11.2.3 Laboratory Methods

The water content (WC) and organic content (LOI) were analysed according to standard procedures (Håkanson and Jansson 1983). Total carbon (TOT-C) and total nitrogen (TOT-N) were analysed in a Leco element analyser following the manual supplied by the manufacturer. Inorganic carbon (IN-C) was determined by the acid-leaching procedure described by Hedges and Stern (1984). Exchangeable (EX-N) and fixed (FIX-N) nitrogen were measured according to the modified methods outlined by Mackin and Aller (1984) and Silva and Bremner (1966), respectively. Mobile phosphorus (MOB-P) was measured according to Carman and Jonsson (1991) without pretreatment with any chemicals. Total phosphorus (TOT-P) and inorganic phosphorus (IN-P) were analysed according to Froelich et al. (1988). The amounts of organic carbon, nitrogen and phosphorus were calculated by the difference between the total and the inorganic amounts. Finally, apatite phosphorus (AP-P) was calculated as the difference between IN-P and MOB-P.

Macrofaunal biomass estimates made from wet weight values are shown to have a large variance, because of methodological differences between laboratories (HELCOM 1991). Very few dry weight and ash-free dry weight data are so far reported. Therefore, the only way to get an acceptable areal coverage of biomass data in the calculations was to use the wet weight data.

In order to calculate biomass as carbon and nitrogen, analyses of carbon and nitrogen content were made for seven common species: *Halicryptus spinulosus, Harmothoe sarsi, Saduria entomon, Pontoporeia femorata, Monoporeia affinis, Mytilus edulis* and *Macoma balthica*. Together these species make up >90 % of the total biomass (measured as wet weight) in all sub-basins and depth strata in the Baltic Sea, except the shallowest zone of the Bothnian Bay (Ankar 1986; Cederwall and Leonardsson 1986, 1987). The animals were collected in the Askö-Landsort area, northern Baltic proper, in June–July 1993. They were frozen alive. After wet weighing of the animals and drying and weighing them again, the analyses were performed in a Leco CHN analyzer. The bivalves were treated with 1 M HCl before analysis, to eliminate the inorganic carbon content of the shell. Correlations were estimated between wet weight and carbon and nitrogen content respectively.

Very few data on carbon and nitrogen content of Baltic Sea species have been published (Cederwall and Jermakovs 1999, Leonardsson, pers. comm.). When calculating the total biomass as carbon and nitrogen, all oligochaetes, polychaetes and nemertines therefore had to be assumed to have the same percentage carbon and nitrogen content as *Harmothoe sarsi*. All isopods remaining on the same grounds are assumed to have a similar content of carbon and nitrogen as *Saduria entomon* while other crustaceans (except *Monoporeia affinis*) were assumed to be similar to *Pontoporeia femorata*. Finally, the proportions of carbon and nitrogen in molluscs (except *Mytilus edulis*) were assumed to be identical to *Macoma balthica*. Considering published data on carbon and nitrogen content of other species from other parts of the world (Beers 1966; Jørgensen et al. 1991), it seems fair to make these assumptions.

Total amounts of wet weight of carbon and nitrogen were estimated by multiplying concentrations per area by the total area of each depth stratum in each sub-basin of the Baltic Sea.

11.3 Results

11.3.1 Sediments

11.3.1.1 Physical Characteristics

The highest water and organic contents are found in the deep offshore areas of the Baltic proper. The corresponding deep offshore areas of the Gulf of Bothnia have on average between 1.5 and 2 times less organic content. Similar results were reported by Jonsson and Carman (1994). The central deep areas of the Gulf of Finland and the Gulf of Riga have on average 20% lower values of organic matter than the Baltic proper (Table 11.3). This observation is generally valid also in nearshore and archipelago deposition bottoms. At these bottoms and at some transportation bottoms the normal pattern is one of decreasing values in the upper part of the profiles followed by relatively constant values beneath. However, it is also common at transportation bottoms that the decreasing trends in the surficial sediments are followed by increasing contents deeper down. At erosion bottoms it is quite normal that the vertical concentration pattern of WC and LOI show increasing values with increasing depth below the seafloor (sand and pebble overlying glacial clay) or low and relatively stable values with increasing depth in the sediment profile.

A common observation in both limnic (Håkansson and Jansson 1983) and marine environments (Akal 1972) is that the water and organic matter contents are closely related to each other. This is also valid for the surface sediments (0–5 cm) of the entire Baltic Sea (Table 11.4).

11.3.1.2 Carbon and Nutrients

Table 11.3 gives the average total, organic and inorganic (by subtraction between the total and the organic content) concentrations of carbon, nitrogen and phosphorus in the uppermost centimetre of the sediment profile. The concentrations are calculated from three different bottom types (Table 11.2) and three or four water-depth intervals (Table 11.1). The average sediment weight is given (for the uppermost centimetre) in Table 11.3 in order to make it possible to calculate the average total amount (0–1 cm) for each bottom type. These total amounts at deposition bottoms are presented in Table 11.4 together with average amounts in the uppermost 5 cm of the sediment. They are calculated as an average amount from all depth intervals since it is unreliable, from the few existing data, to calculate the storage distribution at deposition bottoms within each water depth stratum. Thus, the data given in Table 11.4 were calculated by using the total area of each basin (Table 11.1) multiplied by the percentage deposition bottom area (Table 11.2) and the average amount (g/m^2) at the actual deposition bottoms.

The coefficient of determination (R^2) between organic carbon and organic nitrogen is high in all sub-basins, i.e. $R^2>0.95$ (Fig. 11.4). For both C_{org}/P_{org} and N_{org}/P_{org} the coefficients of determination are considerably lower. The variations in C_{org}/N_{org} ratio at deposition bottoms are relatively small compared to ratios found at transportation and erosion bottoms, both vertically in the sediment profile and within each sub-basin.

11.3.2 Macrofauna

Table 11.7 presents the results of the analysis of carbon and nitrogen content. The highest concentrations were found for *Monoporeia affinis* and the lowest for *Macoma balthica*. Bivalves show low values because their weight includes their shells, while the C and N analyses were performed on shell-free material. The C/N weight ratios (Table 11.7) vary from 4.0 (*Halicryptus spinulosus*) to 9.1 (*Pontoporeia femorata*).

The mean biomass per unit area, for different depth strata and sub-basins, is given in Table 11.5. Note the considerable difference between the oligotrophic Bothnian Bay and the rest of the Baltic Sea (Fig. 11.3). The values for the intermediate depth zone (30–70 m) of the Bothnian Bay (Table 11.5) are especially low. The low concentrations for the deep areas of the Baltic proper (Fig. 11.3, Table 11.5) and the Gulf of Finland (Table 11.5) are caused by an oxygen deficiency that has been present for most of the 1970s and the 1980s (HELCOM 1990).

From the data in Table 11.5 and the proportions of the different depth zones making up the total areas, the mean wet weight of biomass per unit

Table 11.7. Carbon and nitrogen content (as percentage of dry and wet weight including shells; carbonate carbon of bivalves not included), ± standard errors of mean, of some common Baltic Sea macrobenthos species. The ratio of carbon to nitrogen and number of animals (*n*) analysed are also given

Species	Carbon % Dry wt.	% Wet wt.	Nitrogen % Dry wt.	% Wet wt.	C/N Weight ratio	*n*
Halicryptus spinulosus	42.4±0.4	4.3±0.2	10.5±0.2	1.1±0.0	4.0	14
Harmothoe sarsi	43.9±1.2	1.9±0.3	9.0±0.5	0.4±0.1	4.9	6
Saduria entomon	37.2±0.6	13.1±0.7	5.9±0.2	2.1±0.1	6.3	12
Pontoporeia femorata	46.3±0.3	19.8±0.8	5.1±0.2	2.2±0.1	9.1	14
Monoporeia affinis	46.0±1.0	29.0±2.7	5.1±0.3	3.2±0.4	9.0	9
Mytilus edulis	7.7±0.5	3.1±0.2	1.6±0.1	0.6±0.0	4.9	10
Macoma balthica	10.9±0.8	5.3±0.4	1.5±0.1	0.7±0.1	7.4	10

area can also be calculated; 2.4 for the Bothnian Bay, 47 for the Bothnian Sea, 77 for the Gulf of Finland and 54 g/m^2 for the Baltic proper.

Table 11.6 shows estimations of the total biomass of the three depth zones and their sums, in the different subareas of the Baltic Sea. The Baltic proper, being the largest subarea, clearly shows the highest values, while the lowest values are found in the oligotrophic Bothnian Bay. Note that the ratios between the most shallow and the second most shallow depth zones are quite different for wet weights than for carbon concentrations. This is caused by the molluscs (which include the shells, with little organic carbon) dominating the most shallow depth zone (except in the Bothnian Bay).

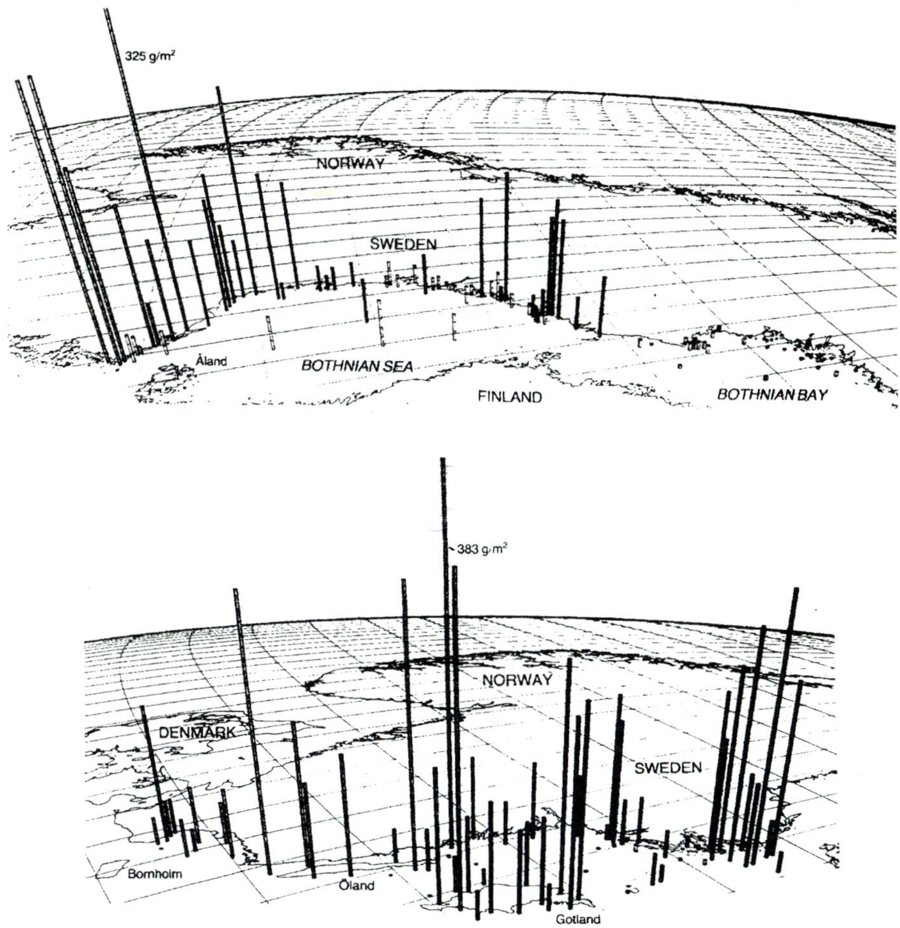

Fig. 11.3. Biomass of benthic fauna in the Baltic Sea

11.4 Discussion

11.4.1 Sediments

11.4.1.1 Physical Characteristics

The close connection between water and organic matter content in the entire Baltic Sea gave a coefficient of determination of at least R^2=0.90 for each of the sub-basins (Table 11.8). The relationship is almost independent of depth in the sediment and physical and biogeochemical characteristics. Consequently, they show the same distribution patterns. The individual amounts can be calculated from each other with good accuracy. The sometimes deviating values may be due to slicing artifacts and/or analytical errors in determination of the water content, especially in samples with very high water content and in samples containing coarse particles.

The general increase in water and organic matter content towards the sediment surface in deposition bottoms is due to the upper parts of the sediment being still in a dynamic condition. Bioturbation, microbial degradation, compaction and resuspension interact, thus causing the increased water and organic matter content in the consolidated sediments further down. The disparity in the vertical distribution pattern at some transport and erosion bottoms is most likely a result of the large alteration in water depths in the Baltic Sea during the Pleistocene period. Thus, some of these bottoms have during different stages of the Baltic Sea history switched between acting as accumulation and acting as transportation-erosion bottoms.

Although LOI is not a chemically defined parameter, it is useful in that it is highly correlated not only to the water content (Fig. 11.4), but also to

Table 11.8. Logarithmic fit between water content and organic matter content (measured as loss on ignition) within the Baltic Sea ($y=a\times\ln(x)+b$; where x is loss on ignition and y the water content)

Area	Coefficients a	Coefficients b	Coefficient of determination (R^2)
Bothnian Bay	24.6	20.6	0.88
Bothnian Sea	27.8	12.9	0.94
Baltic proper	24.4	15.2	0.91
Gulf of Finland	27.7	7.8	0.93
Gulf of Riga	23.5	20.3	0.96
Entire Baltic Sea	23.7	17.9	0.90

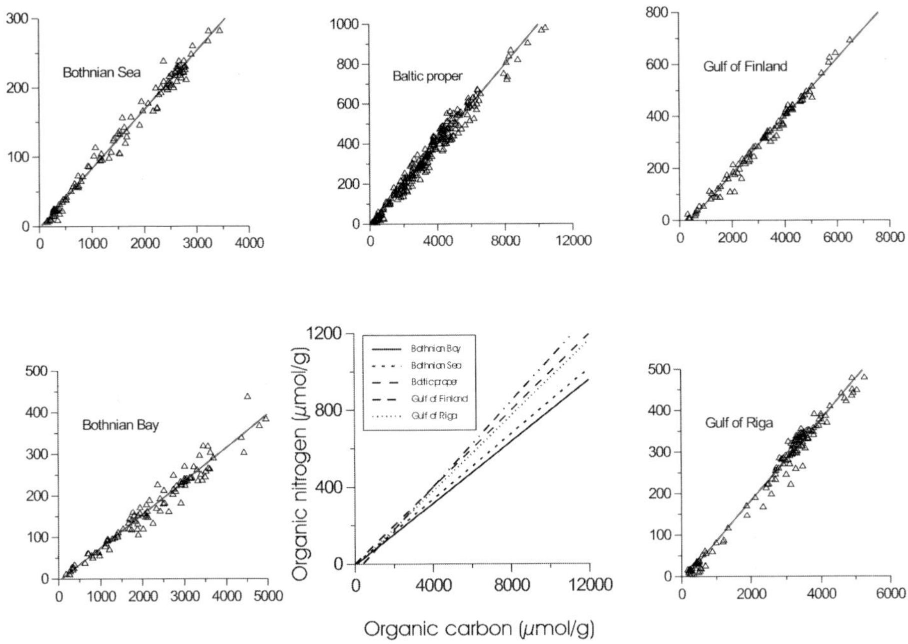

Fig. 11.4. Relationship between organic carbon and nitrogen within the different sub-basins of the Baltic Sea. Coefficients of determination are 0.95, 0.99, 0.97, 0.99 and 0.98 for Bothnian Bay, Bothnian Sea, Baltic proper, Gulf of Finland and Gulf of Riga, respectively

many chemical parameters in the sediment, for instance organic carbon. Thus, since both the water content and LOI have been, and still are, frequently measured in practically all kinds of sediment investigations, they could, with advantage, be used as a covariable for calculating concentrations of selective chemical parameters from other sediment investigations which lack certain chemical data. Furthermore, the measurements of LOI, which is a fast and precise method, could consequently often be a substitute for several time-consuming chemical analyses, e.g. of organic carbon and nitrogen.

11.4.1.2 Carbon

At deposition bottoms the carbonates constitute normally less than 1% dry wt. or less than 5% of the total carbon content (Table 11.3) in all subareas of the Baltic Sea. The low concentration of inorganic C in the surface sediments of the same region is due to the fact that most kinds of carbonates are thermodynamically unstable (Carman and Rahm 1997). The instability is largest in the most intense aerobic mineralization zones due to decreasing pH (Jahnke 1990). Saturation of carbonates (preferably mixed manganese carbonates) is only reached in certain sulphate-reducing pore-water environments (see e.g. Suess 1979; Jakobsen and Postma 1989; Carman 1996). This is due to raised levels of both pH and carbonate alkalinity in such environments (Froelich et al. 1988). Indications of such authigenic precipitation could also be seen in this study since the highest concentrations of inorganic carbon were found at sites situated in the deep, euxinic eastern Gotland Basin area.

The high amount of organic carbon in the deep offshore sediments of the Baltic proper (Fig. 11.5, LOI and Table 11.3) is most likely due to high primary production (Jonsson and Carman 1994) and altered decomposition efficiency. However, differences in water depth and distance from shore may be additional factors influencing its distribution pattern. Another factor of importance is the almost total lack of fauna in the deep anoxic regions of the Baltic proper, fauna that, under oxic conditions, would have been present and would have stimulated mineralization of organic carbon into CO_2.

11.4.1.3 Nitrogen

Besides organically bound nitrogen in the sediments there is also a certain amount of inorganically sequestered nitrogen, that is FIX-N and EX-N. Its concentration in the sediments depends on prevailing redox conditions, surface area of sediment particles, clay mineral compositions and ammonium concentrations (see e.g. Müller 1977; Rosenfeldt 1979; Mackin and Aller 1984). Even though the concentrations of inorganic nitrogen in some sediments lie in the upper range compared to other marine sediments (Müller 1977), it is generally less than 10% of the total nitrogen pool. However, even though it represents a minor portion of the total amount of nitrogen, its concentration could not be neglected, since the ratios of organic carbon to total nitrogen do not reflect the true chemical composition of the sedimentary organic matter.

11.4.1.4 Phosphorus

In contrast to carbon and nitrogen a substantial portion of the phosphorus in the sediments is inorganically bound (Table 11.3). The nature of the sedimentary matrix (grain size, chemistry, etc.) combined with oxygen

Fig. 11.5. Loss on ignition values in offshore deposition sediments in the Baltic Sea. Each unit of scale=2 % LOI

conditions control the extent to which burial of P occurs. The iron cycle is important for the redox-dependent, reversible abiotic sequestering of P within the sediment. In the oxic zone, large amounts of P adsorb to solid ferric complexes (e.g. Berner 1973; Lucotte and D'Anglejan 1988). When those complexes dissolve in the reduced zone the adsorbed amount of P will consequently also be released together with an almost simultaneous release of P from other sources (Gächter et al. 1989; Gächter and Meyer 1993). The released P can either be sequestered as another P-bearing phases (Froelich et al. 1982; Ruttenberg and Berner 1993) or be transported upwards through the redox cline by diffusion where it again can be adsorbed on iron oxyhydroxides (Shaffer 1986) and be incorporated into bacterial biomass (Gächter et al. 1989). Such behaviour obviously also takes place in the sediments of this study where high amounts of MOB-P are found in the oxidized parts of the sediment column whereas the concentrations quickly decrease below the transition zone. MOB-P approaches zero concentration in the reduced zone.

The normal behaviour in the Baltic proper, Gulf of Finland and Gulf of Riga is decreasing MOB-P concentrations with increasing burial depth. However, in the Bothnian Sea and Bothnian Bay, a more complex pattern often occurs. High concentrations of MOB-P are observed even deep down in the sediments. The main explanation for such distribution patterns is of course that most of the sediments in the Gulf of Bothnia have far better oxygen conditions than those of the Baltic proper. The sediments are normally oxidized deep down in the sediment profile. The high peaks of MOB-P at certain levels are most likely due to the abundant occurrence of solid manganese concretions (with high concentrations of iron and manganese) in the fine-grained sediments of this region (Ingri 1985; Carman 1996). The high concentrations of redox-dependent metals thus contribute to enhance the concentration of MOB-P (Carman 1996).

Besides adsorption on solid ferric complexes it is known that clays, Al oxides and Mn oxides also, to varying degrees, adsorb P (see e.g. Froelich et al. 1982). All sorption reactions will be enhanced as a function of decreasing grain size of the sediment. This is clearly reflected by far higher P concentrations in the fine-grained accumulation sediments than in the coarse transportation-erosion sediments. However, in terms of total amount per area, it is not uncommon that bottoms with coarser sediments contain higher amounts (Table 11.3) due to a larger sediment weight per unit area.

The remaining inorganic concentration of P (AP-P; calculated as the difference between the total inorganic amount and the concentration of MOB-P) contains P adsorbed or bounded to all sediment complexes other than the redox-dependent ones, for instance Al oxides, clays, calcite and authigenic and biogenic apatite (Froelich et al. 1982). The vertical variation in AP-P is small compared to that of MOB-P. The highest concentrations are found in the deposition sediments of the Bothnian Bay and Bothnian Sea, probably in connection with deposits of different manganese concretions.

11.4.1.5 Organic C:N:P Ratios in Baltic Sea Sediments

It has been argued that most of the mineralization of the annually produced organic matter in the Baltic Sea occurs during sedimentation in the water mass (Emelyanov and Lakashin 1986; Emelyanov 1988). As a consequence, less than 3% reaches the seafloor (Blashchishin and Pustelnikov 1977). However, the mineralization continues at the seafloor and there the remaining is finally buried. The amounts of organic matter will become less than 0.2% of its production in the photic zone (Blashchishin and Emelyanov 1977; see also Chap. 8). Fractionation processes during halmyrolysis (biogeochemical processes in the water column or at an early diagenetic stage) and diagenesis usually increases the C_{org}/N_{org} and the C_{org}/P_{org} as well as the N_{org}/P_{org}. This is caused by a preferential N and P utilization compared to C (see e.g. Froelich et al. 1988). However, according to the C_{org}/N_{org} ratios found in the sediments of this study (Fig. 11.3) it seems that most of the preferential N fractionation occurs during halmyrolysis or during an early diagenetic stage. This is because stable C_{org}/N_{org} ratios are found, within each sub-basin, almost irrespectively of variation in sediment depth in the profiles and physical, biological and chemical parameters. Further mineralization in the upper 5 cm proceeds thus principally with constant C/N proportionality. Even P is apparently preferentially decomposed during sedimentation in the water column, both with respect to C and N (Table 11.3). The mineralization during diagenesis seems almost independent of the concentration of C in difference to N. Differences in the extent of early diagenesis of organic P depend most likely on differences in sedimentation rates and redox conditions. Another factor is that residual P-rich organic phases form in certain environments (Gächter et al. 1989; Ingall and van Capellen 1990; Ingall et al. 1993; Ingall and Jahnke 1994; Carman et al. 1996).

11.4.1.6 Storage of C, N and P in Deposition Sediments of the Baltic Sea

The successively larger storage of carbon and nitrogen in the uppermost 5 cm of the deposition sediments when going from the Bothnian Bay southwards (Table 11.5) roughly reflects the present eutrophication conditions of respective basins. Almost the same trend can be discerned in the uppermost centimetre even though the largest storage is found in the Gulf of Finland instead of the Gulf of Riga. The large storage of N in the Gulf of Finland explains also the very high flux rates of N reported in Conley et al. (1997). Phosphorus has a different storage pattern. The lowest amounts are found in the Baltic proper (Table 11.3), whereas considerably higher amounts are observed in the Bothnian Sea as well as in the Gulf of Finland and Gulf of Riga. The accepted explanation for this behaviour is that a large proportion of the phosphorus is inorganically bound to elements easily affected by alterations in redox conditions (e.g. iron). In the Baltic proper a large proportion of the deposition sediments are reduced (Jonsson et al. 1990) and thus, to a large extent, incapable of adsorbing phosphorus (Carman and Wulff 1989; Carman and Jonsson 1991). In the other sub-basins far better oxygen conditions normally occur which results in a larger amount of phosphorus stored in the surficial sediments. This is clearly seen in the storage data presented in Tables 11.3 and 11.5.

11.4.1.7 Deposition Rates

According to the literature the sedimentation rate in the Baltic proper varies within a wide range, depending on, among other factors, distance from shore and locality of the accumulation bottom. The sedimentation rate in the offshore areas varies from 1 mm/year (Gotland deep; Ignatius 1958; Niemistö and Voipio 1981) up to 2–4 mm/year at several locations in the southern and northwestern Baltic proper (Erlenkeuser et al. 1974; Suess and Erlenkeuser 1975; Suess 1979; Kögler and Larsen 1979; Jonsson et al. 1990). In the Swedish archipelagos of the Baltic proper much higher rates are common, i.e. up to more than 5 cm/year has been observed (cesium dated; Carman et al., unpubl. research). If we assume an average deposition velocity of 1.5 mm/year in the offshore area of the Baltic proper, approximately 0.38 kg/m^2 accumulates annually (see Table 11.3 for figures). The total accumulation bottom area of the Baltic proper is around 78,000 km^2 (Tables 11.1 and 11.2) of which approximately 70,000 km^2 are located in the offshore region (Carman and Jonsson 1991). Calculated average total concentrations of C, N and P (in the uppermost 5 cm) are 4100, 410 and 28 μmol/g, respectively. Based on an accumulation of 0.38 kg/m^2, one estimates that 1558, 156 and 10.6 mmol/m^2 of C, N and P are buried in the offshore region annually. However, since it is based on the total amounts estimated for the uppermost 5 cm of the sediment, obtained estimates are certainly an overestimation since this part of the sediment is not yet fully remineralized. On the other hand, authigenic processes may increase the concentrations further down the sediment pack. For phosphorus the burial is lower than estimated for the Baltic proper (Wulff and Stigebrandt 1989; recalculated for burial only at accumulation bottoms) but comparable with the burial rates calculated by Jonsson et al. (1990). For C and N it is also close to the results obtained by Jonsson et al. (1990).

11.4.2 Macrofauna

The data for mean biomass can be compared with the data given by Elmgren et al. (1984) and Elmgren (1978). The data presented here are more than three times higher than those by Elmgren. For the Bothnian Bay this can be an effect of more coastal stations being used in this study. This is supported by the fact that the biomass found in the present study for the shallowest zone is 16 times higher than that reported by Elmgren et al. (1984), while the data for the deeper zones are nearly the same. For the Bothnian Sea the comparison shows different results: the biomass estimated in this study is, for the shallowest area, 40 % lower than those found by Elmgren et al.(1984), while the biomass for the deeper areas is 50 % higher in this study. The biomass given for the Baltic proper by Elmgren (1978) is based on stations only from the northern and central parts of the Baltic proper, so those data cannot be compared with the data given here. One should bear in mind that the data given by Elmgren et al. (1984) and Elmgren (1978) were based on a very limited number of sites: 13 from the Bothnian Bay, 32 from the Bothnian Sea and 28 from the Baltic proper.

The figures for carbon and nitrogen content of invertebrates published here (Table 11.7) agree well with figures reported from other areas of the Baltic Sea (Cederwall and Jermakovs 1999; Leonardsson, pers. comm.) as well as for other species from other parts of the world (Beers 1966; Jørgensen et al. 1991).

Out of the C/N ratios given in Table 11.7, the data for *Halicryptus spinulosus*, *Harmothoe sarsi* and *Mytilus edulis* fall well within the limits reported in the literature for benthic animals (Müller 1977). The ratio found for *Saduria entomon* is more like the data reported for zooplankton (Müller 1977), while the ratios for the amphipods are much higher, reflecting their high lipid content (e.g. Hill et al. 1992).

Since the earliest quantitative estimates of the benthic communities in the Baltic Sea were made (e.g. Hessle 1924), the community has changed. The most dramatic change has occurred in the deepest parts of the Baltic proper, where the macrofauna became more or less extinct during the late 1960s because of decreasing oxygen concentrations in the deep water.

During the following decades macrofauna only temporarily succeeded in establishing themselves in this area (HELCOM 1987, 1990). In the more shallower parts of the Baltic proper, the macrofauna was found to have increased in biomass between the 1920s and the 1970s (Cederwall and Elmgren 1980), as a response to eutrophication. The bivalves were responsible for the largest part of this increase. The increase continued until the mid-1980s (HELCOM 1990), but after that no increase has been shown. Also in the Bothnian Sea the biomass of the macrofauna has increased, but in the Bothnian Bay no such change has been detected (Cederwall and Blomqvist 1991).

Acknowledgements. This work was supported by the Swedish Environmental Protection Agency (SNV) and the Swedish Natural Science Research Council (NFR).

References

Akal T (1972) The relationship between the physical properties of underwater sediments that affect bottom reflections. Mar Geol 13:251–266

Andersin A-B, Lassig J, Parkkonen L, Sandler H (1978) The decline of macrofauna in the Baltic proper and the Gulf of Finland. Kieler Meeresforsch Sonderh 4:23–52

Anderson JM, Macfayden A (eds) (1976) The role of aquatic organisms in decomposition processes. Blackwell Scientific Publications, Oxford

Ankar S (1977) The soft bottom ecosystem of the northern Baltic proper with special reference to the macrofauna. Contrib Askö Lab 19:1–62

Ankar S (1986) Övervakning av mjukbottenfauna i Östersjöns kustområde (in Swedish). Rapport från verksamheten 1985. Naturvårdsverket Rapp 3248:1–54

Balzer W (1982) On the distribution of iron and manganese in the sediment/water interface: thermodynamic versus kinetic controls. Geochim Cosmochim Acta 46:1153–1161

Balzer W (1984) Organic matter degradation and biogenic element cycling in a near shore sediment (Kiel Bight). Limnol Oceanogr 29:1231–1246

Balzer W (1986) Forms of phosphorus and its accumulation in coastal sediments of Kieler Bight. Ophelia 26:19–35

Barnett PRO, Watson J, Conelly D (1984) A multiple corer for taking virtually undisturbed samples from shelf, bathyal and abyssal sediments. Oceanol Acta 7:399–408

Beers JR (1966) Studies on the chemical composition of the major zooplankton groups in the Sargasso Sea off Bermuda. Limnol Oceanogr 11:520–528

Berner RA (1973) Phosphate adsorption from sea water by adsorption on volcanogenic ferric oxides. Earth Planet Sci Lett 18:77–86

Birge EA (1921) A second report on limnological apparatus. Trans Wis Acad Sci Arts Lett 20:533–552

Blackburn TH (1986) Nitrogen cycle in marine sediments. Ophelia 26:65–76

Blackburn TH, Sörensen J (eds) (1988) Nitrogen cycling in coastal marine environments. SCOPE 33. Wiley, Chichester

Blashchishin AI, Emelyanov EM (1977) The basic features of geochemistry of the Baltic Sea. In: Geokhemicheskiye issledovaniya i poisk na territorii Belorussi i Pribaltiki (in Russian). Minsk Nauka Tekhnika 6:161–172

Blashchishin AI, Pustelnikov OS (1977) Biogenic components (C_{org} and $SiO_{2amorph}$) in water, suspended matter and bottom sediments of the Baltic Sea (in Russian). Baltica 6:161–172

Blomqvist S, Abrahamsson B (1985) An improved Kajak-type gravity core sampler for soft bottom sediments. Schweiz Z Hydrol 47:81–84

Bloomquist RG (1977) Chemical characteristics of interstitial waters from the southern Baltic Sea. PhD Thesis, Department of Geology, University of Stockholm

Boesen C, Postma D (1988) Pyrite formation in anoxic environments of the Baltic. Am J Sci 288:575–603

Boström K, Burman IO, Boström B, Pontér C, Brandlöf S, Alm B (1978) Geochemistry, mineralogy and origin of the sediments in the Gulf of Bothnia. Finn Mar Res 244:8–35

Carman R (1996) Vertical and spatial distribution of carbon and nutrients in the soft bottom sediments of the Baltic Sea. Proc ICES 1996 Annu Science Conf, Reykjavik

Carman R, Aigars J (1997) Vertical and spatial distribution of biogenic silica in the sediment of the Gulf of Riga, Baltic Sea. Toxicol Environ Chem 60:245–259

Carman R, Jonsson P (1991) The distribution pattern of different forms of phosphorus in some surficial sediment types of the Baltic Sea. Chem Geol 90:91–106

Carman R, Rahm L (1997) Early diagenesis and chemical characteristics of interstitial water and sediments in the deep deposition bottoms of the Baltic proper. J Sea Res 37:25–47

Carman R, Wulff F (1989) Adsorption capacity of phosphorus in Baltic Sea sediments. Estuarine Coast Shelf Sci 29:447–456

Carman R, Aigars J, Larsen B (1996) Carbon and nutrient geochemistry of the surface sediments of the Gulf of Riga, Baltic Sea. Mar Geol 134:57–76

Cederwall H (1978) Long term fluctuations in the macrofauna of northern Baltic soft bottoms. I. 1970–1973. Contrib Askö Lab Univ Stockh 22:1–83

Cederwall H, Blomqvist M (1991) Long-term changes in soft bottom macrofauna in the Gulf of Bothnia – a sign of eutrophication. Acta Ichtyol Pisc 21(Suppl):39–40

Cederwall H, Elmgren R (1980) Biomass increase of benthic macrofauna demonstrates eutrophication of the Baltic Sea. Ophelia Suppl 1:287–304

Cederwall H, Elmgren R (1990) Biological effects of eutrophication in the Baltic Sea, particularly the coastal zone. Ambio 19:109–112

Cederwall H, Jermakovs V (1999) Growth and production of some important macrofauna species in the Gulf of Riga, including comparisons with other areas. Hydrobiologia 393:201–210

Cederwall H, Leonardsson K (1986) Övervakning av mjukbottenfauna i Bottniska Vikens kustområden. Rapport från verksamheten 1985 (in Swedish). Naturvårdsverket Rapport 3249:1–28

Cederwall H, Leonardsson K (1987) Övervakning av mjukbottenfauna i Bottniska Vikens kustområden. Rapport från verksamheten 1986 (in Swedish). Naturvårdsverket Rapp 3338:1–28

Conley DJ, Johnstone R (1995) Biogeochemistry of N, P and Si in the Baltic Sea sediments: response to a simulated deposition of a spring diatom bloom. Mar Ecol Prog Ser 23:45–55

Conley DJ, Stockenberg A, Carman R, Johnstone RW, Rahm L, Wulff F (1997) Sediment-water nutrient fluxes along a eutrophication gradient in the Gulf of Finland, Baltic Sea. Estuarine Coast Shelf Sci 45:591–598

County Administrative Board of Norrbotten (1974) Vattenkvalitet, sedimentkvalitet och vegetationsförhållanden i Bottenviken vid Norrbottenskusten augusti 1974. Länsstyrelsen i Norrbotttens län, naturvårdsenheten, pp 1–48
Dybern BI, Ackefors H, Elmgren R (1976) Recommendations on methods for marine biological studies in the Baltic Sea. BMB Publ 1:1–98
Elderfield H, McCaffrey RJ, Luedtke N, Bender M, Truesdale VW (1981) Chemical diagenesis in Narragansett Bay sediments. Am J Sci 281:1021–1055
Elmgren R (1978) Structure and dynamics of Baltic benthos communities, with particular reference to the relationship between macro- and meiofauna. Kieler Meeresforsch Sonderh 4:1–22
Elmgren R (1989) Man's impact on the ecosystem of the Baltic Sea: energy flows today and at the turn of the century. Ambio 18:326–332
Elmgren R, Rosenberg R, Andersin A-B, Evans S, Kangas P, Lassig J, Leppäkoski E, Varmo R (1984) Benthic macro- and meiofauna in the Gulf of Bothnia (northern Baltic). Finn Mar Res 250:3–18
Emelyanov EM (1988) Biogenic sedimentation in the Baltic Sea and its consequences. In: Winterhalter B (ed) The Baltic Sea. Spec Pap 6. Geological Survey of Finland, Helsinki, pp 127–135
Emelyanov EM, Lukashin VN (1986) Geochemistry of the sedimentary process in the Baltic Sea (in Russian). Nauka, Moscow
Engvall A-G (1978) The fate of nitrogen in early diagenesis of Baltic sediments. A study of the sediment–water interface. Contribution in microbial geochemistry. Department of Geology, University of Stockholm
Erlenkeuser H, Suess E, Willkomm H (1974) Industralization affects heavy metal and carbon isotope concentrations in recent Baltic sediments. Geochim Cosmochim Acta 38:823–842
Fonselius S (1974) Oceanografi (in Swedish). Generalstabens Litografiska Anstalts Förlag, Stockholm, 248 pp
Froelich PN, Bender ML, Luedtke NA, Heath GR, DeVries T (1982) The marine phosphorus cycle. Am J Sci 282:474–511
Froelich PN, Arthur MA, Burnett WC, Deakin M, Hensley V, Jahnke R, Kaul L, Kim K-H, Roe K, Soutar A, Vathakanon C (1988) Early diagenesis of organic matter in Peru continental margin sediments: phosphorite precipitation. Mar Geol 80:309–343
Gächter R, Meyer JS (1993) The role of microorganisms in mobilization and fixation of phosphorus in sediments. Hydrobiologia 253:103–121
Gächter R, Meyer JS, Mares A (1989) Contribution of bacteria to release and fixation of phosphorus in lake sediments. Limnol Oceanogr 33:1542–1558
Graf G (1991) Die Reaktionen des Benthals auf den saisonalen Partikelfluss und die laterale Advektion, sowie deren Bedeutung für Sauerstoff- und Kohlenbilanzen. Rapport Sonderforshungsbereich 313. University of Kiel, pp 1–77
Gorshkova TI (1972) Organic matter in the sediment of the Baltic Sea and its biological importance. In: Bonitet mirovogo okeana (in Russian), part I. All-Union Research Institute of Marine Fisheries and Oceanography (VNIRO), vol 75. Pishchevaa Promyshlennost, Moscow, pp 191–219
Gripenberg S (1934) A study of the sediments of the north Baltic and adjoining areas. Merentutkimuslait Julk/Havsforskningsinst Skr 96:1–231
Hagmeier A (1930) Die Bodenfauna der Ostsee im April 1929. Ber Dtsch Wiss Komm Meeresforsch 5:156–173
Håkanson L, Jansson M (1983) Principles of lake sedimentology. Springer, Berlin Heidelberg New York, 316 pp
Hartman M (1964) Zur Geochemie von mangan und eisen in der Ostsee. Meyniana 14:3–20

Hedges JI, Stern JH (1984) Carbon and nitrogen determinations of carbonate-containing solids. Limnol Oceanogr 29:657–663

HELCOM (1987) First periodic assessment of the state of the marine environment of the Baltic Sea area 1980–1985; background document. Baltic Sea Environ Proc 17B. Swedish Environmental Protection Agency, Stockholm

HELCOM (1990) Second periodic assessment of the state of the marine environment of the Baltic Sea 1984–1988; background document. Baltic Sea Environ Proc 35B. Swedish Environmental Protection Agency, Stockholm

HELCOM (1991) Third biological intercalibration workshop, 27–31 Aug 1990, Visby. Baltic Sea Environment Proc 38. Swedish Environmental Protection Agency, Stockholm

Hessle C (1924) Bottenboniteringar i inre Östersjön (in Swedish). Medd Kgl Lantbruksstyr 250:1–52

Hill C, Elmgren R (1987) Vertical distribution in the sediments of the co-occurring benthic amphipods *Pontoporeia affinis* and *P. femorata*. Oikos 49:221–229

Hill C, Quigley M, Cavletto J, Gordon W (1992) Seasonal changes in lipid content and composition in the benthic amphipods *Monoporeia affinis* and *Pontoporeia femorata*. Limnol Oceanogr 37:1280–1289

Höglund K-J (1984) Kustundersökningen 75 (in Swedish). Länsstyrelsen Västerbottens Län Medd 2:1–67

Holm NG (1978) Phosphorus exchange through the sediment–water interface. Mechanisms studies of dynamic processes in the Baltic Sea. Contrib Microb Geochem 3:1–149

Ignatius H (1958) On the rate of sedimentation in the Baltic Sea. Bull Comm Geol Finland 180:135–145

Ingall DE, Jahnke R (1994) Evidence for enhanced phosphorus regeneration from marine sediments overlain by oxygen depleted water. Geochim Cosmochim Acta 58:2571–2575

Ingall ED, van Cappellen P (1990) Relation between sedimentation rate and burial of organic phosphorus and organic carbon in marine sediments. Geochim Cosmochim Acta 54:373–386

Ingall ED, Bustin RM, van Cappellen P (1993) Influence of water column anoxia on the burial and preservation of carbon and phosphorus in marine shales. Geochim Cosmochim Acta 57:303–316

Ingri J (1985) Geochemistry of ferromanganese concretions and associated sediments in the Gulf of Bothnia. PhD Dissertation, Department of Economic Geology, University of Luleå

Jahnke RA (1990) Early diagenesis and recycling of biogenic debris at the seafloor, Santa Monica Basin, California. J Mar Res 48:413–436

Jakobsen R, Postma D (1989) Formation and solid solution behaviour of Ca-rhodochrosites in marine muds of the Baltic deeps. Geochim Cosmochim Acta 53:2639–2648

Järvekülg A (1979) The bottom fauna of the eastern part of the Baltic Sea. Valgus, Tallinn, 382 pp

Jonsson P, Carman R (1994) Changes in deposition of organic matter and nutrients in the Baltic Sea during the twentieth century. Mar Pollut Bull 28:417–426

Jonsson P, Jonsson B (1988) Dramatic changes in Baltic sediments during the last three decades. Ambio 17:158–160

Jonsson P, Carman R, Wulff F (1990) Laminated sediments in the Baltic Sea – a tool for evaluating nutrient mass balances. Ambio 19:152–158

Jørgensen SE, Nielsen SN, Jørgensen LA (1991) Handbook of ecological parameters and ecotoxicology. Elsevier, Amsterdam

Juškevics V, Steibrins O, Baraskovs V (1994) Deposits of upper Pleistocene and Holocene in the Gulf of Riga. In: Larsen B (ed) Gulf of Riga project. Nordic Environ Res Program, Nordic Council of Ministers, Sweden

Kögler F-C, Larsen B (1979) The west Bornholm basin in the Baltic Sea: geological structure and Quaternary sediments. Boreas 8:1–22

Koop K, Boynton WR, Wulff F, Carman R (1990) Sediment–water oxygen and nutrient exchange along a depth gradient in the Baltic Sea. Mar Ecol Prog Ser 63:65–77

Kullenberg G (1981) Physical oceanography. In: Voipio A (ed) The Baltic Sea. Elsevier, Amsterdam, pp 135–181

Lagzdins G, Saule A, Pallo P (1987) Zoobenthos of the coastal zone of the Baltic proper, Gulfs of Riga and Finland, and their areal division. In: Hydrochemical and hydrobiological characteristics and areal division of the coastal zone of the Baltic proper, the Gulfs of Riga and Finland. Zinatne, Riga, pp 164–180

Larsson U, Elmgren R, Wulff F (1985) Eutrophication and the Baltic Sea: causes and consequences. Ambio 14:9–14

Lithner G (1974) Rönnskärsundersökningen 1973. SNV PM 497. National Swedish Environmental Protection Board, Drottningholm, 70 pp

Lucotte M, d'Anglejan B (1988) Processes controlling phosphate adsorption by iron oxides in estuaries. Chem Geol 67:75–83

Mackin JE, Aller RC (1984) Ammonium adsorption in marine sediments. Limnol Oceanogr 29:250–257

Manheim FT (1961) A geochemical profile of the Baltic Sea. Geochim Cosmochim Acta 25:52–70

Martens CS, Berner RA, Rosenfeld JK (1978) Interstitial water chemistry of anoxic Long Island Sound sediments: 2. Nutrient regeneration and phosphorus removal. Limnol Oceanogr 23:605–617

Müller P J (1977) C/N ratio in Pacific deep-sea sediments: effect of inorganic ammonium and organic nitrogen compounds sorbed by clays. Geochim Cosmochim Acta 41:765–776

Niemistö L, Voipio A (1981) Studies on the recent sediments in the Gotland Deep. Geologi 3:43–46

Niemistö L, Tervo V, Voipio A (1978) Storage of iron and phosphorus in the sediments of the Bothnian Sea. Merentutkimuslait Julk 244:36–41

Olausson E, Gustafsson O, Mellin T, Svensson R (1977) Current level of heavy metal pollution and eutrophication in the Baltic proper. Medd Maringeol Lab, Göteborg

Persson L-E (1982) Macrobenthic associations of the Hanö Bight, southern Baltic. Sarsia 67:93–106

Petersen CGJ (1913) Havets bonitering II. Om havbundens dyresamfund og om disses betydning for den marine zoogeografi. Beretn Till Landbrugsmin Fra Den Danske Biol Stn 21:1–42

Pustelnikov OS, Yu Tamoshyunaite N, Jankovski H (1982) Geochemistry of suspensions and bottom deposits of the Baltic basin. Trudy AN Lit SSR 2(129):57–66

Raudsep R, Kask J, Valing P (1994) Distribution and lithological composition of modern (Holocene) marine bottom sediments in the Gulf of Riga north of 58°N. In. Larsen B (ed) Gulf of Riga project. Nordic Environ Res Program, Nordic Council of Ministers, Sweden

Rosenfeld JK (1979 Ammonium adsorption in near shore anoxic sediments. Limnol Oceanogr 24:356–364

Ruttenberg KC, Berner RA (1993) Authigenic apatite formation and burial in sediments from non-upwelling continental margin environments. Geochim Cosmochim Acta 57:991–1007

Sandén P, Rahm L, Wulff F (1991) Non-parametric trend test of Baltic Sea data. Environmetrics 2:263–278
Seire A (1992) Benthic macrofauna of the deep areas of the Gulf of Finland in 1989/1990 compared with earlier investigations. In: Bjørnestad E, Hagerman L, Jensen K (eds) Proc 12th Baltic Marine Biologists Symp. Olsen and Olsen, Fredensborg, pp 151–156
Shaffer G (1986) Phosphate pumps and shuttles in the Black Sea. Nature 321:515–517
Silva JA, Bremner JM (1966) Determination of isotope-ratio analysis of different forms of nitrogen in soils. 5: Fixed nitrogen. Soil Sci Am Proc 30:587–594
Sternbeck J, Sohlenius G (1997) Authigenic sulfide and carbonate mineral formation in Holocene sediments in the Baltic Sea. Chem Geol 135:55–73
Stigebrandt A, Wulff F (1987) A model for the dynamics of nutrients and oxygen in the Baltic proper. J Mar Res 45:729–759
Suess E (1979) Mineral phases formed in anoxic sediments by microbial decomposition of organic matter. Geochim Cosmochim Acta 43:339–352
Suess E, Djafari D (1977) Trace metal distribution in the Baltic Sea ferromanganese concretions: inferences on accretion rates. Earth Planet Sci Lett 35:49–54
Suess E, Erlenkeuser H (1975) History of metal pollution and carbon input in the Baltic Sea sediments. Meyniana 27:63–75
Thulin G (1922) Bottenboniteringar i södra Östersjön i samband med fisktrålningar. Svenska Hydrogr Biol Kommn Skr 6:1–9
Vink S, Atkinson M (1985) High dissolved C:P excretion ratios for large benthic marine invertebrates. Mar Ecol Prog Ser 21:191–195
Winterhalter B, Flodén T, Ignatius H, Axberg S, Niemistö L (1981) Geology of the Baltic Sea. In: Voipio A (ed) The Baltic Sea. Elsevier Oceanography Series 30. Elsevier, Amsterdam, pp 1-121
Wulff F, Stigebrandt A (1989) Time-dependent budget model for nutrients in the Baltic Sea. Global Biogeochem Cycles 3(1):63–78
Wulff F, Stigebrandt A, Rahm L (1990) Nutrient dynamics of the Baltic Sea. Ambio 19:126–133
Zmudzinski L, Ostrowski J (1982) Baltic Sea zoobenthos of the sixties. Wyzsza Szkola Pedagogiczna Slupsku, Slupsk, pp 1–92

12 Statistical Analysis of Spatial and Temporal Variations in the Baltic Sea

L. RAHM and Å. DANIELSSON

12.1 Introduction

Various signs of a changing Baltic Sea (e.g. increased occurrence of green algae blooms in the coastal zone, reduced transparency in the water column and more frequent periods of anoxic conditions in the deeper sub-basins) have made people aware of the vulnerability of the Sea (see e.g. Rosenberg et al. 1990). To restore the Baltic Sea to a pre-eutrophied state, the decision-makers have advocated remedy actions (HELCOM 1991), similar to what successfully has been done for limnic ecosystems. There are, however, several problems with this. The major problem is that the Baltic has a most heterogeneous character. While limnic systems usually are phosphorus limited, the Baltic proper generally has a nitrogen-limited primary production. Phosphorus limitation may be found in the coastal zone and the Bothnian Bay. Only the Bothnian Sea has more balanced conditions in its interior (Granéli et al. 1990). The dominant sources of nitrogen are, in addition, usually of diffusive character, which will obstruct any remedy action.

To make a large-scale analysis, long time-series of data are needed. For a few variables, like salinity and oxygen, data are available from the end of the nineteenth century (Fonselius et al. 1984). For other variables, like phosphorus, reliable data have only been available from the beginning of the 1960s and for nitrogen data did not become reliable before the end of the same decade. This hampers any trend analysis with focus on possible long-term changes in nutrient dynamics, a necessity in a system with characteristic time scales of decades (see Chap. 13).

Larsson et al. (1985) have made a bold estimate of the gross changes in nutrient load since the beginning of this century. They claimed that a four- and eightfold increase in nitrogen and phosphorus load, respectively, have taken place, with a major increase since the 1950s. While the detailed estimates have been questioned, no serious opposition has been raised against their major conclusions.

Ecological Studies, Vol. 148
F. Wulff et al. (eds.)
A Systems Analysis of the Baltic Sea

In this chapter, changes in hydrography, nutrient concentrations and nutrient ratios over time are presented together with spatial variations. Also, we discuss monitoring programs and the question of representativity of stations.

12.2 Time Series Analysis

When analyzing time series, the aim is to detect changes and describe development over time. Often the result is used to predict the nearest future. Analysis of the nutrient dynamics is based on information concerning the nutrient load and forcing functions (see Chaps. 3–5) as well as the response of the marine ecosystem itself.

The main interest here lies in the nutrient concentrations and ratios, but also surface salinity, halocline depth and strength are presented. Since primary production is difficult to estimate (Smith 1991), secondary variables like O_2 supersaturation (Stigebrandt 1991) and Secchi disc depth (Sandén and Håkansson 1996) have been used.

12.2.1 Methods

Environmental data are usually not very well suited to traditional time series analysis, e.g. ARIMA and linear regression. There are several reasons for this: highly skewed data, seasonality, missing data, censored data and a pronounced autocorrelation.

Nutrient concentration data usually come from a truncated (cut at concentration zero), positively skewed distribution (most data at low concentrations together with a few at very high levels). This violates the assumption of normality, which often is demanded in statistical analysis. Sampling expeditions are only of a few days or weeks duration. Observations are therefore irregularly spread over time as well as space, leading to long, numerous and infrequent gaps in the time series. This is especially true for the northern parts of the Baltic Sea which, due to the severe ice conditions, only have a few winter observations. Censored data are data under detection limit. The range of observation is known, but not its actual value, making it difficult to include in any further analysis. Sometimes there is also more than one detection limit, due to different methods used. There are, however, methods to overcome this problem (e.g. Danielsson 1998). Finally, autocorrelation occurs when observations within a variable are highly correlated with each other.

To reduce the problems of the data mentioned above, a non-parametric monotonic trend test was used. It was developed by Hirsch et al. (1982) for hydrological needs and later modified to account for autocorrelation (Hirsch and Slack 1984). Seasonal Kendall's slope estimator (see Hirsch et al. 1982) was used to obtain the magnitude of the slope. It can be described as the median annual change, adjusted for seasonal variation. Hipel et al. (1988), Berryman et al. (1988) and Reckhow and Stow (1989) have recommended these tests for analyzing water quality in time series. It is robust against deviations of normality. It uses the median value for each season (making missing data of no concern) and the seasonality is then no problem as it calculates the trend for each season separately and then combines the results. As long as there only exists one limit of detection, this is no problem either.

Before testing for trend, the data were arranged in different groups of depth and time. In some of the analysis the entire range of depths was of interest. Therefore, the depths were treated as groups, according to the intrablock method proposed by van Belle and Hughes (1984), i.e. the trend was calculated for each depth interval. To account for seasonality, the data were divided into four seasons (January–March, April–June, July–September and October–December). As both increasing and decreasing trends were of interest, two-sided tests were performed.

As the method only accounted for monotonic trends, problems could occur if a trend changed from increasing to decreasing (or vice versa) within the investigated period. There is also a possibility that no trend will be observed which is why the time series plots must be carefully examined for drastic changes. If a change exists, the series is divided assuming a monotonic trend within each part. The test was then performed for each of these subintervals and compared with the results from the entire time period. One example of these drastic changes occurred in 1976–1977 (Huber et al. 1994), due to a large deep-water inflow to the Baltic proper. In the silicate study by Sandén et al. (1991), the investigation period was for this reason divided into two separate periods (see Sect. 12.2.3). Detailed results of the statistical analysis are presented in the respective works and only the major achievements are presented below.

12.2.2 Salinity

Salinity distribution dominates the density field and hence governs such physical processes as mixing and circulation, with consequences for both nutrient dynamics and recycling rates. During this century the long-term variation in salinity in the Baltic has been the concern of many oceanographers. A catchphrase has been the "oceanisation of the Baltic Sea", focusing on the salinity increase from the 1930s to the large deep-water inflow of 1976–1977. A decline in salinity, a "brackisation" (Schulz 1993), has been reported since then (see Gerlach 1994). This phenomenon is elucidated by trend analysis of surface salinity for the investigation period. Due to this great inflow the time series were split (1970–1976 and 1977–1994). Results from the trend analysis are presented in Table 12.1. The observations between the surface and 15 m are pooled to get an overview picture of the surface waters. Not dividing the time series in two, there are no significant monotonic trends for most stations and seasons. However, using the two sub-periods another picture is revealed. Generally, the latter of the two subperiods gave significantly negative trends, irrespective of independence assumed or not, while only BY15 and BY5 show significant trends for the former sub-period. The danger of not taking trend shifts into account is shown for BY5 in Fig. 12.1.

The halocline is the transition between the brackish surface water and the more saline deep water. The strength and position of the halocline generally govern the diapycnal mixing processes, i.e. the exchange between the stratified deep water and the well-mixed surface layer (see e.g. Matthäus 1980, 1990; Stigebrandt 1987). Wind-induced erosion, deep-water inflow and diapycnal mixing processes influence the halocline position.

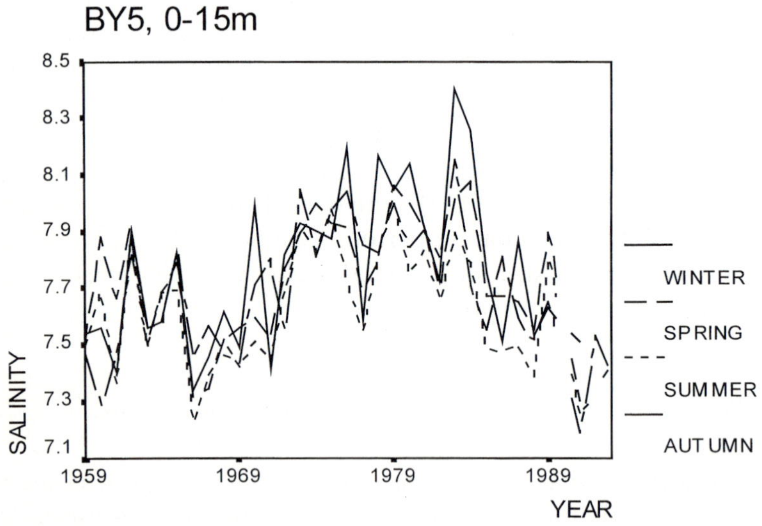

Fig. 12.1. Time series of surface salinity from station BY5 in the Bornholm basin in the southern Baltic proper. Series are partitioned according to season

Table 12.1. Trend analysis of surface salinity. Average value, *p*-value (two-sided test, dependent/independent), and slope

Station and periods	Number of observations	Mean salinity (PSU)	p_{dep}/p_{indep}	Slope (PSU year^{-1})
BO3				
1970–1994	63	3.52	0.6793/0.7592	–0.0013
1970–1976	10	3.42	1.0000/1.0000	0.0006
1978–1994	50	3.53	0.0379/0.1247	–0.0124
US5b				
1962–1994	73	5.61	0.4766/0.4661	–0.0035
1962–1976	17	5.45	0.6123/0.5793	–0.0108
1978–1994	53	5.66	0.0457/0.0592	–0.0155
SR5				
1959–1994	102	5.87	0.1972/0.4024	–0.0016
1959–1976	39	5.84	0.0728/0.1936	0.0066
1978–1994	59	5.89	0.0000/0.0004	–0.0230
BY28				
1959–1994	101	7.06	0.3715/0.5249	–0.0030
1959–1976	48	6.98	0.2686/0.1574	0.0068
1978–1994	53	7.11	0.0000/0.0015	–0.0492
BY31				
1959–1994	117	6.87	0.0000/0.0001	–0.0027
1959–1976	69	6.89	0.1479/0.2887	0.0143
1978–1994	44	6.84	0.0000/0.0014	–0.0404
BY15				
1962–1994	116	7.43	0.9032/0.9423	0.0002
1962–1976	52	7.41	0.0000/0.0003	0.0366
1978–1994	60	7.43	0.0000/0.0008	–0.0365
BY5				
1959–1994	131	7.70	0.0504/0.2803	0.0044
1959–1976	72	7.66	0.0000/0.0213	0.0223
1978–1994	55	7.75	0.0000/0.0020	–0.0429

Stations from both the Baltic proper and the Bothnian Sea were used in the trend analysis (Table 12.2). The time period was, for most stations, 1959–1993. As mentioned above, the time series were divided around 1976–1977. Only a few stations showed significant trends for the two sub-periods. Salinity profiles were estimated by applying cubic spline functions to all observations at each station, season and year. The position and strength of the halocline was then defined using the level of the largest gradient of the spline function. These estimates defined two time series,

Table 12.2. Mean position, trend and *p* value for halocline movement (1959–1993)

Station	Average halocline position (m)	Max. gradient (PSU/m)	Number of observations	Halocline movement (m/year)	*p*-value	Rate of change in halocline gradient (PSU m^{-1} $year^{-1}$)	*p*_value
US5b	57.3	0.016	53	–0.36	0.2904	0.000	0.0029
SR5	74.5	0.019	93	+0.108	0.5679	0.000	0.0124
BY28	79.7	0.060	95	+0.338	0.0001	–0.002	0.0000
BY15	75.3	0.053	113	+0.906	0.0000	–0.001	0.0000
BY31	–	0.046	117	–	–	–0.001	0.0005
BY5	79.7	0.243	95	+0.037	0.2471	–0.002	0.0164
BY2	42.0	0.459	114	–0.015	0.0097	–0.004	0.1451

which could be analyzed with respect to trend. Only the annual estimates are shown in Table 12.2.

Except for BY15 and, but to a lesser degree, BY28 the depth of the halocline did not change significantly over the period. It is also evident that the strength of the halocline (i.e. the stability of the stratification) generally decreased. The two subperiods showed that this decrease was more pronounced after the deep-water inflow of 1976–1977, while the former period was characterized by only one significant negative trend (BY28). To summarize, the stability of the water masses gradually decreased and the conditions for turbulent exchange across the halocline increased.

12.2.3 Nutrients

The occurrence of trends has been studied for several nutrient substances: NO_3-N, NH_4-N, PO_4-P, P_{tot}-P and N_{tot}-N (Sandén and Rahm 1993) as well as N/P ratios (Rahm et al. 1994). The organic form of nitrogen and phosphorus has been calculated as the difference between the total and inorganic forms. This was done to see if qualitative changes have occurred in the nutrient dynamics, changes that may alter the ecosystem in a significant way. Studies of N/P ratios (calculated from pairwise measurements of N and P) are based on the concept of nutrient limitation of primary production, originally suggested by Redfield (1934) and re-evaluated by Redfield et al. (1963) and Sambrotto et al. (1993).

Sandén et al. (1991) also studied trends in DSi-Si (dissolved silicate) concentrations, as this may be used as an indicator of trends in the primary production. Schelske and Stroemer (1971) argued that an increased primary production should lead to an increased organic matter deposition. Hence, if the silica supply is not affected by human activity, its pelagic pool should also decrease. Further, DSi/N_{inorg} ratios have been studied (Rahm et al. 1995b). These ratios cover a slightly shorter period than those for nutrients, 1970–1990. Nine stations were included, as they were the only ones to have a sufficient amount of data for the analysis (Fig. 12.2).

12.2.3.1 Nitrogen

Figure 12.3a shows the results of the inorganic nitrogen trend test, with care taken for season. Ammonium (Fig. 12.3b) has downward trends in the surface water for the entire Baltic Sea. In the Bothnian Bay the trend is in

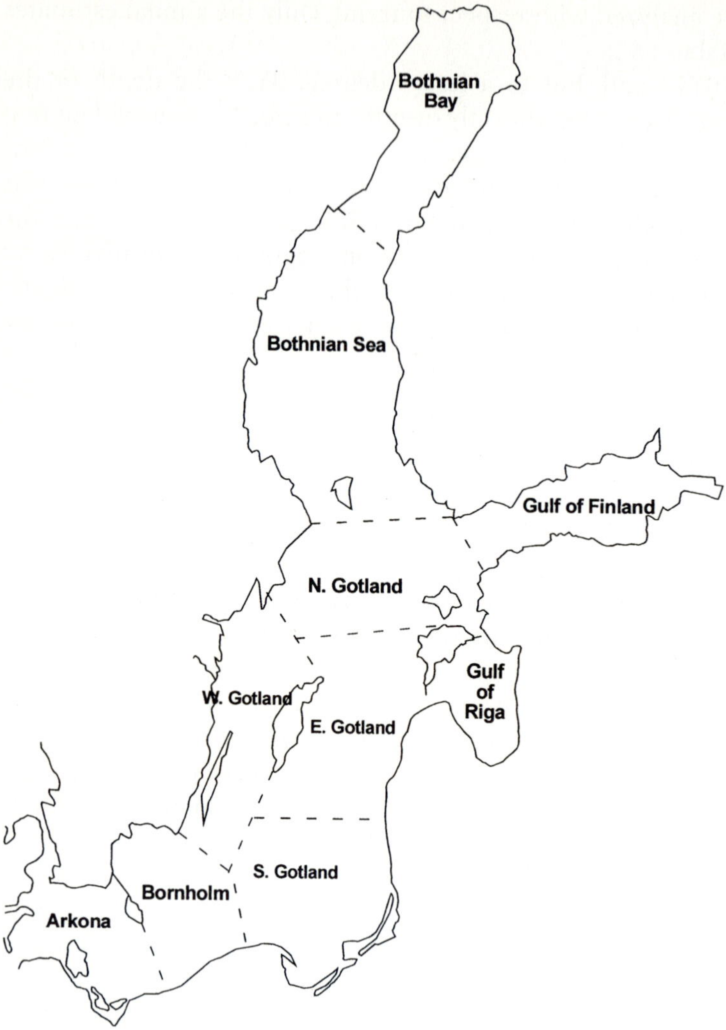

Fig. 12.2. Schematic map of the Baltic Sea with its sub-basins

the order of 4% year^{-1} in the surface water and even steeper in the deeper layers. The only trend detected in the Bothnian Sea is in the surface layer (5–6% year^{-1}). In the Baltic proper the falling trends (3–5% year^{-1}) are found not only in the surface layer, but also close to the bottom. The decreasing ammonium concentrations are attributed to the assumed increase in net primary production.

Nitrate generally shows increasing trends (4–5% year^{-1}) for all depths in the Bothnian Bay and below 20 m in the Bothnian Sea. In the Baltic proper

a
b

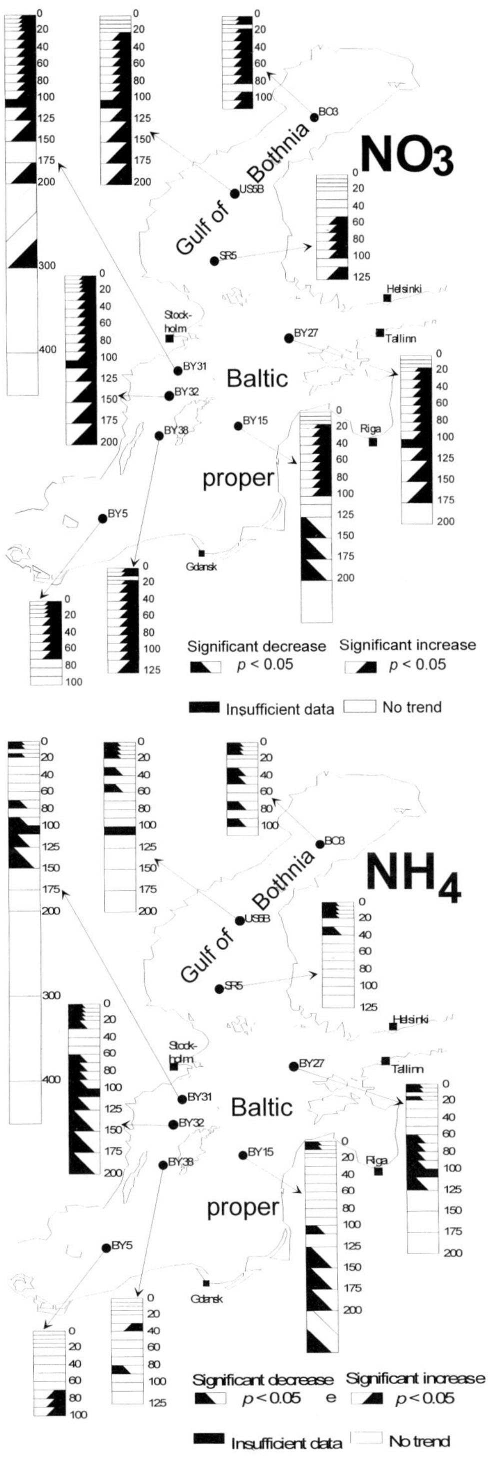

Fig. 12.3. Significant trends of **a** nitrate and **b** ammonium (two-sided test) at nine stations in the Baltic Sea at different depths (m). (After Sandén and Rahm 1993)

most stations and depths show increasing trends. One exception is BY15, where the trends are decreasing (4–6% year^{-1}) for a depth of 125 m and deeper (this level roughly coincides with the redoxcline). The increasing nitrate concentrations are probably caused by increased nitrogen load from both the atmosphere and some rivers (see Chaps. 4 and 5).

Total nitrogen shows similar patterns as NO_3-N, with increasing trends in the order of 1–4% year^{-1}. It is closely related to nitrate and therefore its increase can be explained by the same causes as for nitrate.

12.2.3.2 Phosphorus

Phosphate generally shows significant increasing trends with a magnitude of about 2% year^{-1} (Fig. 12.4a). In the Bothnian Bay, however, the only trends detected are negative. Total phosphorus follows the pattern of phosphate, but, for the Bothnian Bay, with increasing trends of 2% year^{-1} in the surface and bottom waters. In the Bothnian Sea there are increasing trends (2–3% year^{-1}) for all depths, except for the surface water. In the Baltic proper, there are similar trends, at least down to 60 m. The rising trends of phosphate may be due to increased recycling from the sediments. Anaerobic conditions, which occur in the Baltic proper, should facilitate such processes.

12.2.3.3 Silicate

Silica is assumed to be an important indicator of ongoing eutrophication, since its load is not primarily affected by human activities. Further, the diatoms are generally the dominant species in the marine spring bloom. A qualitative change in species succession may be a result of changes in silica over time. A lower limit of the DSi-Si concentration for production is reported as around 1–4 μM (Egge and Aksnes 1992). Silica-limited primary production has also been reported for the southern Baltic proper (Munk-Sörensen and Nielsen 1992). Dissolved silicate shows falling trends for the entire area (Fig. 12.4b). This does not come as a surprise as it is a common phenomenon in the northern hemisphere (Conley et al. 1993).

12.2.3.4 N/P

Redfield et al. (1963) suggested a general nitrogen-phosphorus ratio of 16:1 (by atoms) in plankton composition. Typical inorganic N:P ratios of

a, b

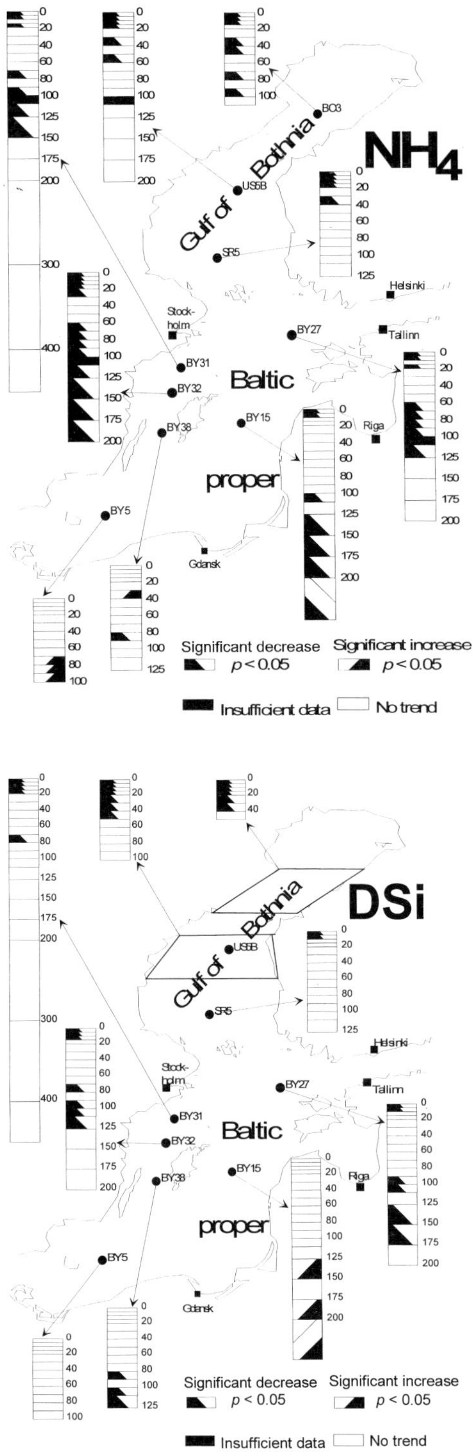

Fig. 12.4. Significant trends of **a** phosphate and **b** dissolved silicate (two-sided test) at nine stations in the Baltic Sea at different depths (m). (After Sandén et al. 1991; Sandén and Rahm 1993)

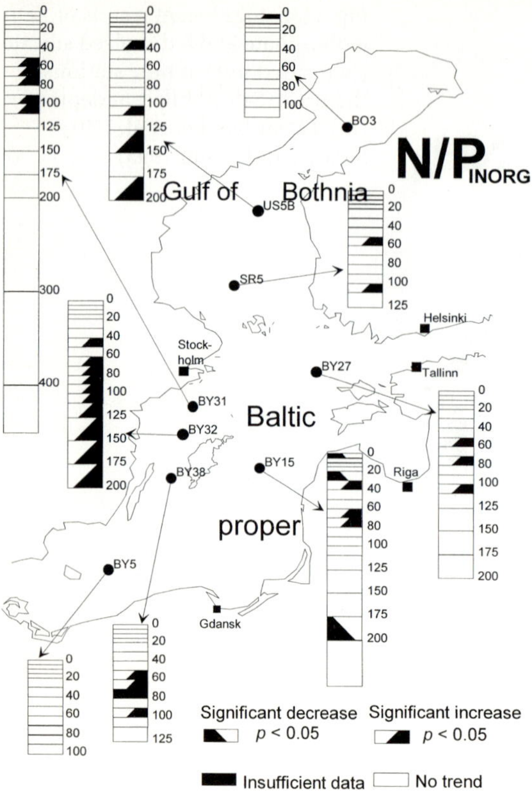

Fig. 12.5. Significant trends of inorganic nitrogen to phosphorus ratio at nine stations in the Baltic Sea at different depths (m)

the Baltic proper surface layer indicate nitrogen as the limiting factor for primary production (Granéli et al. 1990). By the same reasoning, phosphorus should be the limiting nutrient in the Bothnian Bay. The Bothnian Sea is in an intermediate state, where neither nitrogen nor phosphorus is a limiting factor (Elmgren 1989).

In the Baltic proper the inorganic N:P ratio is low (Fig. 12.5), but with increasing trends (3–7 % $year^{-1}$) in the deep waters (Rahm et al. 1994). This may be caused by the, previously mentioned, long stagnation period. In the trophogenic zone there are only a few falling trends, which are in the range of 3–5 % $year^{-1}$. N_{tot}/P_{tot}, as well as N_{org}/P_{org}, present increasing trends of 1–6 % $year^{-1}$. It should be noted that due to lack of observations there are only few significant results from the Gulf of Bothnia. To conclude, despite the pronounced changes reported for N and P, their ratios are unchanged in the trophogenic layers.

12.2.3.5 DSi/N

Silicate is interesting due to its potential as a limiting nutrient without direct anthropogenic links. The Redfield ratio for silica to nitrogen is 1:1, where a lesser ratio indicates silicate-limited growth conditions. In the Baltic Sea this ratio has clearly been above one due to the massive inputs of silica from the rivers, especially in the northern parts. In a recent study (Rahm et al. 1995b) decreasing silica trends are presented. When the ratio goes below unity it can be seen as both a sign of ongoing eutrophication and possible qualitative changes in the ecosystem.

In the water masses there are decreasing trends (5–10%) at all stations (Fig. 12.6). These trends are primarily coupled to the spring bloom. The Baltic proper has generally been nitrogen limited. However, as silicate recycles slower than both nitrogen and phosphorus (Wulff and Stigebrandt 1989), a silica-limited primary production may occur above ratios of 1:1. Since the monotonic trend tests indicate that the ratio at most stations gets close to one at the end of the investigation period, there are good reasons not to exclude the possibility of a DSi-limited spring bloom in the near future.

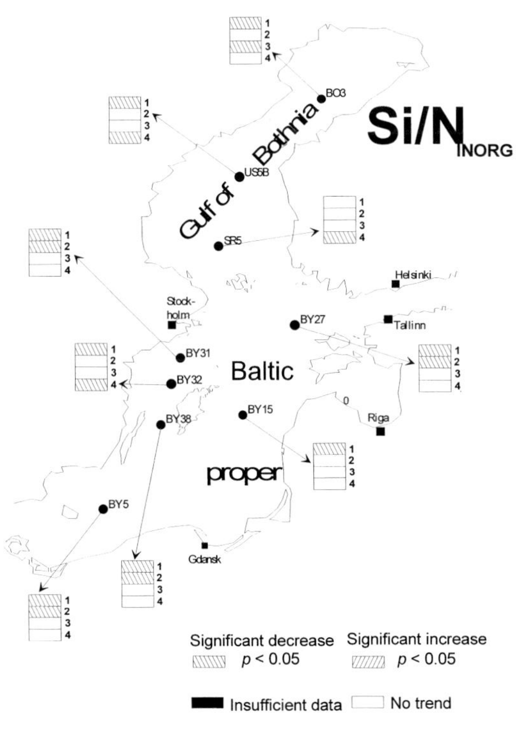

Fig. 12.6. Significant trends of dissolved silicate to inorganic nitrogen ratio at nine stations in the surface waters of the Baltic Sea for the four seasons (Rahm et al. 1995b)

It is difficult to draw any certain conclusion about the river input since only a few significant trends were detected during the actual period, some of them increasing (Rahm et al. 1995b). The few decreasing trends found in river input support the hypothesis that a decreasing DSi:N ratio in the water mass cannot be explained by a corresponding decrease in river input. Further, the dissolution of silica shales is not primarily redox-dependent. Hence, the observed decrease in DSi:N ratio in the water mass is probably a consequence of other factors such as increased N load and increased primary production.

12.2.4 Primary Production

Measurements of potential primary production are carried out within several monitoring programs in the Baltic. Due to analytical difficulties and high diurnal variations, direct observations of primary production are more difficult to obtain. Dugdale and Goering (1967) have suggested oxygen supersaturation in surface layers as an indirect measure of net primary production, something also used by Craigh and Hayward (1987). Stigebrandt (1991) has presented a modified approach. In order to obtain a reliable quantitative estimate, the gas exchange between the eutrophic zone and both the atmosphere and the deep water as well as changing salinity, temperature and wave action are considered. Empirical studies gave an upper limit of 102.5 % saturation, due to abiotic factors like wave-induced air bubbles in the surface layer. Levels above this were assumed to be caused by net primary production. The short-term estimates have been corroborated by independent estimates of net primary production (i.e. budget calculations by Stigebrandt and Wulff 1987).

Significant increasing monotonic trends in the order of 0.1–0.2 % $year^{-1}$ were reported (Stigebrandt 1991). Rahm et al. (1995a) repeated the analysis, but for a larger data set. They found that several of the basic assumptions for linear regression used in Stigebrandt (1991) were not fulfilled. Their parallel analysis, based on the non-parametric method described in Section 12.2.1 above, gave only one significant trend – at Arkona. Hence, the available oxygen data do not support any assumption about long-term changes in primary production, while Elmgren (1989) claims an increase in annual primary production of 30–70 % over the period 1976–1985 for the northwestern Baltic proper.

Secchi disc depth as a secondary variable to primary production has been suggested, for example, by Launiainen et al. (1989) and Sandén and Håkansson (1996). It primarily measures the degree of light attenuation in the water column. Particulate matter like plankton, but also dissolved

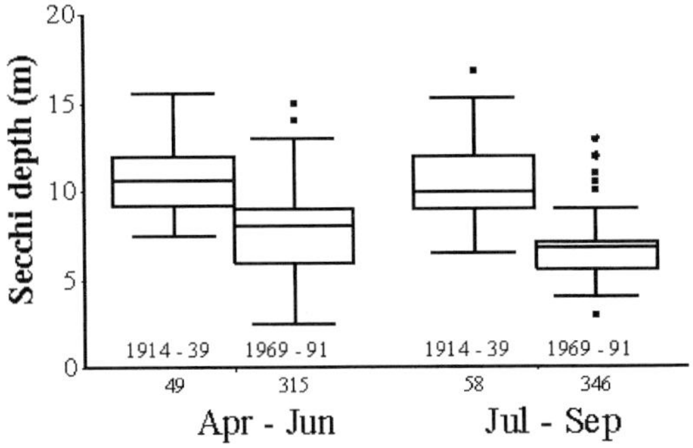

Fig. 12.7. Box plot of Secchi depth for two time periods and spring and summer. Number of observations are shown below time axes. (Sandén and Håkansson 1996)

substances like humic and fulvic acids, influence it – a real problem in the Baltic Sea. A large uncertainty is introduced, according to Sandén and Håkansson (1996), when one tries to couple changes in Secchi depth via chlorophyll (Falkowski and Wilson 1992) to changes in primary production (Eppley et al. 1985). Nevertheless, the extensive data sets obtained during various monitoring programs are analyzed. The investigation concerned the period 1969–1991 and five major stations in the Baltic proper. The authors reported statistically significant changes in Secchi depth of about 0.05 m $year^{-1}$, which should correspond to a change in gross primary production in the range of 0.4–1.0 % $year^{-1}$ (Fig. 12.7). This is somewhat lower than what Elmgren (1989) estimated, but the time period is more than double his periods and the relation between depth, chlorophyll and production is unclear. Trzosinska (1992) has reported the same tendencies of decreasing Secchi depths from the Polish part of the Baltic proper.

12.3 Spatial Analysis

Not only is it important to study changes in time in order to describe and understand the Baltic Sea, it is also important to study spatial variation and changes in concentrations depending on location. It may reveal point sources, concentration distributions, be helpful in improving monitoring programs, etc. (see e. g. Danielsson 1998).

Sampling programs have up to now often been decided subjectively, i. e. where and when to sample. This is partly due to more or less well-founded assumptions about the spatial distribution of the concentration fields and partly to various causes of interest. The insight has grown during this project that the use of subjectively chosen "representative stations" may lead the study astray because inherent structures cannot be resolved.

12.3.1 Differences Between Stations/Basins

Sandén and Danielsson (1995) made an attempt to study differences between basins and to test representativity of stations. Transects were used to find differences between stations, three in north–south direction and three in east-west direction (Fig. 12.8). Their positions were chosen using two criteria: sufficient number of data for the tests and their positions should fit as close as possible to a straight line. As for time period, 1980–1989 was used, where the data were divided into the same four seasons used in the time series analysis. Only the inorganic fraction of the nutrients was studied, i. e. nitrate, phosphate and silicate.

For all seasons and nutrients there are significant differences between the sub-basins. For silicate the Gulf of Bothnia showed significantly higher concentrations than the rest of the sub-basins, where Bothnian Bay had an extremely high median value. Phosphorus shows significantly higher concentrations in the Gulf of Bothnia as well as the Gulf of Finland. Nitrate shows significantly higher concentrations in the Gulf of Finland and Bothnian Bay (see Fig. 12.9). These differences are mainly apparent during autumn and winter, but a tendency can also

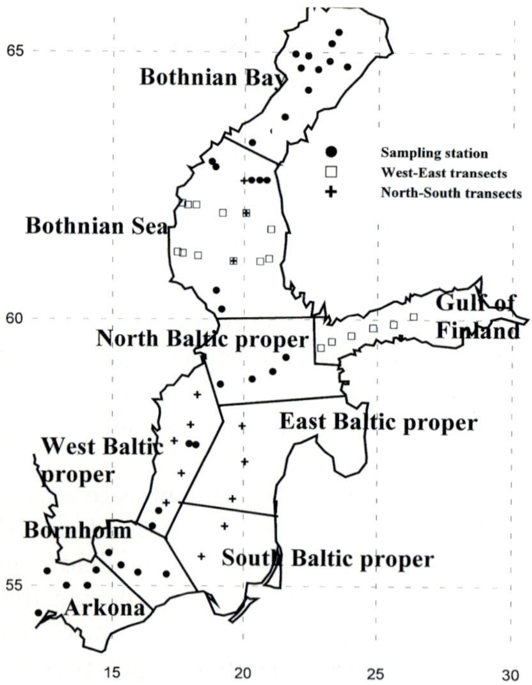

Fig. 12.8. The Baltic Sea with sub-basins, stations and transects used in the spatial analysis of Sandén and Danielsson (1995)

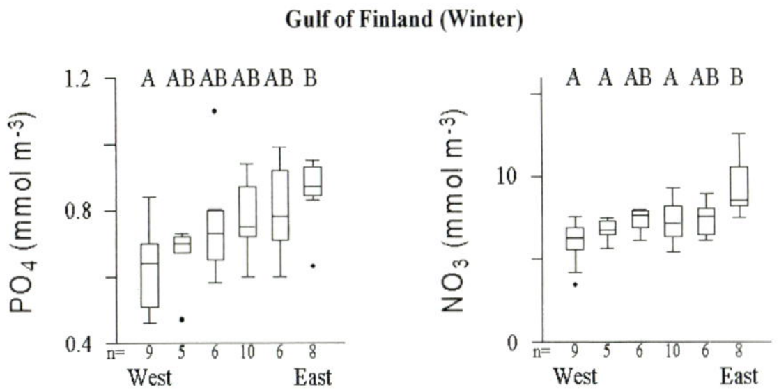

Fig. 12.9. Box plots and significant differences between stations on an east-west transect in the Gulf of Finland for phosphate and nitrate. Stations with the same *letter* for a nutrient and season are not significantly different at the 5% significance level (A<B<C...). (Sandén and Danielsson 1995)

be found for spring and summer. During the latter two, the biological production is high, causing low nutrient concentrations (sometimes even below detection limit), making it harder to reveal any patterns.

An analysis of the conditions within each sub-basin gave significant differences for some seasons and nutrient fractions, but only for the transects in longitudinal direction. For example, the Gulf of Finland revealed a gradient for nitrogen, as well as for phosphorus, with the lowest concentrations at the mouth (Fig. 12.10). The high loads in the innermost parts of the Gulf can explain this. In the Bothnian Sea phosphorus showed significantly higher concentrations at the Finnish coast, while the opposite was true for silicate and nitrate. This can be explained by nutrient loads, but also by the large-scale anticlockwise currents and transport limitation apparent in the sea.

12.3.2 Spatial Correlation

Toompuu and Wulff (1995) investigated the spatial correlations between sites in one sub-basin, the Gulf of Finland, in order to interpolate the concentration fields over a 3D area. When the correlation structure is known, a few stations are sufficient to give the conditions in the sub-basin (see e.g. Danielsson 1998).

The data were the same as for the trend tests above. The water column was divided into layers following the standard sampling depths and the time period was 1972–1991, which was further divided into four seasons. The analysis was carried out for each season and variable separately. Correlations were calculated in both vertical and horizontal directions. For all variables but one (salinity in summer), the correlations are positive. For phosphorus the correlation for Tot-P is slightly higher than that for PO_4-P, with decreases with distance. Tot-P, NO_3-N, NO_2-N as well as temperature show seasonal variation, with highest values during winter (decreasing in

Deep has stayed aerobic during the entire stagnation period. Despite these changes, the position of the halocline has not changed significantly, but has gradually become weaker. This has probably facilitated the turbulent mass exchange across the same layer.

During the two decades of this investigation, most nutrient concentrations increased. Ammonium and dissolved silica are the only exceptions. Nevertheless, a decrease in concentration of these two is attributed to the ongoing eutrophication of the Baltic Sea (Sandén et al. 1991; Sandén and Rahm 1993). The marked increase in primary production since the beginning of this century (Elmgren 1989) has lead to an increase in the oxygen consumption, caused by degradation of this organic matter. Due to the anoxic conditions in the deep water from 1977 and onwards, the deposition, remineralization and recycling of nutrients have changed character. A large pool of ammonium and silicate has been formed in the now anaerobic deep layers.

The changes in nutrient load do not contradict the assumption of an increased primary production. According to Grimvall and Stålnacke (Chap. 4), nutrient concentrations in the rivers have not changed significantly. However, because of the increased river flow, during the latter part of the investigation period, the riverborne load has increased. This results in increases in the water mass of organic and certain inorganic nutrients. The previously increasing atmospheric deposition seems to be leveling off (see Chap. 5). The significant decreases found in the DSi/N_{inorg} ratio for the Baltic also fit into this description. Rahm et al. (1995b) has shown that no general trend in DSi/N_{inorg} ratio is found in the river supply during the investigation period. A qualitative change in the spring bloom, i.e. a switch to less silica-dependent species, may be a consequence if the ongoing eutrophication is not arrested. There is, however, a possibility of changing riverborne silica load but before the investigation period due to river regulation and silica retention in dams. Due to the long characteristic time scales of the Baltic Sea, this could still be of consequence for the nutrient dynamics as the development of hydroelectric power occurred mainly during the 1950s and 1960s (Humborg et al. 1999).

The observed parameters used in this study rely on monitoring programs and single sampling events. Thus the study's outcome is dependent on both the quality of the data, the observation frequency and the representativity of the stations and depths used. Sandén and Danielsson (1995) showed the inherent bias in the existing monitoring programs, which neglected the lateral properties of the different sub-basins. Toompuu and Wulff (1995) presented a method based on spatial correlations to optimize future station locations. It is obvious that here is ample scope for future improvements in the existing monitoring programs. It also points out dangers with existing budget and model efforts. There are weaknesses in time series analysis as a consequence of the properties of existing data sets. More robust alternatives have been presented (see Sect. 12.2.1). To conclude, there is a eutrophication of the Baltic Sea. It is important to follow up the progress of the sea, by improvement of both monitoring program and the statistical analysis. This is a necessity in any future environmental assessment work (see e.g. Rahm and Danielsson 1995).

Acknowledgements. The Swedish Environmental Protection Agency and the Swedish Natural Science Research Council financially supported this work. The authors also thank Mr. G. Yesper for unselfish support.

References

Bergström S, Carlsson B (1995) River runoff to the Baltic Sea: 1950–1990. Ambio 23:280–287

Berryman D, Bobee B, Cluis D, Haemmerli J (1988) Non-parametric tests for trend detection in water quality time series. Water Resour Bull 24:545–556

Conley DJ, Schelske CL, Stoermer EF (1993) Modification of the biogeochemical cycle of silica with eutrophication. Mar Ecol Prog Ser 101:179–192

Craigh H, Hayward T (1987) Oxygen supersaturation in the ocean: biological versus physical contributions. Science 235:199–202

Danielsson Å (1998) Spatial modeling in sediments. Thesis, Department of Water and Environmental Studies, Linköping University

Dugdale RC, Goering JJ (1967) Uptake of new and regenerated forms of nitrogen in primary production. Limnol Oceanogr 12:196–206

Egge JK, Aksnes DL (1992) Silicate as regulating nutrient in phytoplankton competition. Mar Ecol Prog Ser 83:281–289

Elmgren R (1989) Man's impact on the ecosystem of the Baltic Sea: energy flows today and at the turn of the century. Ambio 18:326–332

Eppley RW, Stewart E, Abbott MR, Heyman U (1985) Estimating ocean primary production from satellite chlorophyll. Introduction to regional differences and statistics for southern California. J Plankton Res 7:57–70

Falkowski PG, Wilson C (1992) Phytoplankton productivity in the North Pacific Ocean since 1900 and implications for adsorption of anthropogenic CO_2. Nature 358: 741–743

Fonselius SH, Szaron J, Öström B (1984) Long-term salinity variations in the Baltic Sea deep water. Rapp P-v Réun Cons Int Explor Mer 185:140–149

Gerlach SA (1994) Oxygen conditions improve when the salinity in the Baltic Sea decreases. Mar Pollut Bull 28:413–416

Granéli E, Wallström K, Larsson U, Granéli W, Rosenberg R (1990) Nutrient limitation of primary production in the Baltic Sea area. Ambio 19:142–151

HELCOM (1991) The Baltic Sea joint comprehensive programme. Interim report of the HELCOM ad hoc High Level Task Force, August 1991, Helsinki Commission, Helsinki

Hipel KW, McLeod AI, Weiler RR (1988) Data analysis of water quality time series in Lake Erie. Water Resour Bull 24:533–544

Hirsch RM, Slack JR (1984) A nonparametric trend test for seasonal data with serial dependence. Water Resour Res 20:727–732

Hirsch RM, Slack JR, Smith RA (1982) Techniques of trend analysis for monthly water quality data. Water Resour Res 18:107–121

Huber K, Kleine E, Lass H-U, Matthäus W (1994) The major Baltic inflow in January 1993 – measurements and modelling results. Dtsch Hydrograf Z 46:103–114

Humborg C, Conley D, Rahm L, Wulff F, Cociasu A, Ittekkot V (1999) Silica retention in river basins: far-reaching effects on biogeochemistry and aquatic food webs in coastal marine environments. Ambio 29:45–50

Larsson U, Elmgren R, Wulff F (1985) Eutrophication and the Baltic Sea. Ambio 14:9–14

Launiainen J, Pokki J, Vainio J, Niemiaa J, Voipio A (1989) Näkyösyvyyden vaihteluista ja muuttumisesta pohjoisella Itämerellä [Long term changes in the Secchi depth in the northern Baltic Sea] (in Finnish with English Abstract). Proc XIV Geofysiikan Päivät, 3–4 May, Helsinki

Matthäus W (1980) Is the Baltic halocline really rising? Ophelia Suppl 1:157–163

Matthäus W (1990) Mixing across the primary Baltic halocline. Beitr Meereskd 6:21–312

Munk-Sörensen H, Nielsen K (1992) Åhus bugt 1990–1991; Hydrografi, närsalter og plankton. Åhus Amt, Åhus

Rahm L, Danielsson Å (1995) Making interlinkages in environmental assessment explicit. Proc 50th Session of International Statistics Institute, pp 199–215

Rahm L, Sandén P, Wulff F, Danielsson Å (1994) Trends of nitrogen to phosphorus ratios in the Baltic Sea (manuscript)

Rahm L, Sandén P, Sturesson L, Danielsson Å (1995a) Oxygen saturation trends in the Baltic Sea – problems with statistical inference of marine environmental data. Environ Monit Assess 35:13–25

Rahm L, Sandén P, Wulff F, Stålnacke P, Conley D (1995b) A time series analysis of nutrient inputs to the Baltic Sea and changing DSi/N ratios. Mar Ecol Prog Ser 130: 221–228

Reckhow KH, Stow C (1989) Monitoring design and data analysis for trend detection. Lake Reservoir Manage 6:49–60

Redfield AC (1934) On the proportions of organic derivatives in sea water and their relation to the composition of plankton. James Johnstone Memorial Volume, Liverpool, pp 177–192

Redfield AC, Ketchum BH, Richards FA (1963) The influence of organisms on the composition of seawater. In: Hill MN (ed) The sea, vol 2. Wiley, New York, pp 26–77

Rosenberg R, Elmgren R, Fleischer S, Jonsson P, Persson G, Dahlin H (1990) Marine eutrophication case studies in Sweden. Ambio 19:102–108

Sambrotto RN, Savidge G, Robinson C, Boyd P, Takahashi T, Karl DM, Langdon C, Chipman D, Marra J, Codispoti L (1993) Elevated consumption of carbon relative to nitrogen in the surface ocean. Nature 363:248–250

Sandén P, Rahm L (1993) Nutrient trends in the Baltic Sea. Environmetrics 4:75–103

Sandén P, Håkansson B (1996) Long-term trends in Secchi depth in the Baltic Sea. Limnol Oceanogr 41:346–351

Sandén P, Danielsson Å (1995) Spatial properties of nutrient concentrations in the Baltic. Environ Monit Assess 34:278–307

Sandén P, Rahm L, Wulff F (1991) Non-parametric trend test of Baltic Sea data. Environmetrics 2:263–278

Schelske CL, Stroemer EF (1971) Eutrophication, silica depletion and predicted changes in algal quantity in Lake Michigan. Science 173:423–424

Schulz S (1993) The Baltic Sea on the way to a freshwater lake? Consequences for the Baltic ecosystem. Lecture given at the 13th Baltic Marine Biologists Symp, Riga, 1–4 Sept

Smith SV (1991) Stoichiometry of C:N:P fluxes in shallow-water marine ecosystems. In: Cole J, Lovett G, Findlay S (eds) Comparative analyses of ecosystems: patterns, mechanisms and theories. Springer, Berlin Heidelberg New York, pp 259–286

Stigebrandt A (1987) A model for the vertical circulation of the Baltic Deep Water. J Phys Oceanogr 17:1772–1785

Stigebrandt A (1991) Computation of oxygen fluxes through the sea surface and the net production of organic matter with application to the Baltic and adjacent seas. Limnol Oceanogr 3:444–454

Stigebrandt A, Wulff F (1987) A model for the dynamics of nutrients and oxygen in the Baltic proper. J Mar Res 45:729–759

Toompuu A, Wulff F (1995) Spatial large-scale correlations for optimal interpolation of temperature, salinity and nutrient concentrations in the Gulf of Finland. Environment 6:55–72

Trzosinska A (1992) Water transparency in the Polish zone of the Baltic Sea. Oceanologia 33:203–209

van Belle G, Hughes JP (1984) Nonparametric tests for trend in water quality. Water Resour Res 20:127–136

Wulff F, Stigebrandt A (1989) A time-dependent budget model for nutrients in the Baltic Sea. Global Biogeochem Cycles 3:63–78

13 A Nutrient Budget Model of the Baltic Sea

F. Wulff, L. Rahm, A.-K. Hallin, and J. Sandberg

13.1 Introduction

The loads of nutrients to the Baltic Sea have increased during this century, especially since the 1950s (Larsson et al. 1985). Since the Baltic is a semi-enclosed estuary with a narrow entrance to the North Sea, it has a long hydraulic residence time (e.g. Wulff and Stigebrandt 1989; Chap. 2). One consequence is a high degree of nutrient accumulation in the water mass (see Chap. 12). Empirical modeling of the nutrient dynamics in this region by Wulff and Stigebrandt (1989) gave long residence times and a low degree of nutrient export to the North Sea, a common feature of coastal systems with long water residence time (Nixon et al. 1996). The environmental effects of eutrophication in the Baltic Sea are qualitatively well documented (Rosenberg et al. 1990; HELCOM 1997). Despite this, the quantitative relationships between biogeochemical and physical processes, nutrient loads, climatic variations and eutrophication are not well described. Nevertheless, they are crucial in any scientific analysis of this ecosystem as well as from the point of view of coastal zone management. A first step towards this is to construct nutrient budgets where the relationships between loads, concentrations as well as exports and imports by advective or biogeochemical processes are estimated. This is the purpose of this chapter.

Nutrient fluxes to and from the various basins of the Baltic Sea have been estimated by budget models based on simple continuity assumptions using salt as a conservative tracer to estimate the various fluxes in the system. Wulff and Stigebrandt (1989) reported calculations based on two 5-year means (1977–1981 and 1982–1986) of winter concentrations. These estimates compared favorably with an annual estimate from a highly detailed study based on monthly observations carried out during the Gulf of Bothnia in 1991 and integrated over that year by Wulff et al. (1996).

The approach we will use here is to develop budgets by using existing data. The aim is to determine the average nutrient fluxes and sinks in the Baltic Sea. Reliable nutrient data are only available from the end of the

Ecological Studies, Vol. 148
F. Wulff et al. (eds.)
A Systems Analysis of the Baltic Sea

1960s, hence the period chosen is 1970–1991. The budget calculations are a complementary approach to mechanistic modeling (see Chap. 14).

The main questions to be addressed are: (1) what is the relative importance of nutrient inputs, advective and diffusive fluxes as well as of internal biogeochemical processes (sources and sinks) in controlling nutrient concentrations? and (2) what is the regional variation of hydrographic and biogeochemical conditions in the Baltic sub-basins?

13.2 Material and Methods

13.2.1 Boundaries and Flows

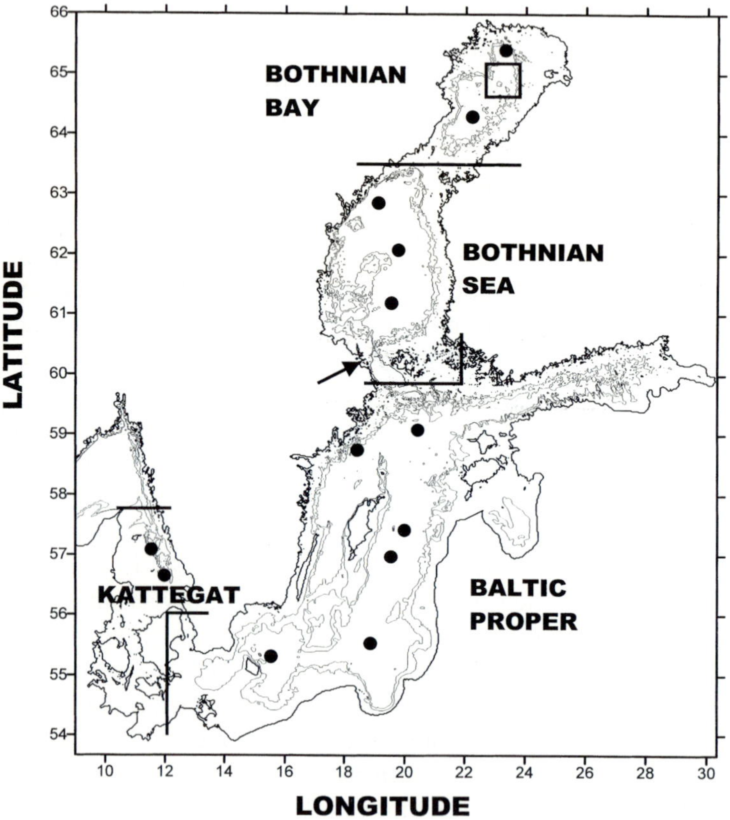

Fig. 13.1. Map of the Baltic Sea. Boundaries between sub-basins are shown as well as positions of stations from which data were used in the budget calculations. *Arrow* indicates position of Forsmark from where water levels were used

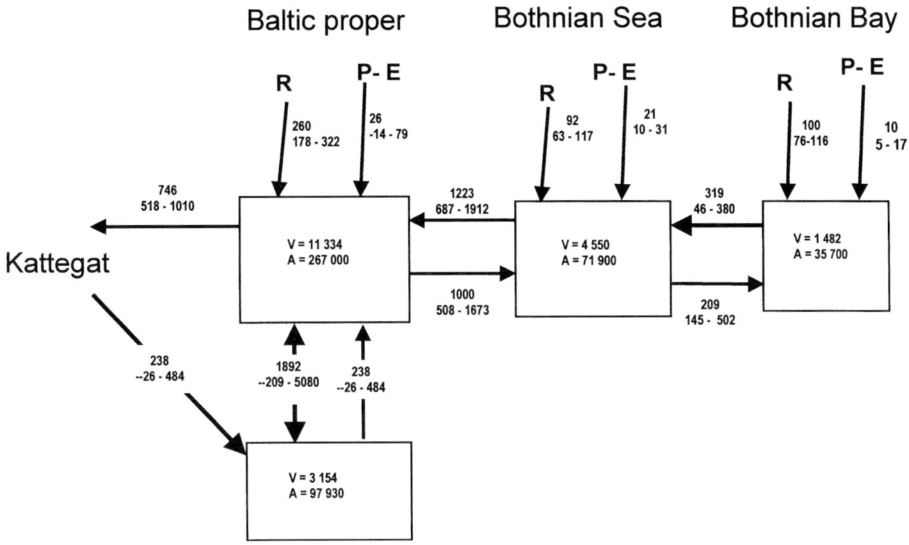

Fig. 13.2. Water budget averaged for 1971–1990 of the Baltic Sea region. Flows are expressed in km^3 $year^{-1}$ and mean values as well as range (maximum–minimum) are shown. Freshwater inputs are via rivers (*R*) and precipitation minus evaporation (*P–E*)

Three major basins separated by shallow sills constitute the Baltic Sea (see Fig. 13.1). They form three distinct hydrographic and biogeochemical areas (Sandén and Danielsson 1995; Chap. 2), which are also reflected in the levels of primary production (Elmgren 1984). The connections to the Kattegat and the North Sea are shallow, something which together with the relatively large freshwater input (Bergström and Carlsson 1994) gives the brackish character of the Baltic Sea. It also forms the two-layer structure of the Baltic proper with a strong halocline well below the entrance sills. This is considered in our modeling effort.

A four-box model is formulated (Fig. 13.2) where the Gulf of Bothnia is represented by the Bothnian Bay (BB) and Bothnian Sea (BS) without any vertical resolution due to their lack of a pronounced halocline. (In fact, this part is identical to the one used in Wulff et al. 1996.) The Baltic proper (BP), on the other hand, is simulated by a two-box configuration, representing the water above and below the permanent halocline. The flows are defined as positive (Fig. 13.2) in the direction of the arrows.

The deep-water renewal takes place as a bottom current from the Kattegat to the deep basins of the BP (see Chap. 2). The diapycnal mass exchange is modeled by advective fluxes and turbulent mixing across a fixed level halocline. The diffusive diapycnal flux is modeled as Fickian diffusion.

13.2.2 Numerical Methods

13.2.2.1 Calculation of Water Flows

For a semi-enclosed water body, it is possible to use salt as a conservative tracer to calculate water flows. This is achieved by requiring volume and mass continuity. It is called Knudsen's hydrographic theorem. This was originally developed by Knudsen (1900) for the Baltic Sea. It has since then been used in many budget calculations of various water bodies. Officer (1980) provides a comprehensive review on how to use this approach.

The first step in the analysis is to determine the net flows of water between the basins. It is formulated using the following assumptions: (1) the model has a "free surface" which takes into account the long-term changes in water volumes of the basins, which in the Baltic can be substantial; (2) the freshwater input includes river transport, diffusive input from coastal strips not included in the river catchments, precipitation and evaporation. The latter two components have previously been assumed to balance and have therefore been ignored in budget calculations. However, we found that we could not account for some of the interannual salinity variations if we did not include an additional freshwater source. Several studies have shown that excess precipitation compared to evaporation may give a net input corresponding to up to 15% of the annual river input (Omstedt et al. 1996).

The mass balance calculations are based on total amounts of salt within each basin. This means that we do not require the basins to be basin-wide homogeneous. In fact, this will allow us to use the "actual" salinity in the outgoing water from a basin instead of its volume-weighted average salinity. Consequently, we could take care of the existing inhomogeneities within the basins. The extent of these "buffer" zones is, however, questionable. Wulff and Stigebrandt (1989) used a constant factor (10%) in their budget calculations to adjust for the difference in salinity between the volume-weighted mean and those at the boundaries. The present work uses corrections based on actual observations (Alfa, see Table 13.2) that are averaged over the model period. By estimating the salinity of the outgoing water as a given fraction of the volume-weighted mean salinity for the entire basin, the inherent flexibility is sacrificed to achieve a simpler model formulation: (1) the basins are assumed homogeneous with regard to the diapycnal fluxes; (2) the turbulent exchange across the basin boundaries is included in the advective flows.

The water flow equations are formulated as difference equations and thereafter solved algebraically. Note that one must start with the innermost basin and with salinity as a dependent variable.

13.2.2.2 Nutrient Fluxes

The resulting water flows are then used as forcing functions for the nutrient fluxes in the same box configuration. The lack of data close to the boundaries made estimates of nutrient gradients within each basin very uncertain. Therefore, we used averaged nutrient concentration in these calculations. The fluxes are estimated using a commercial program, STELLA. The solutions are evaluated numerically using our forcing functions with a time step of 1 year.

13.2.2.3 Interpolation Scheme

The interpolation scheme is a fine-grain version of the one described in Wulff and Rahm (1988). The bathymetric database used has a resolution of 5′ × 5′ in latitude–longitude direction. The total amounts of salt and nutrients are monthly averaged and then summed up into an annual average. Data for missing months have been estimated by linear interpolation.

In an earlier study, Wulff and Rahm (1989) found that data from only a few strategically placed monitoring stations were necessary to obtain a reasonable estimate of total amounts of salt and nutrients (defined as the results obtained by using all available observations). We found that these estimates could vary within 10 % depending on resolution. Thus, we have to develop a scheme where a minimum number of stations, representing typical concentrations in various offshore regions, must exist for a valid interpolation. The stations that had to be included are (see Fig. 13.1) either ANHOLT E or FLADEN for the Kattegat; BY15, BY31, BY29, BY10, BY5 and BCSIII10 for the Baltic proper; either US3, MS8 or SR5 in the Bothnian Sea; and either BO3, F2 or either one of Cv1, F1–F5 and F8 in Bothnian Bay. There was also a requirement of at least ten observations each month in order to be chosen. The number of observations above and below the halocline was set to a minimum of six for each of the sub-basins in the Baltic Sea and four in the Kattegat. The halocline depth in the Baltic proper was fixed but at different depths ranging from 40 to 70 m from south to north. Only the upper 20 m of the Kattegat was included due to the shallow entrance to the Baltic proper.

13.2.3 Data

13.2.3.1 Freshwater Inputs

The period under consideration is 1971–1991. The annual freshwater inputs from the drainage areas to the three basins were estimated from regular monitoring data obtained from the hydrological agencies of all Baltic countries and, but to a lesser degree, from simulations (Bergström and Carlsson 1994; Chap. 3).

Dahlström (1986) estimated the long-term (1931–1960) net difference between precipitation and evaporation to be 141 mm/year over the Baltic Sea, i.e. an input of the same magnitude as the River Neva. Omstedt et al. (1996) have made a similar study (1981–1994) but based on a high-resolution meteorological database within the Baltic Environmental Database (BED) at the Department of Systems Ecology, Stockholm University. In their study they also considered the effect of sea ice. This gave 151 mm/year, hence the net input seems stable though the interannual and regional differences are large. For reasons of consistency the regional estimate by the latter investigation is used.

13.2.3.2 Water Level Variations

Stigebrandt (Chap. 2) discusses sea level variations in the Baltic. Interannual variations could be large enough to affect total volumes (and total amounts of salt and nutrients) in our budget calculations. A change in water level of 1 cm corresponds to 2.6, 0.7 and 0.4 km^3 for the BP, BS and BB, respectively. The annual mean water level is estimated from the gauge at Forsmark in the southern BS. This station is assumed to represent the entire Baltic Sea, a reasonable assumption with regard to the time scales involved (see also Carlsson 1997).

13.2.3.3 Nutrient Inputs

Time series of monthly nutrient load (1974–1991) were compiled for the three sub-basins. The nutrient river load includes water-quality data obtained from over 100 near river-mouth sampling stations and simulated nutrient transport from 80 non-monitored rivers/coastal areas around the Baltic (Stålnacke 1996). Water quality parameters include NH_4, NO_2, NO_3, Tot-N, PO_4, Tot-P and DSi. Missing river data were filled in by using the

Table 13.1. Long-term averages and range (thousand tons) of annual load of nutrients to the Baltic Sea for the period 1974–1991 for all except atmospheric loads (Atm). Some sources from Germany, Denmark and Poland are calculated separately

	Tot N	Tot P	Inorganic N	PO_4^{2-}	SiO_4^{2-}
Bothnian Bay					
Atm[a]	13.9 ±1.45				
Coast	6.74±1.42	0.434±0.99	2.19±0.544	0.159±0.045	31.0± 6.33
Point	3.20	0.211			
River	36.9 ±6.60	2.36 ±0.439	10.6 ±1.91	0.711±0.166	203 ±33.8
Total	**60.7**	**3.00**	**12.8**	**0.869**	**234**
Bothnian Sea					
Atm[a]	46.8±4.50				
Coast	11.9±2.95	0.786±0.223	5.80±1.72	0.323±0.100	56.8±21.7
Point	6537	582			
River	35.5±6.44	1.70 ±0.298	13.7 ±2.32	0.515±0.108	145 ±25.4
Total	**101**	**3.07**	**19.5**	**0.839**	**202**
Baltic proper					
Atm[a]	213 ±17.4				
Coast	97.7± 29.9	3.64±0.931	62.2±19.3	1.83±0.560	64.2
Point	6.34	8.45			
River	431 ±111	17.6 ±5.02	285 ±75.1	8.84±2.65	132 ±46.9
Germany and Denmark[b,c]	23.8± 4.48	1.65±0.302	16.1± 3.14	1.07±0.194	8.97± 1.95
Poland[b]					122±28.2
Total	**829**	**31.3**	**363**	**11.7**	**327**

[a] Wet and dry fraction of inorganic nitrogen; [b] Coast and river loading of nutrients;
[c] Time period 1970–1990

long-term monthly mean contribution (1970–1990) of individual flows to the total basin load.

Monthly atmospheric NH_4 and NO_3 deposition (wet and dry) have been estimated for a $1 \times 1°$ grid (Chap. 5) covering the entire Baltic. Good regional coverage in sampling stations was available only for 1986–1990. The atmospheric deposition of phosphorus, which is assumed to be insignificant compared to river input, was assumed to be 16 μmol m^{-2} $month^{-1}$ for the entire region and investigation period (Areskoug 1993). This is a questionable rate; an areal mean based, for example, on Rosenberg et al. (1990) will reach more than double this value. However, presently a lack of empirical data from the actual region will make all estimates questionable which is why we prefer to rely on this conservative value.

The contributions of Tot-N and Tot-P from coastal point (urban, industrial) sources have been taken from the Second Baltic Sea Pollution Load Compilation (HELCOM 1993). A complete summary of Tot-P and Tot-N loads existed only for 1990 and consequently these data were used for the entire period (see also Table 13.1). Nitrogen fixation by cyanobacteria is not included in the model as no reliable estimates are at hand.

13.2.3.4 Salinity

The annual averaged salinities, estimated from monthly linear interpolations of selected monitoring stations, are shown in Table 13.2. These estimates are within expected bounds with a maximum variance less than 13 % of the means, except for the BP deep water. This water has a slightly larger variation, probably due to the small volumes and the large infrequent inflows of saline water.

13.2.3.5 Horizontal Gradients

The salt concentrations at the sills between the basins were accounted for by first forming annual ratios between the spatially interpolated mean salinity in the respective border area (see Fig. 13.1) and the average for the whole basin. These averages are then estimated for the entire period 1970–1991. Temporal variability in salt concentrations at these borders in relation to the whole basin was ignored by this procedure.

Table 13.2. Long-term averages of salinities of different sub-regions of the Baltic Sea accounted for in the budget

	Salinity (min–max)	Alfa (SD)
Bothnian Bay		
Average (C3)	3.54 (3.32–3.76)	
At the southern border (C3s)	3.49 (3.32–3.63)	0.99 (0.021)
Bothnian Sea		
Average (C2)	5.81 (5.64–6.00)	
At the northern border (C2n)	5.29 (5.00–5.65)	0.91 (0.030)
At the southern border (C2s)	5.80 (5.59–5.95)	1.00 (0.014)
Baltic proper		
Average above halocline (C1a)	7.50 (7.11–7.74)	
Average below halocline (C1b)	9.52 (8.50–10.21)	
At the northern border (C1an)	7.12 (6.58–7.42)	0.95 (0.023)
At the southern border (C1as)	7.64 (7.20–7.96)	1.02 (0.014)
Kattegat		
Average (C0)	23.08 (21.42–25.16)	

13.2.3.6 Nutrients

Interpolation of the water mass data was done in the same way as for salt, i.e. the observations were interpolated over each month and basin and then averaged for each ear. Our attempt to correct mass fluxes across the sills due to gradients within the basins could not be carried out due to lack of observations close to the boundaries and basin-wide averages were used instead.

13.2.3.7 Results

The average water balance for the period is presented in Fig. 13.2. The interannual variations are also shown. The mean net freshwater inputs are higher (~7 %) in all three basins compared to those presented by Wulff and Stigebrandt (1989) for the period 1971–1981. While the water exchange between BB and BS has increased, the flow rates between the other basins and with the Kattegat are lower (12–20 %). The deep-water inflow to the BP is much smaller (~50 %). This is probably a reflection of the long "stagnation period" after the major deep-water inflow of 1976 and up to the end of the investigation period (Matthäus 1995) during which insignificant amounts of high saline water reached the deeper parts of the Baltic proper.

The interannual variations in the advective flows are considerable. The largest variations are, however, found in the diffusive diapycnic flows within BP. "Negative flows" occurred during 1 year (1985) and were caused by a violation of the maximum adjustment rate of the system. The reason for this may be interpolation errors due to erroneous or scarce observations, but low salinities are observed in both the Baltic proper and the North Sea during this year. In fact, the latter region shows the lowest maximum salinity observed for over six decades during 1984–1985 (Anonymous 1993).

13.2.3.8 Model Validity and Uncertainty

Our model formulation assumes, as mentioned above, non-negative flows. This will put constraints on the rate of change of total mass content (i.e. the salt pools). Since the salinity decreases successively from the entrance to the northern end of the Bothnian Bay, one can safely assume that the salinity in an upstream basin is always lower than in its downstream neighbor.

This means that, due to the model formulation, the rate of decrease is limited. If these constraints are violated due to erroneous data or processes not included in the model, the model will respond in an unphysical way, i.e. generate "negative flows".

In order to test the uncertainty in our model predictions, two types of error estimates were used. The first is the common "relative maximum error estimate". The uncertainties used were the standard deviations from annual means based on monthly calculations. Due to the inherent structure of the model, solved sequentially from north and southwards, the estimated "maximum" flow errors will propagate to the nearest downstream basin and further on. This will lead to an error accumulation southwards, with unreasonable estimates as a consequence. A more realistic way to test the model's uncertainty was by generating random values for the individual variables from given rectangular randomization distributions. The previous error propagation was then eliminated as well as the rather unrealistic "maximum error estimate". One hundred runs were done for each set of distributions, after which mean and standard deviation were estimated. We defined unacceptable errors as those flows that exceeded ±100 % of the mean. It turned out that maximum error for individual variables should not exceed ±5 %. This puts severe restrictions on the estimates of the "forcing functions", uncertainties we presently do not know about. Hence our results are marred by errors inescapable with a monitoring program not dedicated to the specific requirements of budget calculations. A consolation is that our results are in reasonable agreement with other attempts.

13.3 Nutrient Budgets

We first calculated budgets for inorganic and totals of N and P separately but merged these into common budgets where we distinguish between inorganic and "organic" fractions, the latter being the differences between total amount and the inorganic fractions, as shown in Fig. 13.3. The average amounts for 1975–1990 found in each basin are shown within each box. The mean changes in these pools are also shown in the boxes for the same period and so are the fluxes. Downward flows from above represent input from the drainage basin and atmosphere. Internal sinks, shown as downward flow below the boxes, are calculated by difference, i.e. the amounts not explained by the other flows and changes in the pools. We assume here that these sinks occur from the pools of organic matter (see also Wulff et al. 1996).

13.3.1 Nitrogen

The long-term average budgets for inorganic nitrogen (DIN) and organic nitrogen (DON) are shown in Fig. 13.3A. For Tot-N, there is a net flux from BB to BS and BP. The net input from the northern basins to the BP is only about 18×10^3 tons year^{-1} (this mainly as Org-N), i.e. only about 10% of the total inputs to the Gulf of Bothnia. The internal sinks and accumulations in the northern basin are small while the larger loss in the next basin (BS) accounts for the very small net export that occurs to the BP. The net export to the Kattegat is about 14% of the Tot-N load, but for DIN the export is only 3%. The internal sinks are of the same order of magnitude as the storages. The residence times (Table 13.3) vary between 2 and 6 years for Tot-N and between 1 and 4 years for DIN in the different basins based on calculations using all inputs to each basin. The residence time is generally shorter for the inorganic fractions, something that reflects the large utilization of DIN compared to the storages. Only the BB differs in that the low productivity and the P limitation leads to a weak utilization of the available N storage. The average residence time for the entire Baltic Sea is 2

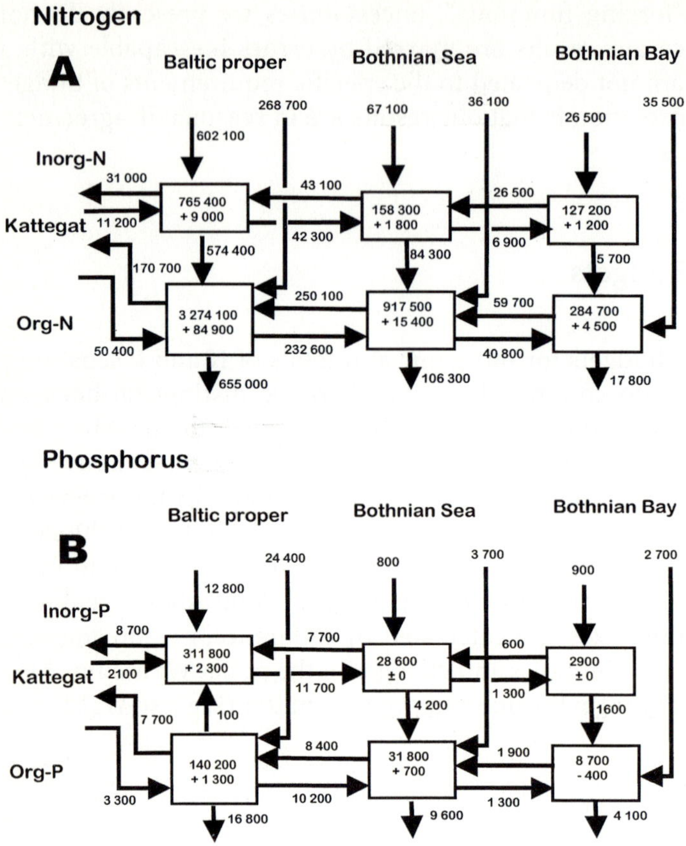

Fig. 13.3. Phosphorus (**A**) and nitrogen (**B**) budgets for 1975–1991 of the Baltic Sea region (total amounts in tons). Flows and changes in pools within each basin are expressed in tons year^{-1}

and 6 years, respectively. The latter is comparable (5.5 years) to that found by Wulff and Stigebrandt (1989). If one only bases these estimates on external loads, the residence times for the northern basins increase substantially while the others remain almost the same (Table 13.3).

13.3.2 Phosphorus

The long-term average budget for PO_4 and DOP is shown in Fig. 13.3B. The net exchange of Tot-P between BB and BS is insignificant while the BS imports 5.8×10^3 tons year^{-1} from the BP (40 % of the total import to the Gulf of Bothnia). Twenty five percent of the Tot-P load is exported to the

Table 13.3. Residence times τ of different nutrient fractions, based on long-term average flows and concentration as shown in Figs. 13.3 and 13.5. Times are given for both total inputs (including fluxes from adjacent basins) and only external load, respectively

System	Silica	Tot-P	Org-P	PO_4-P	DIN	Org-N	Tot-N	Water
Bothnian Bay	3.5/ 4.7	1.8/ 3.2	1.9/3.2	1.6/ 3.2	3.9/4.8	3.8/ 8.1	3.7/ 6.6	4.7/13.5
Bothnian Sea	2.6 /9.8	2.9/13.4	2.8/8.6	2.2/35.8	1.4/2.4	3.0/25.4	2.9/10.4	3.2/40.3
Baltic proper upper	2.4	1.7	3.7	1.1	0.6	5.1	2.4	3.1
Baltic proper lower	0.9	2.1	1.7	1.7	2.1	8.4	6.0	1.5
Entire Baltic proper	6.7/19.3	7.9/12.2	4.0/5.8	15.1/24.4	1.2/1.2	6.6/12.2	3.3/ 4.6	8.3/50.7
Total Baltic	9.8/12.5	11.2/11.6	5.6/5.7	20.6/23.7	1.5/1.5	11.5/13.2	5.0/5.3	27.5/40.3

Kattegat compared to almost half the DIP load; this is considerably higher than the estimates by Wulff and Stigebrandt (1989). The residence times vary between 1 and 15 years and 1 and 4 years for the DIP and Tot-P, respectively. The residence time for the entire Baltic Sea is 21 and 11 years, respectively.

The differences in residence times between sublayers and the entire BP are considerably larger for phosphorus than for nitrogen (Table 13.3). This illustrates that there is a much larger internal recycling of P, compared to N.

13.3.3 N/P Ratios

The N/P ratios for the different terms in the nutrient budgets are shown in Fig. 13.4. The ratios for loads and concentrations of total N/P are almost identical to those found by Wulff and Stigebrandt (1989). There is a decreasing trend in both total and inorganic N/P going from north to south, in terms of concentrations as well as in fluxes between basins. The lowest ratio is found in the deep basin of the BP, illustrating the preferential loss of N compared to P. Usually there are lower ratios for inorganic than for organic N/P. The high ratios in the phosphorus-limited BB stands in contrast to the low ratios in the generally nitrogen-limited BP. The high ratios of inorganic inputs are drastically reduced in the BS and BP while the opposite occurs in the BB, which is again an illustration of preferential N loss for the two southern basins compared to a more efficient P loss in the northernmost basin.

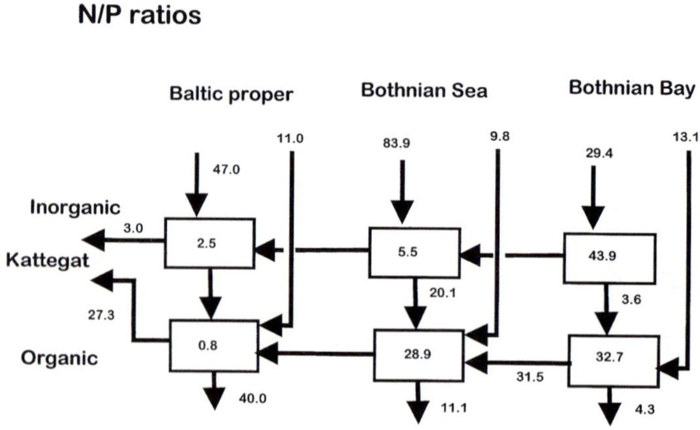

Fig. 13.4. N/P ratios of the budgets, from calculations shown in Fig. 13.3. For simplicity, ratios are only shown here, where net transport of both N and P are in the same direction

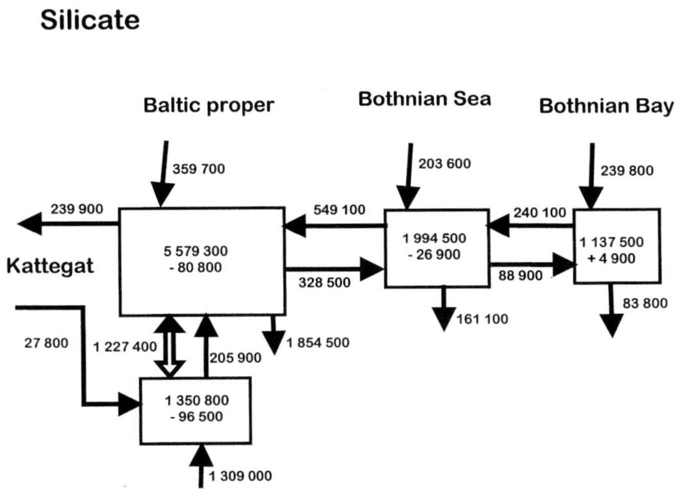

Fig. 13.5. Silica budget for 1974–1991 of the Baltic Sea region (total amounts in tons). Flows and changes in pools within each basin are expressed in tons year^{-1}

While the inorganic N/P ratio for the sink in the upper BP basin is close to that of Redfield (≈7.2), the return flow from the lower basin gets close to unity, indicating not only an efficient N sink but also efficient P recycling. This is in contrast to the efficient P sink and poor N sink in the BB, illustrated by the low ratio in the loss term compared to all other terms in this basin.

13.3.4 Silica

The long-term average budget for silica is shown in Fig. 13.5. The net export from the Baltic Sea is about 26% of the external load. While the annual decrease in the water mass (BP and BS) amounts to 2% of the storage, the annual net sink is about 8% for each basin. The residence time varies from 4, 3, 2 and 1 year going from BB, BS to BP upper and lower layers. The average residence time for the entire BP is 7 years, considerably longer than for each layer, illustrating the large internal recycling.

13.4 Conclusions and Discussion

Our qualitative knowledge about the eutrophication effects in the Baltic Sea is substantial, but the complex interplay between biogeochemical and physical processes, nutrient loads, climatic variations and eutrophication is not well understood.

This is illustrated in Fig. 13.6, where the total load of phosphorus and nitrogen to the BP (from the modeling efforts) is compared to their respective concentration development in the basin. While the N and P load has remained essentially unaltered during the investigation period (Stålnacke 1996) the concentration of N and P has increased in the water mass (see also Sandén and Rahm 1992). Consequently, the sinks have not increased simultaneously with the load and the system is obviously not in a steady state.

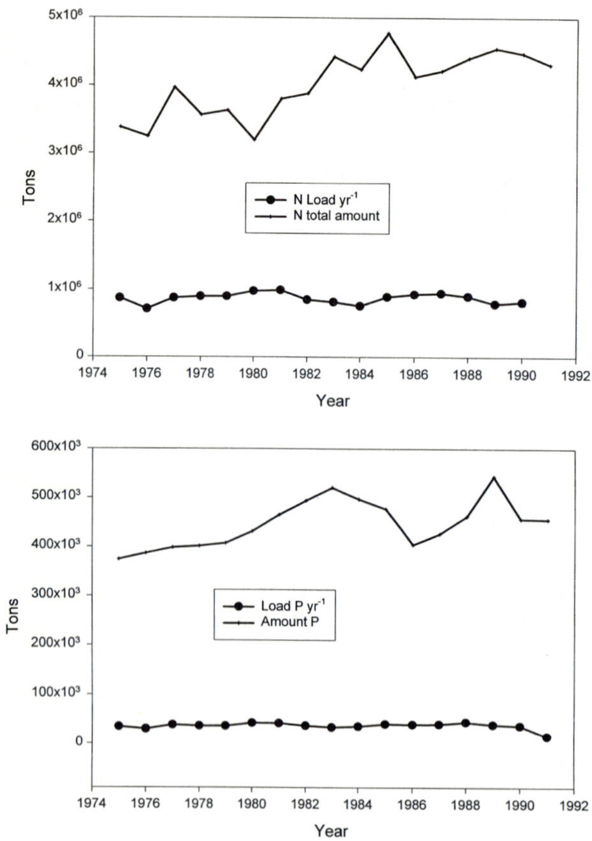

Fig. 13.6. Annual inputs of nitrogen (*above*) and phosphorus (*below*) vs. total amounts in the entire Baltic Sea between 1975 and 1991

Residence times based on external input and total amounts within each basin (Table 13.3) indicate a system with relatively short residence times for inorganic nutrients, especially for PO_4 in the BB and DIN in the BP. This clearly reflects the P-limited BB and the N-limited BP and BS. While the residence times for Org-N are substantially longer than for DIN because the Org-N load is much smaller but its pool is larger than its inorganic part, generally the opposite is true for Org-P. A comparison with the corresponding numbers for the entire Baltic Sea by Wulff and Stigebrandt (1989) reveals a high degree of similarity.

If one includes the interbasin exchange in the calculations, the residence times become much shorter, indicating the relative importance of this process. Moreover, PO_4 gets even a longer residence time than for water, which may seem impossible but may be due to internal sources not included as well as uncertainties in mass flux estimations (PO_4 concentrations are low compared to both Si and N).

Generally, an efficient sink, probably denitrification, as shown already by Rönner (1985), matches the load of nitrogen, but not completely since the concentration of N increases. Phosphorus, on the other hand, is to a large degree recycled before it is sequestered in the sediments. The long period of hypoxia in the BP has probably enhanced this recycling. Nevertheless, 86 and 76 % of the total terrigenous and atmospheric load of Tot-N and Tot-P is trapped in the Baltic Sea. Only in the well-mixed and oligotrophic P-limited BB has phosphorus a shorter residence time than nitrogen. While there are substantial nutrient loads to the basins, the exchange between the Baltic proper and the Kattegat is rather limited. The exchange between the Bothnian Sea and the Baltic proper is more intense with regard to water but there is little net exchange of N and even a northward flux of P. This has management implications since even substantial

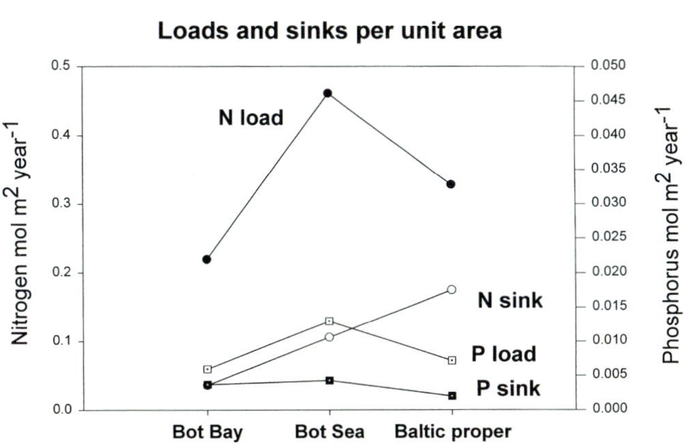

Fig. 13.7. Area-specific nitrogen and phosphorus loads, sinks and concentrations for the Baltic Sea sub-basins

reductions in nutrient loads to the BS will have only minor effects on the total amounts in this basin and in the BP.

The area-specific load of phosphorus (Fig. 13.7) to the three basins is rather similar. In fact, the load to the BS is the largest. The regional differences in the phosphorus sink are small but decrease going from north to south. Nitrogen, on the other hand, shows larger regional differences, but the load to the BS is by far the largest, calculated per unit area. On the other hand, the corresponding nitrogen sink exhibits a marked increase from the oligotrophic BB to the eutrophic BP.

The N/P ratios (based on total amount, in molar units; Fig. 13.8) further emphasize the characteristics of the three basins. The differences between the basins in terms of ratios in loads (calculated on external as well as on external plus internal loads) are small. On the other hand, the N/P ratio of the sinks varies by roughly a factor 8, from the BP to the BB, while the ratio in the water mass varies by a factor 4, but in the opposite direction. Hence it seems fair to regard the BP on average as N limited, the BB as P limited, while the BS seems to be in some sort of intermediate position. This is in line with what Granéli et al. (1990) reported, based on additional studies of primary production.

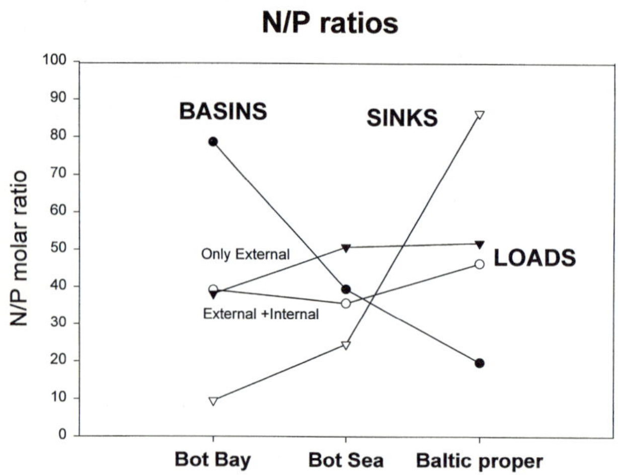

Fig. 13.8. Ratios between total nitrogen and phosphorus expressed in molar units for the Baltic Sea sub-basins. Comparison of ratios calculated on concentrations, of sinks and of loads. Ratios of loads are based on only external (from drainage basins and atmosphere) and with loads from adjacent sub-basins included

Acknowledgements. We thank Ulf Hinrichssen and Per Stålnacke for help with the model calculations and nutrient load estimates respectively. This work was funded within the BASYS project through the MAST program (contract MAS3-CT96-0058) and by the Swedish Environmental Protection Agency.

References

Anonymous (1993) North Sea Task Force 1993. North Sea Quality Status Report 1993. Oslo and Paris Commissions, London. Olsen and Olsen, Fredensborg, Denmark, p 22

Areskoug H (1993) Nedfall av kväve och fosfor till Sverige, Östersjön och Västerhavet (in Swedish). Rapport 4148. Swedish Environmental Protection Agency, Solna

Bergström S, Carlsson B (1994) River runoff to the Baltic Sea: 1950–1990. Ambio 23:280–287

Carlsson M (1997) Sea level and salinity variations in the Baltic Sea – an oceanographic study using historical data. PhD Thesis, Department of Oceanography, Gothenburg University, 18 pp

Dahlström B (1986) Determination of the areal prcipitation for the Baltic Sea. Reports Meteorology and Climatology, no 54, Swedish Meteorological and Hydrological Institute, Norrköping, 71 pp

Elmgren R (1984) Trophic dynamics in the enclosed, brackish Baltic Sea. Rapp P V Réun Cons Int Explor Mer 183:152–169

Granéli E, Wallström K, Larsson U, Granéli W, Elmgren R (1990) Nutrient limitation of primary production in the Baltic Sea area. Ambio 19:142–151

HELCOM (1993) Second Baltic Sea pollution load compilation. Baltic Sea Environ Proc 45. Helsinki Commission, Helsinki, 161 pp

HELCOM (1997) The third periodic assessment of the Baltic Sea 1989–1993: background document. Baltic Sea Environ Proc 64B. Helsinki Commission, Helsinki, 252 pp

Knudsen M (1900) Ein hydrographischer Lehrsatz. Ann Hydrogr Maritimen Met 28:316–320

Larsson U, Elmgren R, Wulff F (1985) Eutrophication and the Baltic Sea. Ambio 14:9–14

Matthäus W (1995) Natural variability and human impacts reflected in long-term changes in the Baltic deep water conditions – a brief review. Dtsch Hydrogr Z 47: 47–65

Nixon SW, Ammerman JW, Atkinson LP, Berounsky VM, Billen G, Boicourt WC, Boynton WR, Church TM, Ditoro DM, Elmgren R, Garber JH, Giblin AE, Jahnke RA, Owens NJP, Pilson MEQ, Seitzinger SP (1996) The fate of nitrogen and phosphorus at the land-sea margin of the North Atlantic Ocean. Biogeochemistry 35:141–180

Officer CB (1980) Box models revisited. In: Hamilton P, McDonald KB (eds) Estuarine and wetland processes. Plenum, New York, pp 65–114

Omstedt A, Meuller L, Nyberg L (1996) Interannual, seasonal and regional variations of precipitation and evaporation over the Baltic Sea. Ambio 26:484–492

Rosenberg R, Elmgren R, Fleischer S, Jonsson P, Persson G, Dahlin H (1990) Marine eutrophication case studies in Sweden. Ambio 19:102–108

Rönner U (1985) Nitrogen transformations in the Baltic proper: denitrification counteracts eutrophication. Ambio 14:1241–1243

Sandén P, Danielsson Å (1995) Spatial properties of nutrient concentrations in the Baltic Sea. Environ Monit Assess 34:289–307

Sandén P, Håkansson B (1996) Long-term trends in Secchi depth in the Baltic Sea. Limnol Oceanogr 41:346–351

Sandén P, Rahm L (1992) Nutrient trends in the Baltic Sea. Environmetrics 4:75–103

Stålnacke P (1996) Nutrient loads to the Baltic Sea. PhD Thesis, Department of Water and Environmental Studies, Linköping University, 78 pp

Wulff F, Rahm L (1988) Long-term, seasonal and spatial variations of nitrogen, phosphorus and silicate in the Baltic. An overview. Mar Environ Res 26:19–37

Wulff F, Rahm L (1989) Optimizing the Baltic sampling programmes: the effects of using different stations in calculations of total amount of nutrients. Beitr Meereskd 60: 61–66

Wulff F, Stigebrandt A (1989) A time-dependent budget model for nutrients in the Baltic Sea. Global Biogeochem Cycles 3:63–78

Wulff F, Pertillä M, Rahm L (1996) Monitoring, mass-balance calculations of nutrients and the future of the Gulf of Bothnia. Ambio Spec Rep 8:28–35

14 A Model of the Biogeochemical Cycles of Nitrogen and Phosphorus in the Baltic

O. Savchuk and F. Wulff

14.1 Introduction

The effects of pollution on the Baltic ecosystem are now well described in many scientific studies and have been followed by extensive monitoring programs. However, most of the causal relationships behind eutrophication are described only qualitatively. The complex interplay between physical and biogeochemical processes, nutrient inputs and climatic variations can only be understood by using models. Such models are also necessary in order to develop a management strategy on where and how much the nutrient inputs should be reduced.

Two different modeling approaches have been used, i.e. 'empirical' and 'mechanistic' models. Empirical models are built by using the large data sets from this region to develop nutrient budgets. The budget models are then used to calculate how changes in nutrient inputs will affect concentrations of nitrogen, phosphorus, and silica. The models demonstrate principal differences in the behavior of these nutrients and between the different subbasins of the Baltic Sea. The differences to a large extent depend on the varying efficiency of the internal sinks (and sources) of phosphorus compared to nitrogen, at different levels of eutrophication (see Chap. 13).

Mechanistic models describe various physical and biogeochemical processes controlling nutrient levels and productivity of the sea. These models have been useful tools to test various hypotheses about the relative importance of different processes. In this chapter, we are expanding earlier versions dealing with only nitrogen, to also include phosphorus. We take advantage of an improved understanding of nutrient inputs to the Baltic Sea by rivers (Chap. 4) and by the atmosphere (Chap. 5). Our understanding of critical processes, i.e. the pelagic community (Chap. 7), sedimentation (Chap. 8) and the sediments (Chaps. 9 and 11), as well as of physical processes (Chaps. 2 and 3) has also improved and could be used to develop and verify the model.

Ecological Studies, Vol. 148
F. Wulff et al. (eds.)
A Systems Analysis of the Baltic Sea

14.2 Objectives and Modeling Strategy

The development of a process-oriented model will test our current understanding of the relative importance of various critical processes. If the verification of the model is successful, it could then be used to forecast possible responses of the Baltic Sea to different scenarios of future variations in climate (see, e.g., Barrow 1993; Rotmans et al. 1994) and different management strategies to reduce pollution (HELCOM 1993b).

To reach these objectives the model must be capable of simulating relevant fluxes created by physical, chemical and biological processes and external forcing. Our model must thus be able to describe the production of organic matter and the factors that control the temporal and spatial distributions of organic matter and those nutrients that limit production in this ecosystem. We need to know how and what amounts of nutrients are entering the sea, how much is assimilated and recycled in the water column, how much is accumulated in the sediments or returned back to the water, and what the losses are due to internal sinks and by water exchange.

Consequently, a detailed ecological description of the ecological communities is by no means an end in itself for this model. From all possible ecological relationships and interactions, we would try to choose only those quantitatively important for large-scale and long-term fluxes of nutrients and organic production.

We have tried to formulate a model that can be related and compared to field and experimental data. Data are extensively used for the parameterization and scaling of various processes and for initial and boundary conditions. However, not only does a model imply a simplification of the real world in terms of the number of state variables and processes, but also there is a reduction in temporal and spatial scales. Thus, "real" data have to be aggregated on relevant time and space scales in order to be used as inputs to the model and to become comparable to its outputs.

We have also tried to keep our model as simple as possible and only as complex as necessary to simulate long-term and large-scale phenomena. Modeling of the nutrient cycles in the Baltic is a test of our understanding of what are the most important properties and mechanisms responsible for the biogeochemical fluxes. What level of complexity is needed to hindcast and, finally, to make forecasts with an appropriate accuracy?

14.3 Formulations and Realizations

"Entities should not be multiplied unnecessarily" (Occam's razor)

All the formulations presented below have been described in detail by Stigebrandt and Wulff (1987) and Savchuk and Wulff (1993, 1996). For the sake of completeness, these will be repeated here briefly. We start with our choice of spatial and temporal scales since it is mutually interconnected with the selection of variables and interactions.

14.3.1 Space and Time Scales

The residence times of nutrients in the Baltic Sea have earlier been calculated to be in the order of 5–15 years and an adjustment of the system to changing inputs might take 20–25 years (Wulff and Stigebrandt 1989; Savchuk and Volkova 1990). Thus, we need a model that can simulate the system over decades. Seasonal and climatic long-term variations in weather, fresh and salt water and nutrient inputs will affect the integral values of fluxes caused by production and sedimentation of organic matter and recycling of nutrients. Irregular events, like salt-water inflows, occur on weekly to monthly scales but influence the conditions in the Baltic for years. For these reasons, we have chosen an internal time scale in the model of 1 day (24 h).

The horizontal gradients are small within the major sub-basins of the Baltic Sea compared to the seasonal and vertical variations (Wulff and Rahm 1988). The production/recycling of nutrient in the trophogenic upper layer is strongly connected with day-to-day fluctuations in the mixed layer depth, superimposed on the seasonal variations. In the deep waters, sporadic penetrations of salt- and oxygen-rich water into different layers will change both physical transports and biochemical transformations. Sharp vertical gradients of salinity, oxygen, and nutrients are a characteristic of the Baltic. Therefore, we need a high vertical resolution in order to describe the processes that both create these gradients and are affected by them. The model has a vertical resolution of 1 m but is horizontally integrated. Thus, we ignore possible modifications of horizontally averaged fluxes, which could be caused by local inhomogeneities and gradients connected with coastal-offshore interactions, upwelling, eddies, plankton patchiness, etc.

14.3.2 State Variables and Interactions

To be able to describe the most important transport and translocation processes occurring at these time and space scales we have made the following choice of variables and interactions (Fig. 14.1).

Nitrogen and phosphorus in the water column are cycled through seven pelagic state variables: heterotrophs or zooplankton (ZOO), autotrophs or phytoplankton (PHYTO), suspended organic (detritus) nitrogen and

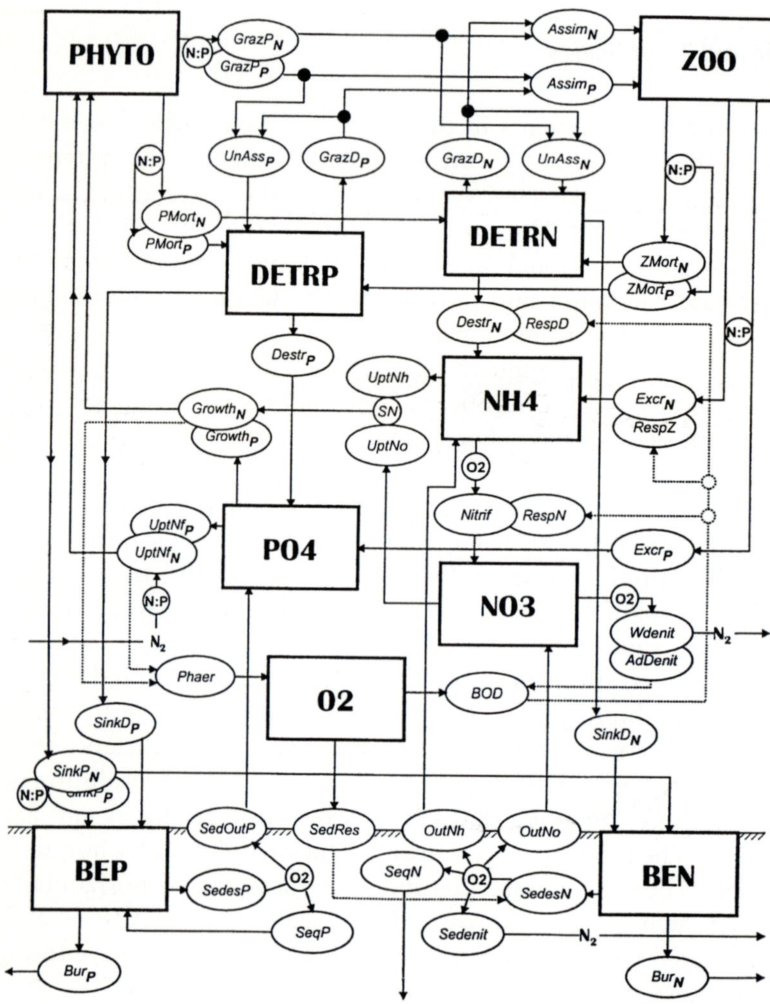

Fig. 14.1. State variables (*boxes*) and processes (*ellipses*) described in the coupled benthic–pelagic submodel

phosphorus (DETRN and DETRP), reduced (NH_4) and oxidized (NO_3=NO_2+NO_3) inorganic nitrogen and phosphorus (PO_4). In order to describe redox-dependent processes, dissolved oxygen (O_2) is also included in the model. Hydrogen sulfide (H_2S) is accounted for as "negative oxygen" (Fonselius 1969, 1981).

The biomasses of the two living components are described in units of nitrogen with constant but different N:P ratios for heterotrophs and autotrophs. Thus, it is not necessary to follow the phosphorus content in biomasses as separate state variables in the model.

Sediment–water interactions play an important role in the closure of the biogeochemical cycles in the shallow Baltic Sea. We have therefore introduced two benthic state variables – nitrogen (BEN) and phosphorus (BEP). These variables are considered as the total amount of labile nitrogen and phosphorus in the upper ("active") layer of the sediments.

The three groups of processes that drive the biogeochemical cycles in the overall model are *translocations* (between state variables), (physical) *transports*, and *exchanges* (across system boundaries).

14.3.3 Biogeochemical Translocations

Fluxes of nitrogen, phosphorus and oxygen between model variables within each 1-m layer are caused by the following processes (see Fig. 14.1). Heterotrophs or zooplankton (ZOO) utilize organic matter through feeding on ($GrazP_N$ & $GrazP_P$)* autotrophs or phytoplankton (PHYTO) and on ($GrazD_N$ and $GrazD_P$) detrital nitrogen and phosphorus (DETRN and DETRP). A part of the consumed food is assimilated ($Assim_N$ and $Assim_P$), while unassimilated particles ($UnAss_N$ and $UnAss_P$) are returned to the suspended detritus pools of organic nitrogen and phosphorus. The ZOO biomass is decreased due to natural and predatory-induced mortality ($ZMort_N$ & $ZMort_P$) and due to excretion ($Excr_N$ and $Excr_P$) producing inorganic nitrogen and phosphorus. The heterotrophs are assumed to be able to keep their vertical position against the gravity force. Thus we disregard diurnal vertical migrations.

The PHYTO biomass is increasing as a result of growth through uptake of inorganic nitrogen and phosphate ($Growth_N$ & $Growth_P$). Nitrogen fixation ($UptNf_N$ & $UptNf_P$) occurs under certain conditions, forming an "internal source" of nitrogen. The mortality of autotrophs ($PMort_N$ & $PMort_P$) increases the concentrations of DETRN and DETRP.

* The designation & means that the nitrogen and phosphorus fluxes are strictly proportional to the N:P ratio of the donor compartment.

The DETRN and DETRP concentrations are increased by the processes mentioned above and are decreased due to sedimentation ($SinkD_N$, $SinkD_P$) and to microbial decomposition which forms fluxes of mineralized nitrogen ($Destr_N$) and phosphorus ($Destr_P$) to ammonium (NH_4) and phosphate (PO_4). The concentrations of nutrients in the water column are further increased by outputs from the sediments (*OutNh*, *OutNo* and *SedOutP*). Nutrient uptakes (*UptAm*, *UptNo* and *GrowthP*) including phosphate utilization by nitrogen fixers ($UptNf_P$) decrease the concentrations. Ammonium (NH_4) is nitrified (*Nitrif*) to nitrate (NO_3) under oxic conditions, while NO_3 is denitrified under anoxic conditions (*Wdenit*), forming an "internal sink" of nitrogen.

Dissolved oxygen is produced by photosynthetic processes in the water column (*Phaer*) and is consumed through respiration of heterotrophs (*RespZ*), microbial mineralization of organic matter (*RespD*), and nitrification (*RespN*) as well as by mineralization in the sediments (*SedRes*). Since denitrification in the water column is due to anaerobic decomposition of organic matter it is described in the model as a process equivalent to oxygen production (*AdDenit*).

Sediment pools of nitrogen and phosphorus are increased by sedimentation of autotrophs and detritus ($SinkP_N$ & $SinkP_P$, $SinkD_N$ and $SinkD_P$) and are decreased due to mineralization ($Sedes_N$ and $Sedes_P$) and burial (Bur_N and Bur_P). Eventually, depending on the redox state, a proportion of mineralized nitrogen is lost due to sediment denitrification of nitrate (*Sedenit*), and as ammonium adsorbed onto the particles of sediments (*SeqN*). A proportion of the mineralized phosphorus is also sequestered by sorption (*SeqP*).

14.3.4 Physical Transports

Four physical processes are taken into account in the model: (1) dynamics of the upper well-mixed layer, (2) dense bottom currents caused by inflows of saline water from the Kattegat, (3) vertical diffusion, and (4) sinking or sedimentation of particulate matter (see also Chap. 2). Sedimentation is computed in the biochemical submodel while the other three fluxes are resolved in the physical submodel that is still in the form originally developed by Stigebrandt (1985, 1987) and coupled with a biochemical submodel by Stigebrandt and Wulff (1987) and Savchuk and Wulff (1996).

A seasonal pycnocline model (Stigebrandt 1985) is used to simulate the seasonal thermocline formation and disappearance with a corresponding entrainment and detrainment of water containing all the ecosystem components into or out of the mixed layer. Being entrained, variables are "instantly" (within one time-step) homogeneously distributed throughout the water column down to the mixed layer depth (concentrations are averaged taking into account the different water volumes of each 1-m layer). Being detrained, concentrations are redistributed by the other transport processes. The variations in the mixed layer depth (MIXLD) are subjected to both seasonal and day-to-day fluctuations. This is ensured by the statistical procedure that generates time series of boundary conditions (Stigebrandt 1985, 1987).

Sporadic salt-water inflows to the Baltic generate dense bottom currents that pass along the bottom slope. The salt water is diluted by entrainment of ambient, less dense Baltic water. At a certain depth this mixture becomes neutrally buoyant and interleaves into a layer with the same density. Intrusions of dense water in the deeper parts of the basin push lighter water upwards, thus inducing vertical advective flows. In real time it takes months for the inflowing water to arrive, e.g. from the Belt Sea to the eastern Gotland Basin. In this horizontally integrated model there is no time delay.

Any gradients in the vertical distributions of model variables generated in the interior layers, by either physical or biochemical processes, are subjected to smoothing by vertical turbulent diffusion, dependent on density stratification. Gravitational sinking is another physical process that will redistribute autotrophs and detrital matter. These particle fluxes are divided into two components in correspondence to the ratios between areas and volumes of consecutive layers. One fraction sediments out from the water column onto the bottom occurring within that layer. The remaining part ($FSINKP_N$ & $FSINKP_P$, $FSINKD_N$ and $FSINKD_P$) will increase concentrations of the autotrophs and detritus in the layer below.

14.3.5 Exchanges Across System Boundaries

The present version of the model comprises the whole Baltic proper from the Danish Straits to the Åland Sea. This is reflected by the corresponding vertical distributions of areas and volumes (Fig. 14.2). Nutrients enter this system from land sources, via the atmosphere, and with waters inflowing from the Kattegat. Nitrogen is also added by nitrogen fixation. Nutrients exit the system with out-flowing water or via internal sinks due to denitrification and burial in the sediments.

The exchanges with the Gulf of Bothnia are neglected. To a certain degree such simplification may be justified by the relative insignificance of the nitrogen and phosphorus exchanges in comparison with total amounts of nutrients in the Baltic proper (Wulff and Stigebrandt 1989; Wulff et al. 1994a).

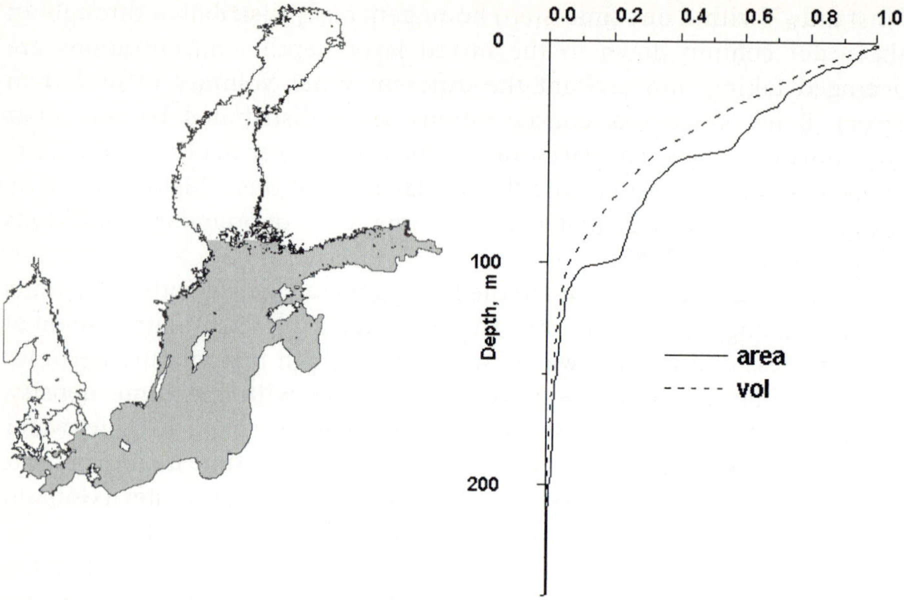

Fig. 14.2. Map of the Baltic (*left*) and hypsographic curve (*right*) that shows depth distributions of water volumes and sediment areas within the Baltic proper ($A_0=255\times10^3\,km^2, V_0=13.5\times10^3\,km^3$)

14.3.6 The System of Equations

The time–space evolution of pelagic state variables is described by a system of horizontally integrated vertical diffusion equations for non-conservative substances (Stigebrandt and Wulff 1987):

$$\frac{\partial C_j}{\partial t}=\frac{1}{A}\left\{\frac{\partial}{\partial z}\left(A\cdot K_Z\frac{\partial C_j}{\partial z}\right)+\frac{\partial}{\partial z}\left(A\cdot w\cdot C_j\right)\right\}+\Phi_j\left(C,z,t\right) \tag{1}$$

In Eq. (1), C(*z*,*t*) is a vector-function of the state variables that consist of $C_j(z,t)$ elements denoting volume concentrations of variables [mg N (or P) m^{-3}], *A(z)* is the depth distribution of horizontal surface areas (i.e. a bathygraphic curve), $K_z(z,t)$ is the horizontally averaged vertical diffusion coefficient, *w*(*z*,*t*) is the horizontally averaged velocity of upward vertical advection induced by the dense bottom current, Φ_j (C,*z*,*t*) is a biogeochemical translocation function for the *j*th state variable which is an algebraic sum of all local biogeochemical fluxes described above (Fig. 14.1).

14.3.7 Parameterization of Biogeochemical Pathways

The formal presentation and details of the parameterizations are found in Savchuk and Wulff (1993, 1996). Here, we will only give a general description of the assumptions behind these formulations.

The photosynthetic increase in the autotroph biomass in this model is equivalent to net primary production. The rate of phytoplankton growth is dependent on temperature, nutrient concentrations and light. The uptake of nitrogen and phosphorus follows the Redfield ratio. Effects of nutrient concentrations are described by Michaelis-Menten-type expressions and non-competitive inhibition of nitrate uptake by ammonium is also taken into account. The light extinction will change with concentrations of plankton and detritus.

The nitrogen-fixing capacity is dependent on the deficit of inorganic nitrogen, compared to phosphate. Since the development of the species capable of utilizing nitrogen gas is highly dependent on water temperature, we made the nitrogen fixation negligible when the temperature is below 12 °C, with a rapid increase between 13 and 15 °C. The growth of cyanobacteria is also dependent on phosphate availability with a Michaelis-Menten kinetic. The cyanobacterial development affects trophodynamic interactions and sedimentation. We describe these effects as being dependent on the contribution of nitrogen fixation to total primary production.

Zooplankton feeding is described with a filtration rate dependent on temperature and food, which consists of both autotrophs and detritus, in relation to their relative concentrations. The ingestion efficiency of zooplankton decreases with an increase of cyanobacteria.

The excretion by heterotrophs is used in the model for the stoichiometric adjustments of the nitrogen and phosphorus cycles (Savchuk 1986; Savchuk et al. 1988). The heterotrophs can, to a certain degree, maintain a constant body nutrient ratio (Corner and Davies 1971; Mayzaud and Conover 1988; Wiebe 1988; Sterner 1990; Sommer 1992; Elser and Hassett 1994). We can therefore define the phosphorus flux as the sum of two components. The first component is the phosphorus being excreted with the same constant N:P ratio of heterotrophs. The second component is the excessive phosphorus excreted when the consumed food has a lower N:P ratio than the consumers. The net growth efficiency is assumed to be inversely related to the water temperature to reflect higher excretion rates of small heterotrophs more abundant in summer.

The mortality of heterotrophs is a closure term in the model (Anderson et al. 1993; Totterdell et al. 1993), with a combination of density- and time-dependent mortality. Mineralization of detritus nitrogen and phosphorus is described as first-order temperature-dependent reactions. Nitrification or microbial oxidation of ammonium to nitrate is described as a temperature-dependent process with photo-inhibition in the surface waters and inhibition at low oxygen concentrations.

Denitrification or reduction of nitrate to molecular nitrogen occurs in the water column at low oxygen concentrations. It is described as a process dependent both on nitrate concentrations and on availability of electron donors such as organic matter or hydrogen sulfide. Assuming as a first approximation that reserves of electron donors in the vicinity of the oxic–anoxic interface are not limiting in the Baltic deep basins (Brettar and Rheinheimer 1991), we will simulate a rather sharp start of denitrification at low oxygen concentrations being dependent on nitrate concentration and temperature. Oxygen production and consumption by autotrophs and heterotrophs and for microbial mineralization and nitrification as well as "oxygen reimbursement" due to denitrification are set proportional to the corresponding nitrogen fluxes.

The natural mortality and sinking of the autotrophs are described as temperature-dependent processes inversely related to the limiting function that is used in the description of autotroph growth. During cyanobacterial blooms, both natural mortality rate and sinking velocity are reduced. Nitrogen and phosphorus fluxes due to sinking of detritus are described with a temperature dependence similar to that for autotrophs.

Any model that attempts to describe the biogeochemical cycles of a shallow sea like the Baltic must include recycling benthic feedback due to the sediment mineralization (Chap. 9). It is also here that most of the internal nitrogen losses occur through denitrification.

The fate of ammonium, initially produced by the mineralization of organic nitrogen, is defined by the redox state of the sediments. All the inorganic nitrogen produced in the model is channeled into four pathways with dynamically changing proportions defined by the oxygen conditions in the bottom water. At low oxygen concentrations a certain part of produced ammonium is adsorbed onto sediment. This will reduce the exchangeable part of NH_4. In oxic sediments, ammonium is nitrified and a certain fraction of nitrate is released to the bottom water while the remaining fraction is denitrified. This formulation allows for co-occurrence of nitrification and denitrification. With higher oxygen concentration in the water above the sediment the oxidized sediment layer is thicker and, consequently, more nitrogen will be released as nitrate. With less oxygen and a redoxcline closer to the sediment surface, a larger fraction of nitrate will diffuse downward into the reduced sediment to be denitrified. We also assume that at high oxygen concentrations, the oxygen flux into sediments is controlled by boundary layer resistance rather than by concentration gradients.

Oxygen consumption by the bottom sediments is due to aerobic decomposition of organic matter and nitrification. Oxygen consumption in the sediment by reduced compounds such as hydrogen sulfide, iron and manganese has not been taken into account.

The fate of mineralized phosphorus is also dependent on the redox state of sediments. Under aerobic conditions, one part of the phosphate produced is adsorbed onto solid particles. The remaining fraction is released to the overlying water. Under anaerobic conditions, no adsorption occurs and some of the previously adsorbed particle-bound P is released to the water column as well.

Since we want to follow the fate of nutrients over decades, we have to take into account a small but permanent system "sink" of nitrogen and phosphorus through the lower boundary of the active sediment layer due to burial. In the upper 5-cm layer of unconsolidated sediments this boundary is moving upward at a rate of about 1.5–4.2 mm $year^{-1}$ in the Baltic (Jonsson et al. 1990; see also Chap. 11).

14.3.8 Boundary and Initial Conditions

For meteorological and hydrological forcing functions the original sets described in Stigebrandt (1985, 1987) and Stigebrandt and Wulff (1987) were used. Wind speeds for each simulation day were randomly selected from monthly frequency distributions obtained from a 30-year time series of observations. The same approach was used for the daily mean air temperature, but the random selection was made on every fifth day. The volume and salinity of inflowing water were for each fifteenth day randomly selected from distributions presented in Stigebrandt (1987). Finally, the temperature of inflowing water was given as a repeated sinusoidal annual cycle with an amplitude of 3 °C around a mean of 8 °C. The described procedures were used to calculate time-series of boundary conditions that were the same in all the numerical experiments described below. Thus, the modeled hydrographic variations and internal transport processes were the same from experiment to experiment.

The seasonal variations in atmospheric precipitation and river runoff of water and nutrients are computed by half-day linear interpolation between mean monthly values. The estimates for wet and dry atmospheric deposition of ammonium and nitrate onto the Baltic proper area were made by Ann-Katrin Hallin (Department of Systems Ecology, Stockholm University) using computations by Granat (Chap. 5) for the whole Baltic Sea area with a $1 \times 1°$ resolution. Taking into account the spatio-temporal coverage and reliability of the primary data, we have chosen 1979–1991 as a basic time interval to calculate mean monthly sums of nitrogen loads from the atmosphere.

Annual phosphorus atmospheric load onto the entire Baltic Sea has been estimated at 6000–7000 tons (Davidavichiene and Sopauskiene 1989; HELCOM 1989). Ignoring geographical differences and taking into consideration the ratio between areas of the Baltic proper ($255 \times 10^3\,km^2$) and the whole Baltic Sea ($415 \times 10^3\,km^2$), we calculated that 3700 tons are deposited annually onto our modeled domain. This amount was then distributed throughout the year.

All the inputs from land were divided into two sources: land runoff (see Chap. 4) and coastal point sources. The data on land runoff were extracted and combined by Ann-Katrin Hallin (Department of Systems Ecology, Stockholm University). We have used monthly integrated values of ammonium, nitrate and phosphate as well as organic nitrogen and phosphorus, averaged for 1979–1990. Organic fractions were calculated as the difference between total and inorganic nitrogen and phosphorus. The load from the point sources has been taken from HELCOM (1993a) as annual inputs of ammonium, total nitrogen and phosphorus for 1990. Organic nitrogen was calculated by difference. Organic phosphorus and phosphate were each assumed to be 50% of total phosphorus. These inputs were evenly distributed throughout the year. In accordance with the model formulation we will only account for the "labile" fraction of organic matter. In the "standard case", 30% of organic nitrogen and 60% of organic phosphorus entering the sea from the land were assumed to be biodegradable.

Salt water flowing into the Baltic will also bring in nutrients and oxygen. We have used average monthly concentrations calculated for the surface layer of the Belt Sea from all measurements collected in the BED database between 1970 and 1990.

The total nitrogen and phosphorus inputs from all described sources are shown in Fig. 14.3. On average, 739×10^3 tons of nitrogen and 38×10^3 tons of phosphorus are added to the Baltic proper annually, with an average "labile" N:P molar ratio of 43. Small year-to-year variations, less than 20,000 tons for nitrogen and 3000 tons for phosphorus, are caused by fluctuations in both the volume of water flowing in from the Kattegat and the time of the year when these events occur.

The initial concentrations of the state variables have been computed in the following way. First, some plausible first guess-estimates of horizontally averaged vertical distributions at the end of winter for the Baltic proper were used to run the model for 20 years. The final vertical distri-

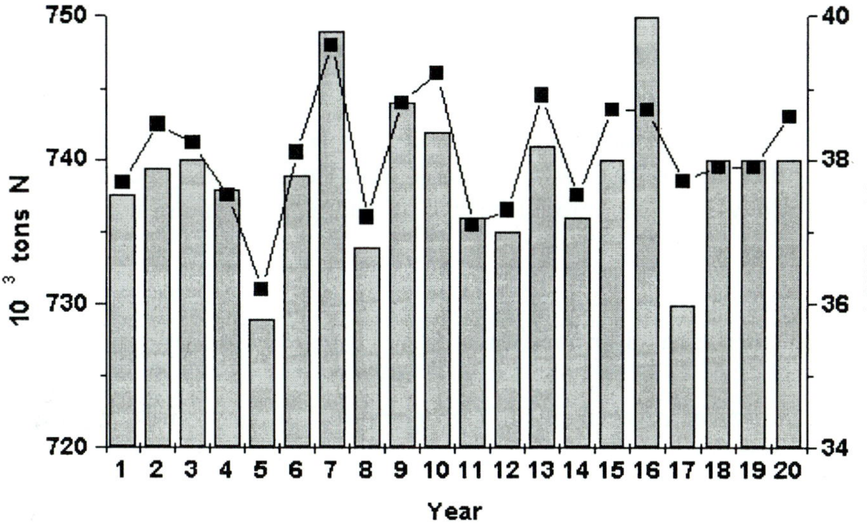

Fig. 14.3. Total loads of nitrogen (*columns*) and phosphorous (*line*) to the Baltic proper, used in the model as boundary conditions. Annual sums of all sources (rivers, point sources, atmosphere and inflows from the Kattegat)

Table 14.1. Total amounts of inorganic nutrients (10^3 tons) given as initial conditions and estimated from field measurements for the Baltic proper. (Wulff et al. 1994b)

Model variables	Initial conditions	March 1982/1986	March 1987/1991
NH_4	79	83	–
NO_3	970	987	1080
PO_4	610	458	483

butions, on 1 March of year 21, were then used as initial conditions for the "standard case". The corresponding total amounts of inorganic nutrients are comparable with similar estimates from field measurements (Table 14.1). The latter represent total monthly amounts computed from interpolated, volume-weighted monthly means of nutrient concentrations for 5-year periods with a 1° horizontal resolution and for 22 depth intervals (Wulff et al. 1994b). This procedure gives estimates that are rather compatible with the model variables during the last 15 years. Later calculation by Wulff et al. (Chap. 13) gave for 1991 lower estimates: 855,000 and 335,000 tons of inorganic N and P, respectively.

14.4 Results and Discussion

Here, we first present the results of numerical experiment with our model with the initial and boundary conditions described above ("the standard case") and compare them with available data from the last decades. Next, we will show how the model reacts to changes in external inputs of nutrients.

14.4.1 Results of the Standard Case

14.4.1.1 Long-Term Variations in Hydrography, Nutrients and Oxygen

Twenty years of time–depth variations for those state variables that can be directly compared to long-term field observations are presented in Fig. 14.4. The variations in water temperature and salinity reproduced by the physical submodel (Fig. 14.4A,B) are quite comparable to the seasonal cycle and year-to-year variations "typical" for the Baltic proper, at least in the 1960s and 1970s (Stigebrandt and Wulff 1987).

Stagnation periods, with oxygen (Fig. 14.4E) decreasing to -5 ml l^{-1} (or 2.5 ml l^{-1} H_2S) resulted in accumulation of ammonium and phosphate (Fig. 14.4D) as well as in total depletion of nitrate (Fig. 14.4C). The combined effect of new inflows of high salinity and oxygenated water into the deep layers changes the redox conditions and ammonium and phosphate concentrations drop down, while nitrate concentration increases. It can also be seen in the figures how the "old" stagnant water, depleted of nitrate and enriched in phosphate, is pushed up into the intermediate layers.

The described features are quite similar to the variations that have actually occurred in the Baltic proper, at least qualitatively. However, the large data sets of nutrient and oxygen concentrations compiled from studies of the Baltic allow us to statistically compare field measurements with the similarly treated results of 20-year simulation. Since a statistical procedure has been used to give hydrological and meteorological forcing to simulate hydrographic features typical for the Baltic proper, such a comparison is more relevant.

Apparently, we cannot expect that a statistically driven model would generate the time-sequence of inflow events that had actually occurred in the Baltic during the last decades, the long stagnation of 1978/1993 in particular. We will therefore compare the model outputs with statistical estimates based on international data collected in the 1960s to 1980s at the

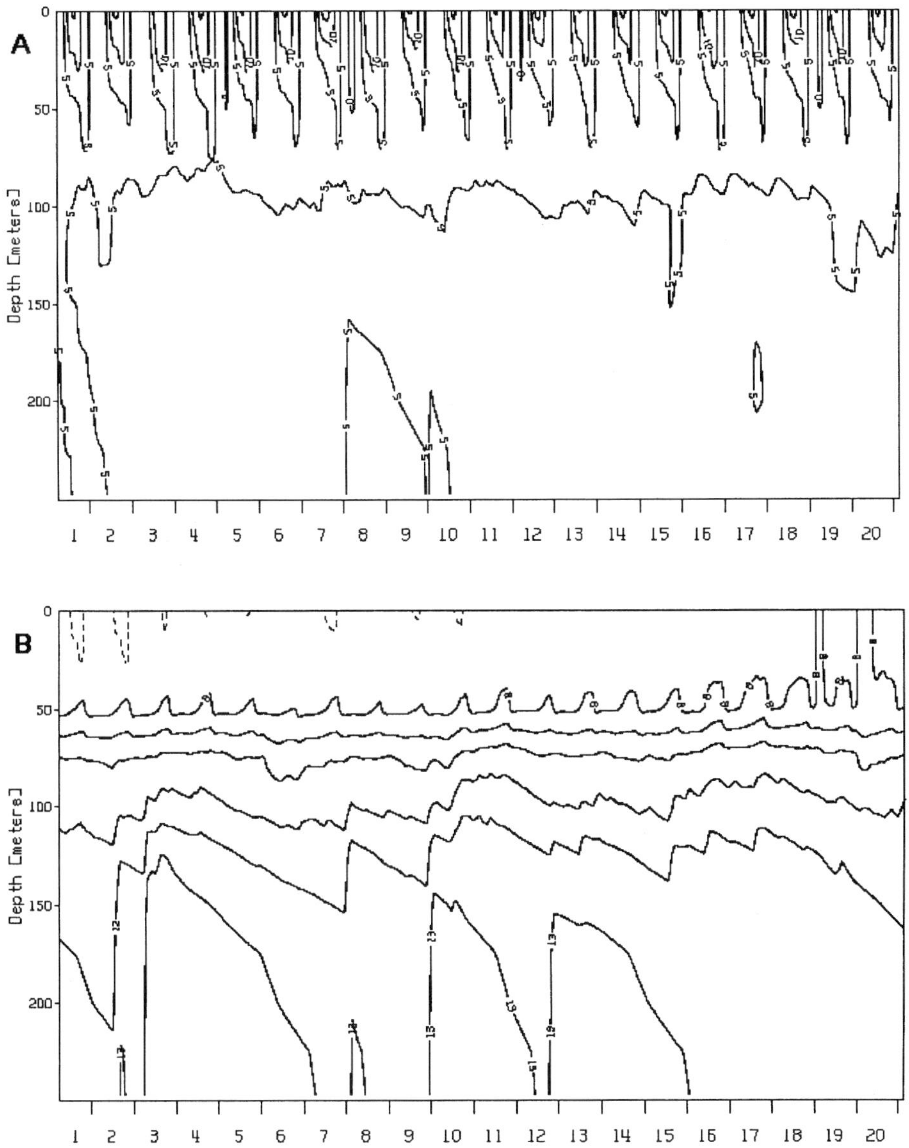

Fig. 14.4. Simulated time–depth variations of **A** temperature (°C), **B** salinity (PSU), **C** nitrate (μM N), **D** phosphate (μM P), and **E** oxygen (ml l^{-1}) from the standard case (anoxic zone is emphasized by *dashed isolines*). The 20-year contour plots are made from 30-day averages calculated for each 25th meter

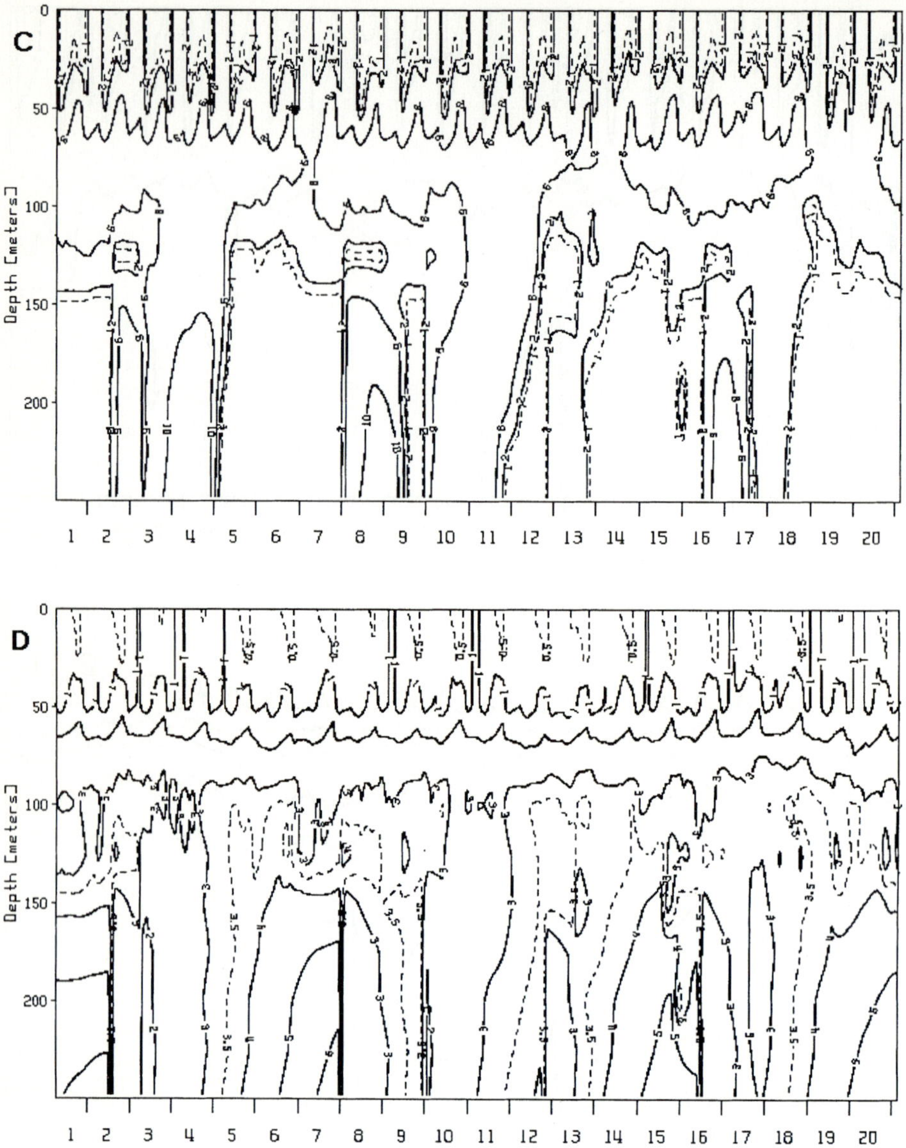

Fig. 14.4. *(continued)*

central Gotland Deep station (Savchuk et al. 1994). At the same time, the "standard case" describes a quasi-periodic steady state, formed on a long-term scale under current permanent high nutrient inputs. Therefore, the winter nutrient accumulation has to be compared with recent observations.

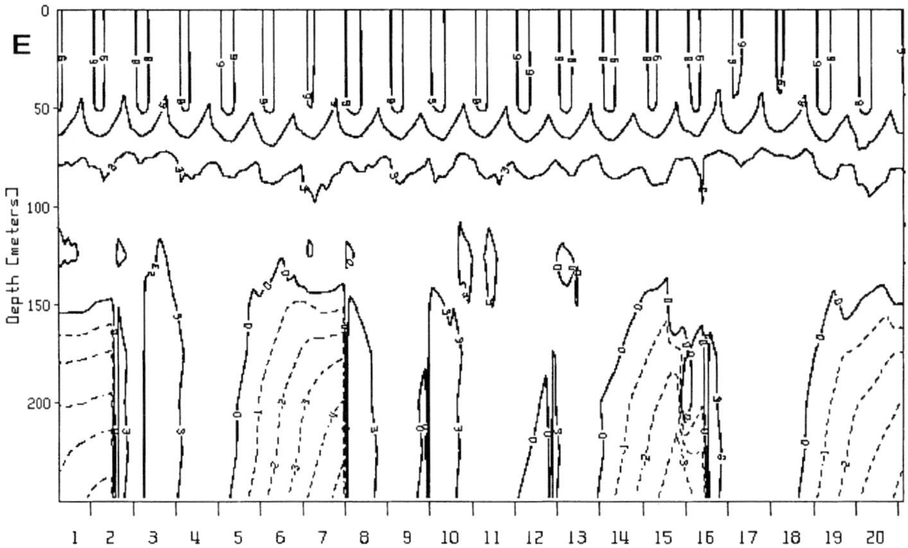

Fig. 14.4. *(continued)*

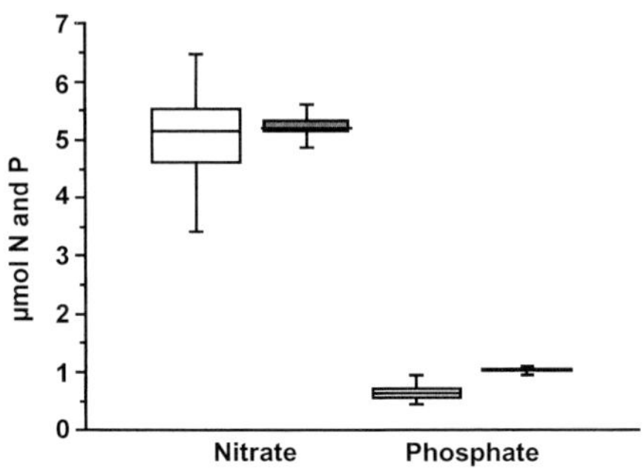

Fig. 14.5. Comparison of statistical estimates (median, 25 % and 75 % quartiles, and extremes) for observed (*open boxes*) and simulated (*shaded boxes*) variations in winter surface nutrients (µM)

In the model, the surface winter nutrient maximum occurs in March with an N:P molar ratio varying between 4.5 and 5.5. The comparison of maximum values of nutrient surface concentrations found in the BED database (February–April, 1990–1996) with those retrieved from the model shows a fairly good fit for nitrate, while phosphate concentrations are higher (Fig. 14.5). The summer minimum phosphate concentrations, between 0.15 and 0.30 µM P in August–September are also higher than most typical field values. Some limitations in the present model formulations will result in the elevated phosphate concentrations in the oxygenated layers. The reasons for this are discussed below.

The simulated variations in nutrient concentrations in deep layers are also quite comparable with the corresponding statistical estimates for the

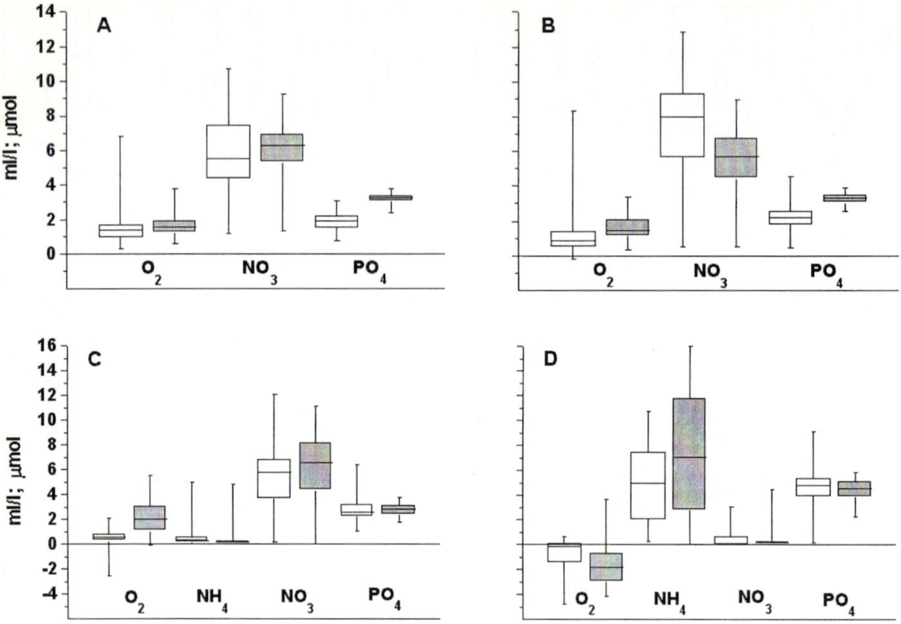

Fig. 14.6. Comparison of observed (*open boxes*; Savchuk et al. 1994) and simulated (*shaded boxes*) variations in oxygen (ml l^{-1}) and nutrients (µM) at 100 m (*upper graphs*) and 200 m (*lower graphs*). **A** and **C** show statistical estimates (median, 25 and 75% quartiles, and extremes) for periods with oxic conditions in the bottom layer; **B** and **D** correspond to anoxic periods

Gotland Deep (Fig. 14.6). The most pronounced difference is that major inflows result in higher oxygen concentrations in the model. This is a consequence of the one-dimension model formulation used here. The new inflow of oxygenated water is instantly distributed into the whole horizontally integrated water column. The observations presented here are from a central station in the Baltic, to where the new salt water will reach after several months' delay, passing several smaller basins while exposed to oxygen consumption.

14.4.1.2 Seasonal and Interannual Variations in the Pelagic Subsystem

To show the seasonal variations in more detail we have chosen three specific time intervals – years 3–4, 7–8 and 15–16 – of the 20-year simulation. Typical examples of seasonal dynamics for some pelagic variables are presented in Fig. 14.7.

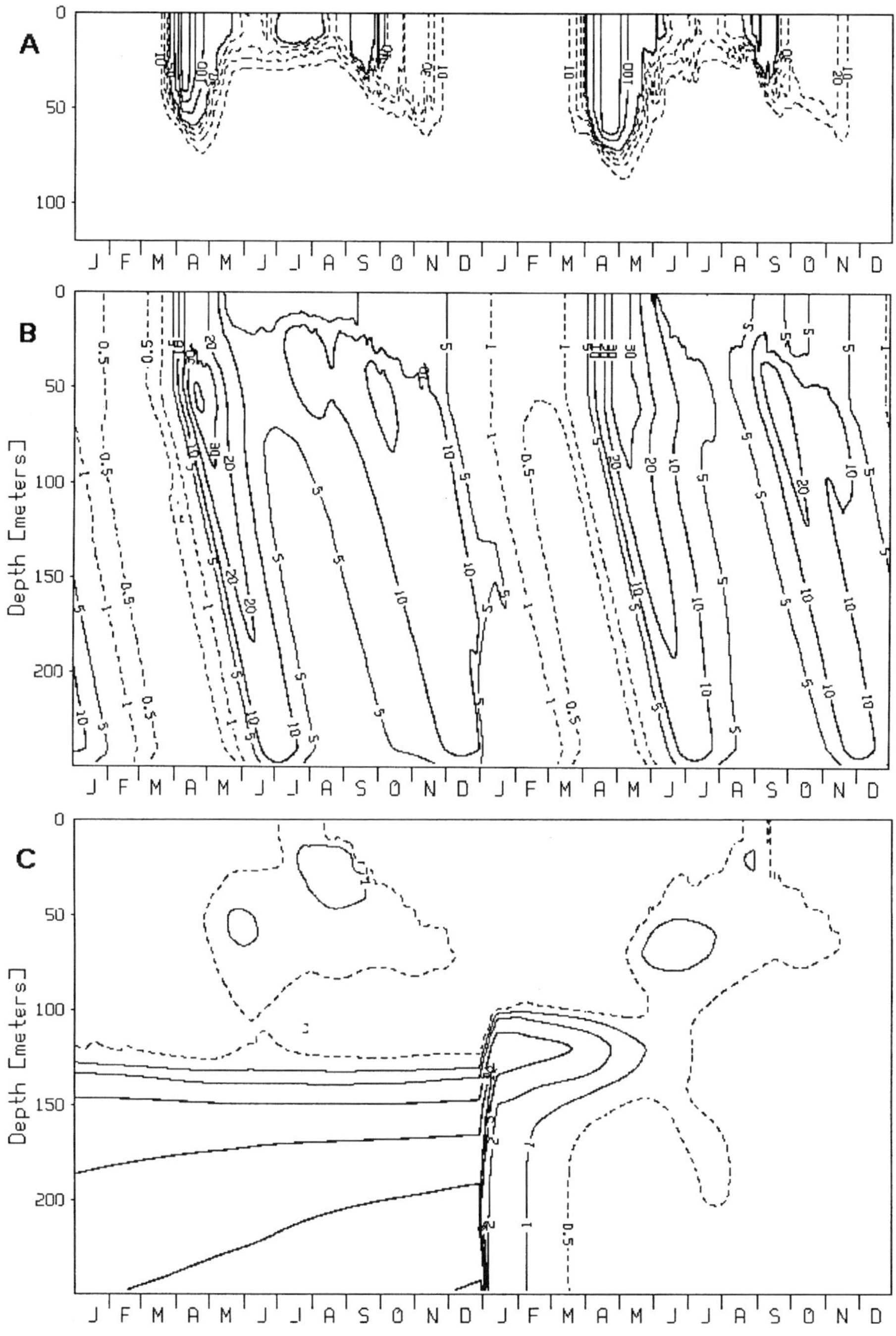

Fig. 14.7. Simulated time–depth distributions of **A** autotrophs (mg C m^{-3}), **B** detritus (mg N m^{-3}) and **C** ammonium (μMN) as exemplified for years 7 and 8 of the "standard case". Contour plots are made from 2-day averages, calculated for each 10th meter

14.4.1.2.1 Primary Production and Autotroph Biomass

In the model, the spring phytoplankton bloom starts in the beginning of March and culminates in April with biomass up to 250 mg C m^{-3} (Fig. 14.7A)* or 8–9 g C m^{-2} (Fig. 14.8B). There is a spring primary production maximum of 110–140 mg C m^{-3} day^{-1} or about 1–1.2 g C m^{-2} day^{-1} (Fig. 14.8A). The autotroph biomass decreases after the bloom to about 20–30 mg C m^{-3} in June. It then fluctuates below 100 mg C m^{-3} (or 2 g C m^{-2}) during the rest of

* Here, and further onward in the text, we assume a Redfield C:N ratio of 6 (by weight) to express simulated nitrogen values as carbon units.

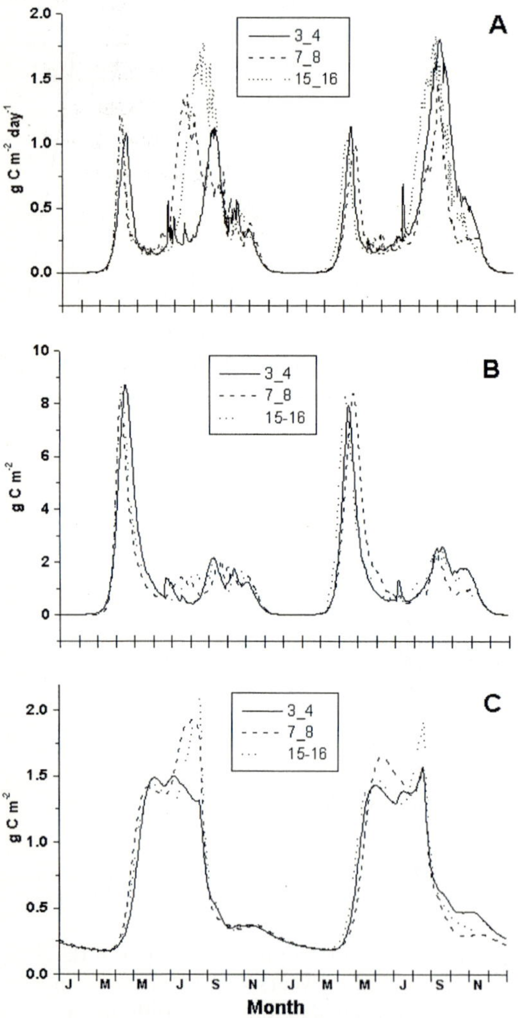

Fig. 14.8. Seasonal variation in depth-integrated primary production (**A**), autotroph (**B**) and heterotroph (**C**) biomasses generated by the model and exemplified for two consecutive years and for three periods (years 3–4, 7–8 and 15–16) of the 20-year simulation

Table 14.2. Some estimates of biomass and primary production of phytoplankton in the Baltic proper

Season, location, time	Biomass, C ($mg\ m^{-3}$)	($g\ m^{-2}$)	Primary production, C ($mg\ m^{-3}\ day^{-1}$)	($g\ m^{-2}\ day^{-1}$)	Reference
Spring bloom					
Arkona Sea, 1981	200–250[a]	4–6[a]	100–400	2–3	Schulz (1985)
Bornholm Sea, 1975	100–150		100–200	1–1.5	von Bodungen et al. (1981)
Northern Baltic, 1982	250–370		120–280	1.8–2.8	Leppänen and Kononen (1988)
Northern Baltic, 1988		7.1		1.4–1.8	Lignell et al. (1993)
Askö area, 1975		4–6	50–100	0.5–0.7	Larsson et al. (1986)
Askö area, 1977/1988		2–8[a]		0.4–1.2[a]	Johansson et al. (1992)
Summer					
Arkona Sea, 1978	20–80		40–80	0.35–0.7	Schulz and Breuel (1984)
Arkona Sea, 1981/1988	80–120		60–180		HELCOM (1990)[a,b]
Bornholm Sea, 1981/1988	60–100		60–120		HELCOM (1990)[a,b]
Gotland Sea, 1981/1988	60–180		50–150		HELCOM (1990)[a,b]
Landsort Deep, 1982	40–50		30	0.25–0.6	Wulff and Cederwall (1984)
Northern Baltic, 1981/1988	40–140		50–150		HELCOM (1990)[a,b]
Gulf of Finland, 1981/1988	80–160		50–150		HELCOM (1990)[a,b]
Askö area, 1977/1988		0.8–1.7[a]		0.3–1.4	Johansson et al. (1992)

[a] Chlorophyll "a" values converted into carbon ones with C/Chl=40

[b] Hourly values of the PPC were multiplied by 12 to convert into daily ones

the summer, despite high primary production (40–140 mg C m^{-3} day^{-1} or 0.6–1.7 g C m^{-2} day^{-1}). An autumn bloom occurs in late August–October.

When comparing these values with some estimates based on field observations in the Baltic (Table 14.2), one should bear in mind that the simulated "primary production" corresponds to or is even lower than measured net primary production (using the ^{14}C method). In the model we are calculating the net growth rates and no account is made for exudates.

Autotroph mortality exceeds primary production only in the senescent phase of the spring blooms and in late autumn. The sedimentation is also playing a noticeable role during these periods (Figs. 14.7B and 14.13 below). It is the heterotroph grazing that keeps autotroph biomass at low levels despite a high primary production during summer.

14.4.1.2.2 Heterotrophs

In the model, the increase in the heterotroph biomass starts in April following the vernal bloom as it usually occurs in the Baltic (Johansson et al. 1992; Lignell et al. 1993; Viitasalo 1994). The highest biomasses of 40–70 mg C m^{-3} or 1.5–2 g C m^{-2} occur in August–September (Fig. 14.8C). The secondary production is about 2.0–6.0 mg C m^{-3} day^{-1} or 50–120 mg C m^{-2} day^{-1}, calculated as the difference between assimilation and excretion.

These values seem to be rather high, even for composite variable "heterotrophs" in comparison with some estimates from the 1980s for net zooplankton (Table 14.3). They are closer to the values found in a eutrophicated bay (Johansson et al. 1992), especially with respect to a shift of maximum development towards August.

Table 14.3. Estimates of summer zooplankton biomasses in the Baltic proper

Location, time	Biomass, C[a] (mg m^{-3})	(g m^{-2})	Reference
Arkona Sea, 1978	30–50		Schulz and Breuel (1984)
Southern Baltic, 1981/1983		0.7–1.0	Wulff et al. (1986)
Landsort Deep, 1982	17.7	0.75	Wulff and Cederwall (1984)
Northern Baltic, 1988		0.9[b]	Lignell et al. (1993)
Northern Baltic, 1980s	26–52		Viitasalo (1994)
Askö area, 1977/1988		0.45	Johansson et al. (1992)
Eutrophic Bay, 1977/1988		0.9	Johansson et al. (1992)

[a] Wet weight values converted into carbon units with C/ww=0.05
[b] Post-bloom amount, May 26

Annual secondary production in the model varies between 7.5 and 10.1 g C m^{-2} or 1.9–2.6 10^6 tons C. This is quite comparable with or even lower than estimates for the coastal Askö area (mean for 1977/1988 of 6.4 with a range of 2.0–9.7 g C m^{-2}; Johansson et al. 1992) and for the whole Baltic Sea (13 g C m^{-2}; Elmgren 1989).

14.4.1.2.3 Role of Different Nitrogen Sources

As shown in Fig. 14.9A, nitrate uptake dominates in spring when it supports up to 95 % of primary production and becomes again significant in autumn. During the summer, 60–90 % of dissolved inorganic nitrogen requirements are met by ammonium. Occasionally the proportion of nitrate uptake exceeds 20–40 %, due to sporadic deepening of the mixed layer during summer. Few field studies have been made in the Baltic region. Karlson et al. (1996) showed in a study made during different seasons in Skagerrak that nitrate contributed with 21–30 %, ammonia with 47–55 %, and urea with 21–23 % of the nitrogen uptake. In our simulations,

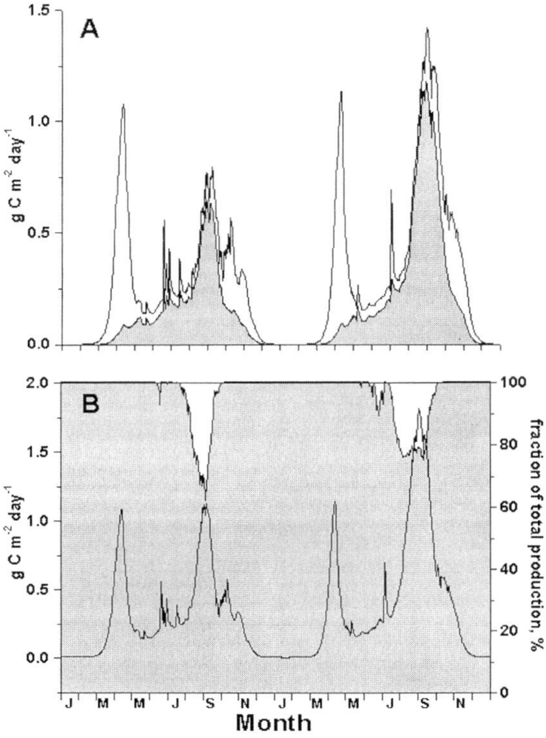

Fig. 14.9. Simulated seasonal variations in (**A**) new (*open*) and regenerated (*shaded*) primary production and (**B**) contributions (*right axis*) of dissolved inorganic nitrogen uptake (*shaded*) and nitrogen fixation (*open*) to total primary production (*left axis*), as exemplified for years 3–4 of the simulation

regenerated inorganic nitrogen (only ammonia and no urea in our model) contributes with 49–63% of the 4–5.6×10[6] tons required annually as dissolved inorganic nitrogen.

Nitrogen gas fixation seldom exceeds 30% of total daily primary production even at summer maximum (Fig. 14.9B). Thus, in the model, molecular nitrogen contributes only 8–11% (Table 14.4) of the total annual nitrogen requirements of 4.4–6.2 10[6] tons.

Simulated mineralization rates are decreasing with depth from a maximum around 0.2 mg N m^{-3} day^{-1} at 20- to 30-m depths to 0.05–0.15 mg N m^{-3} day^{-1} at 240–250 m. The seasonal maximum of nitrification, about 1.5 mg N m^{-3} day^{-1}, is related to a subsurface maximum of ammonium; quite a common phenomenon in the Baltic proper (see, e.g., Gundersen 1981; Yurkovskis 1983; Savchuk et al. 1994), reproduced by the model (cf. Figs. 14.7C and 14.10). Higher nitrification rates, up to 2 mg N m^{-3} day^{-1}, occur occasionally in the deep layers when waters containing oxygen are mixed with stagnant waters enriched with ammonium. These values are approximately in the middle of the broad range (0.2–34 mg N m^{-3} day^{-1}) compiled by Kaplan (1983) for estuaries and coastal bays. They are also quite comparable with the maximum of 3.9 mg N m^{-3} day^{-1} observed by Enoksson (1986) at the intermediate depths in the western Gotland Basin and with 2.8 mg N m^{-3} day^{-1} measured by Berounsky and Nixon (1985). The latter value is from studies made in mesocosms with Narragansett Bay waters containing ammonium and nitrite in concentrations similar to those usually found in the Baltic Sea.

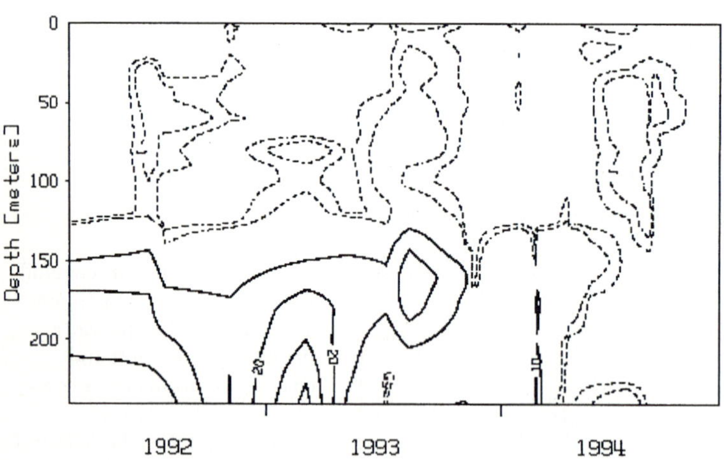

Fig. 14.10. Time–depth variations in ammonium (μM N) in the Gotland Deep (BY-15) during 1992–1994. (Data from BED database combined with those kindly provided by A. Yurkovskis)

Table 14.4. Rates of nitrogen fixation simulated by the model and recalculated from literature sources

Units	Model	Estimates	Location, time	Reference
Mg N m^{-3} day^{-1}	4.4–6.8[a]	0.1–1.6[b]	Northern and central Baltic, 1974/1975	Rinne et al. (1978)
		0.3–1.2[b]	Northern Baltic, 1978/1979	Rinne et al. (1981)
		0.5–2.0	Askö area, 1982	Wulff and Cederwall (1984)
		17–27[c]	Arkona Sea, 1974/1975	Hübel and Hübel (1987)
% PP (daily N uptake)	30[a]	18 (13)	Askö area, 1982	Wulff and Cederwall (1984)
		16[d]	Northern Baltic(?) 19(??)	Sahlsten and Sörensen (1989)[e]
Mg N m^{-2} day^{-1}	50–65[a]	0.7–9.0[b]	Northern Baltic, 1978/1979	Rinne et al. (1981)
		42 (80)[f]	Askö area, 1976	Lindahl et al. (1978)
		5–34	??, 1980/1984	Niemistõ et al. (1989)
Mg C m^{-2} day^{-1}	300–390[a]	250[g]	Northern Baltic, 1978	Larsson and Hagström (1982)[e]
% PP (on C base)	30–40[a]	23[g]	Northern Baltic, 1978	Larsson and Hagström (1982)[e]
g N m^{-2} $year^{-1}$	1.5–2.8[h]	0.3–1.1	Askö area, 1976	Lindahl et al. (1978)
10^3 tons $year^{-1}$	356–703[h]	170	Baltic proper+Gulf of Finland	Leppänen et al. (1988)
g C m^{-2} $year^{-1}$	8–17[h]	13[i]	Entrance to Gulf of Finland	Kuparinen et al. 1984[e]
% Total annual PP	8–11[h]	15[i]	Entrance to Gulf of Finland	Kuparinen et al. (1984)[e]

[a] Range of absolute maximum for each of simulated years
[b] Recalculated from molar C_2H_2 units into weight N units with a factor 56=6 (2 h/day) × 1/3 (mol N_2/mol C_2H_2) × 28 (g N/mol N_2)
[c] Maximum values recalculated from hourly into daily with a factor 12
[d] Highest % of total nitrogen uptake
[e] As cited by Kononen (1992)
[f] Recalculated from maximum values of acetylene reduction with a factor 28=3 (4 h/day) × 1/3 (mol N_2/mol C_2H_2) × 28 (g N/mol N_2)
[g] Maximum values for size group >10 μm
[h] Range from 20 years of simulation
[i] Estimated for size group >3 μm

Maximum denitrification rates, up to 2–4 mg N m^{-3} day^{-1} in the vicinity of the redoxcline, are comparable with 3–4 mg N m^{-3} day^{-1} calculated by Rönner and Sörensson (1985) for the western Gotland Basin as well as with 0.6 and 1.5 mg N m^{-3} day^{-1} found by Brettar and Rheinheimer (1991, 1992) in the Gotland Deep.

The statistical forcing has also resulted in pronounced year-to-year variations in the physical and chemical environment and, consequently, in intensity of biogeochemical cycling (Fig. 14.11). The years 3, 5, 8, 13 and 19 with markedly lower biological activity are generally colder than others. Also, these years are slightly more windy and the upper, trophogenic layer is more mixed during spring and summer.

Due to the highly temperature-dependent parameterization of nitrogen fixation, the nitrogen fixation is considerably lower during cold years. Similar relationships between cyanobacteria development and external driving forces are, in nature, probably the reason for a good interannual correlation between areal cyanobacteria accumulation estimated from satellite images and sunshine duration (Kahru et al. 1994; Kahru and Rud 1995) as well as with mean surface water temperature in August averaged from field observations for the corresponding years (Fig. 14.12).

In colder years, less N_2 is involved in summer metabolism via nitrogen fixation. Recycling is also less efficient due to lowered metabolic activity of heterotrophs, caused both by decreased filtration rate and increased net growth efficiency. A combination of decreased nitrogen fixation and regenerated production with enhanced summer mixing leads to an increase in the relative significance of new production by nitrate uptake.

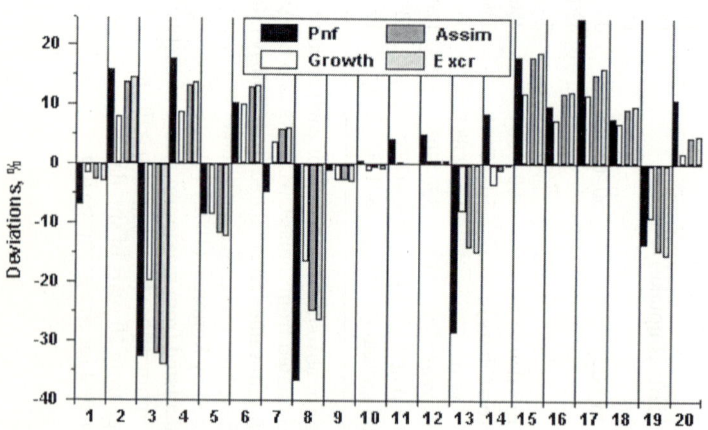

Fig. 14.11. Relative deviations of annual nitrogen fixation (*Pnf*), inorganic nitrogen uptake (*Growth*), heterotroph assimilation (*Assim*) and excretion (*Excr*) fluxes from 20-year averages

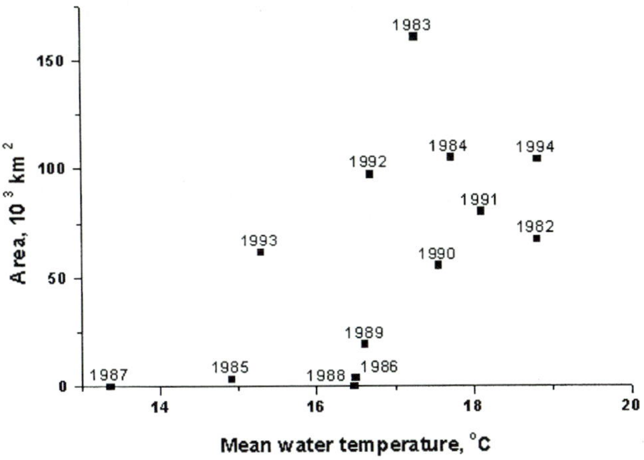

Fig. 14.12. Cyanobacteria areal accumulation (Kahru and Rud 1995) compared to average surface water temperature in August (1982–1994) calculated from BED

14.4.1.2.4 Sedimentation

The sedimentation of organic matter has two clear seasonal maxima following the spring and autumn blooms (cf. Fig. 14.7B). While in the upper 50–60 ms sedimentation is taking place nearly simultaneously, the delay is increased and the total depositing amount is decreased with depth (Fig. 14.13A). The proportion of live autotrophs in sedimenting organic matter is also decreasing with depth, from about 50 % as a seasonal maximum at 50 m (Fig. 14.13B) to a few percent at 125 m. The composition of detritus is changing both in time and with depth. The N:P ratio of detritus is close to 15.5 (the value prescribed in the model for autotrophs) both in winter and during spring blooms. It rarely exceeds 17 during summer in the upper layers, and increases with depth in the autumn up to 20 due to larger contribution of dead heterotrophs with a prescribed N:P ratio of 24.4 (Fig. 14.13C). Thus, relatively less phosphorus is transported to deep sediments in comparison with nitrogen.

The maximal sedimentation rates at 25–50 m during the spring bloom vary within 0.5–0.6 g C m^{-2} day^{-1}, assuming a C:N weight ratio of 6. This is quite comparable with the ranges reported, e.g. for the Northeast Atlantic: 0.1–0.6 (0.8) g C m^{-2} day^{-1} (Wassmann et al. 1991) and for the entrance to the Gulf of Finland in 1983 as phytoplankton carbon: 0.3–0.4 g C m^{-2} day^{-1} (Heiskanen and Kononen 1994). However, they are lower than maximum estimates for the Kiel Bight: 0.5–1.5 g C m^{-2} day^{-1} (Smetacek 1980, 1984), the Bornholm Deep: 1.6–2.0 g C m^{-2} day^{-1} (von Bodungen et al. 1981), and the entrance to the Gulf of Finland: 0.7–1.9 g C m^{-2} day^{-1} (Lignell et al. 1993). Comparison with estimates for the entire spring bloom period gives similar discrepancies: the simulated range for 2 months in spring at

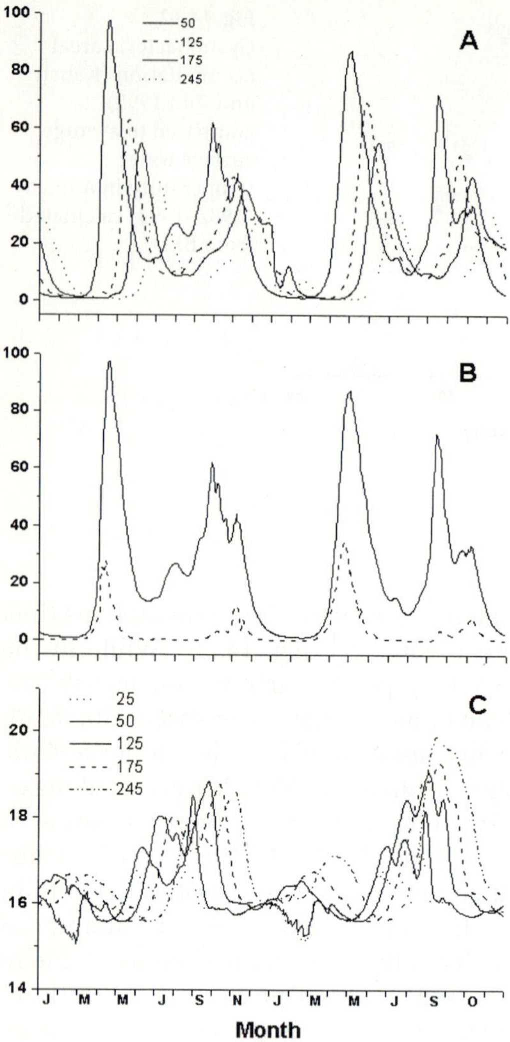

Fig. 14.13. Simulated deposition of organic matter during years 7 and 8: **A** total deposition (mg N m^{-2} day^{-1}) at different depths; **B** total (*solid*) and living algae (*dotted*) deposition (mg N m^{-2} day^{-1}) at 50 m; **C** N:P ratio of detritus deposited at different depths

2.8–3.0 g N m^{-2} (17–18 g C m^{-2}) is close to the 2.5 g N m^{-2} reported for the Northern Baltic (Leppänen et al. 1986) and lower than 32.8 g C m^{-2} at the entrance to the Gulf of Finland (Lignell et al. 1993).

The reasons for these discrepancies are probably explained both by the model parameterizations and with the field measurement techniques. In the model, we are dealing primarily with an internal nutrient cycling, since total inputs (see Fig. 14.3) account for only about 15 and 5% of nitrogen and phosphorus requirements by autotrophs, respectively.

A number of uncertainties are also connected with the sedimentation measurements per se, for instance, trapping efficiency of sediment traps, decomposition of settled material within traps and lateral transport of resuspended material (Wassmann and Heiskanen 1988). Resuspension is shown to be considerable in the Baltic (Blomqvist and Larsson 1994). When some of these effects have been taken into account, the measured and corrected values of 2.2 g N m^{-2} and 18.8 g C m^{-2} for the spring bloom period at the entrance to the Gulf of Finland (Heiskanen and Leppänen 1995) seem to be quite close to the model.

The modeled late summer sedimentation rates of 40–80 mg N m^{-2} day^{-1} (0.25–0.5 g C m^{-2} day^{-1}) are close to those found in the Kiel Bight after summer-autumn blooms (Smetacek 1980) but are definitely higher than 4–6 mg C m^{-2} day^{-1} (Heiskanen and Kononen 1994) or 15–30 mg N m^{-2} day^{-1} (Heiskanen and Leppänen 1995) measured at the entrance to the Gulf of Finland in 1983. So far no account has been made in the model for a seasonal species succession and corresponding changes in the detritus composition. Consequently, the large summer sedimentation in the model is caused both by high mortality rates defined by nitrogen limitation of unified autotroph community and by high sinking velocities of autotrophs and detritus defined by high water temperatures, both parameterizations being invariable during the year.

The simulated time–depth variations in the N:P ratio of sedimenting detritus are at least qualitatively similar to measurements at the entrance to the Gulf of Finland (Heiskanen and Leppänen 1995). Here the N:P molar ratio of suspended particulate material increases from 13–15 in the surface (0–15 m) layer during the spring bloom to 53 in the deeper layer (15–40 m) in summer.

14.4.1.3 Seasonal and Interannual Variations in the Sediment Subsystem

In general, depth distribution of mineralization rates is similar to the sediment content of nitrogen and phosphorus, with the highest rates (25–35 mg N and 13–22 mg P m^{-2} day^{-1}) found at 20–30 m (Fig. 14.14). The seasonal dynamics is caused by variations in both water temperature and sedimentation, with small but distinct year-to-year differences.

The total range of model ammonification rates (10–35 mg N m^{-2} day^{-1}) is quite comparable with those estimated both by Enoksson et al. (1990) for the SE Kattegat (10–40 mg N m^{-2} day^{-1}) and by Lohse et al. (1993) for the North Sea [2–32(167) mg N m^{-2} day^{-1}] but lower than the range (30–150 mg N m^{-2} day^{-1}) compiled by Henriksen and Kemp (1988) for shallow sediments from more eutrophied environments. The annual averaged sediment oxygen consumption of 140–150 g O_2 m^{-2} corresponds rather well with the range of 46–176 g O_2 m^{-2} $year^{-1}$ compiled by Canfield

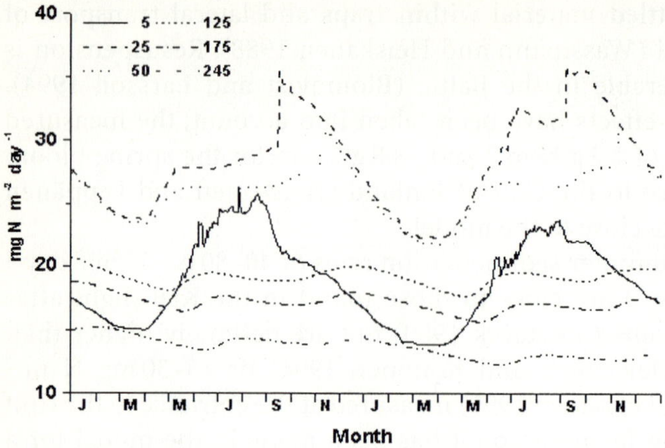

Fig. 14.14. Sediment nitrogen mineralization rates at different depths during years 7 and 8

(1989) for similar environments. Simulated rates of other sediment processes are also comparable with those estimated from field measurements, although they tend to be closer to the lower parts of the reported ranges (Table 14.5).

The simulated subsurface maximum of labile sediment phosphorus and nitrogen (BEN and BEP) is mainly due to the fact that the largest part of the sedimenting particles is trapped and/or mineralized on the vast shallow areas. The model does not account for erosion and resuspension of sediments, which were proposed as major downward transport mechanisms (Jonsson et al. 1990). Unfortunately, we do not have enough proper field observations to draw a definite conclusion as to what degree the simulated accumulation rates are a pure model artifact or to what degree they describe natural distributions.

The comparison with historical data available on depth distributions even for total (combining organic and inorganic, labile and stable) amounts of nitrogen and phosphorus in the Baltic sediments is hampered by the necessity to compare sparse measurements made at specific sites with simulated horizontally integrated distributions. A few data available (Carman and Jonsson 1991) show a similar pattern with the model. Maximum concentrations of both total and, especially, potentially mobile phosphorus were found at intermediate depths, while there is a sharp decrease towards greater depths and accumulation bottoms. The same general features of depth distribution have recently been shown by Carman and Cederwall (Chap. 11). The depth distributions seem to be rather similar to the simulated distributions, if the observations are recalculated into the same units (Table 14.6).

Table 14.5. Rates of sediment processes [mg N (P) (O_2) m^{-2} day^{-1}] simulated by the model and recalculated from literature for similar marine environments

Process	Simulation	Field estimates	Reference
Nitrification	0–37	0–60	Kaplan (1983)
		11–48	Henriksen and Kemp (1988)
		0.7–8.7	Lohse et al. (1993)
Denitrification	0–20	0–34	Hattori (1983)
		17–85	Seitzinger (1988)
		5–188	Billen and Lancelot (1988)
		6–45	Henriksen and Kemp (1988)
		0–112	Koike and Sorensen (1988)
		1.3–32	Enoksson et al. (1990)
		0–2.8	Lohse et al. (1993)
		2.2–3.7	Conley et al. (1996)
Ammonium release	0–8	12–66	Klump and Martens (1983)
		10–78	Billen and Lancelot (1988)
		0.5–10.2	Enoksson et al. (1990)
		0–11.8	Koop et al. (1990)
		0.7–28	Lohse et al. (1993)
		0–170	Cowan and Boynton (1996)
		3.1–11.7	Conley et al. (1996)
Nitrate release	0–21	3–73	Klump and Martens (1983)
		0–58	Billen and Lancelot (1988)
		0.2–2	Enoksson et al. (1990)
		0–7.7	Lohse et al. (1993)
		0–11.8	Cowan and Boynton (1996)
		0–4.2	Conley et al. (1996)
Phosphate release	0.5–8	0.6–2.0	Enoksson et al. (1990)
		0–27.5	Koop et al. (1990)
		0–110.1	Cowan and Boynton (1996)
		0.4–1.0	Hall et al. (1996)
		1.2–21.2	Conley et al. (1996)
Oxygen uptake	180–640	206–435	Enoksson et al. (1990)
		130–922	Koop et al. (1990)
		50–860	Cowan and Boynton (1996)
		355–404	Conley et al. (1996)

The sediment phosphorus in the model has no permanent sink except for a small burial, in contrast to nitrogen, where most is denitrified. Consequently, phosphorus is accumulated in the sediments and may eventually be released into the water column. The excess of phosphorus will further promote N limitation of primary production and positive P feedback in the overall biogeochemical cycle.

Table 14.6. Depth distributions of nitrogen and phosphorus areal concentrations in sediments of the Baltic proper as recalculated from Carman and Cederwall (Chap. 11) and simulated by the model. Data on total nitrogen and phosphorus concentrations (g/g) for the uppermost centimeter of accumulation, transportation, and erosion types of the bottom are recalculated as areal concentrations with corresponding sediment weights. Values for different types of bottoms were further combined as weighted averages with weights given as proportions of bottom types found within corresponding depth intervals. Finally, area-weighted averages for the simulated domain (Baltic proper+Gulf of Finland+Gulf of Riga in the cited paper) were calculated with corresponding area proportions of depth intervals within each water body. All mentioned values are taken from the cited paper

Depth intervals (m)	Nitrogen (mg N m^{-2})	Normalized	Model	Phosphorus (mg P m^{-2})	Normalized	Model
0– 30	12,597	1.02	0.81	3842	0.71	0.86
30– 70	12,326	1.00	1.00	5445	1.00	1.00
70–120	10,889	0.88	0.82	3235	0.59	0.29
120 – bottom	8 378	0.68	0.71	1064	0.20	0.18

14.4.1.4 Annual and Seasonal Budgets of the Entire System

Generalized annual budgets of nitrogen and phosphorus, averaged from the 20-year simulation, are shown in Fig. 14.15. In the model, we are dealing only with the labile (mobile) fractions of organic matter and inorganic nutrients. This prevents a direct comparison of absolute values from field data where total amounts of nitrogen and phosphorus are used. However, it is still possible to analyze both the relative significance of different biogeochemical processes and the coupling between nitrogen and phosphorus. On an annual basis the system is at nearly steady state.

The budgets show that internal cycling between pelagic and benthic subsystems is far more important for phosphorus than for nitrogen. For

Table 14.7. Residence times (years) of nitrogen and phosphorus for the model of the Baltic proper calculated from total amounts (in winter) divided by arithmetic means of sources and sinks

System	N	P
Baltic proper	4.6	87.0
Pelagic subsystem	0.5	1.9
Benthic subsystem	2.3	7.6

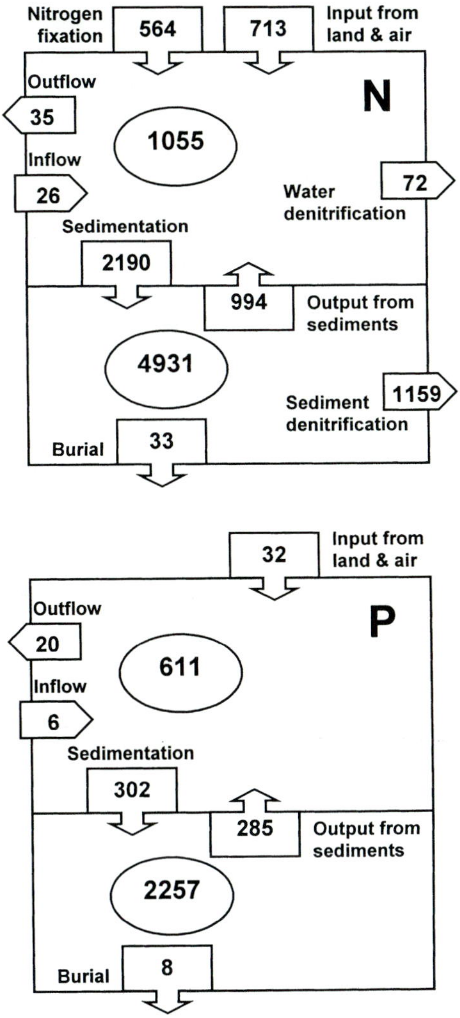

Fig. 14.15. Simulated average annual budgets of nitrogen and phosphorus for the Baltic proper (in thousand tons)

nitrogen, N_2 fixation is almost of the same magnitude as all other external sources. Denitrification is nearly balancing external inputs.

These differences between the nitrogen and phosphorus cycles are reflected in the residence (turnover) times (Table 14.7). On a subsystem scale, both nitrogen and phosphorus are similarly active in a biogeochemical sense, as shown by similar residence times. However, on an overall system scale, phosphorus is much more conservative than nitrogen due to insignificant external sinks. This is the lesson to be taken into account in long-term mass balance studies.

Denitrification and underutilization of phosphorus in the pelagic nitrogen-limited subsystem are the major reasons for the drastic changes in integral N:P ratios. They change from 50 (or even 89 accounting for nitrogen fixation) in the total input to very low values both in pelagic (3.8) and benthic (4.8) subsystems as well as for the overall system (4.6).

14.4.2 Effects of Nutrient Load Reductions

There are many properties of a large-scale ecosystem model that could be evaluated by sensitivity analysis. Here, we have chosen to examine how the model will react to changes in nutrient loads in three different scenarios. All the numerical experiments reported below have been conducted with the same initial conditions and with the same subset of physical forcing functions that were used in the standard case. Consequently, the biogeochemical transport and translocation fluxes are affected by the same physical environment.

To simplify the evaluation, we drastically reduced the inputs of nutrients and prescribed a 50 % reduction in land-based sources each 5 years. This reduction scheme means that in year 20, only 38 % N and 32 % P remains (Fig. 14.16). Consequently, the N:P ratio of inputs has increased from 43 to 53, since we have not reduced the atmospheric deposition.

Effect of Only N Reduction. When only nitrogen load was reduced, nitrogen fixation is increased up to 40–50 % compared to the standard case (Fig. 14.17A). The changes in total primary production are barely noticeable at all.

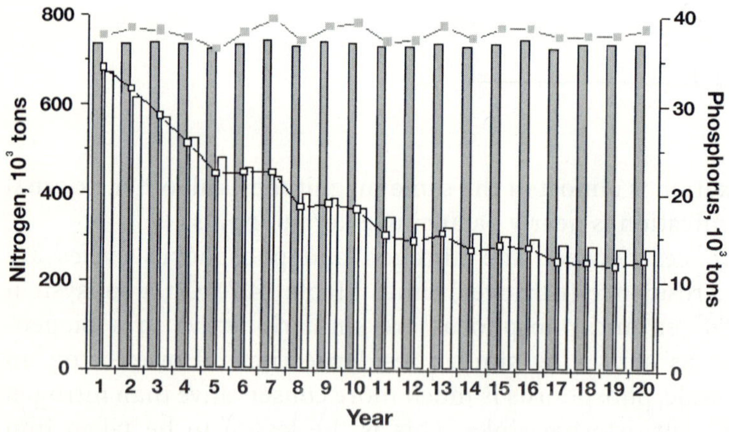

Fig. 14.16. Comparison of annual nitrogen (*columns*) and phosphorus (*curves*) loads between the standard case (*shaded columns* and *squares*) and reduction scenario (*open columns* and *squares*)

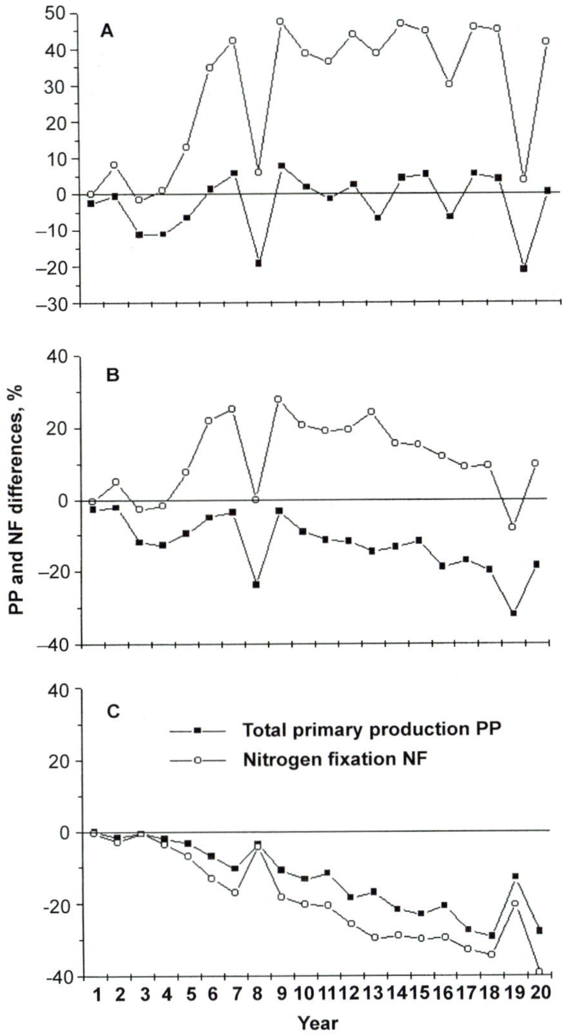

Fig. 14.17. Simulated reactions of the Baltic proper ecosystem following load reductions. Relative differences of primary production (*PP*) and nitrogen fixation (*NF*) between the standard case and N reduction (**A**), "N and P reduction" (**B**), and P reduction (**C**) scenarios

Effect of Simultaneous N and P Reductions. These drastic reductions in loads resulted in a moderate 20–30 % reduction in primary production in 20 years (Fig. 14.17B). In this scenario the general decrease in inorganic nitrogen uptake is also to a certain degree compensated by increased nitrogen fixation and reductions take effect rather slowly.

Effect of Only P Reduction. The resulting long-term decrease in total primary production was approximately the same as in the previous scenario, but nitrogen fixation is significantly lower (Fig. 14.17C).

These experiments confirm our earlier conclusions, based on more simple models (Wulff and Stigebrandt 1989; Savchuk and Volkova 1990), that the recovery of the Baltic Sea to the less eutrophied state would take decades, even with drastic reductions in loads. The argument is that the present productivity level is mainly maintained by the large amounts of nutrients already accumulated in the system rather than by external inputs. According to model estimates from the standard run, the land load supplies only 9 % of nitrogen and 3.5 % of phosphorus utilized annually in primary production.

However, the different reactions of the model to reduction of different nutrients are worth explaining in more detail. For this purpose we will compare the results of the last simulation year when the meteorological and hydrological forcing functions were close to the average conditions and all the reductions had already taken a full effect (Fig. 14.18).

The *reduction of nitrogen load* to this nitrogen-limited and phosphorus-enriched system increases the existing phosphorus surplus (cf. Fig. 14.15). Correspondingly, the nitrogen fixation, dependent in our model both on nitrogen deficit and on phosphate concentration, is also increased. That, in turn, leads to a subsequent increase in regenerated production due to recycling of fixed nitrogen. Thus, the significance of new production decreases even further.

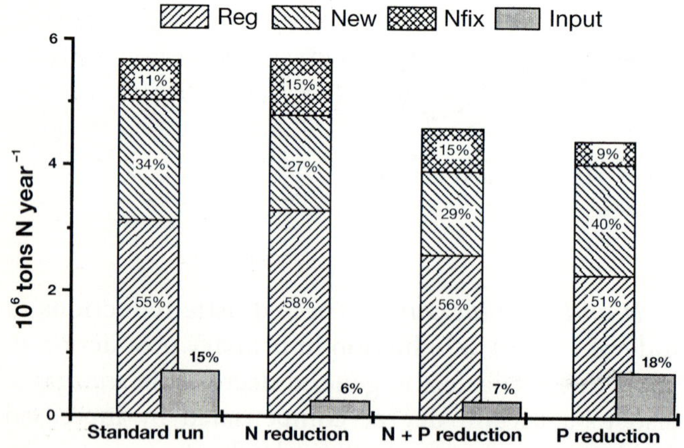

Fig. 14.18. Nitrogen sources of annual primary production in year 20 of the standard run and reduction scenarios, presented as annual amounts and proportions. *Reg* and *New* uptake of ammonium and nitrate, respectively; *Nfix* nitrogen fixation; *Input* sum of all external loads. Note: since nitrogen fixation can also be considered as an external supply, percent values of *Input* were calculated relatively to uptake of dissolved inorganic nutrients (*Reg+New*)

In the case of *phosphorus reduction*, the low inorganic N:P molar ratios (6.2 in March and 3.8 in August as compared to 4.7 and 2.5 in the standard run, respectively) imply that the primary production still remains nitrogen limited even after 20 years of reduced phosphorus load. This is not surprising, taking into account the long residence time of phosphorus in the system (see Table 14.7 above). However, in the P-reduction case, less phosphorus is left underutilized and both the nitrogen fixation and regenerated production are reduced. Decreased contribution of fixed and recycled nitrogen results in increased significance of new production. Thus, the system appears to be phosphorus controlled, although primary production is still nitrogen limited!

The simultaneous load reduction of both nutrients results in an intermediate effect. The N:P ratios were not changed as much as in the phosphorus reduction scenario and nitrogen fixation was rather high but not enough to compensate a decrease in new production due to nitrogen reduction.

The dynamic interaction of the two nitrogen recycling loops, based on nitrogen fixation and on nitrate uptake, is also the mechanism behind the pronounced year-to-year variations demonstrated in Fig. 14.17. Evidently, significant reductions in both nitrogen fixation and regeneration rates in the cold and windy years (cf. Fig. 14.11 and related discussion above) have most pronounced negative effects under nitrogen-reduction scenarios, when the contribution of the "nitrogen fixation–recycling" loop to the biogeochemical cycle is higher. In the only P reduction scenario this loop is to a certain degree suppressed and the significance of "new" production becomes more important. Consequently, the relative increase* of biotic activity in cold years is primarily driven by a "new production–recycling" loop. Primary production, enhanced by nitrate uptake sustains a heterotroph activity that, in turn, would maintain the autotroph community through nutrient regeneration.

14.4 Conclusions

A comparison of simulation results with available data shows that the model is actually capable of reproducing a number of key features of the

* Note, that here we are speaking about relative deviations within each of the scenarios. In the P reduction scenario the absolute values of mentioned fluxes are still lower than in the standard run, while in the N reduction scenario the nitrogen fixation was even slightly higher.

Baltic proper. For instance the model generates a nearly quantitative description of the seasonal, interannual, and long-term dynamics of nutrients and oxygen, a plausible description of the seasonal dynamics of plankton variables and a reasonable imitation of depth distribution of sediment variables, compared to observations. Even more important, the modeled pelagic and benthic biogeochemical fluxes appeared to be quite realistic when compared with the corresponding estimates compiled from the literature.

The residence times of nutrients in the system are useful indicators of the intensity of biogeochemical cycles. However, it is extremely important to make corresponding calculations for the entire system. As our experience shows (Table 14.7), the estimates made only for the water body itself or for the pelagic and benthic subsystems separately might be quite misleading. In the phosphorus case, for instance, the residence time for an overall system is an order-of-magnitude longer than for subsystems due to a much more intensive internal cycling, compared to external exchanges.

The model has also served as a "workbench" where formulations interact with empirical knowledge to test and improve our understanding. In that respect, we have learned some useful lessons.

Even after 100 years of intensive field and experimental studies in the Baltic Sea, the most reliable data to use in the model calibration, verification and validation are still the variations in nutrients and oxygen, and to a lesser extent also biomasses of planktonic variables. There is a clear shortage of data on rates of many important processes and correspondent biogeochemical fluxes. Among existing estimates, the measured rates presented on a per cubic (square) meter and per day (hour) basis seem to be more reliable than extrapolations to the whole water column or for the whole basin made under many different assumptions, sometimes not even clearly stated in the publications. Evidently, more process-oriented and goal-specified studies are needed in order to cover the existing gaps.

During the calibration of the model, we realized even more clearly the existence of a complementary principle. With highly aggregated variables, more complex parameterizations of interactions between them are needed, and vice versa. The splitting into several less aggregated variables would hopefully allow us to avoid some of the more complicated formulations. In the present model, such "over-parameterized" relationships are probably those that describe nutrient uptake and nitrogen fixation by the autotroph community as well as in benthic recycling. However, we found that more detailed model formulations described in the literature and used for other systems could not be supported by experimental or field data from the Baltic, although this region is considered as one of the best-studied marine systems. Therefore, in many cases we decided it would be futile to go into detailed formulations, which could not be supported by data.

It is tempting to speculate how a less aggregated system would affect the system performance, but our experience with this model shows the counter-intuitive behavior of complex systems. Further model analysis, prompting new field and experimental studies using corresponding scales, is the only way to evaluate these hypotheses. However, as Murphy, the Optimist, had once put it: "Anything you try to fix will take longer and cost more than you thought".

Acknowledgements. This project was funded through the BASYS project contract MAS3-96-0058 within the EU-MAST program and through several grants from the Swedish Environment Protection Agency.

References

Anderson NR, Andersen V, Fransz HG, Frost BW, Klepper O, Rassoulzadegan F, Wulff F (1993) Modelling zooplankton. In: Evans GT, Fasham MJR (eds) Towards a model of ocean biogeochemical processes. Springer, Berlin Heidelberg New York, pp 177–192

Barrow EM (1993) Scenarios of climate change for the European Community. Eur J Agron 2:247–260

Berounsky VM, Nixon SW (1985) Eutrophication and the rate of net nitrification in a coastal marine ecosystem. Estuar Coast Shelf Sci 20:773–781

Billen G, Lancelot C (1988) Modelling benthic nitrogen cycling in temperate coastal ecosystems. In: Blackburn TH, Sorensen J (eds) Nitrogen cycling in coastal marine environments. Wiley, New York, pp 341–378

Blomqvist S, Larsson U (1994) Detrital bedrock elements as tracers of settling re-suspended particulate matter in a coastal area of the Baltic Sea. Limnol Oceanogr 39(4):880–896

Brettar I, Rheinheimer G (1991) Denitrification in the central Baltic: evidence for H_2S-oxidation as motor of denitrification at the oxic–anoxic interface. Mar Ecol Prog Ser 77:157–169

Brettar I, Rheinheimer G (1992) Influence of carbon availability on denitrification in the central Baltic Sea. Limnol Oceanogr 37(2):1146–1163

Canfield DE (1989) Sulphate reduction and oxic respiration in marine sediments: implications for organic carbon preservation in euxinic environments. Deep Sea Res 36(1):121–138

Carman R, Jonsson P (1991) Distribution patterns of different forms of phosphorus in some surficial sediments of the Baltic Sea. Chem Geol 90:91–106

Conley DJ, Stockenberg A, Carman R, Johnstone RW, Rahm L, Wulff F (1996) Sediment-water nutrient fluxes in the Gulf of Finland, Baltic Sea. Estuar Coast Shelf Sci 45:591–598

Corner EDS, Davies AG (1971) Plankton as a factor in the nitrogen and phosphorus cycles in the sea. Adv Mar Biol 9:101–204

Cowan JLW, Boynton WR (1996) Sediment-water oxygen and nutrient exchange along the longitudinal axis of Chesapeake Bay: seasonal patterns, controlling factors and ecological significance. Estuaries 19(3):562–580

Davidavichiene LG, Sopauckiene DA (1989) Inputs of nitrogen and phosphorus compounds from atmosphere. In: Davidan IN, Savchuk OP (eds) "Baltica" Project. Problems of studies and mathematical modelling of the Baltic Sea ecosystem, issue 4. Main tendencies of the ecosystem's evolution (in Russian). Gidrometeoizdat, Leningrad, pp 112–118

Elmgren R (1989) Man's impact on the ecosystem of the Baltic: energy flows today and at the turn of the century. Ambio 18(6):326–332

Elser JJ, Hassett RP (1994) A stoichiometric analysis of the zooplankton–phytoplankton interaction in marine and freshwater ecosystems. Nature 370:211–213

Enoksson V (1986) Nitrification rates in the Baltic Sea: comparison of three isotope techniques. Appl Environ Microbiol 51(2):244–250

Enoksson V, Sorensson F, Granéli W (1990) Nitrogen transformations in the Kattegat. Ambio 19(3):159–166

Fonselius SH (1969) Hydrography of the Baltic deep basins III. Fish Bd Swed Ser Hydrogr Rep 23:97

Fonselius SH (1981) Oxygen and hydrogen sulfide conditions in the Baltic Sea. Mar Pollut Bull 12:187–194

Gundersen K (1981) Nitrogen. In: Melvasalo T, Pawlak J, Grasshoff K, Thorell L, Tsiban A (eds) Assessment of the effects of pollution on the natural resources of the Baltic Sea. Baltic Sea Environ Proc 5B. Helsinki Commission, Helsinki, pp 169–191

Hall PO J, Hulth S, Hulthe G, Landen A, Tendberg A (1996) Benthic nutrient fluxes on a basin-wide scale in the Skagerrak (north-eastern North Sea). J Sea Res 35:123–137

Hattori J (1983) Denitrification and dissimilatory nitrate reduction. In: Carpenter EJ, Capone DG (eds) Nitrogen in the marine environment. Academic Press, New York, pp 191–232

Heiskanen A-S, Kononen K (1994) Sedimentation of vernal and late summer phytoplankton communities in the coastal Baltic Sea. Arch Hydrobiol 131(2):175–198

Heiskanen A-S, Leppänen J-M (1995) Estimation of export production in the coastal Baltic Sea: effect of resuspension and microbial decomposition on sedimentation measurements. Hydrobiologia 316:211–224

HELCOM (1989) Deposition of airborne pollutants to the Baltic Sea area: 1983–1985 and 1986. Baltic Sea Environ Proc 32. Helsinki Commission, Helsinki

HELCOM (1990) Second periodic assessment of the state of the marine environment of the Baltic Sea, 1984–1988; background documents. Baltic Sea Environ Proc 35B. Helsinki Commission, Helsinki

HELCOM (1993a) Second Baltic Sea pollution load compilation. Baltic Sea Environ Proc 45. Helsinki Commission, Helsinki

HELCOM (1993b) The Baltic Sea joint comprehensive environmental action programme. Baltic Sea Environ Proc 48. Helsinki Commission, Helsinki

Henriksen K, Kemp WM (1988) Nitrification in estuarine and coastal marine sediments. In: Blackburn TH, Sorensen J (eds) Nitrogen cycling in coastal marine environments. Wiley, New York, pp 207–249

Hübel H, Hübel M (1987) Nitrogen fixation in Baltic coastal waters. In: Proc 4th Symp of the Baltic Marine Biologists, Gdansk, 1975. Morski Instytut Rybacki, Gdynia, pp 27–32

Johansson S (1992) Regulating factors for coastal zooplankton community structure in the northern Baltic proper. PhD Thesis, Department of Zoology, Stockholm University, 30 pp

Johansson S, Larsson U, Skärlund K (1992) Annual dynamics of Baltic Sea coastal zooplankton communities in relation to eutrophication. In: Johansson S (ed) Regulating factors for coastal zooplankton community structure in the northern Baltic proper. PhD Thesis, Department of Zoology, Stockholm University

Jonsson P, Carman R, Wulff F (1990) Laminated sediments in the Baltic – a tool for evaluating nutrient mass balances. Ambio 19(3):152–158

Kahru M, Rud O (1995) Monitoring the decadal-scale variability of cyanobacteria blooms in the Baltic Sea by satellites. In: Proc 3rd Thematic Conference on Remote Sensing for Marine and Coastal Environments, vol 2, pp 76–83

Kahru M, Horstmann U, Rud O (1994) Satellite detection of increased cyanobacteria blooms in the Baltic Sea: natural fluctuation or ecosystem change? Ambio 23(8):469–472

Karlson B, Edler L, Granéli W, Sahlsten E, Kuylenstierna M (1996) Subsurface chlorophyll maxima in the Skagerrak – processes and plankton community structure. J Sea Res 35(1–3):139–158

Kaplan WA (1983) Nitrification. In: Carpenter EJ, Capone DG (eds) Nitrogen in the marine environment. Academic Press, New York, pp 139–190

Klump JV, Martens CV (1983) Benthic nitrogen regeneration. In: Carpenter EJ, Capone DG (eds) Nitrogen in the marine environment. Academic Press, New York, pp 385–457

Koike I, Sorensen J (1988) Nitrate reduction and denitrification in marine sediments. In: Blackburn TH, Sorensen J (eds) Nitrogen cycling in coastal marine environments. Wiley, New York, pp 251–273

Kononen K (1992) Dynamics of the toxic cyanobacterial blooms in the Baltic Sea. Finn Mar Res 261:36

Koop K, Boynton WR, Wulff F, Carman R (1990) Sediment-water oxygen and nutrient exchanges along the depth gradient in the Baltic Sea. Mar Ecol Prog Ser 63:65–77

Larsson U, Hobro R, Wulff F (1986) Dynamics of a phytoplankton spring bloom in a coastal area of the northern Baltic. Contrib Askö Lab Stockh Univ 30:1–32

Leppänen J-M, Kononen K (1988) Cycling of organic matter during the vernal growth period in the open northern Baltic proper. III. Phytoplankton composition and estimation of loss rates of phytoplankton production. Finn Mar Res 255:37–54

Leppänen J-M, Lahdes E, Bruun JE (1986) Cycling of nitrogen during the vernal growth period in the northern Baltic proper in 1982. Ophelia 26:293–303

Leppänen J-M, Niemi Å, Rinne I (1988) Nitrogen fixation of cyanobacteria (blue-green algae) and the nitrogen cycle of the Baltic Sea. Symbiosis 6:181–194

Lignell R, Heiskanen A-S, Kuosa H, Gundersen K, Kuuppo-Leinikki P, Pajuniemi R, Uitto A (1993) Fate of a phytoplankton spring bloom: sedimentation and carbon flow in the planktonic food web in the northern Baltic. Mar Ecol Prog Ser 94:239–252

Lindahl G, Wallström K, Brattberg G (1978) On nitrogen fixation in a coastal area of the northern Baltic. Kieler Meeresforsch Sonderh 4:171–177

Lohse L, Malschaert JFP, Slomp CP, Helder W, van Raaphorst W (1993) Nitrogen cycling in North Sea sediments: interaction of denitrification and nitrification in offshore and coastal areas. Mar Ecol Prog Ser 101:283–296

Mayzaud P, Conover RJ (1988) O:N atomic ratio as a tool to describe zooplankton metabolism. Mar Ecol Prog Ser 45:289–302

Niemistö L, Rinne I, Melvasalo T, Niemi Å (1989) Blue-green algae and their nitrogen fixation in the Baltic Sea in 1980, 1982 and 1984. Meri 17:1–59

Rinne I, Melvasalo T, Niemi Å, Niemistö L (1978) Nitrogen fixation by blue-green algae in the Baltic Sea. Kieler Meeresforsch Sonderh 4:178–187

Rinne I, Melvasalo T, Niemi Å, Niemistö L (1981) Studies on nitrogen fixation in the Gulf of Bothnia. Finn Mar Re 248:117–127

Rotmans J, Hulme M, Downing TE (1994) Climate change implications for Europe. An application of the ESCAPE Model. Global Environ Change 4(2):97–124
Rönner U, Sörensson F (1985) Denitrification rates in the low-oxygen waters of the stratified Baltic proper. Appl Environ Microbiol 50:801–806
Savchuk O (1986) The study of the Baltic Sea eutrophication problems with the aid of simulation models. In: Proc Baltic Sea Monitoring Symp, Tallinn. Baltic Sea Environ Proc 19, Helsinki Commission, Helsinki, pp 51–61
Savchuk OP, Volkova VV (1990) Forecasting the future state of the Baltic Sea eutrophication. In: Proc 17th Conf of Baltic Oceanographers, Norrköping
Savchuk OP, Wulff F (1993) Biogeochemical transformations of nitrogen and phosphorus – a pelagic submodel for the Baltic Sea. Contrib Systems Ecol Stockh Univ 1:50
Savchuk O, Wulff F (1996) Biogeochemical transformations of nitrogen and phosphorus in the marine environment – coupling hydrodynamic and biogeochemical processes in models for the Baltic proper. Contrib Systems Ecol Stock Univ 2:79
Savchuk OP, Kolodochka AA, Gutsabbath ES (1988) Simulation of the matter cycle in the Baltic Sea ecosystem. In: Proc 16th Conf of Baltic Oceanographers, Kiel, pp 921–931
Savchuk OP, Arsenchuk MO, Shpaer IS (1994) Nutrients. In: Terzijev FS, Rozhkov VA, Rimsh EY, Spaer IS (eds) Project "The seas". Hydrometeorology and hydrochemistry of the seas, vol III. The Baltic Sea, issue 2. Hydrochemical conditions and oceanological base of biological productivity (in Russian). Hydrometeoizdat, St Petersburg, pp 307–372
Schulz S (1985) Results of ecological time-series investigation in spring and summer in the Arkona Sea. In: Symp on Ecological Investigations of the Baltic Sea Environment (under the auspices of HELCOM), March 1983, Riga. Valtion painatuskeskus, Helsinki, pp 161–175
Schulz S, Breuel G (1984) On the variability of some biological parameters in the summer pelagic system of the Arkona Sea. Limnologica 15:365–370
Seitzinger SP (1988) Denitrification in freshwater and coastal ecosystems: ecological and geochemical significance. Limnol Oceanogr 33(4/2):702–724
Smetacek V (1980) Annual cycle of sedimentation in relation to plankton ecology in western Kiel Bight. Ophelia Suppl 1:65–76
Smetacek V(1984) The supply of food to the benthos. In: Fasham MJR (ed) Flows of energy and materials in marine ecosystems. Plenum, New York, pp 517–547
Sommer U (1992) Phosphorus-limited *Daphnia*: intraspecific facilitation instead of competition. Limnol Oceanogr 37(5):996–973
Sterner RW (1990) The ratio of nitrogen to phosphorus resupplied by herbivores: zooplankton and the algal competitive arena. Am Nat 136(2):209–229
Stigebrandt A (1985) A model for the seasonal pycnocline in rotating systems with application to the Baltic proper. J Phys Oceanogr 15(11):1392–1404
Stigebrandt A (1987) A model for the circulation of the Baltic deep water. J Phys Oceanogr 17(10):1772–1785
Stigebrandt A, Wulff F (1987) A model for the dynamics of nutrients and oxygen in the Baltic proper. J Mar Res 45:729–759
Totterdell IJ, Armstrong RA, Drange H, Parslow JS, Powell TM, Taylor AH (1993) Trophic resolution. In: Evans GT, Fasham MJR (eds) Towards a model of ocean biogeochemical processes. Springer, Berlin Heidelberg New York, pp 71–92
Viitasalo M (1994) Seasonal succession and long-term changes of mesozooplankton in the Northern Baltic Sea. Finn Mar Res 263:39
von Bodungen B, von Bröckel K, Smetacek V, Zeitschel B (1981) Growth and sedimentation of the phytoplankton spring bloom in the Bornholm Sea (Baltic Sea). Kieler Meeresforsch Sonderh 5:49–60

Wassmann P, Heiskanen A-S (eds) (1988) Sediment trap studies in the Nordic countries. Proc of a Worksh at Tvärminne Zoological Station, Feb, 207 pp

Wassmann P, Peinert R, Smetacek V (1991) Patterns of production and sedimentation in the boreal and polar northeast Atlantic. Polar Res 10(1):209–228

Wiebe PH (1988) Functional regression equations for zooplankton displacement volume, wet weight, dry weight, and carbon: a correction. Fish Bull 86(4):833–835

Wulff F, Cederwall H (1984) Dynamics of a *Nodularia* bloom in the Baltic proper. ICES CM/B 4:11

Wulff F, Rahm L (1988) Long-term, seasonal and spatial variations of nitrogen, phosphorus and silicate in the Baltic. An overview. Mar Environ Res 26:19–37

Wulff F, Stigebrandt A (1989) A time-dependent budget model for nutrients in the Baltic Sea. Global Biogeochem Cycles 3(1):63–78

Wulff F, Aertebjerg G, Nicolaus G, Niemi A, Ciszewska P, Schulz S, Kaiser W (1986) The changing pelagic ecosystem of the Baltic Sea. Ophelia Suppl 4:299–319

Wulff F, Pertillä M, Rahm L (1994a) Monitoring, mass balance calculation of nutrients and the future of the Gulf of Bothnia. Ambio Spec Rep 8:28–35

Wulff F, Rahm L, Rodriguez-Medina M (1994b) Long-term and regional variations of nutrients in the Baltic Sea; 1972–1991. Finn Mar Res 262:35–50

Yurkovskis AK (1983) Nutrients. In: Davidan IN, Savchuk OP, Sustavov Yu V (eds) "Baltica" project. The problems of research and mathematical modelling of the Baltic Sea ecosystem. Issue 1. The ecosystem and its components (in Russian). Hydrometeoizdat, Leningrad, pp 113–133

15 A Multicompartmental, Multi-Basin Fugacity Model Describing the Fate of PCBs in the Baltic Sea

F. WANIA, D. BROMAN, J. AXELMAN, C. NÄF, and C. AGRELL

15.1 Introduction

Persistent organic pollutants (POPs) such as polycyclic aromatic hydrocarbons and organochlorinated compounds reach lakes, estuaries and the oceans through direct discharge, riverine inflow or atmospheric deposition. These generally quite hydrophobic chemicals have a low affinity for the aqueous phase and tend to partition into the organic phases of suspended solids and bottom sediments or the lipid fraction of aquatic biota. Some POPs show a potential to biomagnify and the highest concentrations are often found in the upper trophic levels of aquatic food chains.

The fate of POPs in an aquatic environment is governed by a range of processes such as partitioning between water and suspended solids, transport across the interfaces atmosphere/water and water/sediment, mixing within the water phase and degradation. Multimedia compartmental mass balance models have been shown to be valuable tools for describing, understanding and quantifying this complex and intertwined set of processes (Mackay et al. 1983). Most of the fate modeling of POPs in large water bodies has until now been focused on the Great Lakes of North America (e.g. Mackay 1989 and references therein; Jeremiasen et al. 1994).

The aquatic environment of the Baltic Sea is heavily contaminated with POPs. The presence of DDT and PCBs in marine animals from the Baltic was established as early as 1969 (Jensen et al. 1969) and a myriad of other POPs has been detected since then (Jansson et al. 1993; Järnberg et al. 1993; various references in Johnels 1992). High levels of chlorinated organic compounds have been linked with the decline of seal populations in the Baltic Sea (Addison 1989; Olsson et al. 1994). Concentration and flux data exist for numerous POPs in a whole range of environmental media in the Baltic Sea region. In recent years, measurements have been reported in the atmosphere (e.g. Broman et al. 1991a; see also Chap. 6), water and sus-

Ecological Studies, Vol. 148
F. Wulff et al. (eds.)
A Systems Analysis of the Baltic Sea

pended particulate matter (Broman et al. 1989, 1991b, 1992, 1996; Näf et al. 1992; Schulz-Bull et al. 1994), sediments (Østfeldt et al. 1994; see also Chap. 10), and biota (e.g. Haahti and Perttilä 1988; Jansson et al. 1993; Järnberg et al. 1993; Bignert et al. 1995).

Despite the observation of intense contamination with POPs in the Baltic through the chemical analysis of numerous environmental samples and even though the Helsinki Commission concluded in its first assessment of the pollution of the Baltic Sea that "it would be valuable to construct a model of the flow of halogenated hydrocarbons in the Baltic ecosystem and to determine the budget of these compounds in the Baltic Sea" (Helsinki Commission 1980), it has so far, to our knowledge, not been attempted to describe the fate and behavior of POPs in the Baltic Sea employing a multimedia modeling approach. A notable exception is the recent attempt to describe the mass balance of chlorinated pulp mill effluents by modeling the fate of extractable organic chlorine EOCl (Wulff et al. 1993).

One reason for this may be that the Baltic is a large and very heterogeneous system and therefore not easily amenable to a compartmental mass balance modeling approach, which is essentially based on the assumption of homogeneous environmental characteristics and chemical concentrations. The Baltic Sea has been labeled a prime example of ecosystem variability, which is characterized by a number of physical, chemical and biological gradients (Leppäkoski and Bonsdorff 1989). One major gradient is quite obviously climatic: the Baltic Sea extends over 12 latitudinal degrees and southern and northern portions experience widely different conditions regarding temperature, sea-ice cover and hydrological water balance. Through its connection with the North Sea the Baltic Sea is really a huge and, for its size, relatively shallow estuary and salinity decreases with increasing distance from the Danish Straits. Within the Baltic proper there is additionally a strong vertical gradient of salinity, temperature and chemical composition. Physical and chemical gradients cause similar gradients of biological characteristics such as species composition and biological productivity.

Another reason may be that the Baltic is surrounded by no less than nine countries (Sweden, Finland, Russia, Estonia, Latvia, Lithuania, Poland, Germany, Denmark) and for several decades had been divided by rather tight political boundaries. There has been no common jurisdiction to coordinate the investigation and management of marine contamination in this area and data consistency and availability may have been poor.

In this chapter, we introduce B-POP, a comprehensive steady-state mass balance model aiming to describe the behavior of POPs in the aquatic system of the Baltic Sea environment. A description of the model structure will be followed by results of calculations simulating the fate of PCBs in the

Baltic Sea in the late 1980s and by an outlook on future challenges of POP fate modeling in this heavily contaminated marine area.

15.2 Model Structure

The concept and formulation of the B-POP model builds upon a previously described model of the long-term behavior of PCBs in Lake Ontario (Mackay 1989), which in itself is a modified version of the generic lake model QWASI (Mackay et al. 1983). The description of the model in this chapter will be kept brief and mostly restricted to those features which deviate from these previously described models. Some familiarity with the basics of fugacity modeling and the respective terminology (Mackay 1991) may be required.

Very briefly, mass-balance models based on the fugacity approach comprise a number of compartments, which are assumed to have homogeneous environmental conditions and chemical concentrations. Expressions are defined for the capacity of each compartment to hold a chemical [Z-values in units of $mol/(m^3 \cdot Pa)$] and expressions describing transformation and intercompartmental transport processes [D-values in units of $mol/(h \cdot Pa)$]. Mass-balance equations for all compartments are formulated and solved for the chemical fugacity (f in units of Pa) in each compartment. Concentrations (in mol/m^3) are then calculated as products of Z and f, transport fluxes and transformation rates (in mol/h) as products of D and f.

Adopting a previously presented approach for modeling the fate of POPs in a large water body, which accounts for the heterogeneity of the environment (Diamond et al. 1994), several sub-basins with distinct characteristics are identified within the Baltic Sea environment, described with a specific set of environmental parameters and interconnected through the formulation of water-transport terms. These sub-basins are (1) Bothnian Bay, (2) Bothnian Sea, (3) Gulf of Finland, (4) Gulf of Riga, (5) Baltic proper, (6) Danish Straits, and (7) Kattegat (Helsinki Commission 1986). The Kattegat is in exchange with the North Sea, which is not a model compartment, but has been assigned the index 8. Due to its larger depth and pronounced vertical stratification, the Baltic proper sub-basin is additionally divided into a surface (index 5) and a bottom section (index 9). Each of these eight basins is represented in the B-POP model by a water and a surface sediment compartment, which are considered well mixed and homogeneous in terms of environmental properties and chemical contamination. The water compartments of these sub-basins are connected through advective mixing processes as shown in Fig. 1.

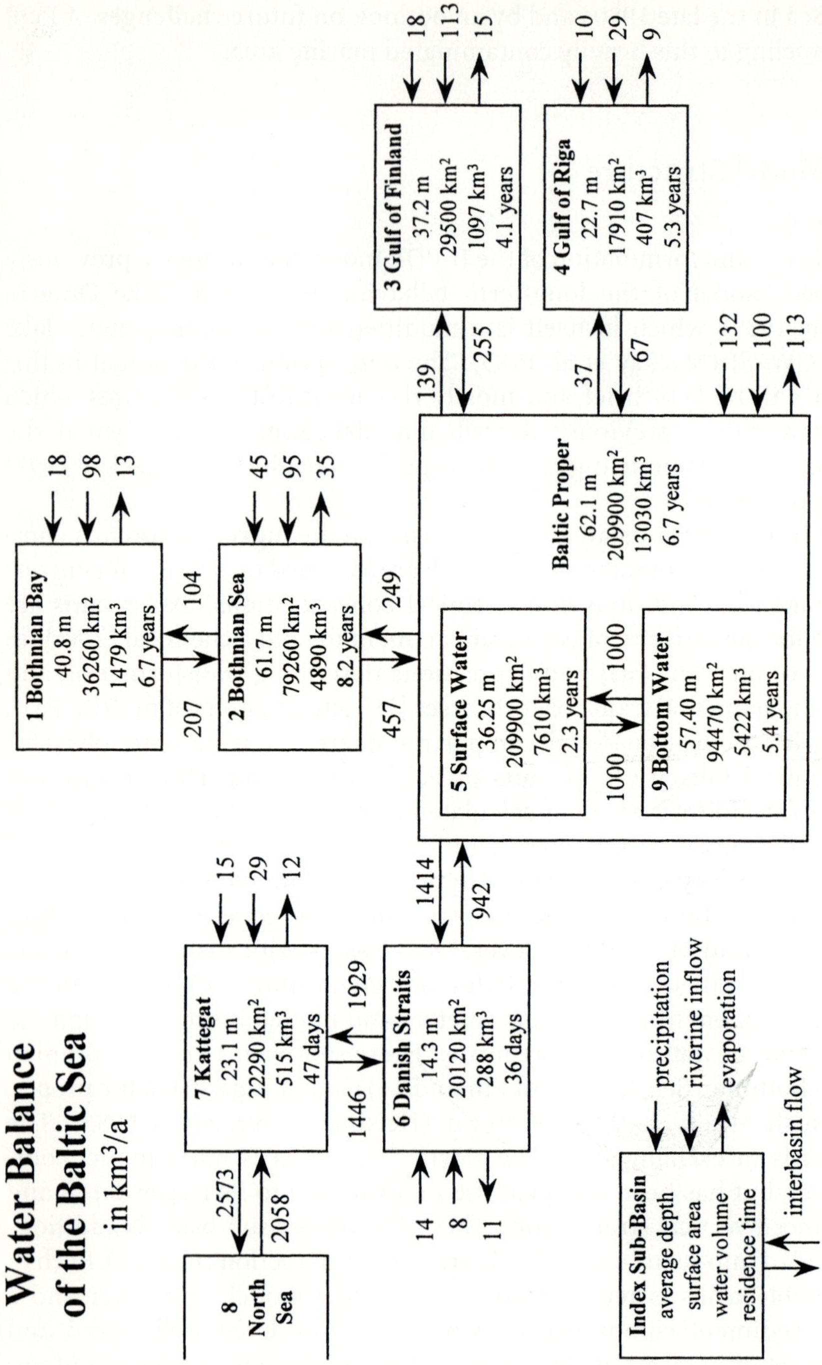

Fig. 15.1. Water balance of the Baltic Sea and its sub-basins based on a study by the Helsinki Commission (1986)

It is assumed that in both water and sediment the POPs are either present in truly dissolved form or attached to particulate organic carbon (POC). There are indications that a considerable fraction of POPs in the water column and in sediment pore water is associated with dissolved and colloidal organic carbon DOC (Brannon et al. 1995). A DOC-bound PCB fraction also occurs in the Baltic Sea (Chap. 10). At the present stage it was considered premature to include partitioning to DOC in the model because crucial information is still missing. Partition coefficients K_{DOC} between water and DOC are very uncertain and seem to vary considerably, depending on the nature of the organic material. Thus K_{DOC} values which are specific for the conditions in the Baltic Sea are required. There is currently also little information available on the concentrations and the dynamics of DOC in the Baltic Sea.

To describe the behavior of an organic contaminant in the aquatic system of the Baltic Sea with the present model it is therefore necessary to establish three mass balances for each sub-basin: I. for water, II. for POC, and III. for the contaminant. These mass balances are interdependent, i.e., the transport of POC will depend on the transport of water, and the fate of the contaminant, in turn, depends on both POC and water movements. Assuming a steady state, all the inputs of water, organic carbon and chemicals to a sub-basin have to match all the outflows and losses.

15.2.1 Mass Balance for Water

A detailed study of the water balance of the Baltic Sea established for all seven sub-basins the mean water depth, water surface area, average precipitation input, evaporative loss and river water inflow as well as the freshwater exchange rates between them (Helsinki Commission 1986). The total water flows between sub-basins can be derived from the freshwater flows using average salinity data in the various sub-basins or, if available, from measured values (e.g., the water exchange through the Danish Straits; Helsinki Commission 1986; see Chap. 8). The average depth of the surface water compartment in the Baltic proper is 36 m, which corresponds to a halocline depth of 50 m (Kullenberg and Jacobsen 1981), and the average vertical water exchange is estimated to be 1000 km^3/a (Hinrichssen, Stockholm University, pers. comm.). Figure 1 summarizes all steady-state water balance parameters including the average water residence times that can be estimated from these flow rates.

15.2.2 Mass Balance for Particulate Organic Carbon

Previous fugacity lake models (Mackay et al. 1983; Mackay 1989) generally included the sorption or partitioning of organic contaminants to solids which are suspended in the water column as well as contaminant transport in connection with the deposition and resuspension of such particles. For hydrophobic organic chemicals the association with the organic matter fraction of the suspended solids is dominant, while sorption to the mineral component tends to be negligible. To simplify model structure and to reduce the number of required input parameters, the B-POP model therefore describes particle-association and particle-mediated transport on the basis of particulate organic carbon (POC) instead of suspended solids.

The rates of mineralization, sedimentation, resuspension and burial of POC are of eminent importance for the fate of POPs in the Baltic Sea. To assure the selection of a consistent set of parameters describing these processes, the B-POP model derives a complete POC mass balance for each sub-basin (Fig. 2). For sub-basins without vertical subdivision the calculation of the POC balance is described in detail in Table 1, giving input parameters and equations. Because of the exchange of POC between surface and bottom water, the calculation for the Baltic proper is more complex and summarized separately in Table 2.

The selected input values listed in Table 3 are based on literature values, personal communications and reasonable guesses. Figure 3 shows the resulting steady-state mass balance for POC in the Baltic Sea as a whole. The

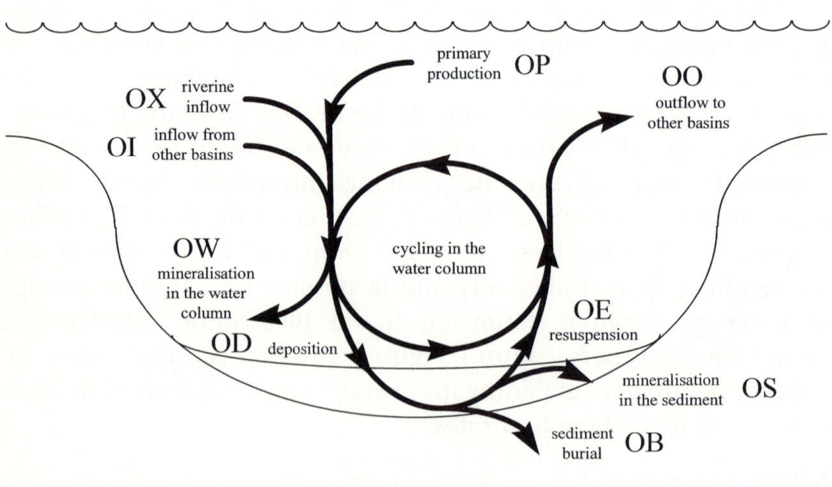

Fig. 15.2. Particulate organic carbon transport fluxes and mineralization rates calculated for each sub-basin of the Baltic Sea

Table 15.1. Calculation of POC mass balance for sub-basins 1 to 4, 6 and 7 (see Appendix)

Input parameters for each sub-basin i
AW_i, GX_i, $G_{i,j}$, $G_{j,i}$ (from water balance, Fig. 1)
$CWpoc_i$, $CXpoc_i$, BP_i, $facOW_i$, $facOE_i$, $facOS_i$

Calculation of POC fluxes and rates in g C/h for each sub-basin i

Autochthonous production	$OP_i = BP_i \cdot AW_i$
inflow from other basins j	$OI_i = \sum(G_{j,i} \cdot CWpoc_j)$
Outflow to other basins j	$OO_i = \sum(G_{i,j} \cdot CWpoc_i)$
Inflow with rivers	$OX_i = GX_i \cdot CXpoc_i$
Mineralization in water column	$OW_i = facOW_i \cdot (OI_i + OP_i + OX_i)$
Resuspension from the sediment	$OE_i = \dfrac{(OI_i + OP_i + OX_i - OO_i - OW_i)}{(1/facOE_i - 1)}$
Deposition to the sediment	$OD_i = \dfrac{OE_i}{facOE_i}$
Mineralization in surface sediment	$OS_i = facOS_i \cdot (OD_i - OE_i)$
Burial to deeper sediment layers	$OB_i = OI_i + OP_i + OX_i - OO_i - OW_i - OS_i$

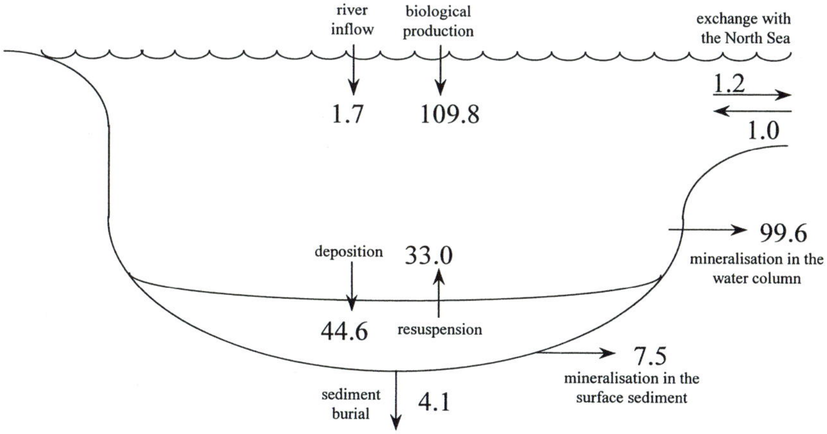

Fig. 15.3. Weighted average steady-state mass balances of particulate organic carbon in the Baltic Sea (including Danish Straits and Kattegat) as defined in the B-POP model. Fluxes are given in units of g POC per m^2 and year

Table 15.2. Calculation of POC mass balance for the Baltic proper (see Appendix)

Input parameters

AW_5, AW_9, GX_5, $G_{5,j}$, $G_{j,5}$ (from water balance, Fig. 1)

$CWpoc_5$, $CWpoc_9$, $CXpoc_5$, BP_5, $facOW_5$, $facOW_9$, $facOE_5$, $facOE_9$, $facOS_5$, $facOS_9$

Calculated as in Table 1

OP_5, OI_5, OO_5, OX_5, OW_5

POC transport from bottom water to surface water

$O_{9,5}=G_{9,5} \cdot CWpoc_i$

Ten unknown parameters

OD_5, OD_9, OE_5, OE_9, OB_5, OB_9, $O_{5,9}$, OS_5, OS_9, OW_9

Ten equations

The following set of equations is solved simultaneously using a numerical method

$OI_5+OP_5+OX_5+O_{9,5}+OE_5=OO_5+OW_5+OD_5+O_{5,9}$ [a]

$O_{9,5}+OE_9=OD_9+OW_9+O_{5,9}$ [a]

$OD_5=OE_5+OB_5+OS_5$ [b]

$OD_9=OE_9+OB_9+OS_9$ [b]

$OE_5=facOE_5 \cdot OD_5$

$OE_9=facOE_9 \cdot OD_9$

$OS_5=facOS_5 \cdot (OD_5-OE_5)$

$OS_9=facOS_9 \cdot (OD_9-OE_5)$

$OW_9=facOW_9 \cdot O_{5,9}$

$OD_5 \cdot AW_9=(O_{5,9}-G_{5,9} \cdot CWpoc_5) \cdot (AW_5-AW_9)$ [c]

---

[a] Mass balance for water compartment

[b] Mass balance for sediment compartment

[c] This equation is based on the assumption that, on an area basis, the POC settling rate to the shallow sediments and to the bottom water is the same

POC balance reveals that autochthonous production by primary producers [110 g C/(m^2·a)] is by far the most important source of POC in the Baltic Sea, riverine inflow accounting only for a small fraction [2 g C/(m^2·a)]. Most of the POC [100 g C/(m^2·a)] is mineralized before it settles to the surface sediment, where further decomposition takes place [8 g/(m^2·a)]. Only a minor fraction [4 g/(m^2·a)] is buried in deeper sediment layers. There is a fairly intense recycling of POC occurring across the sediment-water interface. Figure 4 shows the respective mass balance for the Baltic

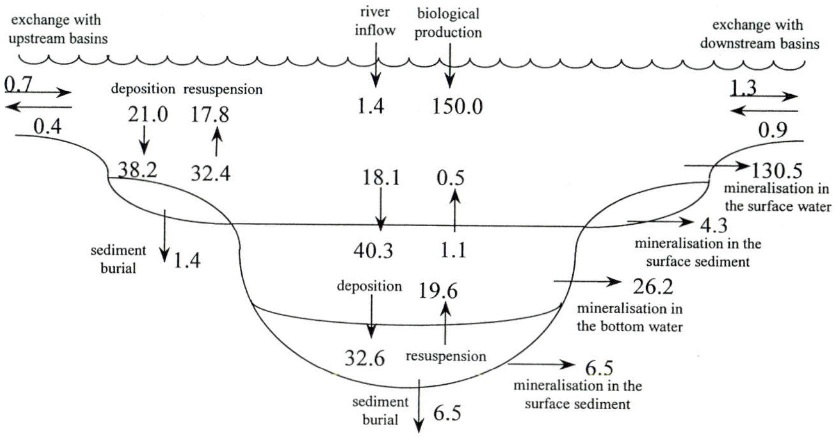

Fig. 15.4. Steady-state mass balance of particulate organic carbon in the Baltic proper as defined in the B-POP model. Fluxes are given in units of g POC per m^2 and year

Table 15.3. Selected input parameters describing concentrations and fate of particulate organic carbon in the Baltic Sea (see Appendix)

	Bothnian Bay	Bothnian Sea	Gulf of Finland	Gulf of Riga	Baltic Prp. (S)	Baltic Prp. (B)	Danish Straits	Kattegat
ORGS[a], g/g	0.05	0.05	0.05	0.05	0.05	0.08	0.05	0.05
CXpoc[b], µg/l	1000	1000	750	2500	3000	n.a.	1000	1000
Cwpoc[c], µg/l	200	200	200	200	200	100	200	200
BP[d], g C/(m^2·a)	20	55	75	85	150	n.a.	125	125
facOW[e]	0.65	0.70	0.80	0.75	0.85	0.65	0.70	0.70
facOS[e]	0.60	0.65	0.70	0.75	0.75	0.50	0.75	0.75
facOE[f]	0.75	0.70	0.70	0.80	0.85	0.60	0.75	0.75

[a] Based on data from Jonsson and Carman (1994), Østfeldt et al. (1994)

[b] Estimated using data on the annual mean inflow of river water and suspended material (Helsinki Commission 1986) and assuming an OC content of the suspended material of approximately 0.1 g C/g.

[c] U. Larsson, Stockholm University (pers. comm.)

[d] Based on data from Lassig et al. (1978).

[e] Based on typical rates of mineralization in the aquatic environment (e.g., Wetzel 1983; Valiela 1984; Larsson and Nellbring 1990; Baker et al. 1991), assuming reduced rates of mineralization in the northern basins and in the bottom water of the Baltic proper, due to low temperatures and partially anoxic conditions, respectively

[f] Based on reports of intense benthic recycling of POC (e.g., Baker et al. 1991; Blomqvist and Larsson 1994); assuming higher resuspension intensity in shallow sub-basins

Table 15.4. Key parameters describing the dynamics of organic carbon in the Baltic Sea as given by Elmgren (1984) and Jonsson and Carman (1994), compared with the weighted average of the values used in the model

	Elmgren	Jonsson and Carman	This model
Region	Baltic Sea inside Danish Straits	Northern Baltic proper	Baltic Sea plus Danish waters
Rates in g C/($m^2 \cdot a$)			
Primary production[a]	139	165	110
Allochthonous input[b]	10		2
"sedimentation"[c]	48	63	45
"sequestration"[d]	1.6	13	11.5
Concentrations in µg C/l			
River water	11,500		750 to 3000
Seawater	4 500		100 to 200

[a] Planktonic and benthic primary production
[b] River inflow, direct discharge, and inflow from other sub-basins
[c] Gross settling flux from water to sediment
[d] Mineralization within surface sediment and burial

proper sub-basin. The shallow coastal areas experience strong resuspension and only minor burial, while the deep-water bottoms have less intense resuspension and a fairly high burial rate.

The description of the dynamics of POC in the model thus reproduces the general picture of energy flow in the Baltic Sea as outlined by Elmgren (1984, 1989) and others (e.g., Jonsson and Carman 1994; Table 4). When comparing the figures in Table 4, it is important to realize that they refer to different areas and possibly different definitions of organic carbon. For instance, the model presented here only considers particulate organic carbon, which may explain the lower carbon concentrations in sea and river water, and accordingly the lower riverine carbon inflow.

15.2.3 Mass Balance for the Persistent Organic Pollutant

15.2.3.1 Phase Capacities Z

In the atmosphere the POPs occur in the vapor phase and attached to aerosol particles. In aqueous phases, i.e., in seawater, river water and surface sediment, the POPs are present either in truly dissolved form or attached to POC. Between these subphases instant equilibrium partition-

Table 15.5. Calculation of Z-values in units of mol/($m^3 \cdot$ Pa) (see Appendix)

Input parameters	
TA_i, P_i, H_i, K_{OW}, $ORGS_i$, DENoc, DENM, $CWpoc_i$, $CXpoc_i$, VAQ_i, VSW_i	
Auxiliary parameters	
$Kpoc=0.41 \cdot K_{OW}$	$DENS_i = \dfrac{1}{\dfrac{1-ORGS_i}{DENM}+\dfrac{ORGS_i}{DENoc}}$
Pure phases	
Water	$ZW_i=1/H_i$
POC	$Zpoc_i=ZW_i \cdot Kpoc$
Sediment particles	$ZS_i=ZW_i \cdot DENS_i \cdot ORGS_i \cdot Kpoc/1000000$
Air	$ZA_i=1/R \cdot TA_i$
Aerosols	$ZQ_i=ZA_i \cdot 6000000/P_i$
Bulk phases	
Bulk water	$BZW_i=ZW_i+(CWpoc_i/DENoc) \cdot Zpoc_i$
Bulk river water	$BZX_i=ZW_i+(CXpoc_i/DENoc) \cdot Zpoc_i$
Bulk sediment	$BZS_i=VSW_i \cdot ZW_i+(1-VSW_i) \cdot ZS_i$
Bulk air	$BZA_i=VAQ_i \cdot ZQ_i+(1-VAQ_i) \cdot ZA_i$

ing is assumed to occur. Table 5 summarizes the calculation of capacity values Z for each subcompartment and bulk phase.

15.2.3.2 Process Terms D

In the atmosphere the aerosol-attached fraction of the contaminant is subject to dry-particle deposition and wet-particle scavenging. POP vapors can diffuse into precipitation and directly into the seawater, whereby rainwater is assumed to be in chemical equilibrium with the vapor phase.

In seawater the truly dissolved fraction can volatilize and diffuse into the surface sediment, while the POC-associated fraction is subject to particle settling and resuspension. Both fractions are subject to degradation processes and advective transport of the water phase. The truly dissolved POPs in sediment pore water can diffuse in the overlying water column, while the POC-attached fraction can be resuspended or buried in deeper sediment layers. Again, both fractions are subject to degradation. Table 6 summarizes the derivation of rate parameters D for these processes.

Table 15.6. Calculation of D-values in units of mol/(h·Pa) (see Appendix)

Input parameters	
UA_i, UW_i, UQ_i, UT_i, SVG_i, HLW_i, HLS_i	
AW_i, AS_i, VW_i, VS_i, GX_i, GM_i, $G_{i,j}$ (from water balance, Fig. 1)	
OD_i, OE_i, OB_i, $O_{5,9}$ (from POC balance, Tables 1 and 2)	
ZW_i, $Zpoc_i$, ZA_i, ZQ_i, BZW_i, BZX_i, BZA_i, BZS_i (see Table 4)	
Riverine Inflow	
Riverine inflow	$DX_i=GX_i\cdot BZX_i$
Interbasin Transport	
Horizontal flow	$D_{i,j}=G_{i,j}\cdot BZW_i$
Vertical upward flow	$D_{9,5}=G_{9,5}\cdot BZW_9$
Vertical downward flow	$D_{5,9}=G_{5,9}\cdot ZW_5+O_{5,9}\cdot Zpoc_5$
Air–water exchange	
Wet deposition	$DM_i=GM_i\cdot (ZW_i+SVG_i\cdot VAQ_i\cdot ZQ_i)$
Dry particle deposition	$DQ_i=AW_i\cdot UQ_i\cdot VAQ_i\cdot ZQ_i$
Diffusive gas exchange	$DV_i=1\Big/\left(\frac{1}{AW_i\cdot UA_i\cdot ZA_i}+\frac{1}{AW_i\cdot UW_i\cdot ZW_i}\right)$
Water–sediment exchange	
Sedimentation	$DD_i=OD_i\cdot Zpoc_i/DENoc$
Resuspension	$DR_i=OE_i\cdot Zpoc_i/DENoc$
Sediment burial	$DB_i=OB_i\cdot Zpoc_i/DENoc$
Water-sediment diffusion	$DT_i=AS_i\cdot UT_i\cdot ZW_i$
Degradation	
In the water column	$DW_i=VW_i\cdot BZW_i\cdot ln2/HLW_i$
In the surface sediment	$DS_i=VS_i\cdot BZS_i\cdot ln2/HLS_i$

15.2.3.3 Mass Balance Equations

Employing these Z and D-values, two mass-balance equations can be set up for each basin i (the index j refers to all basins adjacent to basin i), for the water and the surface sediment. In its current form the B-POP model is assuming a steady-state situation, i.e., the sum of all inputs of contaminant to a compartment matches the sum of the outputs.

The generic mass balances for the water and sediment compartment of a sub-basin *i* read

$$EW_i+DX_i\cdot FX_i+(DM_i+DQ_i+DV_i)\cdot FA_i+(DR_i+DT_i)\cdot FS_i+\sum_j(D_{ji}\cdot FW_j)=$$
$$=(DW_i+DV_i+DT_i+DD_i+\sum_j D_{ij})\cdot FW_i$$

$$FW_i\cdot (DT_i+DD_i)=FS_i\cdot(DR_i+DT_i+DS_i+DB_i)$$

The solution is greatly facilitated by deriving a D-value describing the net flow from the water column to the sediment. This makes it possible to eliminate the sediment fugacity FS from the water mass balance.

$$Dsed_i = \frac{(DT_i + DD_i)\cdot(DB_i + DS_i)}{DB_i + DS_i + DR_i + DT_i}$$

By summarizing all the inputs with the exception of the inflow from neighboring compartments in a parameter I

$$I_i = EW_i + DX_i \cdot FX_i + (DM_i + DQ_i + DV_i) \cdot FA_i \qquad \text{(special case: } I_9 = 0\text{)}$$

and all D-values describing loss from the water column in a total D-value Dtot

$$Dtot_i = Dsed_i + DW_i + DV_i + \sum_j D_{ij}$$

the generic mass balance for the water compartment of sub-basin i becomes

$$I_i + \sum_j (D_{ji} \cdot FW_j) = Dtot_i \cdot FW_i$$

Based on the connection of the sub-basins (Fig. 1) the following set of mass balances for the water compartments can be formulated:

Bothnian Bay	$I_1 = +Dtot_1 \cdot FW_1 - D_{21} \cdot FW_2$
Bothnian Sea	$I_2 = -D_{12} \cdot FW_1 + Dtot_2 \cdot FW_2 - D_{52} \cdot FW_5$
Gulf of Finland	$I_3 = +Dtot_3 \cdot FW_3 - D_{53} \cdot FW_5$
Gulf of Riga	$I_4 = +Dtot_4 \cdot FW_4 - D_{54} \cdot FW_5$
Baltic proper (S)	$I_5 = -D_{25} \cdot FW_2 - D_{35} \cdot FW_3 - D_{45} \cdot FW_4 + Dtot_5 \cdot FW_5 - D_{65} \cdot FW_6 - D_{95} \cdot FW_9$
Danish Straits	$I_6 = -D_{56} \cdot FW_5 + Dtot_6 \cdot FW_6 - D_{76} \cdot FW_7$
Kattegat	$I_7 = -D_{67} \cdot FW_6 + Dtot_7 \cdot FW_7 - D_{87} \cdot FW_8$
Baltic proper (B)	$0 = -D_{59} \cdot FW_5 + Dtot_9 \cdot FW_9$

If EW_i, FX_i, FA_i, FW_8 are input parameters, the I and D terms can be calculated and this system of eight linear equation can be solved numerically for the eight unknown water fugacities FW_i. It is then possible to calculate the sediment fugacities FS_i

$$FS_i = FW_i \cdot \frac{DT_i + DD_i}{DB_i + DS_i + DR_i + DT_i}$$

Once the fugacities in water and sediment are known, concentrations C in mol/m^3, amounts M in mol and rates N of transport and transformation in mol/h are easily calculated employing the Z- and D-values:

$$C = F \cdot Z$$

$$M = F \cdot Z \cdot V$$

$$N = F \cdot D$$

15.3 PCBs in the Baltic Sea

The data on the occurrence and sources of POPs, which were generated during the Project "Large-Scale Environmental Effect and Ecological Processes in the Baltic Sea" and which are presented elsewhere in this book (see Agrell et al., Chap. 6; Axelman et al., Chap. 10), are uniquely suitable input and evaluation parameters for the B-POP model. Because the current model assumes a steady-state situation, a simulation has to focus on a specific time period where temporal changes in the behavior of a contaminant were minor. An attempted is made to use the model to describe the fate of PCBs within the Baltic Sea environment during the late 1980s. Throughout the 1960s and 1970s the use and thus the release of PCBs in northern Europe underwent large changes, while in the 1980s a slower development seemed to be established. Long-term trend studies on PCB levels in biota from the Baltic (Bignert et al. 1993, 1995) confirm relatively stable or only slightly decreasing concentrations of PCBs towards the late 1980s, when most of the measurements of the Baltic Sea Project were performed.

15.3.1 Selection of Input Parameters

15.3.1.1 Chemical Properties

Although ideally the environmental behavior of various PCB congeners should be modeled separately, this first and still preliminary effort uses concentrations for input and evaluation which refer to the sum of the six most commonly reported PCB isomers, 52, 101, 118, 138, 153, 180, thus including one tetra-, two penta-, two hexa- and one heptachlorobiphenyl.

Table 15.7. Physical-chemical properties and degradation half-lives of "PCBs" used in the model calculations

Molecular weight[a]	360.9 g/mol
Melting point[a]	114 °C
Entropy of fusion[a]	45.2 J/(K · mol)
Vapor pressure[b]	log (p in Pa)=14.84–5399/(T in K)
Aqueous solubility[c]	s in mol/m³=–0.154–1497/(T in K)
log K_{OW}[a]	7.0
Degradation half-life in water	50 years
Degradation half-life in sediment	∞

[a] PCB-155, Mackay et al. (1992)
[b] PCB-155, Wania et al. (1994)
[c] PCB-155, Shiu et al. (1997)

The number of chlorine atoms in the PCB molecule strongly affects vapor pressures, aqueous solubilities and octanol-water partition coefficient. According to Falconer and Bidleman (1994), the vapor pressure of the subcooled liquid of PCB-52 is more than two orders of magnitude higher than that of PCB-180. Similarly, the K_{OW} of these compounds covers a range of one order of magnitude (Mackay et al. 1992). The use of a single set of physicochemical properties is thus a major simplification. Table 7 lists the selected chemical property data, which reflect the typical properties of a hexachlorobiphenyl. Since the average temperatures in the Baltic Sea environment are considerably lower than room temperature, a temperature correction for vapor pressure and solubility is employed.

A tricky task is the selection of realistic degradation half-lives for POPs in the environment. Degradation rates established in the laboratory are not easily transferred to real environmental conditions. Two recent investigations on concentration profiles in sediment cores (Axelman et al. 1995; Kjeller and Rappe 1995) suggest that degradation of PCBs in Baltic sediments is very slow or does not occur to any significant extent at all. Therefore, an artificially long degradation half-life in surface sediment was used.

15.3.1.2 Source and Emission Information

The model requires three types of source and emission information: average bulk air concentrations CA_i (vapor and aerosol-bound phase), average bulk river water concentrations CX_i (truly dissolved and POC-bound phase) and an estimate of the direct discharge rate into water EW_i,

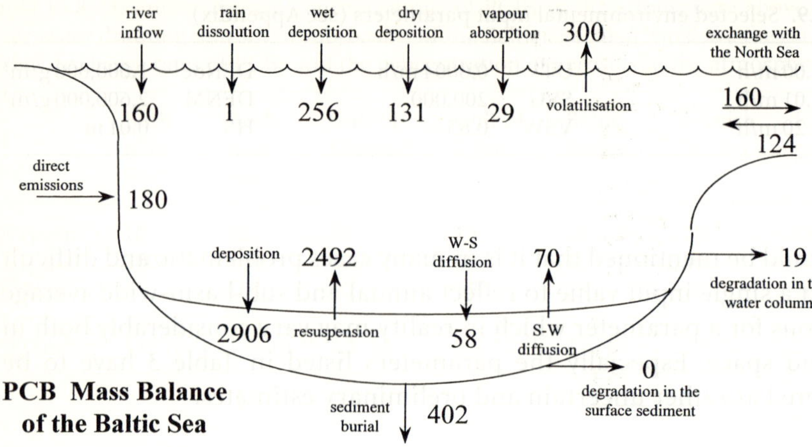

Fig. 15.5. Overall steady-state mass balance of PCBs in the Baltic Sea (including Danish Straits and Kattegat) as calculated by the B-POP model for the end of the 1980s/early 1990s. Fluxes are given in units of kg/year

A closer inspection shows a net annual air–water exchange of 117 kg into the water, i.e., the Baltic Sea is a sink for atmospheric PCBs, which does not exclude the possibility that PCB volatilization exceeds deposition during parts of the year. Major atmospheric deposition processes are wet and dry deposition of aerosol-associated PCBs. The net flux of 402 kg/a between water and surface sediment is also downward and major processes are the deposition and resuspension of POC. The overall pattern of PCB transport in the Baltic Sea environment is thus a downward flux from the atmosphere to the water, to the surface sediment, and finally to the deep sediment. This implies that the Baltic Sea acts as a sink for PCBs. This is in contrast to recent investigations in the Great Lakes of North America, which suggest that volatilization exceeds atmospheric deposition in these lakes and Lake Ontario (Mackay et al. 1994), Lake Michigan (Hornbuckle et al. 1995) and Lake Superior (Jeremiasen et al. 1994) now act as sources of PCBs to the regional atmosphere.

Figure 6 shows the calculated interbasin transport rates, input and loss terms, and concentrations for each of the sub-basins. Not all sub-basins show a net atmospheric input of PCBs. The model calculations suggest that the Gulf of Finland acts as a source of PCBs to the atmosphere and that in the Bothnian Bay and the Gulf of Riga atmospheric deposition and volatilization are almost equal. Like the net water flux, the net flux of PCBs between the sub-basins is almost always directed from the upstream basins towards the North Sea. The exception is that between Baltic proper and Danish

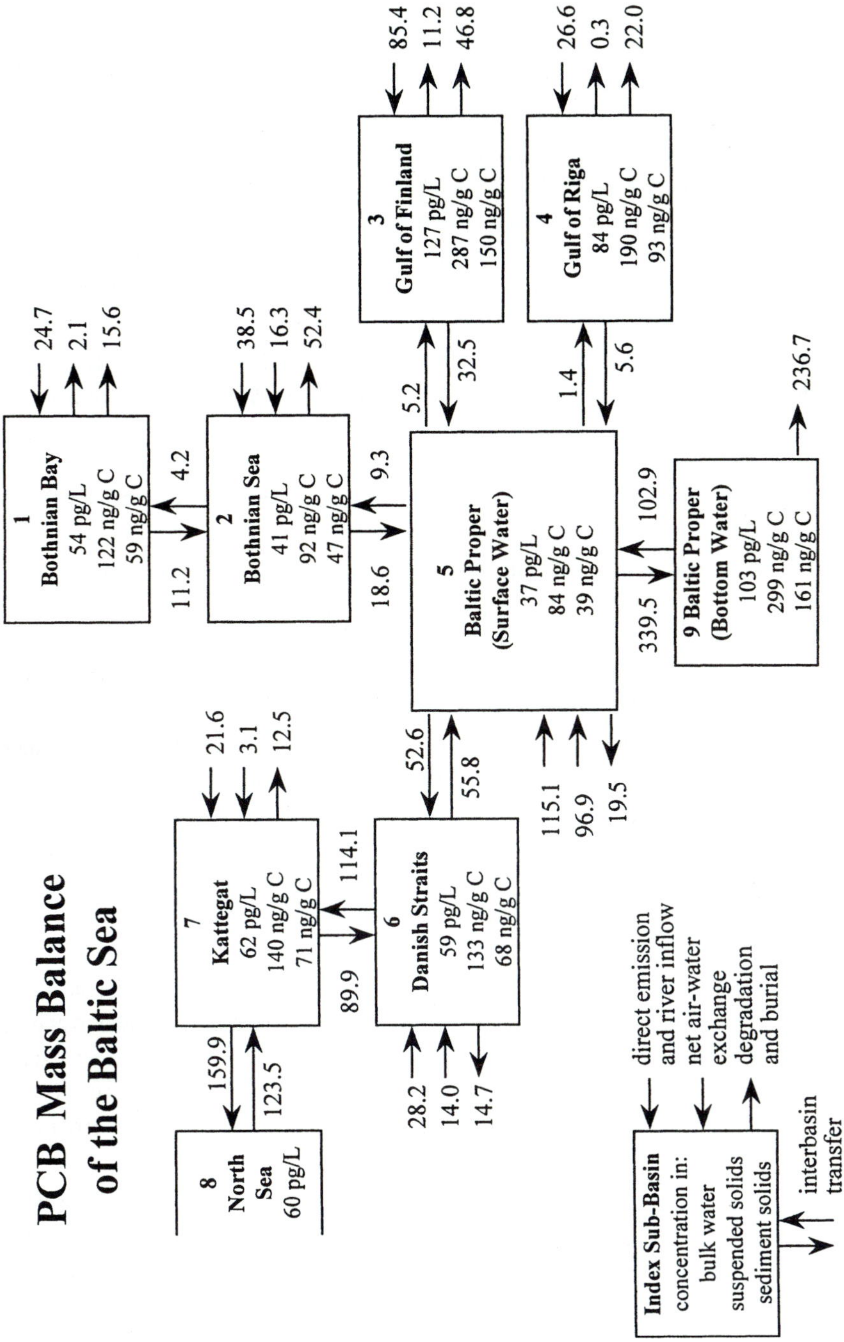

Fig. 15.6. Steady-state concentrations and fluxes of PCBs in the seven sub-basins of the Baltic Sea as calculated by the B-POP model for the end of the 1980s/early 1990s. Fluxes are given in units of kg/year

Straits where virtually no net exchange of PCBs occurs. "Vertical" processes, i.e., those occurring within a basin, tend to be more important than "horizontal", i.e., interbasin transport processes. In the Baltic proper, for example, the calculated gross flux of PCBs to the bottom water is five times as high as the outflow to all other basins together – a result of the very efficient scavenging of PCBs by settling POC. The highly dynamic marine system between Denmark and southern Sweden (basins 6 and 7), in which horizontal exchange processes become dominant, is an exception.

Riverine and atmospheric sources are of the same order of magnitude in most sub-basins, but a detailed analysis reveals that the sub-basins differ considerably in their source characteristics. Table 10 summarizes the relative importance of various input and output processes in the sub-basins. In the large basin of the Bothnian Sea and Baltic proper, PCB input is dominated by atmospheric deposition, while in the smaller and more densely populated Gulfs of Finland and Riga direct emissions and riverine inflow become far more important. Expectedly, the input of PCBs to the Danish Straits and the Kattegat is determined to a large extent by the inflow of water from the Baltic proper and the North Sea. The Bothnian Bay finally has large contributions from both rivers and atmosphere. Differences are less pronounced with respect to PCB loss, as volatilization and sediment burial are the major loss processes for all basins except the Danish Straits and the Kattegat, where interbasin flow is by far the most important loss pathway. In the deep Baltic proper sediment burial becomes relatively more important.

Table 15.10. Relative importance of various input and output pathways for PCBs in the sub-basins of the Baltic Sea (in percent)

	Bothnian Bay	Bothnian Sea	Gulf of Finland	Gulf of Riga	Baltic Proper	Danish Straits	Kattegat
Input							
Direct emissions	11	8	32	31	17	12	4
River inflow	41	25	36	24	9	2	4
Atmospheric deposition	39	49	29	42	49	17	9
Interbasin flow	9	18	4	3	25	69	83
Output							
Volatilization	44	35	37	43	28	10	8
Degradation	2	2	2	1	3	0	0
Sediment burial	30	43	35	44	54	7	4
Interbasin flow	24	20	26	12	15	83	88

Figure 6 also displays regional PCB concentration differences in water and sediment. The model suggests that the Gulf of Finland, due to high riverine inflow and relatively high population/industry density (St. Petersburg, Helsinki, Tallinn, Kohtla-Järve/Narva), should show elevated concentrations in both water and sediment. To a smaller degree this should also apply to the Gulf of Riga. The model also suggests fairly high concentrations in the bottom water and deep sediments of the Baltic proper. For the remaining basins the model calculates rather uniform concentrations between 37 to 62 pg/l in seawater and 40 to 70 ng/g C in surface sediment.

15.4.1 Response Times

Ratios of volume times bulk Z-value over the sum of all D-values contributing to loss from a compartment or a group of compartments are characteristic constants, indicating the response time of this system to loading changes. A detailed discussion of derivation and interpretation of these time constants is given by Mackay (1989). Table 11 lists the time constants for the water column and the entire aquatic system for each of the sub-basins.

The response time of the water column is approximately 6 months, indicating that a reduction in the PCB loading, for example, by reducing atmospheric or river concentrations, would be reflected by decreasing water-column concentrations within a fairly short time period. This is because the residence time of PCBs in the water column is not determined by the hydraulic water residence time which is in the range of several years;

Table 15.11. Calculated characteristic response times of the water column and the total aquatic system (i.e., water column and surface sediment)

	Response time of water in days	Response time of aquatic system in years
Bothnian Bay	182	19.3
Bothnian Sea	160	14.2
Gulf of Finland	105	14.2
Gulf of Riga	32	14.3
Baltic proper	195	16.6
Baltic proper (surface)	113	3.7
Baltic proper (bottom)	193	28.4
Danish Straits	16	2.5
Kattegat	25	1.9
Baltic Sea	151	18.3

see Fig. 1), but by POC sedimentation. (The smaller basins have smaller time constants because here water advection is a more important loss process.) The response time of the entire aquatic system is much longer, in the range of 15 years. This indicates that volatilization, sediment burial, degradation, and outflow to the North Sea are too slow to effectively reduce PCB levels in the surface sediments of the Baltic Sea within a few years. This implies that loading reductions can only be expected to reduce water concentrations to a certain level, which will be maintained for many years by resuspension from surface sediments. Not surprisingly, the deep sections of the Baltic Sea are particularly slow to respond. It can be hypothesized that the sediment concentrations in the Baltic Sea are likely to experience a long time lag of several years in responding to loading changes similar to that observed in Lake Ontario (Gobas et al. 1995). The sediment core from the Baltic proper, analyzed by Kjeller and Rappe (1995), showed the highest concentrations in the slice corresponding to the years 1974–1981, and thus well after the maximum PCB loading in the Baltic had occurred

15.5 Evaluation of Model Results

One of the most interesting questions, of course, is how the calculated concentrations and fluxes and their spatial patterns compare with measured concentration levels and transport rates.

15.5.1 Water Concentrations

Two investigations of PCBs in the seawater of the Baltic Sea during the late 1980s (Schulz-Bull et al. 1994; see also Chap. 10) revealed very heterogeneous concentrations in both space and time. Schulz-Bull et al. (1994) reported concentrations (six PCB isomers, dissolved phase, and fraction adsorbed to suspended particles) between 8.5 and 1,227 pg/l, with most concentration in the lower end of this range (10 to 100 pg/l). Axelman et al. (Chap. 10) measured concentrations between 5.4 and 74.2 pg/l. The calculated water concentrations (see Fig. 6) are thus within the range of measured values. In neither study was it possible to discern clear geographical or seasonal patterns, which would allow the evaluation of the calculated regional differences.

15.5.2 Sediment Concentrations

PCB concentrations in surface sediments of the Baltic Sea (Gaul 1991; Østfeldt et al. 1994; see also Chap. 10) are less variable than the water concentrations. For 1987, Gaul (1991) reported concentrations (six PCB isomers) of 75 to 150 ng/g C in the Kattegat, 35 to 90 ng/g in the southern Baltic proper, but higher concentrations in some estuaries in the Belt region (280 to 460 ng/g C). More recently, Østfeldt et al. (1994) reported values between 67 and 238 ng/g LOI (LOI equals approximately 0.5·C), while Axelman et al. (Chap. 10) measured between 50 and 600 ng/g C. The calculated concentrations (Fig. 6) are thus slightly lower than the measured values. With respect to regional differences, the measured data indicate slightly decreasing concentrations towards the north, but especially in the Baltic proper the measurements show large variations within rather small areas. Levels are less variable in the northern basins and the measurements suggest higher concentrations in the Gulf of Finland compared to the Gulf of Bothnia, though not as clear a difference as suggested by the model.

15.5.3 Concentrations in Precipitation

PCB concentrations in precipitation deduced from the model calculations are between 0.6 ng/l in the Bothnian Bay to 1.6 ng/l in the Danish Straits region. Calculated PCB deposition fluxes (i.e., the sum of wet deposition and dry particle deposition) are between 1.3 and 4.4 ng/(m^2·day). These values are higher than in measured precipitation samples by Agrell et al. (Chap. 6), which for the entire Baltic Sea region, show a median rain concentration of 0.46 ng/l and a median depositional flux of 0.54 ng/(m^2·day). Deposition measurements in the Kattegat by Brorström-Lundén (1995), which included particles sticking to the sampler surface, yielded median values between 1.6 and 2.6 ng/(m^2·day), which compare better with a calculated depositional flux in this area of 3.1 ng/(m^2·day).

15.5.4 Mass Balance of PCBs

In addition to selected concentrations, the overall picture of PCB behavior in the Baltic Sea, as determined by the B-POP model (Fig. 5), can be compared with the semi-empirical mass balance derived from measurements by Axelman et al. (Chap. 10). In general, fluxes and amounts of PCBs predicted by the model and those derived from measurements are within a factor of three of each other. The sediment-related fluxes, namely, the net sedimentation of PCBs to the surface sediments and the burial rates, are similar within a factor of two. Also, the input by land-based sources (riverine and direct input), the relative minor importance assigned to the outflow of PCBs to the North Sea, and the estimated volatilization rates are very similar in both approaches. The largest differences are seen in the relative importance attached to various deposition processes. Whereas the model identifies particle-mediated wet and dry deposition as the major atmospheric deposition pathway of PCBs to the Baltic Sea, the semi-empirical approach assigns more importance to diffusive gas absorption. In general, however, the pictures of the overall environmental fate of PCBs in the Baltic Sea environment painted by the two approaches agree rather well.

15.6 Conclusion

We believe that the B-POP model succeeds in outlining the general picture of PCB behavior in the Baltic Sea at the end of the 1980s by identifying the major environmental pathways and reservoirs within the system. When judging the accuracy and precision of the simulation, it is important to bear in mind the simplifications and assumptions made by the model. The most obvious are:

- the use of a single set of physicochemical properties to describe a mixture of compounds with fairly variable volatility and hydrophobicity,
- the assumption of complete and instantaneous mixing of large and heterogeneous sub-basins,
- the assumption of steady state, i.e., stable conditions in time, including the omission of any seasonal variability in environmental parameters and concentrations,
- the uncertainty of many of the selected environmental input parameters, which in some cases are generic or estimated values.

Considering the limitations inherent in the present model, the agreement between modeled and measured concentrations is surprisingly good. In most cases, the calculated value is within the range and close to the average of the measured values. This agreement lends support to those model results not directly validated by comparison with measurements such as the response times.

15.7 Outlook

There are several possibilities of using such a model beyond obtaining a comprehensive picture of the behavior of POPs in the Baltic Sea. The analysis of time-response characteristics was only hinted at above and should be conducted in far more detail to evaluate the feasibility and potential results of various "clean-up" scenarios and elimination schemes comparable to the analysis for Lake Ontario by Gobas et al. (1995). It may also be possible to quantify and to a certain extent even to predict the possible interaction of eutrophication and POP behavior in the Baltic Sea. It has been suggested that the reduction in nutrient discharge to the Baltic would reduce POC sedimentation and thus the sequestering of POPs to the sediments (Jonsson 1992). In lakes, eutrophication was implicated in lowering levels of POPs in aquatic vertebrates (Larsson et al. 1992; Taylor et al. 1991). The model could be run with several POC mass balances reflecting different trophic conditions to assess the effect on the calculated water, suspended solids, and sediment concentrations.

The B-POP model should also be improved to address the shortcomings and simplifications mentioned above. Further considerations for model development should include the following:

- Model calculations should be performed with individual PCB isomers rather than PCB mixtures to evaluate the influence of physicochemical properties.
- The selection of environmental parameters, especially the description of the dynamics of POC, should be improved.
- Recent measurements (Chap. 10) have shown that a large fraction of POPs in the water column is associated with dissolved organic carbon (DOC). More studies are required to reflect the impact of DOC on the behavior of POPs in aquatic fate models.
- As a result of loading changes, the behavior of PCBs and other POPs in the Baltic Sea underwent large fluctuations throughout the latter half of this century. Only a nonsteady-state model will be able to reflect these temporal changes adequately. A model allowing the change of parameters in time is also required to include the effects of seasonality on POP behavior. Due to its northern location, seasonal influences may be particularly important in the Baltic Sea.
- To make better use of the large amount of POP concentration data available for biotic material, the B-POP environmental fate model should be complemented by a food-chain accumulation model (Gobas et al. 1995).

- In the current model version, the atmospheric POP concentration is an input parameter used to calculate water concentrations. In reality, the concentration in the air above the Baltic Sea is likely to be influenced or to a certain extent even controlled by volatilization from the water column and from the terrestrial surfaces in the catchment area of the Baltic Sea. A catchment model, including compartments for atmosphere and terrestrial environment, is more likely to reflect realistic environmental conditions.
- Where sufficient measurements are available for validation, a higher spatial resolution of the model may be justified.

Acknowledgements. Funding from the Nordic Council of Ministers (NMR) and the Environment and Climate Research Program of the European Commission (Project POCYCLING-Baltic, ENV4-CT96-0214) is gratefully acknowledged.

Abbreviations

(index i refers to sub-basin)

Dimensions

AW_i water surface area in m^2
AS_i sediment surface area in m^2
HS_i sediment depth in m
VW_i volume of water in m^3
VS_i volume of surface sediment in m^3

Transport Rates of Water in m^3/h

GX_i river water inflow
GM_i precipitation
$G_{i,j}$ water flow from sub-basin i to sub-basin j

Transport and Transformation Rates of POC in g C/h

OP_i	autochthonous production
OI_i	inflow to sub-basin i from neighboring basins
OO_i	outflow from sub-basin i to neighboring basins
OX_i	riverine inflow
OD_i	deposition
OE_i	resuspension
OB_i	sediment burial
OW_i	mineralization rate in the water column
OS_i	mineralization rate in the sediment
$O_{5,9}$	transport rate from surface to bottom water in the Baltic proper

Environmental Parameters

$CWpoc_i$	POC concentration in seawater in g/m^3
$CXpoc_i$	POC concentration in inflowing river water in g/m^3
$ORGS_i$	mass fraction organic carbon in surface sediment solids in g/g
BP_i	biological productivity in the water column in g $OC/(m^2 \cdot h)$
$facOW_i$	fraction of the POC input to the water column (i.e., river inflow, inflow from neighboring basins, and biological production), which is mineralized in the water column
$facOE_i$	fraction of gross POC deposited in the sediment, which is resuspended
$facOS_i$	fraction of net POC deposited in the sediment (i.e., deposition minus resuspension), which is mineralized in the surface sediment
DENoc	density of organic carbon in g/m^3
DENM	density of mineral matter in g/m^3
$DENS_i$	density of sediment particles in g/m^3
SVG_i	scavenging ratio
VAQ_i	volume fraction aerosol in air in m^3/m^3
VSW_i	volume fraction water in surface sediment in m^3/m^3
UQ_i	dry deposition velocity in m/h
UA_i	air–water mass transfer coefficient, air side in m/h
UW_i	air–water mass transfer coefficient, water side in m/h
UT_i	water-sediment mass transfer coefficient in m/h
TA_i	air temperature in K
R	ideal gas constant in $J/(K \cdot mol)$

Chemical Properties

H_i	Henry's law constant in $Pa \cdot m^3/mol$
P_i	vapor pressure of the subcooled liquid in Pa
K_{OW}	partition coefficient octanol to water
Kpoc	partition coefficient particulate organic carbon to water
HLW_i	degradation half-life in water in h
HLS_i	degradation half-life in surface sediment in h

Model Parameters

all Z-values	explained in Table 5
EW_i	direct emission rate into water in mol/h
I_i	total input rate of chemical to water (direct, riverine, and atmospheric) in mol/h
$Dsed_i$	D-value describing net transfer from water to sediment in $mol/(h \cdot Pa)$
$Dtot_i$	sum of all D-values describing loss from the water compartment in $mol/(h \cdot Pa)$
all other D-values	explained in Table 6
FA_i	fugacity in air in Pa
FX_i	fugacity in river water in Pa
FW_i	fugacity in seawater in Pa
FS_i	fugacity in surface sediment in Pa

References

Addison RF (1989) Organochlorines and marine mammal reproduction. Can J Fish Aquat Sci 46:360–368

Andrulewicz E, Trzosinska A (1985) Estimation of input and distribution of chlorinated hydrocarbons in the Baltic Sea. International Council for the Exploration of the Sea (ICES/SCOR WG42), Copenhagen, Denmark

Axelman J, Bandh C, Broman D, Carman R, Jonsson P, Larsson H, Linder H, Näf C, Pettersen H (1995) Time-trend analysis of PAH and PCB sediment fluxes in the northern Baltic proper using different dating methods. Mar Freshwater Res 46: 137–144

Baker JE, Eisenreich SJ, Eadie BJ (1991) Sediment trap fluxes and benthic recycling of organic carbon, polycyclic aromatic hydrocarbons, and polychlorobiphenyl congeners in Lake Superior. Environ Sci Technol 25:500–509

Bignert A, Göthberg A, Jensen S, Litzén K, Odsjö T, Olsson M, Persson W, Reutergårdh L (1993) The need for adequate biological sampling in ecotoxicological investigations: a retrospective study of twenty years pollution monitoring. Sci Total Environ 128:121–139

Bignert A, Litzén K, Odsjö T, Olsson M, Persson W, Reutergårdh L (1995) Time-related factors influence the concentrations of sDDT, PCBs and shell parameters in eggs of Baltic Guillemot *(Uria aalge)*, 1861–1989. Environ Pollut 89:27–36

Blomqvist S, Larsson U (1994) Detrital bedrock metals as tracers of settling resuspended particulate matter in a coastal area of the Baltic Sea. Limnol Oceanogr 39:880–896

Brannon JE, Pennington JC, Davis WM, Hayes C (1995) Flouranthene K_{DOC} in sediment pore waters. Chemosphere 30:419–428

Broman D, Näf C, Zebühr Y, Lexén K (1989) The composition, distribution and flux of PCDDs and PCDFs in settling particulate matter (SPM): a sediment trap study in the Northern Baltic. Chemosphere 19:445–450

Broman D, Näf C, Zebühr Y (1991a) Long term high- and low-volume air sampling of polychlorinated dibenzo-p-dioxins and dibenzofurans and polycyclic aromatic hydrocarbons along a transect from urban to remote areas on the Swedish Baltic coast. Environ Sci Technol 25:1841–1850

Broman D, Näf C, Rolff C, Zebühr Y (1991b) Occurrence and dynamics of polychlorinated dibenzo-p-dioxins and dibenzofurans and polycyclic aromatic hydrocarbons in the mixed surface layer of remote coastal and offshore waters of the Baltic. Environ Sci Technol 25:1850–1864

Broman D, Näf C, Zebühr Y (1992) Occurrence and dynamics of polychlorinated dibenzo-p-dioxins and dibenzofurans and other combustion related pollutants in the aquatic environment of the Baltic. Chemosphere 25:125–128

Brorström-Lundén E (1995) Measurements of semivolatile organic compounds in air and deposition, PhD Thesis, Chalmers University of Technology, Göteborg, Sweden

Brorström-Lundén E, Lindskog A, Mowrer J (1994) Concentrations and fluxes of organic compounds in the atmosphere of the Swedish west coast. Atmos Environ 22:3605–3615

Diamond ML, Poulton DJ, Mackay D, Stride FA (1994) Development of a mass balance model of the fate of 17 chemicals in the Bay of Quinte. J Great Lakes Res 20:643–666

Elmgren R (1984) Trophic dynamics in the enclosed, brackish Baltic Sea. Rapp Réun Cons Int Explor Mer 183:152–169

Elmgren R (1989) Man's impact on the ecosystem of the Baltic Sea: Energy flows today and at the turn of the century. Ambio 18:326–332

Falconer RL, Bidleman TF (1994) Vapor pressure and predicted particle/gas distributions of polychlorinated biphenyl congeners as a function of temperature and chlorine substitution. Atmos Environ 28:547–554

Gaul H (1991) Überwachung des Meeres. Bericht für das Jahr 1989. Teil II: Daten. Bundesamt für Seeschiffahrt und Hydrographie, Hamburg, 196 pp

Gobas FAPC, Z'Graggen MN, Zhang X (1995) Time response of the Lake Ontario ecosystem to virtual elimination of PCBs. Environ Sci Technol 29:2038–2046

Haahti H, Perttilä M (1988) Levels and trends of organochlorines in cod and herring in the northern Baltic. Mar Pollut Bull 19:29–31

Helsinki Commission (1980) Assessment of the effects of pollution on the natural resources of the Baltic Sea, 1980, Baltic Marine Environment Protection Commission. Baltic Sea Environ Proc 5B

Helsinki Commission (1986) Water balance of the Baltic Sea, Baltic Marine Environment Protection Commission. Baltic Sea Environ Proc 16:174

Hornbuckle KC, Jeremiasen JD, Sweet CW, Eisenreich SJ (1994) Seasonal variation in air-water exchange of polychlorinated biphenyls in Lake Superior. Environ Sci Technol 28:1491–1501

Hornbuckle KC, Sweet CW, Pearson RF, Swackhammer DL, Eisenreich SJ (1995) Assessing annual water-air fluxes of polychlorinated biphenyls in Lake Michigan. Environ Sci Technol 29:869–877

Jansson B, Andersson R, Asplund L, Litzén K, Nylund K, Sellström U, Uvemo U-B, Wahlberg C, Wideqvist U, Odsjö T, Olsson M (1993) Chlorinated and brominated persistent organic compounds in biological samples from the environment. Environ Toxicol Chem 12:1163–1174

Järnberg U, Asplund L, de Wit C, Grafström A-K, Haglund P, Jansson B, Lexén K, Strandell M, Olsson M, Jonsson B (1993) Polychlorinated biphenyls and polychlorinated naphthalenes in Swedish sediment and biota: Levels, patterns, and time trends. Environ Sci Technol 27:1364–1374

Jensen S, Johnels A, Olsson M, Otterlind G (1969) DDT and PCB in marine mammals from Swedish water. Nature 224:247–250

Jeremiasen JD, Hornbuckle KC, Eisenreich SJ (1994) PCBs in Lake Superior, 1978–1992: decreases in water concentrations reflect loss by volatilization. Environ Sci Technol 28:903–914

Johnels A (ed) (1992) Special issue: seals and seal protection. Ambio 21:493–606

Jonsson P (1992) Large-scale changes of contaminants in Baltic Sea sediments during the twentieth century, PhD Thesis, Uppsala Univ, Uppsala, Sweden

Jonsson P, Carman R (1994) Changes in deposition of organic matter in the Baltic Sea during the twentieth century and notes on sediment dating problems. Mar Pollut Bull 28:417–426

Jonsson P, Carman R, Wulff F (1990) Laminated sediments in the Baltic – a tool for evaluating nutrient mass balances. Ambio 19:152–158

Kullenberg G, Jacobsen TS (1981) The Baltic Sea: an outline of its physical oceanography. Mar Pollut Bull 12:183–186

Kjeller L-O, Rappe C (1995) Time trends in levels, patterns, and profiles for polychlorinated dibenzo-p-dioxins, dibenzofurans, and biphenyls in a sediment core from the Baltic proper. Environ Sci Technol 29:346–355

Larsson U, Nellbring S (1990) Sedimentation av partikulärt material. In: Intensivövervakning av pelagialen vid landsortdjupet. Naturvårdsverket Rapport 3989, Stockholm

Lassig J, Leppänen J-M, Niemi Å, Tamelander G (1978) Phytoplankton primary production in the Gulf of Bothnia in 1972–1975 as compared with other parts of the Baltic Sea. Finn Mar Res 244:101–115

Larsson P, Collvin L, Okla L, Meyer G (1992) Lake productivity and water chemistry as governors of the uptake of persistent pollutants in fish. Environ Sci Technol 26:346–352

Leppäkoski E, Bonsdorf E (1989) Ecosystem variability and gradients. Examples from the Baltic Sea as a background for hazard assessment. In: Landner L (ed) Chemicals in the aquatic environment: advanced hazard assessment. Springer, Berlin Heidelberg New York, pp 6–73

Mackay D (1989) Modeling the long-term behavior of an organic contaminant in a large lake: application to PCBs in Lake Ontario. J Great Lakes Res 15:283–297

Mackay D (1991) Multimedia environmental models: the fugacity approach. Lewis, Chelsea, MI, 257 pp

Mackay D, Joy M, Paterson S (1983) A quantitative water, air, sediment interaction (QWASI) fugacity model for describing the fate of chemicals in lakes. Chemosphere 12:981–997

Mackay D, Shiu WY, Ma KC (1992) Illustrated handbook of physical-chemical properties and environmental fate for organic chemicals, vol I: monoaromatic hydrocarbons, chlorobenzenes, and PCBs. Lewis, Chelsea, MI, 697 pp

Mackay D, Sang S, Vlahos P, Diamond M, Gobas F, and Dolan D. (1994) A rate constant model of chemical dynamics in a lake ecosystem: PCBs in Lake Ontario. J Great Lakes Res 20:625–642

Näf C, Broman D, Pettersen H, Rolff C, Zebühr Y (1992) Flux estimates and pattern recognition of particulate polycyclic aromatic hydrocarbons, polychlorinated dibenzo-p-dioxins, and dibenzofurans in the waters outside various emission sources on the Swedish Baltic coast. Environ Sci Technol 26:1444–1457

Olsson M, Karlsson B, Ahnland E (1994) Diseases and environmental contaminants in seals from the Baltic and the Swedish west coast. Sci Total Environ 154:217–227

Østfeldt P, Gustavson K, Jansson B, Jonsson P, Miettinen V, Ringstad O, Wesén C (1994) Halogenated organic compounds in the marine environment 1989–1990. Nordic Council of Ministers, TemaNord, p 591

Schulz-Bull DE, Petrick G, Kannan N, Duinker JC (1995) Distribution of individual chlorobiphenyls (PCB) in solution and suspension in the Baltic Sea. Mar Chem 48:245–270

Shiu WY, Wania F, Hung H, Mackay D (1997) Temperature dependence of aqueous solubility of selected chlorobenzenes, polychlorinated biphenyls and dibenzofuran. J Chem Eng Data 42:293–297

Taylor WD, Carey JH, Lean DRS, McQueen DJ (1991) Organochlorine concentrations in the plankton of lakes in southern Ontario and their relationship to plankton biomass. Can J Fish Aquat Sci 48:1960–1966

Thompson S (1992) Estimating the loadings of priority pollutants to Lake Ontario. A report for Environmental Protection - Ontario Region, Conservation and Protection, Environment Canada, 103 pp

Valiela I (1984) Marine ecological processes. Springer, Berlin Heidelberg New York

Wania F, Shiu WY, Mackay D (1994) Measurement of the vapor pressure of several low volatility organochlorine chemicals at low temperatures with a gas saturation method. J Chem Eng Data 39:572–577

Wetzel RG (1983) Limnology, 2nd edn. Sanders College Publ, Orlando, 767 pp

Wulff F, Rahm L, Jonsson P, Brydsten L, Ahl T, Granmo Å (1993) A mass balance model of chlorinated organic matter for the Baltic Sea – a challenge for ecotoxicology. Ambio 22:27–31

Subject Index

Ecological Studies

Volumes published since 1995

Volume 108
Ecology and Biogeography of Mediterranean Ecosystems in Chile, California and Australia (1995)
M.T.K. Arroyo, P.H. Zedler, and M.D. Fox (Eds.)

Volume 109
Mediterranean-Type Ecosystems. The Function of Biodiversity (1995)
G.W. Davis and D.M. Richardson (Eds.)

Volume 110
Tropical Montane Cloud Forests (1995)
L.S. Hamilton, J.O. Juvik, and F.N. Scatena (Eds.)

Volume 111
Peatland Forestry. Ecology and Principles (1995)
E. Paavilainen and J. Päivänen

Volume 112
Tropical Forests: Management and Ecology (1995)
A.E. Lugo and C. Lowe (Eds.)

Volume 113
Arctic and Alpine Biodiversity. Patterns, Causes and Ecosystem Consequences (1995)
F.S. Chapin III and C. Körner (Eds.)

Volume 114
Crassulacean Acid Metabolism. Biochemistry, Ecophysiology and Evolution (1996)
K. Winter and J.A.C. Smith (Eds.)

Volume 115
Islands. Biological Diversity and Ecosystem Function (1995)
P.M. Vitousek, L.L. Loope, and H. Adsersen (Eds.)

Volume 116
High Latitude Rainforests and Associated Ecosystems of the West Coast of the Americas: Climate, Hydrology, Ecology and Conservation (1996)
R.G. Lawford, P. Alaback, and E. Fuentes (Eds.)

Volume 117
Global Change and Mediterranean-Type Ecosystems (1995)
J. Moreno and W.C. Oechel (Eds.)

Volume 118
Impact of Air Pollutants on Southern Pine Forests (1996)
S. Fox and R.A. Mickler (Eds.)

Volume 119
Freshwater Ecosystems of Alaska. Ecological Syntheses (1997)
A.M. Milner and M.W. Oswood (Eds.)

Volume 120
Landscape Function and Disturbance in Arctic Tundra (1996)
J.F. Reynolds and J.D. Tenhunen (Eds.)

Volume 121
Biodiversity and Savanna Ecosystem Processes. A Global Perspective (1996)
O.T. Solbrig, E. Medina, and J.F. Silva (Eds.)

Volume 122
Biodiversity and Ecosystem Processes in Tropical Forests (1996)
G.H. Orians, R. Dirzo, and J.H. Cushman (Eds.)

Volume 123
Marine Benthic Vegetation. Recent Changes and the Effects of Eutrophication (1996)
W. Schramm and P.H. Nienhuis (Eds.)

Volume 124
Global Change and Arctic Terrestrial Ecosystems (1997)
W.C. Oechel et al. (Eds.)

Volume 125
Ecology and Conservation of Great Plains Vertebrates (1997)
F.L. Knopf and F.B. Samson (Eds.)

Volume 126
The Central Amazon Floodplain: Ecology of a Pulsing System (1997)
W.J. Junk (Ed.)

Volume 127
Forest Decline and Ozone: A Comparison of Controlled Chamber and Field Experiments (1997)
H. Sandermann, A.R. Wellburn, and R.L. Heath (Eds.)

Volume 128
The Productivity and Sustainability of Southern Forest Ecosystems in a Changing Environment (1998)
R.A. Mickler and S. Fox (Eds.)

Volume 129
Pelagic Nutrient Cycles: Herbivores as Sources and Sinks (1997)
T. Andersen

Volume 130
Vertical Food Web Interactions: Evolutionary Patterns and Driving Forces (1997)
K. Dettner, G. Bauer, and W. Völkl (Eds.)

Volume 131
The Structuring Role of Submerged Macrophytes in Lakes (1998)
E. Jeppesen et al. (Eds.)

Volume 132
Vegetation of the Tropical Pacific Islands (1998)
D. Mueller-Dombois and F.R. Fosberg

Volume 133
Aquatic Humic Substances: Ecology and Biogeochemistry (1998)
D.O. Hessen and L.J. Tranvik (Eds.)

Volume 134
Oxidant Air Pollution Impacts in the Montane Forests of Southern California (1999)
P.R. Miller and J.R. McBride (Eds.)

Volume 135
Predation in Vertebrate Communities: The Białowieża Primeval Forest as a Case Study (1998)
B. Jędrzejewska and W. Jędrzejewski

Volume 136
Landscape Disturbance and Biodiversity in Mediterranean-Type Ecosystems (1998)
P.W. Rundel, G. Montenegro, and F.M. Jaksic (Eds.)

Volume 137
Ecology of Mediterranean Evergreen Oak Forests (1999)
F. Rodà et al. (Eds.)

Volume 138
Fire, Climate Change and Carbon Cycling in the North American Boreal Forest (2000)
E.S. Kasischke and B. Stocks (Eds.)

Volume 139
Responses of Northern U.S. Forests to Environmental Change (2000)
R. Mickler, R.A. Birdsey, and J. Hom (Eds.)

Volume 140
Rainforest Ecosystems of East Kalimantan: El Niño, Drought, Fire and Human Impacts (2000)
E. Guhardja et al. (Eds.)

Volume 141
Activity Patterns in Small Mammals: An Ecological Approach (2000)
S. Halle and N.C. Stenseth (Eds.)

Volume 142
Carbon and Nitrogen Cycling in European Forest Ecosystems (2000)
E.-D. Schulze (Ed.)

Volume 143
Global Climate Change and Human Impacts on Forest Ecosystems: Postglacial Development, Present Situation and Future Trends in Central Europe (2001)
J. Puhe and B. Ulrich

Volume 144
Coastal Marine Ecosystems of Latin America (2001)
U. Seeliger and B. Kjerfve (Eds.)

Volume 145
Ecology and Evolution of the Freshwater Mussels Unionoida (2001)
G. Bauer and K. Wächtler (Eds.)

Volume 146
Inselbergs: Biotic Diversity of Isolated Rock Outcrops in Tropical and Temperate Regions (2000)
S. Porembski and W. Barthlott (Eds.)

Volume 147
Ecosystem Approaches to Landscape Management in Central Europe (2001)
J.D. Tenhunen, R. Lenz, and R. Hantschel (Eds.)

Volume 148
A Systems Analysis of the Baltic Sea (2001)
F.V. Wulff, L.A. Rahm, and P. Larsson (Eds.)

Printing and Binding: Stürtz AG, Würzburg